Comments on this volume

Many people have reviewed this volume. They have diverse backgrounds and a variety of professions, and are at various stages in life. Their comments below touch on the science, the letters, the remarkable men who wrote them, as well as their own experience of PCT and the particular relevance it has had for them.

–The editor

This is a marvelous and extremely important volume. It is extremely important because it records two towering intellects in an extended correspondence concerning a true scientific revolution in psychology. Powers and Runkel theorize, experiment, model, discuss, criticize and advance our understanding of Perceptual Control Theory (PCT) which was given its first extended formulation in Powers' *Behavior, the Control of Perception* (1973). The correspondence is exceptionally well-written, occasionally wandering, sometimes technical (but no more than advanced algebra) often insightful, and always illuminating. It provides an outstanding case study of how science develops when real scientists are involved. There are suggestions, descriptions of experiments, computer modeling, explorations of consequences, criticisms, false starts, new breakthroughs, and throughout it all the sense that this is real science in the making.

The volume is marvelous because the humanity of these two men of science is also abundantly present. They express their pride in their successes, their frustrations at being misunderstood, their growing respect for each other as scientists and their maturing friendship for each other as persons.

It is a must read for anyone who is interested in bringing psychology out of the dark ages and in observing how two outstanding scientists make science really work.

Hugh Petrie, Ph.D. (Philosophy)
Professor Emeritus and former Dean,
Graduate School of Education
State University of New York at Buffalo

Bill Powers is one of the clearest and most original thinkers in the history of psychology. For decades he has explored with persistence and ingenuity the profound implications of the simple idea that biological organisms are control systems. His background in engineering allowed him to avoid many of the traps that have victimized even the best psychologists of the past. I believe his contributions will stand the test of time.

Henry Yin, Ph.D. (Cognitive Neuroscience)
Professor of Psychology & Neuroscience
Duke University, NC

Bill Powers' work in the 20th century will prove to be as important for the life sciences as Charles Darwin's work in the 19th century. By the time this notion has become common knowledge, historians of science will be very happy with this correspondence between two giants.

Frans X. Plooij, Ph.D. (Behavioral Biology)
Director, International Research-institute on
Infant Studies (IRIS), Arnhem, The Netherlands

I am a former Navy Fire Control Technician, charged with operating, maintaining and repairing the systems that control a warship's gunnery and missile systems. I like to think I know a little bit about control. When I first read *Behavior: The Control of Perception* in 1975, PCT immediately struck a chord with me. Most important, it provided a schematic for analyzing, understanding and improving human performance in the modern workplace.

What Peter Drucker called "the shift to knowledge work" was actually a shift from prefigured or 'canned' work routines to configured or 'crafted' responses. Crafted responses entail figuring out what to do so as to achieve and maintain valued results. That requires employees to exercise a considerable degree of discretion and to vary their behavior in ways that get the job done. The old stimulus-response view of human behavior doesn't offer any help with this kind of performance, and neither does the cognitive view.

The bottom line is that employees must be viewed as agents acting in their employer's best interests instead of compliant instruments of managerial will. The only theory of human behavior (and performance) that fits the bill is William Powers' Perceptual Control Theory (PCT). His view of human beings as "living control systems" is precisely what management needs if it is ever again to have any meaningful impact on workplace performance and productivity. Finally, I find PCT very useful as a way of reflecting upon, understanding and managing my own behavior.

Fred Nickols, www.nickols.us
Managing Partner, Distance Consulting LLC
Exec. Dir. Educational Testing Service (1990–2001)

This book provides a wonderful compilation of the historical underpinnings of Perceptual Control Theory (PCT), and includes many communications between Philip Runkel and William Powers during the time period that PCT was being further developed and refined. As part of the book, other authors' contributions are given, including excellently written comments by the editor, Dag Forssell. PCT first became known to my wife and me in 2004, when we attended a class for retired people at the University of Cincinnati. We had two outstanding teachers, Len Lansky (a retired psychology professor) and Robert Summer (a psychotherapist in private practice). Besides learning about the basic simplicity of PCT and about the huge improvement that PCT has over other concepts of psychology in describing the actions of living things, the class members learned how to effectively use PCT to think about and resolve disagreements between two people. Overall, this new book is extremely valuable in understanding PCT.

Raymond E. Sund, Ph.D. (Nuclear Physics)
Former Director of R&D at Toledo Edison Co.

Bill Powers provides a way of understanding living beings that on the surface might appear simple. And yet, once you look from this perspective, everything you thought you knew is brought into question and a process of re-examination and rediscovery begins. I have found this to be the most valuable learning experience I have ever had and the most significant influence on my work as a clinician and researcher. The collection of letters and papers in this book provides a fascinating opportunity to embark upon the journey of discovery and re-examination shared by Bill Powers and Phil Runkel. It provides an experiential process of learning more about PCT in a way that brings everything to life.

Sara Tai, Ph.D. (Clinical Psychology)
Senior Lecturer, Chartered Clinical Psychologist,
Accredited Cognitive Behavioural Therapist
University of Manchester, UK

It is hard to overstate the importance of this work yet its significance will probably not be fully realized for many years to come. The contribution of Powers' insights to the life sciences are so profound and far-reaching that virtually none of our current knowledge will remain intact in the new paradigm where the phenomena of control and circular causality are the foci of attention. Some current concepts will require only a slight tweaking while others will need a major revision. Still others will become entirely irrelevant.

Being able to follow along while a highly respected scholar such as Runkel spring cleans all that he knows in order to understand PCT accurately is a rare privilege. While Runkel learns more about PCT, the reader cannot help but to benefit from Runkel's searching queries and astute insights.

This book will become an important resource for any serious student of PCT which, in time, will be anyone who seeks to rigorously understand the fundamental elements of the process of living.

Tim Carey, Ph.D. (Clinical Psychology)
Associate Professor in Mental Health, Centre for
Remote Health, a joint Centre of Flinders
University and Charles Darwin University and
Central Australian Mental Health Service,
Alice Springs, Northern Territory, Australia

While Director of Systems Manufacturing at a division of Intel, I had the opportunity to innovate specific management processes and communications strategies based on my understanding of PCT. The experience spanned about 14 months. The result was a level of plant performance that had never been attained before. The plant won numerous accolades for on time delivery, line linearity, and quality. Also, teamwork between departments, which in the past had been less than ideal, improved significantly.

As a consultant, I created programs applying PCT to problems that managers encounter often.

At Apple, I taught managers how they might deal with problem performers in a more effective manner. At Hughes Bipolar Semiconductor, I applied PCT to building teamwork in a production area where performance was so poor that the material cycle time in the area was over 24 weeks. When I completed my work, the line had already reduced cycle time to less than 8 weeks. Morale was the best managers had seen in memory.

I worked with Intel's PC Enhancement Division on Constructive Confrontation communications skills and taught managers how they could aggressively confront problems with others, but in such a way that they did not have to get into angry, stressful arguments. The managers told me that I had given them a new set of tools for dealing effectively with others, even when a problem might get emotional.

Upon reflection, it is not surprising that applying PCT in a number of practical and skillful ways would produce results. PCT is the most comprehensive and accurate model of the human operating system I have ever encountered. I am satisfied that in the future organizations will realize more repeatable results, more efficient and effective problem resolution, a far greater sense of teamwork and esprit de corps, all with managers experiencing far less stress.

Jim Soldani
Intel: Director, Systems Manufacturing,
Memory Systems, Phoenix, AZ. 1978–81
Director, Corporate Training 1981–83
Director, Systems Group WW Materials 1991–94
Author: *Effective Personnel Management: An Application of Control Theory*

I never "bought" the linear determinism of the stimulus-response psychologies because it was inconsistent with my experience, and because promoters of these views seemed always tacitly to exempt themselves from being subject to them. That's probably why learning perceptual control theory was for me in the early 1990s not a tumultuous overthrow of old ideas, such as Dr. Runkel reports in the earliest of the letters here, so much as it was an exploration of the ramifications.

These ramifications, as he also attests, are challenge enough. PCT is unprecedented in its breadth and reach. Its grounding in physics and physiology sets it far above the speculations that prevail in the psychological and social sciences. Its requirement for explicit working models that conform closely to recorded behavior sets standards of excellence that are without parallel in these fields. This, together with the restoration of purpose to the center of the sciences of living things, separates what is essential from what is incidental. The perceptual variables controlled by an individual are the essential matters to be identified and measured. The countless other variables that an investigator might perceive and statistically correlate are revealed as disturbances to or incidental side effects of control, obscured by the aggregation of data from many trials and many individuals.

These characteristics of PCT—its scope, its rejection of IV–DV hand-waving and 'models' that don't work, its demand for hard-science specificity and for correlations near 100%—can make it a hard sell. Those who wish to curry favor among today's makers and breakers of reputation might well steer clear. The essay *Three "Dangerous" Words* in Part II will tell you why. But those who want to do something of lasting value should pay close attention. Our colleague Phil Runkel has gone this way before us under the guidance of our mentor Bill Powers, and their 22 years of wise, articulate, witty correspondence lays out bright lights and signposts for our benefit.

Bruce Nevin, Ph.D. (Linguistics)
Editor, *The Legacy of Zellig Harris: Language and information into the 21st century*
Program Manager & Information Architect,
Cisco Systems Inc.

Dialogue Concerning the Two Chief Approaches to a Science of Life lays out a fascinating, behind-the-scenes and historical look at why a highly regarded academic psychologist, Phil Runkel, came to abandon mainstream psychology principles and adopt Perceptual Control Theory originated by Bill Powers.

PCT will assuredly have a huge impact on the social sciences due to the fundamental insight that people do not respond to stimuli, they act to oppose disturbances to their controlled variables.

I know PCT has significantly improved my thinking about human behavior in general, and about research in psychology and economics in particular. I am currently using PCT principles to investigate my investment process for buying and selling stocks. Gaining a deeper understanding of the higher level reference perceptions underlying my formation of an "investment thesis" for a stock and my interpretation of new information, such as company news announcements, is a lot more complex and difficult than I thought.

Bartley J. Madden, BSME, MBA
Former Managing Director, Credit Suisse,
Author of
Wealth Creation: A Systems Mindset for Building and Investing in Businesses for the Long Term

As a psychologist, Runkel approaches ideas about how human behavior functions that originate from Powers' technical world of engineering. Although Runkel could consider himself to be an expert on human behavior, he puts himself into the position of a student who open-mindedly learns a promising new theory about the nature of psychological processes. As a psychologist, for me it is very exciting to witness the written dialogue between Runkel and Powers, and to put myself into the position of a student as well. Perceptual Control Theory (PCT) has a lot to tell about human behavior, a perspective that contrasts with mainstream psychological reasoning. *Dialogue* is not merely a discussion between two scientists, not just a discourse on PCT—it is a challenge to the fundamental concepts of modern psychology.

Michael Cramer, Dipl.-Psych. (Psychology)
Head of Addictions Department,
Clinic for Forensic Psychiatry and Psychotherapy,
Kaufbeuren, Germany

I was pleasantly surprised by the importance of the contents. Often, correspondence like this gets lost in the shuffle, yet is also often some of the most enlightening material on the sources of the correspondents' thinking—particularly when the correspondents' thinking is as cutting edge as Bill's and Phil's.

The period of time during which I participated in CSG meetings and had conversations with Bill and others was among the most significant of my life. My interests have been and continue to be at the societal level rather than the psychological; of course, as a member of the social systems in which I have interest, it has been important that I develop some form of understanding of myself, my desires and my thinking and how those play into collective processes of social design and transformation, which must include consideration of the desires and thinking of others.

I am an engineer by training, as is Bill, and it was always refreshing to have someone with whom to talk who was first and foremost interested in what is useful, i.e., that which might be employed to alleviate the misery that pervades much of the current human condition. Bill showed tremendous patience with my interests and found ways to discuss control systems theory that continues to inform my search for social design theory and method when I am a part of the very systems I design.

These letters remind me of the richness of those conversations.

Larry Richards, Ph.D. (Operations Research)
President, American Society for Cybernetics, 1986-88.
Executive Vice Chancellor for Academic Affairs
and Professor of Management & Informatics
Indiana University East

Learning about PCT during my undergraduate studies certainly forced me to think again. Now that I have, a world where organisms execute responses to stimuli just does not add up anymore. Grasping Bill Powers' revolutionary idea is not necessarily easy, but that is because it challenges you to review your very fundamental assumptions about how we function as humans. I am glad that I had the chance to do so early on in my career, because it certainly does not get easier later on.

Oliver Schauman, BSc (Hons) Psychology,
University of Manchester

I read this book with great interest and enjoyment. Phil Runkel and Bill Powers are two of the most intelligent people I've ever met, and reading this collection of letters felt like listening in as two great minds engage in intense and highly productive discussions. As in all really good discussions, the parties to this exchange confront conflicts and differences of perception head-on, and both come away with new insights. It was particularly fascinating to watch Phil Runkel reorganize his perceptions of what PCT is all about, and I found it equally intriguing to watch Bill Powers enlarge his view of how the social sciences can work.

Because this exchange pushed both men to the limits of their understanding of the newly developing science of control of perception, readers of the book can also draw fresh insights from their discussions. I expect that every serious student of PCT will want to read this book. Although both Powers and Runkel are superbly clear writers, some of their letters do require close reading, particularly the opening exchange that dissects the 1978 Powers article on *Quantitative analysis of purposive systems*. To figure out what they were talking about, I had to go back and reread that article, but the article was well worth rereading, and the commentary in the letters illuminated points that I had missed the first time around. Again and again through the rest of the book, I came across ideas that suggested answers to questions I've had as I apply PCT in my own sociological work.

Finally, I found it particularly poignant to observe the way that friendship and affection grew between these two men over the course of their correspondence. The two show themselves in these letters not only as great scientists but also as men of compassion, warmth, and humanity. When someday historians of science are writing biographies of Bill Powers, this book will be an invaluable resource.

Kent McClelland, Ph.D. (Sociology)
Professor of Sociology, Grinnell College, IA

B.F. Skinner famously posited that our behavior is caused by what we perceive. William Powers caught my attention when he turned that proposition on its head: What we perceive is caused by our behavior. That is, in fact, what human behavior is—action that creates a change that we perceive; if the perception conforms to what we intended, the action was a success. In a stroke, Powers put purpose at the heart of the human condition. As someone trying to understand the law and its insistence that humans are responsible for their actions, seeing humans as the authors of their behavior made sense of the legal assignment of responsibility, where Skinner's proposition would rob the law of its moral force, making it simply another form, albeit sloppy, of behavior control.

Better than that, Powers went on to illuminate the process by which humans exert their control over perception: the negative feedback process. As it has become articulated in the work that followed *Behavior: The Control of Perception*, Powers's theory grew capable of revealing the inner dynamic of all fields of human behavior, from law to ergonomic design to learning, emotions, and the behavior of crowds. The current volume puts you in the heart of this conversation about human nature, in the hands of two lucid thinkers.

Hugh Gibbons, J.D. (Legal Theory)
Professor of Law Emeritus
New Hampshire School of Law,
University of New Hampshire

As a researcher and practitioner of cognitive behavioural therapy (CBT), I have seen its powerful effects in aiding people's recovery, yet I have been aware of the limitations of, and contradictions between, the cognitive and behavioural theories that inform CBT. When I discovered PCT in the late 1990s, I saw immediately a theory that could bridge the gaps between cognition, behaviour, and motivation by considering them as integral components of a single unit—the negative feedback loop. When I read Powers (1973) further, I realised that these units could be configured in such a way as to model learning, memory, planning and mental imagery. I was 'sold', and since this time I have endeavoured to test and apply PCT within my research and clinical work. It is often difficult for therapists to grasp the notion that there can be a precise, empirical and quantitative model of purposive, humanistic psychology—but here it is.

Warren Mansell, Ph.D. (Clinical Psychology)
Senior Lecturer, Chartered Clinical Psychologist,
Accredited Cognitive Behavioural Therapist
University of Manchester, UK

William Powers' great contribution has been to explain exactly what it means, in scientific terms, to say that people have purposes, and to follow through the logic of this basic idea to build a testable theory of human behaviour. This book shows him explaining these ideas to a colleague in a correspondence over many years, infused throughout with his characteristic warmth, clarity, and vigour. I have found Powers' approach to control systems fruitful in my own work in robotics and in computer-generated human animation.

He is also one of the wisest people it has ever been my good fortune to know.

Richard Kennaway, Ph.D. (Mathematics)
Senior Research Associate
School of Computing Sciences
University of East Anglia, Norwich, UK

Can you imagine that in an hour and a half the course of your life would change?

That's what happened to me in the fall of 1957 when I wandered into a free seminar at the University of Chicago Counseling Center training program, where I was an intern. It changed my life forever.

The topic was: *A General Feedback theory of human behavior*. It left me quivering with excitement. This is it, I felt. Here was a description of how behavior really works—something I had been yearning to find all the way through my graduate courses.

Since that day, I have been striving to draw useful applications from Bill Powers writings and teachings. I have an earnest desire to see the world come to a realistic understanding of human behavior, and a conviction that such knowledge will affect the course of human life for good.

Richard Robertson, Ph.D. (Psychology)
Professor emeritus of Psychology,
Northeastern Illinois University
Co-author and editor,
Introduction to Modern Psychology;
The Control-Theory View

In his 1933 magnum opus, *Science and Sanity*, Alfred Korzybski presented his system of applied epistemology (labeled "general semantics") as his contribution to the foundations of a "science of man." As early as his first 1921 book *Manhood of Humanity*, he had expressed the importance of non-linear, circular ("spiral") causation for understanding human behavior. But his rough working intuition of circular mechanisms didn't line up with the psychology of his day, which mainly operated within the stimulus-response paradigm. So although he regretted the lack of what he considered a "scientific psychology" (an exact theory of the circular mechanisms of behavior didn't exist) he was forced when formulating his own work to make use of the best, though inadequate, studies of his day.

Almost as soon as he became aware of the notion of feedback, which began to rise into public awareness after World War II under Norbert Wiener's rubric "cybernetics," Korzybski leaped on it as "a turning leaf in the history of human evolution and socio-cultural adjustment." But it took a long time after Korzybski's death in 1950 before William T. Powers' 1973 book *Behavior: The Control of Perception* actually showed how negative feedback control, long touted by cyberneticists, might function as the core for an exact and overarching scientific theory for psychology.

Powers is not just a theorist—as an engineer he had intimate contact with the 'guts' of actual mechanical servomechanisms. He's had lots of experience with human servomechanisms too. He and his colleagues have elaborated a detailed research program for psychology, called Perceptual Control Theory (PCT), which emphasizes human autonomy, a phenomenological perspective, and the rigorous modeling of behavior. Their program has already begun to get carried out, although acceptance by the larger social/behavioral science community has been slow going indeed, since much of modern (2010) behavioral/social 'science' still operates under the burden of the outdated but still pervasive stimulus-response, linear cause and effect, paradigm. I am convinced that PCT is at the forefront of a major and needed paradigm shift in the human sciences, part of the non-aristotelian socio-cultural-scientific revision that Korzybski long hoped to foment.

This book of correspondence between Powers and his close colleague, the late Philip J. Runkel, will give the interested reader an irreplaceable inside view (and a very human one) of the developing work-in-progress in PCT over the last several decades and into the new millennium. It seems well nigh certain that Korzybski would have felt delighted to see the substantial standing and growing structure that Powers and Runkel, two early and serious students of his work (see the Name Index), have produced on such korzybskian foundations.

Bruce I. Kodish, Ph.D. (Applied Epistemology)
Author of *Korzybski: A Biography*

For readers possessing prior knowledge of Perceptual Control Theory (PCT), *Dialogue Concerning the Two Chief Approaches to a Science of Life* will provide a fascinating and intimate back story and commentary to PCT as developed by Bill Powers and encountered and understood by Phil Runkel.

Those with little or no knowledge of PCT will find the book to be an enticing, if sometimes challenging, introduction to this revolutionary way of making sense of behavior. Editor Dag Forssell's preface, the correspondence between Powers and Runkel, and a collection of other writings by PCT pioneers provide a rich and often colloquial context in which to encounter and reflect on a perspective that turns behavioral science upside down and inside out.

Sometime in the future, mainstream behavioral scientists will understand behavior as the control of perception. For them, this book will document this paradigm shift's initial diffusion to and further development by a small group of early adopters. It will also serve as a reminder of how difficult and slow such a process can be, even when the availability of personal computers made it possible to simulate in detail the hierarchical perceptual control systems that are at the heart of this new understanding of behavior.

PCT has provided a foundation that has offered me new insights into my interests in evolution, human nature, learning, and education. This book documents the building of this foundation that has broad application to all other disciplines and sub-disciplines in the life sciences.

Gary A. Cziko, PhD. (Psychology)
Professor Emeritus of Educational Psychology
University of Illinois at Urbana-Champaign
Author of *Without Miracles* and *The Things We Do*

At first glance I could not see a reason for this book, but after reading it I realized that it is a brilliant way to guide the reader into an understanding of what Perceptual Control Theory is, and why it is necessary and useful. Next to having one's own prolonged exchange with Powers, one can hardly find a better way to learn PCT and its ramifications than to see how Runkel's understanding develops through thoughtful questioning and equally thoughtful answers. This book should be on the reading list of every student of PCT.

Martin Taylor, Ph.D., P.Eng. (Engineering Physics and Experimental Psychology perception)
Scientist Emeritus, Defence Research and Development Canada – Toronto

Dialogue resonates with us. We have benefited so much from the study and application of PCT.

One benefit, most significant to us at a very personal level, flows from the basic PCT realization that behavior is the control of perception. Action is only a part of that process, and a rather automatic one at that. More significant is intent, and PCT makes it clear that you cannot tell what a person is doing (intent) by watching what the person is doing (action).

Over time, as we studied and internalized PCT, we understood that actions, what we sense and observe, are ***not*** the whole story. As each of us interprets and attaches meaning to the actions of the other, that meaning originates in each of us and does not necessarily yield a valid understanding of the other person's intent. So we have learned to ask what the other intended, was trying to achieve—not to criticize action. This has made us slow to blame or to anger. Of course, we may start by sharing how we felt about or experienced the actions of the other, but then we shift the conversation to a dialogue about intent. That habit has led us to a better understanding and acceptance of each other. Our marriage was good before we discovered PCT. It is even better now!

We also learned from PCT what we can't do. We can't make another person do what we want. In fact, if we try, if we coerce, people resist more often than not. So we proceed invitationally, asking each other and others if they would like to do such and such, participate in this or another project. We are not offended if our invitations are not accepted. Live and let live!

We treasure this dialogue of letters because it reminds us of Bill's and Phil's lively and informative conversations at the annual Control System Group meetings. It has been a singular privilege to know two such exceptional gentlemen.

Lloyd Klinedinst, Ph.D. (Middle English)
Barbara Bollmann, M.A. (Counseling)

As an engineer, I have always considered psychology to be an unreliable fuzzy science. PCT is different. It is a well structured approach that's easy to understand and just makes sense.

I work with very large computer systems. Users come to me with their problems and wish lists. With a very basic understanding of PCT, I learned not to focus on "What are you trying to do?" and get caught up in the user's proposal for changes to the system, but instead get to "What are you trying to achieve?", which meant encouraging the user to spell out the desired end result. That has worked for me as it made it easier for me to suggest alternative, much easier solutions to my users' problems.

Björn Leffler, M.Sc. (Computer Science)
Senior Software Developer
Animal Logic Studios, Australia

In this volume, Bill Powers shows Phil Runkel the way from experimental psychology towards a psychology where perceptions are controlled through negative feedback. I found it most interesting to read Phil Runkel's questions and comments along with Bill Powers' convincing answers. I imagine that Phil Runkel experienced Bill Powers' explanations in much the same way as I have experienced Bill Powers' many books and essays. Phil Runkel's *People as Living Things—The Psychology of Perceptual Control* is further evidence of the way Bill Powers explains things and events.

Bill Powers has meant more to me than any other person as I have developed my understanding of people and human behavior.

Bjørn Simonsen
Former Professor of Chemistry
Bergen University College, Norway

Dialogue Concerning the Two Chief Approaches to a Science of Life

Word Pictures and Correlations versus Working Models

William T. Powers   Philip J. Runkel

Living Control Systems Publishing
Menlo Park, CA

Numerous introductions, explanations and articles in PDF format, tutorials and simulation programs for Windows computer, links to other PCT web sites and related resources are available at the publisher's website, www.livingcontrolsystems.com.

Library of Congress Control Number: 2010930214

Publishers Cataloging in Publication
Powers, William T. 1926–2013
Runkel, Philip Julian, 1917–2007
Dialogue Concerning the Two Chief Approaches
to a Science of Life :
Word Pictures and Correlations versus Working Models
/Dag Forssell, 1940–, Editor
xlvii, 558 p. : ill. ; 28 cm.
Includes name index.
978-0-9740155-1-4 (softcover, perfect binding)
978-1-938090-00-4 (hardcover, case binding)
1. Control (Psychology). 2. Perceptual control theory.
3. Cybernetics. 4. Organizational Development.
5. Psychology—Research—Methodology. 6. Science
7. Control theory. I. Title.
II. Title: Word Pictures and Correlations
versus Working Models

BF455. 2011

♾ The paper used in this book meets all ANSI standards for archival quality paper.

On the back cover: Picture of Phil Runkel in 2003; picture of Bill Powers in 2004

How we deal with other people determines our effectiveness and satisfaction as friends and lovers, managers, parents, sales associates, teachers and counselors, both in the workplace and in our personal lives. With Perceptual Control Theory (PCT), effectively dealing with people no longer has to be confusing, a matter of luck, a gift, or something best left to specialists.

In one sense, every person alive is a psychologist. People studying management want to acquire good people skills so they can be successful. Couples want to understand each other so they can maintain a good relationship. Parents want to know how to teach their children well so they become capable adults. Teachers want to know how to inspire their students. Politicians want to know how to negotiate agreements and lead well. Counselors want to know how to help others resolve conflict. . .

PCT provides a new concept of how living organisms function, which turns much of what we think we know on its head and lays a foundation for psychology to become a science with the accuracy and reliability we expect in the physical and engineering sciences.

PCT is for everyone—a basic understanding is sufficient to achieve new and practical insights about human behavior.

This book is dedicated to those willing to investigate the scientific underpinnings of PCT. Directions to other resources, from basic introductions to instructive computer simulations, are included.

This book

Dialogue Concerning the Two Chief Approaches to a Science of Life
Word Pictures and Correlations versus Working Models

is available as a free PDF download from the publisher's website, www.livingcontrolsystems.com, as well as the free online libraries www.archive.org and www.z-lib.org, which will help ensure that this book and others on the subject of ***Perceptual Control Theory, PCT,*** will be available to students for many decades to come.

File name: DialoguePowersRunkel2011.pdf

The file is password protected. Changes are not allowed. Printing at high resolution and content copying are allowed. Before you print, check the modest price from your favorite Internet bookstore.

For related books and papers, search *Perceptual Control Theory*

For drop ship volume orders, mix and match, contact the publisher.

Minor updates and this note added in 2020.

This PDF file is big, about 30 MB, because most pages are images. Adobe Acrobat Pro has run OCR, so this PDF file can be searched.

Contents

PART II holds 11 papers / statements on PCT, science and revolutions, expanding on the editor's preface.

Much more is included in

Perceptual Control Theory:
An Overview of the Third Grand Theory in Psychology
Introductions, Readings, and Resources

This is a Book of Readings, available the same as this work.

. . . .

You did not invent the loop. It existed in a few mechanical devices in antiquity, and came to engineering fruition when electrical devices became common. Some psychologists even wrote about "feedback." But the manner in which living organisms make use of the feedback loop—or I could say the manner in which the feedback loop enabled living creatures to come into being—that insight is yours alone. That insight by itself should be sufficient to put you down on the pages of the history books as the founder of the science of psychology. I am sure you know that I am not, in that sentence, speaking in hyperbole, but in the straightforward, common meanings of the words.

In a decade or two, I think, historians of psychology will be naming the year 1960 (when your two articles appeared in *Perceptual and Motor Skills*) as the beginning of the modern era. Maybe the historians will call it the Great Divide. The period before 1960 will be treated much as historians of chemistry treat the period before Lavoisier brought quantification to that science.

. . . .

Philip J. Runkel, October 13, 1999

Editor's preface

About these letters

These letters represent much more than 500 pages of correspondence between two lucid gentlemen—the creator of PCT, William T. (Bill) Powers, and Philip J. (Phil) Runkel. The significance lies in the subject matter, Perceptual Control Theory (PCT).

The letters are part of a larger whole. This preface and Part II are intended to provide context.

About the book title

Galileo Galilei is known for *Dialogue Concerning the Two Chief World Systems*, which challenged the old and introduced a new approach to astronomy. For his heresy, church leaders sentenced him to house arrest for life, where he wrote *Dialogues Concerning Two New Sciences*, a discussion of math, physics, and scientific method. For this, Galileo is considered the father of modern physical science. Runkel makes it clear in his letter of October 13, 1999 that he thinks of Bill Powers as the father of a modern science of psychology.

The title of this volume is similar to Galileo's book titles because, just the same, these letters become a dialogue that challenges the old and introduces a new approach—this time in psychology and life science.

You can read these letters as

—questions, answers, and comments on the life sciences in general and psychology in particular
—an account of the gut-wrenching upheaval Phil experienced as his understanding of PCT grew
—an account of what is wrong with methods in psychology
—a prequel to Phil's books *Casting Nets and Testing Specimens* as well as *People as Living Things*
—a tutorial in Perceptual Control Theory (PCT)
—a glimpse into the minds of two intellectual giants
—a partial history of Perceptual Control Theory

This preface and Part II provide

—a brief introduction to PCT (p. 509)
—notes regarding PCT and scientific revolutions
—a guide to resources for your study of PCT

What you will realize

Once you have studied this volume and some of the other PCT resources, especially the tutorial programs, it will be clear to you that psychologists have not provided an understanding of individuals. You will realize that other disciplines which deal with the makeup of individuals and their interactions, such as management, sociology, education, economics, and neurology, suffer due to this lack of understanding.

Specifically, you will realize that:

—Recognizing and understanding control lays a foundation for psychology to become a science with the accuracy and reliability we expect in the physical and engineering sciences.

—Failure to recognize, study, and understand control correctly is crippling the life sciences.

—The Scientific Method has been employed for the study of living organisms without regard to the fact that they control their environment, not the other way around. As a result, psychologists have studied the wrong thing, the wrong way.

—A scientific revolution in psychology is underway, just as upsetting, historic, and productive as the revolution in astronomy 400 years ago.

—The idea of an upsetting scientific revolution in psychology will appear inconceivable, absurd, insulting, and outrageous to people who "know" that progress in science is a matter of an indefinite accumulation of facts.

—To become a true science, psychology will have to start over. Related life sciences will also benefit from a recognition and understanding of control.

—Anyone who chooses to study PCT will understand psychology as well as, or better than, existing experts do, because as Will Rogers said:

> *It isn't what we don't know that gives us trouble, its what we know that ain't so.*

Revolutions happen to sciences

The invention of PCT causes a scientific revolution, yet scientific revolutions are little known or understood. In his book *The Structure of Scientific Revolutions*, Thomas S. Kuhn makes it clear that scientific revolutions are infrequent and, once they have occurred, are rendered invisible. Textbooks are rewritten, obscuring the fact that earlier concepts were not compatible with the new.

While my education was technical, in fields where numerous revolutions have occurred, I was unaware of the idea of scientific revolutions until I read Kuhn. I too took for granted that science was a matter of steadily accumulating facts, building indefinitely on prior research. Not so. There have been numerous upheavals in the physical sciences. The Copernican revolution in astronomy is well known. Chemistry started over when oxygen was discovered in the 1780s. Just over a hundred years ago, light was still propagating through aether, which filled the universe.

If you are not aware of our history of scientific revolutions, it must seem inconceivable that there can be such a thing as psychology starting over. Among other things, this means reconsidering a huge body of research—not necessarily all observations, but certainly conclusions and explanations. In his major work *People as Living Things*, Runkel provides an overview of psychology, reconsidered in light of PCT.

I find that much of what I want to say here I have already written, so why reinvent the wheel? My colleagues in PCT have also written about various aspects of this revolution. That is why I have added Part II, a collection of papers and notes that deal with science and revolutions in general, and psychology in particular. Additional notes regarding revolutions follow below.

About psychology and the life sciences

In the realm of science, psychology is perhaps the most important discipline. Much of our health and satisfaction depends on our ability to live well and get along with others.

It makes good sense that, along with management, psychology is the most popular major in our universities. Several other related disciplines take cues from psychology: sociology, education, economics, neurology, anthropology, psychiatry, management and organizational behavior, political science, social work, counseling …

In one sense, every person alive is a psychologist. People studying management want to acquire good people skills so they can be successful. Couples want to understand each other so they can maintain a good relationship. Parents want to know how to teach their children well so they become capable adults. Teachers want to know how to inspire their students. Politicians want to know how to negotiate agreements and lead well. Counselors want to know how to help others resolve conflict.

You would think that the science of psychology will show us how to live well and be effective, but problems persist at all levels of society; within and between individuals, within and between organizations, within and between nations. The popularity of newspaper cartoons such as *Dilbert*, which portrays bad management and morale in the workplace, is but one symptom of the problems people face in their daily lives.

Several psychologists have pointed out that psychology is not scientific. But until now, nobody has been able to offer an alternative. All have been effectively ignored by the large number of people in this discipline.

I have long been aware that William James is quoted as saying: "This is no science, it is only the hope of a science". I just looked up the context of that quote by purchasing a recent republication. James' statement is much more powerful and aligned with the message of this volume than I expected. I want to share it with you. First some context from the back cover:

> In 1890, after 12 grueling years of writing, thought, and research, the great American psychologist and philosopher William James (1842-1910) finally published his two-volume

Note:
Most documents mentioned in this preface can be found at "the website" meaning either www.PCTresources.com or www.livingcontrolsystems.com. Each has a Google search bar to help you locate the file or document. www.PCTresources.com is my site focused on archives, while www.livingcontrolsystems.com is my publishing site, featuring a wealth of introductory documents, tutorial programs, videos, and numerous PCT-related books. Both sites will change with time, so I do not want to specify at which site any one resource can be found. Files relating to this work, such as enclosures and "About Phil Runkel", show on this volume's web page.

> *Principles of Psychology*—which, in the exhaustion of the moment, James himself characterized as "a loathsome, distended, tumefied, bloated, dropsical mass, testifying . . . that there is no such thing as a science of psychology." More neutral observers immediately recognized James' monumental work as innovative, definitive, and brilliant. Unfortunately, at 1400 pages, it was much too weighty to serve as a text, as James had intended it to be. So in the next two years, he condensed, reworked, and rewrote it as *Psychology: The Briefer Course.* (In academic circles, Principles came to be known simply as "James"—and The Briefer Course as "Jimmy.")
>
> . . . An enormous amount of what James wrote in the fledgling days of psychology is still true, relevant, and thought-provoking today. Students, psychologists, and general readers will welcome this new edition of one of the great—and most readable—classics of psychology.

Here is the last page of the book:

> **Conclusion.**—When, then, we talk of 'psychology as a natural science,' we must not assume that that means a sort of psychology that stands at last on solid ground. It means just the reverse; it means a psychology particularly fragile, and into which the waters of metaphysical criticism leak at every joint, a psychology all of whose elementary assumptions and data must be reconsidered in wider connections and translated into other terms. It is, in short, a phrase of diffidence, and not of arrogance; and it is indeed strange to hear people talk triumphantly of 'the New Psychology,' and write 'Histories of Psychology,' when into the real elements and forces which the word covers not the first glimpse of clear insight exists. A string of raw facts; a little gossip and wrangle about opinions; a little classification and generalization on the mere descriptive level; a strong prejudice that we have states of mind, and that our brain conditions them: but not a single law in the sense in which physics shows us laws, not a single proposition from which any consequence can causally be deduced. We don't even know the terms between which the elementary laws would obtain if we had them. This is no science, it is only the hope of a science. The matter of a science is with us. Something definite happens when to a certain brain-state a certain 'sciousness' corresponds. A genuine glimpse into what it is would be the scientific achievement, before which all past achievements would pale. But at present psychology is in the condition of physics before Galileo and the laws of motion, of chemistry before Lavoisier and the notion that mass is preserved in all reactions. The Galileo and the Lavoisier of psychology will be famous men indeed when they come, as come they some day surely will, or past successes are no index to the future. When they do come, however, the necessities of the case will make them 'metaphysical.' Meanwhile the best way in which we can facilitate their advent is to understand how great is the darkness in which we grope, and never to forget that the natural-science assumptions with which we started are provisional and revisable things.

The situation has not changed in the last 120 years. Psychology remains an art, not a science. Robyn Dawes, with his book with the telling title *House of Cards; Psychology and Psychotherapy Built on Myth,* is one of many who have sounded the alarm.

Here is the Library Journal review by Mary Ann Hughes and P.L. Neill, posted at Amazon.com:

> Dawes (social and decision sciences, Carnegie Mellon Univ.) presents a strong argument, based on empirical research, that psychotherapy is largely a shill game. He argues that while studies have shown that empathetic therapy is often helpful to people in emotional distress, there is no evidence that licensed psychologists or psychiatrists are any better at performing therapy than minimally trained laypeople. Nor are psychologists or psychiatrists any better at predicting future behavior than the average person—a disturbing conclusion when one contemplates the influence such "experts" have on the U.S. judicial system. While other books have criticized the psychologizing of our society, none has been so sweeping or so convincingly argued. This book raises such important societal issues that all academic and public libraries have a duty to make a permanent place for it on their shelves.

To me, this says in plain English that any wise person is on par with educated psychologists or psychiatrists regarding effective therapy. That includes a caring and thoughtful parent or manager, wise village elder, zen master, rabbi, and caring friend. Psychology as taught in our universities is not helping.

Tim Carey, author of *The Method of Levels: How to do Psychotherapy Without Getting in the Way,* provides additional detail. You can access his observations easily. See *A Look At Where We Are*, listed on page 507.

Runkel, whom you will get to know in this volume, said in the foreword to his major work *People as Living Things*:

> I will disagree in serious ways with most of the widely accepted psychological theories you encounter in popular literature, in textbooks (of whatever discipline), and in the halls of academe. I will agree with the other theories at some points, but the underlying assumptions of the theory here (Perceptual Control Theory) are not those you will find either printed or implied on many of the pages printed about psychology. In that sense, this book is disputatious. I do not, by the way, claim that those other authors and lecturers are immoral or mentally deficient. I claim only that they are wrong.

About Perceptual Control Theory, PCT

Developed by William T. (Bill) Powers, Perceptual Control Theory (PCT) is a quantifiable, testable model of how living systems work. In time, PCT will help us understand living organisms with the accuracy and reliability we expect in the physical sciences.

Understanding PCT starts with understanding control systems. We use all kinds of mechanical control systems regularly, such as thermostats and cruise controls. We set a desired temperature, and if there is a difference between that setting and the temperature sensed by the thermostat, it turns on the heater or the air conditioner. We set the speed we want to drive, and if the car notices that we slow down, it automatically steps on the accelerator.

Bill Powers explains:

> Control is a process of acting on the world we perceive to make it the way we want it to be, and to keep it that way. Examples of control: standing upright; walking; steering a car; scrambling eggs; scratching an itch; knitting socks; singing a tune. Extruding a pseudopod to absorb a nanospeck of food (all organisms control, not only human beings).
>
> The smallest organisms control by biochemical means, bigger ones by means of a nervous system. Whole organisms control; the larger ones have brains that control; most have organs that control; if they are composed of many cells, their cells control; the DNA which directs their forms and functions controls; even some molecules, certain enzymes, control by acting on the DNA to repair it when it's damaged. Control is the most basic principle of life and can be seen at every level of organization once you know what to look for.
>
> ...The problem is not that the life sciences got everything wrong; it's just that they got the most important things wrong: what behavior is, how behavior works, and what behavior accomplishes.

Full disclosure:
I refer frequently to Kuhn. His opinion of PCT appeared on the book jacket when Bill's major work, *Behavior: The Control of Perception* was published. (In discussions, this title is often abbreviated B:CP).

THOMAS S. KUHN, Professor of the History of Science, Princeton University; author of *The Structure of Scientific Revolutions.*

> "Powers' manuscript, *Behavior: The Control of Perception,* is among the most exciting I have read in some time. The problems are of vast importance, and not only to psychologists; the achieved synthesis is thoroughly original; and the presentation is often convincing and almost invariably suggestive. I shall be watching with interest what happens to research in the directions to which Powers points."

How is PCT different

Once you understand PCT, you gain a perspective on contemporary psychologies.

Bill Powers portrayed stimulus-response thinking as well as cognitive psychology from a control theory perspective at a Control Systems Group conference. The following is based on his discussion.

Let us start with a control diagram. It is not my intention here to explain PCT, only to identify the variables and functions considered in a control diagram, and how they interact. A convenient summary is featured in *Once Around the Loop*, a paper posted at the website and included in the *Book of Readings*.

(By *Book of Readings*, I mean *Perceptual Control Theory; Science & Applications—a Book of Readings*, credited to Powers and updated from time to time.)

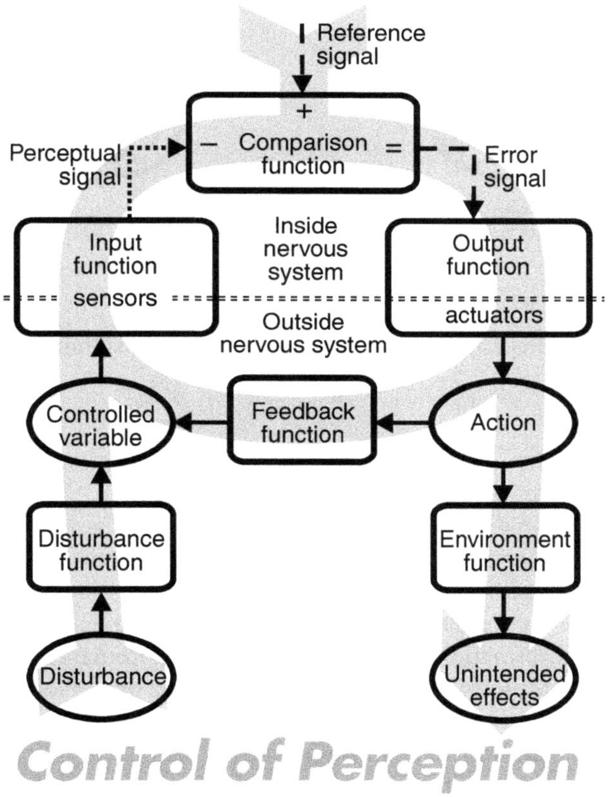

Fig. 1 Perceptual Control Theory, PCT
Closed–loop Psychology
The basic summary control diagram.
The grey overlay highlights the closed-loop flow.

Please review Figure 1: Disturbance is something going on in the environment that affects whatever the organism cares about, the Controlled variable, as represented to the brain by the Perceptual signal.

What the organism wants in regard to the Controlled variable is represented by the Reference Signal. Comparing the two results in the Error signal (a difference signal, not a value judgement) which affects actuators, whether muscles or physiology, so that the Perceptual signal representing the Controlled variable is brought to (or kept) close to the Reference signal.

And no, we do not say it is this simple. This diagram represents an entire hierarchy of control systems—by the millions—at work throughout your nervous system at all times, controlling a multitude of variables inside and outside your body, all simultaneously.

For a conceptual sketch of the proposed hierarchy, see *Perceptual Control—Details and Comments* in the *Book of Readings*. (That and similar illustrations, including the pattern on the cover of this book, draw on Mary Powers' sketch on page 405.)

Behaviorism

The idea of stimulus-response seems intuitively obvious. For example, if you stand on the deck of a ship during a storm, the heaving deck makes you do things (but only if you want to stay upright ☺).

René Descartes formalized the concept of stimulus and response in the mid-1600s. Behaviorists have worked hard to build a science based on this, and while some psychologists will claim that behaviorism is out of fashion, it is very much with us and Experimental Analysis of Behavior (EAB) is alive and well.

Applications permeate our culture. Surely you have heard of gold stars, incentive programs, and one minute management.

Figure 2 shows the control diagram overlaid with an interpretation of what researchers invested in stimulus-response thinking are looking at: Disturbances in the environment and Action by the organism.

As you can see, psychologists studying a stimulus (Independent Variable) and the response to that stimulus (Dependent Variable), creating statistics galore (and mistakenly presuming that correlation implies causation and that statistics tells us about individuals–see Kennaway (1998)), are studying only that which is visible in the environment, thus looking at a very small subset of the whole. It is not possible to build a science based on such a limited understanding of what is going on.

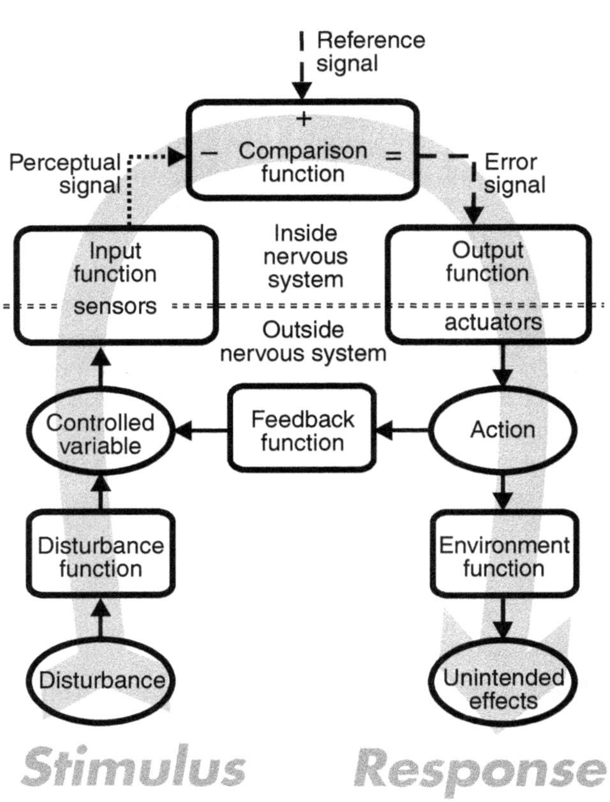

Fig. 2 Behaviorism:
Stimulus–Organism-Response
Illustrated in terms of a basic control diagram

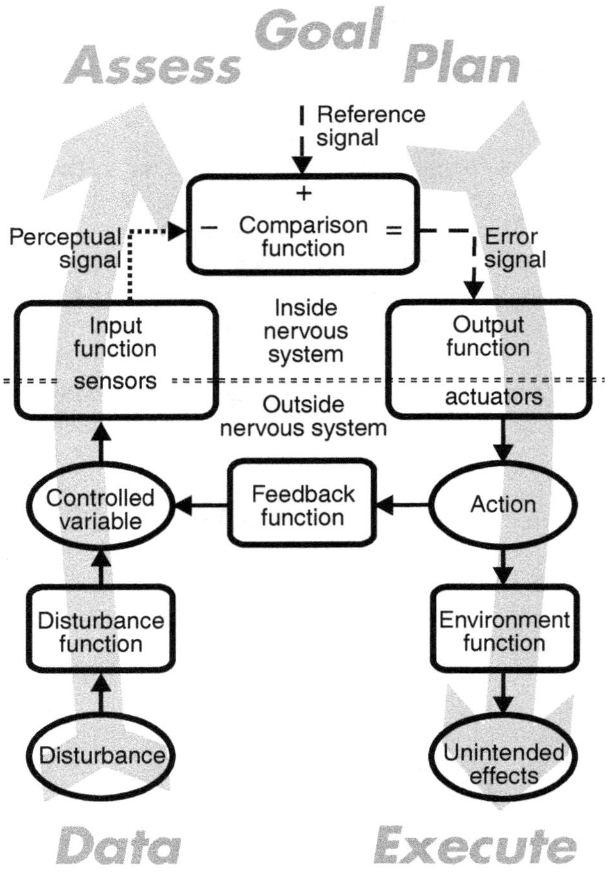

Fig. 3 Cognitive Psychology
Data-Assess-Goal-Plan–Execute
Illustrated in terms of a basic control diagram

Cognitive psychology

It also seems intuitively obvious that your mind issues commands to your muscles.

For example, if your ship is at rest, the environment does not make you walk across the deck to the other side. You just decide to walk. So now we study how the mind can evaluate the environment and plan action, then issue commands to our muscles. Engineers have demonstrated (using a laborious approach called Inverse Kinematics) that it is very possible to precompute commands to muscles and motors so limbs move just so—provided you have a powerful computer and provided that there are no disturbances at all. Muscles must not tire, and the environment must not change. This is the case for robots in repeatable circumstances and for animated 3-D figures in computers, but never for living organisms in the real world.

Figure 3 shows the control diagram overlaid with an interpretation of what researchers in the discipline of cognitive psychology are focusing on.

The intuitively obvious idea that the brain processes information, plans action, and issues commands to our muscles lies at the heart of cognitive psychology, and psychologists are working hard to sort out the complexities of our minds on this basis.

Not so intuitively obvious is the fact that neither the concept of behaviorism nor that of cognitive psychology is sufficient to explain how you can make your way across that heaving deck during the storm, or how a swallow can fly right into the small opening of her nest, without fail, on a windy day.

Discussion

While many psychologists recognize purpose and feedback in general, a detailed, correct understanding of negative feedback control is missing. Without a detailed understanding it is not possible to create a science of psychology.

This situation in psychology today is not very different from the situation in astronomy 400 years ago. At the time, astronomy was well developed with extensive observations and elaborate explanations based on the intuitively obvious idea that the earth is the center of the universe and everything revolves around the earth.

As anyone who spends night after night observing the heavens can see, from time to time Mars and the other planets change course relative to the stars, moving forward, then back, then forward again. A prominent feature of earth-centered astronomy was the explanation that Mars and the other planets not only move in circles around the earth, but at the same time in little circles, epicycles, around a point on the big circle as it progresses around the earth.

Once people reviewed the evidence and understood the mechanism of the solar system, the explanations that went with the earth-centered astronomy crumbled. Mars and the planets never move backwards. It just looks that way. The phenomenon turned out to be an illusion.

Just the same, once you understand the mechanism of control and review the evidence, the explanations that go with stimulus-response and/or plan-execute psychologies crumble.

As illustrated in Figures 2 and 3, Perceptual Control Theory shows that the intuitively appealing explanations in terms of stimulus-response and plan-execute are incomplete at best. The phenomenon of stimulus-response is an illusion. Organisms do not respond to stimuli, they oppose disturbances to their controlled variables. Organisms do not plan actions, they simply change reference signals and the hierarchy of control systems acts as necessary to bring perceptual signals in line with reference signals. PCT is a larger, complete, more all-encompassing explanation than either behaviorism or cognitive psychology. Therefore, PCT cannot be integrated into these limited approaches any more than it was possible to integrate the idea of the solar system into the then existing earth-centered astronomy.

In Figure 1, you can see that Perceptual Control Theory considers disturbances in the environment, a rapidly varying reference signal (think speech, what you want to hear from your mouth varies rapidly and you can control the sound quite well), tiring muscles, changes in how your limbs affect the environment and the controlled variable. Because of the nature of negative feedback control, organisms can deal with rapidly varying reference signals, disturbances, functions and variables.

Clearly, attempting to correlate any two variables is not enough. While cognitive psychologists are fond of talking about a cognitive revolution in psychology, the mistaken application of the scientific method has not changed[1]. Research is still based on correlating an Independent Variable with a Dependent Variable. Neither behaviorists nor cognitive psychologists realize that it has been a profound mistake to focus on Action/Behavior. What is of interest to the organism is the state of its Controlled variable. Conducting research informed by PCT you would look for a very low correlation between any Disturbance and the Controlled variable rather than a high correlation between Disturbance and Action[2]. This is the point of the demonstration Bourbon relates in his paper *Three Dangerous Words.* (See Part II, page 530, right column)

For more on psychological theorizing, see Runkel's *People as Living Things*, Part III Science.

About scientific revolutions

The movie Avatar provides a nice, very personal introduction to scientific revolutions.

A Na'vi girl called Neytiri has just saved our hero Jake from snarling beasts. The Na'vi are natives living on the moon Pandora, resisting the human intruders (Sky people) who are mining their incredibly valuable mineral Unobtanium without regard for the natives or their environment.

As they walk along, Jake asks Neytiri why she saved him. She answers that he has a strong heart and no fear, but that he is *ignorant,* like a child. So he suggests that she should teach him. She answers that Sky people cannot learn, they do not *see*, and that nobody can teach them to *see*.

---

1 For a discussion, see Marken (2009)

2 See Marken (1992, 2002)

Jake is brought to the village gathering and the matriarch Mo'at examines him. She asks why he came; he answers that he came to learn. Mo'at says that the Na'vi people have tried to teach other Sky People, but that it is hard to fill a cup which is already full.

Jake responds that his cup is empty; he is no scientist. Mo'at assigns Neytiri the responsibility of teaching Jake the Na'vi ways, and they will see if his insanity can be cured. How Jake and Neytiri come to appreciate each other, and then to love, illustrates what Ed Ford calls *Quality time* in his book *Freedom From Stress*—an introduction to PCT based on his experience as a counselor.

Scientific revolutions are personal

In the clash between cultures that the movie depicts, what matters are personal understandings. That is why it is significant that Jake's cup is empty. Unlike his colleagues, Jake does *not* have a Ph.D. and has *not* spent three years or more studying the human occupiers' documentation of the Na'vi culture, language, and environment. Thus he does not have an investment in a particular understanding and it is much easier for him to come to see the Na'vi world through Na'vi eyes and appreciate its beauty.

All scientific revolutions are personal. As Clark McPhail, a sociologist and student of PCT, makes very clear in *The Myth of the Madding Crowd,* there is no such thing as a group mind. All individuals are thinking and acting separately.

Thus this revolution in psychology is an issue for each individual who undertakes to study PCT. Bill Powers is the first to point out that none of the people who have looked into PCT so far were taught PCT at an early age. Everyone has a cup that is full already, making the transition that much more difficult.[3] Bill considers himself to be a student of PCT, not a guru, both as a matter of attitude and because much remains to be figured out and researched. It follows that everyone else in the PCT sphere is a student too. This is one reason Bill is tolerant and supportive of anyone who makes an effort to learn PCT.

Given that everyone who is exposed to the concept and explanation that PCT offers has already created a personal web of understandings based on personal experiences and interpretations from an early age, supplemented with teachings at whatever level in school, nobody has a cup that is empty. But there are degrees of fullness and there are variations in how a person thinks, as a result of what the person has experienced and what conclusions the person drew from those experiences.

I hope that telling my story, my journey to PCT and experience to date, will provide useful context —an overview of progress to date, where and how anyone can learn more.

Dag's story

My wife Christine and I grew up, met and married in Sweden. We traveled to the U.S. together in 1967 to see the world before we would settle down in Sweden. We never returned.

I got jobs as an engineer and engineering manager with marketing responsibilities. Christine, while she had worked as a physical education teacher during our first years together in Sweden, worked at home to raise our two daughters.

In 1975, Christine got involved in direct sales of nutritional products—in line with her interest in good health—and I accompanied her to events featuring motivational speakers. I was intrigued. I listened to the speaker spin a story of how she would tell the customer this, and the customer would think that, and then she said the other, complete with detailed explanations of what went on in the customer's mind, the customer's spouse's mind, and their circumstances. Of course the customer would buy the product package. One can get motivated by this kind of imagining, but the euphoria is fleeting. The problem is that while you buy into the story the speaker relates, this scenario is not likely to happen in the real world. Nevertheless, I enjoyed numerous tape recordings of well known motivational speakers, such as Earl Nightingale. I was open to suggestions from various directions and found some of the advice useful.

In my search for insight into what makes people tick, I continued reading books on topics such as listening and character education, one recommending the other. I found *Reality Therapy* by William Glasser, liked it, and read most of his writings. I found his book *Stations of the Mind* fascinating. Here, Glasser explained and illustrated PCT in order to provide theoretical support for *Reality Therapy.* A foreword by Powers discussed the origins of PCT.

3 For some ideas on how we all fill our cups, see *Are All Sciences Created Equal,* starting on page 535 in this volume.

Discovering PCT

Curious about the foreword to *Stations of the Mind,* I overcame the high price, purchased and read Powers' *Behavior: The Control of Perception (B:CP).* I found an elegant, very physical explanation of how our nervous systems can work. To me, the text is clear and well illustrated. I found it easy to visualize interactions between neurons as shown in chapter 3, Premises (featured in the *Book of Readings*). I saw that there are significant differences between Powers' original and Glasser's embellished, very personal interpretation (currently called Choice Theory). But I am glad that through Glasser I found the real thing.

In early 1989 I asked a member of Glasser's faculty about Powers. I was directed to Ed Ford and visited him in Phoenix. Ed supplied me with his book *Freedom From Stress* and told me about Powers' Control Systems Group (CSG) and its forthcoming conference at the Indiana University of Pennsylvania.

Getting involved—conferences, archives, email, teaching, books and recommendations

Traveling to the conference, I met Gary Cziko while waiting for the bus at the Pittsburgh airport. He told me right away that his focus was evolution, and he has since written two excellent books on evolution and PCT—*Without Miracles: Universal Selection Theory and the Second Darwinian Revolution* and *The Things We Do: Using the Lessons of Bernard and Darwin to Understand the What, How, and Why of Our Behavior.* See also *The Origins of Purpose: The First Metasystem Transitions* among Bill's introductions at the website.

Arriving at the conference, I met Bill Powers and many others who have become good friends.

One was Tom Bourbon, who was teaching psychology and PCT at the University of Texas at Austin.

Another was Greg Williams. He and his wife Pat, both MIT engineers, saw the historic significance of Powers' work a few years before I came into the picture. Greg edited the newsletter *Continuing the Conversation* (CC) from 1985 through 1991. CC started out as a forum for conversations about Gregory Bateson, but shifted to cybernetics and PCT in 1986 once Greg discovered PCT. From 1986 through 1989, CC served as the official newsletter of the American Society of Cybernetics (ASC) and is now archived at the website as well as at ASC's website. You will find CC discussed in letters on pages 280 and 345 in this volume. Greg recorded CSG meetings starting in 1987. He also edited *Closed Loop* from 1991 through 1994. At the outset this newsletter featured threads from the mailing list Control Systems Group Network (CSGnet), later complete articles. Closed Loop is archived at the website. The last issue of Closed Loop features a 54-page catalog of CSG archive materials held by Greg and Pat at their home in Kentucky. The extensive list includes items such as all 15 Masters theses by Tom Bourbon's students. While serving as archivist for the Control Systems Group (CSG), Greg made selections from Bill's unpublished papers, edited and typeset Bill's anthologies (1989) and (1992) and the college textbook by Robertson (1990).

I have become a second archivist for CSG. Greg and I will work with educational institutions to make CSG archives available to students and researchers and to ensure that they are duplicated so they will not be lost to history if any one location suffers a catastrophic loss.

One 1989 presentation that has stuck in my mind and that any reader can replicate was Wayne Hershberger's illustration of saccades. Wayne held up a red LED, such as was common on digital clocks. These LEDs actually blink at 60 Hertz because of the AC current. Wayne darkened the room and asked us to look at the red light, then shift our gaze suddenly far to the left. It was easy to see blinks an equal distance to the right, before the light was again stationary in Wayne's hand. If you sit still and move your eyes around, the image of the room in front of you does not shift, or shifts very little, even though obviously the image falls on a different place on your retina. It made sense to me to think that the control hierarchy postulated by PCT would shift the retina's coordinate system as it shifts the directions of the eyeballs, and that the neural coordinate system would shift faster than the physical eyeballs. Thus the blink off to the right. (With a solid light you see a streak instead of discrete dots). At the time, Wayne was editing the anthology *Volitional Action, Conation and Control,* which features 25 chapters. Half relate to PCT.

Jim Soldani, formerly Director of Systems Manufacturing at an Intel plant in Phoenix, contributed the chapter *Effective Personnel Management: An Application of Control Theory.* Jim reported spectacular results from applying his understanding of PCT. For more, see Jim's recent paper *How I applied PCT to get results.* In the fall of 1990 Jim came to the Phoenix airport to spend 45 minutes with me. He shared with me that he had spent six years following his success at Intel developing a consulting business teaching PCT to industry. Despite pockets of considerable success, he found it a hard sell and had to give up.

Purposeful Leadership

I was at a crossroads at the time. I had met Mike Bosworth who was teaching a sales training program called *Solution Selling* and served as a student group coach a few times. I explained to Mike that the basics of his approach might have been based on the way PCT would suggest that you focus on how the customer wants to solve his problem, not on what the salesman wants to sell. Mike encouraged me to teach PCT to sales managers. He explained that a persistent problem is that a star salesperson gets promoted to sales manager and falls flat on his face. While solution selling was a good program for teaching salespeople how to sell, a good program teaching sales managers how to manage would be invaluable.

In spite of Jim Soldani's warnings, I undertook in early 1991 to teach PCT to captains of industry and registered the trademark *Purposeful Leadership*. I put a program together and mailed thousands of letters to executives. Most of these no doubt ended up in wastebaskets, but one technology company allowed me in 1992 to present my program on three consecutive Wednesdays. About 15 people in marketing and 15 engineers signed up. By the third Wednesday, most of the marketing people had dropped out, leaving the engineers. The Human Resources manager told me afterwards I was not as entertaining as she expected but the feedback I received from the engineers was positive. Two of the engineering managers wrote me a year and a half later to report on how they were using what I had taught them and their results.

Much later I assembled articles and an outline of my program along with the feedback in *Management and Leadership: Insight for Effective Practice*.

In the early 1990s, the Deming Management Philosophy was new and Total Quality Management (TQM), was emerging as a management tool. I saw a connection, so I attended Deming's seminar and wrote Dr. Deming, who graciously responded:

> Dear Mr. Forssell, 15 June 1991
>
> I thank you for your letter of 8 June 1991. Yes, Profound Knowledge is not what people are looking for. They seek procedures and formulas. It is a hard broad jump. I agree, psychology is the weak link.
>
> Sincerely yours,
> *W. Edwards Deming*

With the experience of my seminar under my belt, I presented a two-hour introduction to PCT to Deming Users Groups in early 1993. See the video *PCT supports TQM* at the website.

Dr. Deming's note supports the conclusion I drew from my experiences that many people do not expect understanding from seminars; people expect entertainment and prescriptions.

My efforts to develop a teaching and consulting practice failed to generate income, so by 1994 I had to give it up. I found a new profession translating technical texts between English and Swedish.

Staying involved

Gary Cziko sponsored an email discussion group, Control Systems Group Network (CSGnet) in September, 1990. CSGnet became a forum for lively, wide-ranging discussions about PCT, with Bill Powers patiently teaching all comers. Bill is still going strong 20 years later.

Because I was focusing on PCT full time, I immediately began saving all the CSGnet correspondence and have continued to do so. This archive is available at the website. The earliest record consists of Word files, but as of March 1992, mailboxes created by the Eudora e-mail program are available as well. Eudora, now in the public domain, features excellent Boolean search capability, which means that you can search the many megabytes for comments on any topic.

In 1993-1994 I also undertook to assemble about 100 threads from CSGnet. Needless to say, these too are posted at the website. Threads discussing stories, belief and knowledge are particularly relevant to this discussion of scientific revolutions[4].

I had purchased a video camera and editing tape deck for the purpose of teaching PCT, so I brought it to the 1993 CSG conference and have taped most conferences since. More than 300 hours of camera tapes have now been digitized and I will be happy to provide this material to institutions and serious students.

CSG conferences are very informal indeed. That is the way Bill wants it. Participants organize a schedule of presentations on the first evening of a three-day conference and anyone is welcome to present most anything, even where the relation to PCT is tenuous at best.

As I find time to edit video and create flash files, I will post a selection of presentations at the website.

---

4 See especially the threads called Gullibility.pdf, Religion.pdf and the last entry in Clarity.pdf.

Presentations in 1993 that have stayed with me include Bill Powers presentation *The dispute over control theory*, which inspired my illustrations in this preface, and sociologist Clark McPhail telling how researchers in the past never studied crowds, but voiced opinions anyway, and these now dominate textbooks. Clark had recently published his book on crowds. A few sentences from the Foreword tell the story:

> A most peculiar thing has happened. A few scholars of crowd behavior actually have begun to observe and describe systematically the empirical features of crowds. Disconcertingly, this represents a radical development in the annals of crowd analysis. Clark McPhail is the intellectual leader without peer in chronicling and categorizing temporary gatherings before trying to explain them. As a result, his accounts of their variable features have virtually no counterpart.

Kent McClelland gave his first of several presentations on *Conflictive cooperation*, later published as *The Collective Control of Perceptions: Constructing Order from Conflict*. Kent's work is very suggestive about how large groups of people, even while bickering among themselves, control for a set of outcomes with great collective force. This helps explain resistance to new ideas by groups of scientists, as well as the glacial pace of political process involving large groups.

In his presentation, Clark McPhail mentioned that sociologists are interested in purpose. Clark is one of the many contributors to the recent book *Purpose, Meaning, and Action*, which Kent co-edited.

Other 1993 presentations I remember were Tom Bourbon on *Person-Model Interactions: Interference, Control of Another, Countercontrol & Conflict* and his student Michelle Duggins-Schwartz on the topic *When is helping helping?*

In 1994, I presented my interpretation of how *Memory* might be continuously active in the hierarchy and an attempt to sort out *Explanations*, which became *Are All Sciences Created Equal?* on page 535 in this volume.

Ed Ford and collaborators provided an overview of their development in inner city schools in Phoenix of their *Responsible Thinking Program* (RTP); a program designed to resolve discipline problems in schools in a way that is supportive of both students and staff.

In 1995 a group of five PCTers presented at the American Educational Research Association (AERA), conference in San Francisco. The demonstration / workshop was led by Hugh Petrie who at the time was Dean, Graduate School of Education, State University of New York at Buffalo. Once I met Hugh, I ordered his book *The Dilemma of Enquiry and Learning* which spells out how he found that PCT resolves Meno's classic quandary. For more on Hugh and his involvement with PCT, see his Intellectual Autobiography at the website.

In 1997, Wolfgang Zocher, an engineer, presented *Simulating eye movement*, a simulation he had carried out using an analog computer. He later demonstrated his computer when he hosted a CSG conference in his home town of Burgdorf, Germany. Bill Powers presented *Artificial Cerebellum* and *Little Man*, fruits of his increasingly realistic modeling efforts. Tom Bourbon presented *Interactive control, a survey of where PCT has been tested in social interactions.*

Also in 1997, Bill Powers presented prints of the original draft for what became *Making Sense of Behavior*. Many people have expressed appreciation for this slim volume for its simple, basic, easy-to-understand introduction to PCT—featuring neither equations nor graphics.

1998 saw two conferences. The first, at Schloss Kröchlendorff north of Berlin, Germany, featured a fascinating presentation by Bill of his new program *Inverted Pendulum*. We have all balanced a broom in our hand, moving the hand around to keep the broom upright. Well, as we walk about we are our own brooms, so this demo is all about us. Bill explained that he achieved the splendid performance of his model using just five nested control systems.

Frans Plooij presented *PCT and infant research, an 11 year overview*. I consider the work of Hetty van de Rijt and her husband Frans Plooij, now available in English as *The Wonder Weeks*, to provide some of the most compelling, tangible evidence available that Powers' suggestions for a hierarchical arrangement of control systems is much more than hypothetical. When you read their book, you are reading about the mental development of infants in stages of progressively more complex perceptual capability. At the same time you are reading about how your own brain is working right now, a hierarchical layer cake of control systems, with each successive class of perceptions building on those that were developed before it. www.thewonderweeks.com features information about their research as well as supportive research by other behavioral biologists.

In Vancouver, BC, that same year, one of the new developments was Rick Marken's report on how baseball players catch balls by keeping certain perceptual variables under control.

1999 featured a separate two-day pre-conference on *The Method of Levels*, anchored by Bill Powers and Tim Carey. During the regular conference that followed, Tim Carey gave presentations on *Bullying* and *Counter-control.*

At the 2000 conference in Boston, Bill Powers introduced his recent simulation program *14 degrees of freedom* featuring an entire arm, and Hugh Gibbons, professor at the Franklin Pierce Law Center, presented a theory of rights. Hugh is the author of *The Death of Jeffrey Stapleton: Exploring the Way Lawyers Think*, in which he uses PCT to explain the structure of law.

Significant to me, this was also the conference where Phil Runkel tugged at my shirtsleeve during a break and asked me to review his manuscript (which I published in 2003) for technical accuracy. I began development of livingcontrolsystems.com in 2004 to support Phil's work. It has grown into a PCT reference site.

At the 2001 conference in Burgdorf, Germany, Richard Kennaway presented his six-legged bug named Archie, with full physical dynamics, using control systems to operate the legs and a pair of odor-sensing antennae to detect food locations. Archie can walk over uneven terrain, all without using any inverse kinematic or dynamic calculations, any analysis of the terrain, or any plans of action. Richard also presented his work on an *Avatar* that translates simple code into sign language for TV programs, moving smoothly from one sign to another in a very natural way.

During the 2003 conference in Los Angeles, we celebrated the 30th anniversary of the publication of B:CP. A delegation from South China Normal University attended. Tributes to Powers were offered. Lloyd Klinedinst unveiled the web-based Festschrift he had organized as a tribute to Powers' genius.

Bart Madden, an independent researcher, found PCT in early 2005. Bart is focused on market-based solutions to public policy issues. He recently published *Wealth Creation; A Systems Mindset for Building and Investing in Businesses for the Long Term.* The first chapter, *A Systems Mindset*, features a discussion of the importance of considering the purposes of managers as well as employees, shareholders, and customers. Bart correctly introduces the basics of PCT and adapts his insight to his presentation.

A most significant recent development is Powers' *Living control systems III: The Fact of Control.* Runkel read B:CP. I did. Many others have. But it is not all that easy to grasp PCT from the written presentation by Powers, however lucid, or from any other written description. Words get in the way. Our understandings of words are necessarily influenced by our personal experiences, so the meanings of words can never be exactly the same for any two people.

Understanding control and PCT has now become much, much easier. The 13 Windows programs that this book explains in its nine chapters are control systems. By changing parameters of these control systems, you can experience the nature of control directly, in diverse ways. These personal experiences will enable you to understand the intended meaning of the words about control that you will read in this book and in the other works that we have cited.

Shelley Roy's book *A People Primer: The Nature of Living Systems* is a welcome addition to the PCT literature. This book is an easy read, yet portrays PCT correctly as Shelley discusses common problems.

In 2009, Bill Powers wrote an outline for a TV program designed to introduce PCT. The program did not come to pass, but Bill's paper explaining *PCT in 11 Steps*, followed by *Reorganization and MOL*, an overview of how control systems may come into being, change, cause internal conflict, and ways to resolve internal conflict, is an excellent introduction to and summary of PCT.

Final comment

As I have studied PCT, participated on CSGnet, and attended conferences, I have come to understand what turns out to be a frequent problem for PCT—people read about it, figure they understand it because the terminology sounds familiar, and proceed to publish their own distorted versions that cannot work. In his recent intellectual autobiography, posted at the website, Hugh Petrie writes in a note:

> Those familiar with the educational literature will recognize that William Glasser has written extensively in education utilizing a concept he calls "control theory." Although there are superficial resemblances to Powers' perceptual control theory, Glasser completely fails to appreciate that what is controlled are perceptions, not actions or behaviors. This renders Glasser's version of control theory no more insightful than most cognitivist theories in psychology.

I have come to understand that we all make new information fit what we already think we know. If our cups are already full, and depending on how we think, this means that the new may be interpreted and distorted so it will fit, like forcing a square peg into a round hole, even if the result is turning PCT upside down and backwards. PCT itself explains how this works.

Kuhn points out that new ideas are typically resisted by people already steeped in a science, that the new ideas often come from outsiders, and that younger scientists, whose cups are not as full, are the ones who weigh the merits of the new compared to the old and make the choice to go with the new. This is why scientific revolutions tend to take a long time to play out. They require a generation change.

You will find examples and discussions of this phenomenon in this volume and at the website under *Controversy, Comparisons and Acceptance.*

Glasser's mistake is common. The holy grail of psychology has long been the prediction and control of behavior. The idea that we control our behavior permeates our culture. Many control engineers think so too—control systems control their output, right? Wrong! It is not the movement of a motor or the position of the machine that is controlled. It is the reading from (the perceptual signal from) the sensor that reports on the position of the machine that is controlled. This becomes very clear if the sensor is poorly calibrated. Simple control systems have no knowledge of their output/actions, the only thing they "know" is what they sense, their input.

As humans, we can pay attention to and remember our actions, but for the most part we do not. We pay attention to outcomes and whether they match what we intend. People do not control their action/behavior. People control for what they want to experience, outcomes, their sensory input.

It follows that most people alive today, including control engineers, talk about control and presume they understand it, but have never realized that their understanding is deeply flawed. Our cups are full—full of mistaken interpretations—and as a result almost everybody is profoundly ignorant about how and why we all behave as we do.

There is much more available than I have touched on here, at my web sites and at those of other PCTers. More will develop as the world catches on to PCT.

As you can see, this volume, while extensive, is only the proverbial tip of the iceberg of information that is available to you. Enjoy! I hope you find this introduction and the references useful for your studies.

Dag Forssell
Hayward, California

References

Bosworth, Michael (1993) *Solution Selling: Creating Buyers in Difficult Selling Markets.* New York, NY: McGraw-Hill

Bourbon, W. Thomas and William T. Powers (1993) Models and their worlds. *Closed Loop*, 3(1), 47–72. Reprinted as Ch. 12 in Runkel (2003)

Carey, Timothy A. (2006). *The Method of Levels: How to do Psychotherapy Without Getting in the Way.* Menlo Park CA: Living Control Systems Publishing.

Cziko, Gary (1995). *Without Miracles: Universal Selection Theory and the Second Darwinian Revolution.* Cambridge MA: MIT Press.

Cziko, Gary (2000). *The Things We Do: Using the Lessons of Bernard and Darwin to Understand the What, How, and Why of Our Behavior.* Cambridge MA: MIT Press.

Dawes, Robyn M. (1994). *House of Cards: Psychology and Psychotherapy Built on Myth.* New York: Free Press.

Ford, Edward E. (1989, 1993) *Freedom from stress; most people deal with symptoms... this book, based on perceptual control theory, deals with causes.* Scottsdale, AZ: Brandt publishing

Forssell, Dag C. (2008). *Management and Leadership: Insight for Effective Practice.* Menlo Park CA: Living Control Systems Publishing.

Gibbons, Hugh (1990) *The Death of Jeffrey Stapleton: Exploring the Way Lawyers Think.* Concord, NH: Franklin Pierce Law Center. Second edition (2013), updated, Menlo Park CA: Living Control Systems Publishing

Glasser, William, MD. (1981) *Stations of the Mind; New Directions for Reality Therapy.* New York, NY: Harper & Row.

Hershberger, Wayne A. (Ed) (1989) *Volitional Action, Conation and Control.* Amsterdam: North-Holland.

James, William (1892). *Psychology: The Briefer Course.* New York: Holt. Abridged republication (2001) Mineola, NY: Dover Publications

Kennaway, Richard (1998). *Population statistics cannot be used for reliable individual prediction.* School of Information Systems, University of East Anglia. 17 April 1998. [Posted at the website]

Kuhn, Thomas S. (1970) *The Structure of Scientific Revolutions,* second edition, enlarged. Chicago: University of Chicago Press (1996 edition identical except for the addition of an index and smaller print).

Madden, Bartley J. (2010) *Wealth Creation; A Systems Mindset for Building and Investing in Businesses for the Long Term.* Hoboken, NJ: Wiley

Marken, Richard S. (1992). *Mind Readings: Experimental Studies of Purpose.* Gravel Switch KY: Control Systems Group

Marken, Richard S. (2002). *More Mind Readings: Method and Models in the Study of Purpose.* St. Louis MO: newview.

Marken, Richard S. You say you had a revolution: Methodological foundations of closed-loop psychology. *Review of General Psychology.* Vol 13(2), June 2009, 137-145.

McClelland, Kent A. (2004). *The Collective Control of Perceptions: Constructing Order from Conflict.* International Journal of Human-Computer Studies 60:65-99

McClelland, Kent A. and Farraro, Thomas J. (Eds.) (2006) *Purpose, Meaning, and Action; Control Systems Theories in Sociology.* New York, NY: Palgrave Macmillan

McPhail, Clark (1991). *The Myth of the Madding Crowd.* New York: Aldine de Gruyter.

Petrie Hugh G. (1981) *The Dilemma of Enquiry and Learning.* Chicago: University of Chicago Press. Second edition (2011), updated, Menlo Park CA: Living Control Systems Publishing

Powers, William T. (1973). *Behavior: The Control of Perception (B:CP).* Chicago: Aldine. Second edition (2005), revised and expanded, Bloomfield, NJ: Benchmark Publications.

Powers, William T. (1989). *Living Control Systems: Selected Papers of William T. Powers.* Bloomfield, NJ: Benchmark Publications

Powers, William T. (1992). *Living Control Systems II: Selected Papers of William T. Powers.* Bloomfield, NJ: Benchmark Publications

Powers, William T. (1998). *Making Sense of Behavior: The Meaning of Control.* Bloomfield, NJ: Benchmark Publications

Powers, William T. (2008). *Living control systems III: The Fact of Control.* Bloomfield, NJ: Benchmark Publications

Powers, William T. (Creator) (Updated regularly) *Perceptual Control Theory; Science & applications—a book of readings.* Menlo Park CA: Living Control Systems Publishing [Posted at the website]

Powers, William T. (2009). *PCT in 11 Steps.* [p. 509 and posted at the website]

Powers, William T. (2009). *Reorganization and MOL.* [Posted at the website]

Robertson, Richard J. and Wm. T. Powers (Eds) (1990). *Introduction to Modern Psychology.* Bloomfield, NJ: Benchmark Publications.

Roy, Shelley A.W. (2008) *A People Primer: The Nature of Living Systems.* Chapel Hill, N.C.: New View Publications

Runkel, Philip J. (1990). *Casting Nets and Testing Specimens.* New York: Praeger. Second edition (2007), revised and updated, Menlo Park CA: Living Control Systems Publishing

Runkel, Philip J. (2003) *People as Living Things: The Psychology of Perceptual Control.* Menlo Park, CA: Living Control Systems Publishing

Soldani, James (1989). *Effective Personnel Management: An Application of Control Theory.* [Posted at the website]

Soldani, James (2010). *How I applied PCT to get results.* [Posted at the website]

van de Rijt, Hetty, and Plooij, Frans (2010). *The Wonder Weeks: How to stimulate your baby's mental development and help him turn his 10 predictable, great, fussy phases into magical leaps forward.* Arnhem, The Netherlands: Kiddy World Promotions B.V.

Excerpts from the letters

Phil, July 23, 1985

Dear Dr. Powers:

I hope this letter reaches you. Some years have passed since your article "Quantitative analysis of purposive systems" was published in the Psychological Review in 1978. I was captivated by it when it first came out, but I have only recently got round to studying it with care. I am still captivated by it.

Some interpretations of your article are deep, such as the matter of what we allow ourselves to learn from experiments, especially highly controlled experiments. Other interpretations are simple (or so they seem to me just now), such as the common observation that workers on the assembly line can be more influenced by possible accusations from their fellow workers of rate busting than they are by exhortations to increase production, since the sensory input of accusations from fellow workers can usually be kept close to the desired rate (the reference standard is typically zero) without jeopardizing the receipt of wages—another desired input. Or the interpretation that many students pay more attention to signals that a good grade is forthcoming than to signals that their understanding of subject matter is increasing. Am I right about those simple interpretations?

. . . .

I am setting out to write a book on life in organizations. It will be a sort of list and explanation of what you need to know to live a half-way decent life as a member of an organization—what you need to know about individuals, dyads, groups, interfaces of groups, and organizations. Acting to control input is of course one of the vital things to know about individuals—and the other levels of human system also. So I want very much not make a botch of how I talk about the process.

. . . .

P.S.. How come I don't hear my psychological colleagues taking about controlling input? I don't talk with them much, but they still seem to be talking about responses. And I don't think I've read anything in the psychological literature in years that cites your paper or book or even distinguishes between controlling input and output.

Bill, July 29, 1985

Dear Dr. Runkel,

Your letter implies a pleasant project, which will undertake immediately—that is now. More or less in the sequence of your letter:

. . . .

Your final comment was, I hope, intended to be wry. I've been sort of wondering, too, when they'd start talking about controlling input. For about 30 years.

Bill, November 8, 1985:

My argument with [James Grier] Miller [(1978) *Living Systems*] is similar to the argument I have with most theoreticians in psychology, flavored to an extent I am in no position to assess by my own professional jealousy. In my defense, I try to be honest and keep a fine strainer over the drain, but what I find after the last gurgle is usually just a wad of hair.

Miller, like many others, says things with which I can agree. But that isn't enough for me. Before they came to understand what I am about, even strong supporters used to send me reams of useful material showing that so-and-so back in 1937 (e.g., Tolman) stuck his neck out and insisted that behavior is, e.g. purposive. I would write back and say thanks, but I would also explain that thousands of people have had the feeling that behavior is purposive, and have said so, and I can't possibly acknowledge them all. Nor am I inclined to: if all I had to say was that I, too, think behavior is purposive I might as well have stuck to engineering, So my friends caught on, and I no longer get such materials unless the author also offers an explanation of what a purpose is and some attempt to say how purpose works, from which the conclusion follows irresistibly. Needless to say, I don't get much of that stuff any more.

It's easy to make proposals to the effect that this or that phenomenon exists or occurs. Most "theories" in the life sciences do no more than that. To me, however, such proposals are just the start of a theoretical effort: the real question is not what happens, but HOW IT WORKS. Anybody can guess about properties of behavior, and find both data and other people to agree with the guess (given a friendly interpretation in both cases). But to find an explanation that not only fits the data but is internally consistent, rigorously defined, non-statistical, and plausible in terms of what we know about the physical capabilities of an organism—that is the real problem. That's the only problem I consider worth the effort to solve. I don't care if other people agree or disagree. That's a side-issue to me. All I want is a model of behavior that I can't poke holes through, a model I can test, a model that doesn't depend on my faith in it or on unspoken assumptions. I am my own worst critic: I put questions to my own efforts that few others even know how to ask. This is not because I'm smart, but because I KNOW SOMETHING THEY DON'T KNOW: control theory.

Behind essentially every theory of behavior I have ever seen, Millers included, is a basic assumption about the nature of behavior. It's expressed under various names: stimulus-response, input-output, antecedent-consequent, dependent variable-independent variable, and so on. The assumption is that behavior results from influences acting on organisms. This is the only model of a behaving system that most life scientists understand. It underlies EVERYTHING they say, Let me quote Miller, p. 448:

> "Some individuals are stronger, larger, healthier, more talented, better educated, or more disposed toward a certain activity than others. [Who could argue with that?]. Consequently, within the range of species norms for different processes, individual organisms differ in their characteristic input-output relationships."

Aside from the fact that the "consequently" could just as well go with the first sentence (moved to be the second one), this quote shows how the old input-output model is almost invisibly taken for granted. My first reaction to sayings like this is not to the substance, but to the assumption: who says organisms have any characteristic input-output relationships in the first place? I can prove, in fact, that they don't (all you have to do is consider the role of reference signals—or just look at behavior). This results in my losing interest in whatever conclusions follow.

Miller, of course knows a little about control processes, but like most others who do, he relegates them to homeostatic systems; p. 448, title of section 5.2: "Adjustment processes among subsystems or components, used in maintaining variables in steady states." The idea of controlling through varying a reference signal has never occurred to him, or if it has, he hasn't seen what it means.

Looking higher on page 448 I see " ...when different messages arrive at the two eyes or ears simultaneously, a number of factors influence a person's ability to respond appropriately to them...". The embedding paragraph isn't even about S-R theory—that's assumed without defense. It's concerned with information theory and the peculiar idea that "messages" are always clamoring to get into the brain which has to filter out what it can use to avoid being overwhelmed. The tricky term "appropriately" isn't explored at all—just lucky for the organism, I guess.

And so it goes, sentence after sentence, paragraph after paragraph, page after page, book after book. The life sciences are in the grip of a wrong model of behavior, a model that has never been tested, a model that is based on blind faith in a few basic assumptions that aren't even recognized as being testable theoretical assumptions. I don't care how many guesses agree with my conclusions if the basis for them is simply wrong, or worse, non-existent. That doesn't make me right, of course, but why pursue what we know is wrong?

In school, I was always the guy who raised his hand during the introductory lecture. If I can't swallow the basis for an argument, I just can't see any point in hearing the whole tedious thing worked out. I am as certain as I can be that Miller's fundamental assumptions about the very nature of organisms are false to fact. I'm willing to stipulate that his logic is impeccable—but so what? Garbage in, garbage out. Sorry.

I'm sure this testy essay hasn't convinced you of the vacuity of Miller's book, but we'll get back to that sort of thing, without doubt. If I know you, you'll call my bluff.

. . . .

Phil, January 8, 1986

. . . .

Your letter shook me up something awful.

. . . .

Bill, February 21, 1986:

Seems to me, judging from your letter to Rick and your latest to me, that you've reached Stage 2 of the Pilgrim's Progress Toward Control Theory, which is called "So What?"

The foot is poised for the next step, the old road is abandoned, and you're ready to go—only where is the signpost? For that matter, where are the other roads? How come everything looks just the way it did before?

It finally dawns on one that there aren't any other roads.

. . . .

We have to build the base first, just as physics did when the most complicated thing Galileo knew how to do was to run balls down inclined slopes or time pendulums with his pulse. Before he did that, nobody understood about acceleration and gravity —NOBODY. Galileo and a lot of others had to go painfully through all the stuff that is now taught as boring simple laboratory exercises to freshmen—but if they hadn't gone through it, there wouldn't be any physics. The laws of gravity would still be rules of affinity.

We are now exactly where Galileo was. The life sciences have never gone through that development that took place after Galileo. The life sciences still think that events can cause other events, that tendencies mean something, that statistical generalizations are of some use in understanding nature. They think that what happens to organisms makes them behave. Practically everything that is really known about organisms is not "life science" at all—it's just physics and chemistry done inside organisms. As Rick would say, physics and chemistry have been doing just fine, thank you. But the life sciences are still in the Dark Ages.

Galileo got into a lot of trouble, and so did many of the scientists who tried to follow the new lead. They had trouble with established religion. We're going to have, are having, the same problem: the religion we're fighting is called Science, the brand practiced by biologists, neurologists, behaviorists, sociologists, linguists, and the others. All the others who think that behavior is an effect of prior causes. There is nobody around to hold our hands, help us out when we're puzzled, show us what all this is going to mean, or take our side against all the misunderstanding, criticism, and hostility that will come our way. Nobody is going to give us a million dollars to establish a control theory institute where we can work in peace. We are revolutionaries, like it or not, and we are finding out what that means.

Phil, June 11, 1986

. . . .

The reinforcement people nowadays seem to say that to make reinforcement work, you have to find out what is reinforcing for the subject. That seems to be saying that you look to see what condition the subject will work to maintain. That seems to be the same as saying that people have purposes, something I thought reinforcement theorists were not supposed to say. Of course, Hull and his followers, I forget how many years ago, postulated that the organism was motivated by "drives."

I remember vaguely something about maintaining certain conditions in the "tissues." At the time I was reading that stuff, I didn't think of "drive reduction" as a goal. But now it seems to me that they did put purpose into the theory. Tolman did explicitly.

. . . .

Bill, June 17, 1986

. . . .

In some regards, common sense is aligned with control theory. We don't expect the flight attendant to spill the drinks. In fact, I stopped telling my co-workers, mostly blue-collar, about my theoretical work after enough of them had responded to the basic idea by asking, "But doesn't everyone know that?" Of course we have intentions. Of course we resist disturbances. You have to know a lot more about what science believes to understand that control theory throws a monkey-wrench into the works, or as you say, sand into the gears. Most ordinary people greet a description of what scientific psychology believes with incredulity.

. . . .

Bill, July 1, 1986

. . . .

Theories based on nothing but words can't be wrong if you don't want them to be wrong.

. . . .

Bill, September 6, 1986

. . . .

I think the key to the Raven problem is to be found in the reluctance of "anyone in the science division" to sit still for an explanation. Despite what they think on the soft end of the campus, the hard sciences just don't use generalizations, induction, and so on. They make models: if the underlying reality contained such and such entities with such and such properties (very precisely stated), then we would observe so and so, which is precisely what we do observe (if not, change the model until this statement is true: a control process).

. . . .

You will notice that the hard sciences have done a lot better with their subject-matter than the soft ones. The soft scientists attribute the difference to the excessive difficulty in working with living systems. I think the problem is their method. When your only model is "If something happens n times, it is likely to happen n+1 times," you don't have much to work with. "Similarity" is not a property of nature: it is an observer's opinion, based mainly on the habit of categorizing and aided by the fact that perception has limits of discrimination. If you look closely enough at any two things, similarities disappear and variables become continuous. The raven paradox, if there really is one, is caused by categorizing.

. . . .

Bill, September 28, 1986

. . . .

That paper you sent me, although it does drop me a reference-crumb, is disgusting. If there is any virtue in it, it's that Denker et. al. are presenting a model that at least does run. That's the first step toward honest modeling. Most models are simply proposals about the internal organization of some system. There's no proof, however, that the model drawn on the paper would actually behave in the same way as the system being modeled: the idea of running a model is confined to a very few people outside engineering. When you commit your hypotheses to specific functional representations and simulate the consequences on the computer (or otherwise), at least you find out whether your model behaves at all like what you had hoped.

. . . .

Bill, October 18, 1986

Your research on generalization was very interesting. My comments on its use in the hard sciences were made on the basis of general impressions and experience, but not from having searched the literature. I guess I wasn't too far off the track.

Your little project got me to thinking about the subject again, and once again asking myself why I feel that things are done so differently (as your last paragraph comments) in the two divisions of science. It's not easy to put one's finger on such impressions. On the surface, the life sciences seem VERY scientific, with all the trappings of experimentation, objective analysis, cautious advancement of hypotheses, and so on. But why does science work so well in physics and chemistry, and so poorly in psychology? Maybe, I'm thinking, it's something like this:

. . . .

When we have found some regularities, we can start working on theories. Here, I think, is where the two approaches diverge. The basic question that follows finding a regularity is, "Why does this regularity appear?" There are two directions in which we can search for the answer: one leads to workable answers and the other leads to delusion.

. . . .

That's the way physics works, if for buzz-makers and timers and counters we substitute electrons, fields, charges, masses, atoms, and so on. This approach works mainly because we demand that the model behave EXACTLY as the real thing behaves under all circumstances.

Now the other approach, the one that doesn't actually work. [two-page discussion of method in social science]

. . . .

So the physicist demands that his models not only match their behavior to real behavior, but that everything we can find out about the parts of the model by any means at all check out with experimentation and remain internally consistent. It isn't considered good form to propose models in which most of the parts are in principle unobservable directly or indirectly. Nor is it considered good form to let a model go public while there are still observations of any kind that contradict what the model implies. Such observations indicate that the model isn't finished yet.

What I'm getting at, I guess, is that there is no mystery behind the success of physical models, or behind the failure of models—"intervening variables"—in the life sciences. It's Just a matter of where you set your standards for acceptance of a model. If a physicist is baffled by failure of his model to predict correctly in just one important situation, he doesn't say "oh, well, it works most of the time," and publish it anyway. Not my ideal physicist at least. He says "Oh, shit!" and goes back to work. If life scientists demanded that their models work with a high degree of precision in all known circumstances and take into account all known facts, they, too, would generate highly successful models—or, quite properly, admit ignorance. You don't get anywhere by insisting that a model MUST work and that if it doesn't the data must be wrong, and you don't get anywhere by lowering your standards to let models go when they still don't work all the time. But that is exactly what has happened in the life sciences.

Most generalizers I have met object to models. I think they object, without knowing it, to BAD models, models that don't work, because in their fields they have seen nothing else. Unfortunately, when a good model comes along that does work, it doesn't impress the generalizers, because they simply don't expect models to work, to add anything to what observation tells them.

. . . .

Phil, January 6, 87

. . . .

It impresses me that good consultants can hold very different theories. I hear them explaining their behavior in terms of theories for which I have very little respect. I think what is happening is that the theories (mine, too, no doubt) serve more as mnemonic devices than as guides to action. That is, the consultant acts mostly from intuition, and then keeps track of the course of events by hanging memory on the rack of the theory, rarely asking about the logical fit. It is enough that the consultant can say to other members of the team or to himself or herself, "So what happened then was ..., so now I think we are ready to ...." And from knowing the lingo, the other members of the team can get a pretty good notion of the kind of bare action that took place, regardless of the kind of theoretical frame the person is using to call up the picture. The handbooks for organizational consultants are full of the most disparate theoretical viewpoints you can imagine. "This exercise illustrates how ...." and then the author will spill out a theory I think is nonsense. But I use the exercise anyway, because I can see how it will pull participants into awareness of some dynamics I want them to be aware of, and I can hang the events on my theory as I guide the participants through the exercise. A consultant can be a nincompoop according to the standards of the academic experimenter, can espouse and proclaim a theory that the academic experimenters have long ago shown not to hold water, and yet be a very competent consultant. I suppose the people who painted those wonderful pictures on the rimy walls of caves in Spain and France had some pretty wild theories about pigments. They must have had some pretty wild theories about light, too, to paint so many of the pictures in places where they could work only by torchlight. In the middle ages, people had some very wrong notions about ballistics. But they managed to batter down a lot of walls with their cannons.

. . . .

Bill, January 9,1987

. . . .

It seems to me that our knowledge of the world consists of empirically-discovered relationships among perceptions, and nothing else. We are the bellringers, tugging at the ropes, feeling and seeing how they behave under our efforts, but limited forever to that bellringers' room that belongs to human beings. We act and we sense; what we act upon many have immensely more degrees of freedom than what our senses report. We experience a version of the universe, the version created when all the degrees of freedom that actually exist are projected into the space defined by the degrees of freedom of our human senses.

. . . .

Phil, July 4, 1987

. . . .

But I do worry about all the energy you are putting into battling the behaviorists. I know they are still thriving, and shouldn't, but they are not the largest sect among psychologists. Of course, I must admit that their underlying rationale permeates the ordinary talk of other psychologists and of the public. It turns my stomach. Educated people in other fields talk as if reinforcers, rewards, and punishments were facts to be accepted the way we accept water running down hill.

. . . .

Bill, December 4,1987

. . . .

Chapter 11 Specimens: Control theory

The front part of this chapter needs strengthening. I think it is essential to follow the course that Marken set. First we must establish control as a phenomenon. This is not a theoretical matter. We have to show that organisms actually do stabilize external variables of all degrees of complexity against disturbances, maintaining them recognizeably near reference conditions that we can identify experimentally. We have to show that the relationship among controlled variables, disturbances, and actions is a real relationship, a directly observable fact of nature. No theory is needed in order to do this. The fact is that organisms do behave in this way. This observation has nothing to how they could behave this way and still be physical systems.

This is precisely where psychology went astray. Psychologists observed this phenomenon, although they didn't observe it very competently, and chose to disbelieve what they saw because it went against principles they had decided to treat as holy and superior to the data. Essentially all the contortions of psychological theories and philosophies of science have been generated exactly to explain how it is that behavior can appear purposive yet not actually be purposive. I think the miserable record of the life sciences hinges on this fateful choice to ignore the data.

In any case we control theorists have to establish the reality of the observations first. Then we can raise the question of finding a theory that makes sense of them. Fortunately, this theory exists in mature forms it is called control theory. Control theory is the body of analytical methods that has been developed specifically to help us understand the operation of systems that behave as organisms do in relationship to their environments: closed-loop systems of causation. This theory, in turn, leads us to new interpretations of old data, and suggests new ways of exploring both behavior and the nervous system. It suggests a model of the nervous system that is consistent with the many levels of apparent organization that we see in behavior.

So first we have the phenomenon of control. Then we have the theory of control systems. Then we have the model build on that theory to account for more and more of behavior. Control theory is not simply the proposition that organisms control things. That proposition must be treated as a report on a phenomenon, different from the theory that illuminates the phenomenon. Control theory explains control behavior.

. . . .

Bill, November 21, 1988

. . . .

In some piece of writing, I commented that there are actually people who think that invented realities and imagined models are more real than simple silent experience. In your letter to Carol you quote what seems a direct example, in Bogen and Woodward. "For the most part, phenomena cannot be perceived ..."—what an extraordinary statement! They are redefining "phenomenon" to mean "what we imagine or deduce to be the case" as opposed to "what we observe." This usage, I think, defines what is wrong with intellectuals.

It may be that the difficulty lies, as I think you suspect, in their pejorative term "epistemologically privileged status." They talk about beliefs and explananda, justification of beliefs about the natural world, belief that something is the case, claims about existence, evidence, and "phenomena of scientific interest." All these terms speak to me of a person so busy *talking* about experiences that the experiences themselves are just a springboard from which one can reach higher levels of verbal abstraction. One bounce and we're done with *that* (scented handkerchief brushing away the traces).

The concept of levels of perception is probably, as you say, one factor that is missing. But I have always suspected that before the lower levels can even be seen as perceptions, its necessary to get out from under words, language, reasoning, deduction, all that ponderous machinery of thinking. I think we have to become aware of the way we push our patterns of thought toward preselected conclusions, slipping cleverly from one meaning of a word to a different meaning, skipping blithely over holes, switching the train of thought around difficult spots as much as following it to its foreordained destination. Only then can we see that models and other kinds of explanations are no more then plausible imaginings, some more plausible than others. When plausible imaginings are carefully constructed, and when they are tested against nonverbal experience as frequently as possible, they can become powerful tools: viz, physics, at least prior to quantum mechanics. Well, I suppose even after, although I'm reluctant. When imaginings lose their anchors in experience, they turn into intellectual games and we lose the ability to choose the best imaginings.

. . . .

Bill, February 18, 1989

. . . .

Shepard's article reminded me of the many psychologists who are perfectly willing to let perception happen with no relationship to the brain or nervous system. The metaphor of "resonances" is Shepard's way of bypassing the lower levels of perception, letting the higher levels somehow tremble to the Aeolian touch of reality without ever existing as crass neural impulses.

Telling stories is OK if you plan to check up on them somehow (unless they're meant just as entertainment, in which case you wouldn't check up on them). Gibson's story is that the real reality is really there and our brains simply pick it up. Fine, good, OK. Now how are you going to find out if that is true? To see if that is a true statement, you would have to have some way of checking to see if the "optic array" gives us a picture of reality that is just like the actual reality. That means you need a way to know about reality that doesn't depend on your own or anyone else's optic array. Gibson put a lot of store in "tangibility", as if touch weren't a sense. If you can touch it, it's real. But how do you know you're touching the same thing you're looking at? You don't. Gibson doesn't. Nobody does. We just assume, and try to make our senses cohere in terms of each other. There aren't any other terms.

Once in a while I wake up and look at all these solemn people posturing and pronouncing and making up their tales, and I think, "Why, you're nothing but a pack of cards!"

. . . .

Phil, July 20, 1989

. . . .

I know you are not much interested in method, but to me (as you know) theory and method are inseparable. With theory (or metatheory) I include the low-down assumptions lots of researchers never think about, such as linear versus circular causation.

. . . .

Phil, March 16, 1990

. . . .

Here is another salvo in my battle to get you to cease battling with the behaviorists.

Maybe you will want to read only the parts I have marked in red. Or maybe not even those.

. . . .

Bill, March 20, 1990

I'm not battling with behaviorists—only with what they believe. Behaviorists probably wouldn't say "only," although they're perfectly capable of slipping into the use of terms like "believe" (as in "superstitious beliefs").

. . . .

The argument between cognitivists and behaviorists (Bolles/Amsel review) is not the same as mine between control theory and behaviorists. Maybe cognitivists have "polluted" the meaning of behaviorism, but I haven't. I think I understand behaviorism very well. The cognitivists denigrate the "dependent variables" of behaviorism as "colorless movement" and "glandular squirts," which is to say that they object to them on aesthetic grounds. I object to them because they don't exist. They do not "depend" in the assumed way. You don't even need control theory to prove that.

I have read Watson, as Amsel recommends; Watson's works are based on the assumption that behavior is a dependent variable, and that environmental events are the independent variables. Says Bolles/Amsel, "...the major message of behaviorism, conveniently ignored by cognitivist critics because of the questions it raises and problems it poses for rejecting behaviorism, is that knowledge claims of psychology cannot meet the standards of natural science methodology unless behavior is employed as the dependent variable."

There's the problem laid out plain. The behaviorists have always claimed to have a lock on the only true natural science methodology (they assume, incorrectly, that it's the same methodology that physicists use). This is the methodology, of course, that says you vary the independent variable and look for a correlation with the dependent variable. What the behaviorists conveniently overlook (aside from the initial false assumption) is that this way of viewing behavior doesn't meet the methodological standards of the natural sciences either (by which I mean physics and chemistry). The predictions made on this basis don't predict worth a damn.

. . . .

It's the use of abstract constructs that stands in the way of psychologists when they try to understand the control-system model. They simply can't grasp the idea that it is not an abstract construct. Because they treat control theory as just another construct, they compare it with existing constructs using the same criterion they always use: verbal plausibility. They don't ask how well it works, because the idea of a theory working is all but unknown to them. And they certainly don't ask whether the components of control systems physically exist—they never claim that even for their own constructs.

But control theory is a literal description of how an organism works. Never mind whether it's a correct description: that's another subject, the subject of testing models. It's a literal description because every component of a proposed control organization, including boxes and arrows, inside and outside the organism, is supposed to represent the operation of some observable thing. When I speak of reference signals, I'm not just talking about an arrow in a diagram or an algebraic variable. I'm proposing that inside the brain there are real neural signals that we could measure, that act in neural circuits to establish reference levels just as they are established in real electronic devices that we can take apart and study. When I draw a line and label it "perceptual signal," I'm proposing not only that such signals could he found in the brain, but that the signals are identically what we experience when we experience perceptions. These are strong and falsifiable propositions about a real physical system.

. . . .

I consider this standard scientific view of both animal and human consciousness to be not just confused, but pathological. I think it has crippled the life sciences by making the begging of questions a formal part of scientific reasoning.

I do agree with Bolles on one point. The cognitivists have not shown anything wrong with behaviorism; they have simply abandoned it. That leaves all the phenomena discovered by behaviorists in limbo—neither explained nor explained away. The cognitivists have done nothing more than shrug off what they can't explain. I find this attitude irritating beyond support. People have criticized me for spending so much time thinking about operant conditioning; they say, "Nobody thinks that's important any more, why are you wasting your time on that old stuff?" My answer is that it's important until we understand the phenomena; just turning to something else is no answer. Control theory can explain all the substantive phenomena that behaviorists have explained in terms of drives, reinforcements, and so on. But it's not enough to say that we can explain it. We have to do it in such a way as to leave no room to doubt that control theory does a far more convincing job than any behaviorist explanation has done.

I don't buy this "orthogonal" garbage. It's all one system. If we're to understand it, we have to bring the whole thing and all the phenomena associated with it under a single consistent theory. Otherwise we go back to doing our own thing and not worrying about thinking six contradictory thoughts before teatime. "Microtheories" are for dilettantes.

So it isn't just behaviorist ideas that I battle, Phil. It's a whole history of hubris and dishonesty in the life sciences, and the resulting failure to develop even the rudiments of a real science of behavior. As far as I'm concerned, we're starting from zero.

So that's my answering salvo. Now you have to try to guess whether you hit a battleship or a destroyer and where to aim the next round.

Phil, Oct 13, 1999

. . . .

As you know, I have been reading your writings and those of your followers since 1985. I have told you before how, as I strove to understand your view of perception and action, I found my own accustomed views undergoing wrenching, unsettling, unhinging, even frightening changes. I found myself having to disown hundreds, maybe thousands of pages which at one time I had broadcast to my peers with pride. I found, too, that as my new understanding grew, my previous confusions about psychological method, previously a gallimaufry of embarrassments, began to take on an orderliness. Some simply vanished, as chimeras are wont to do. Others lost their crippling effects when I saw how the various methods could be assigned their proper uses—this is what I wrote about in "Casting Nets." For me, the sword that cut the Gordian knot—my tangle of methodological embarrassments—was the distinction between counting instances of acts, on the one hand, and making a tangible, working model of individual functioning, on the other. That idea, which in retrospect seems a simple one, was enough to dissipate (after some months of emotion-fraught reorganization of some cherished principles and system concepts) about 30 years of daily dissatisfaction with mainstream methods of psychological research.

. . . .

You did not invent the loop. It existed in a few mechanical devices in antiquity, and came to engineering fruition when electrical devices became common. Some psychologists even wrote about "feedback." But the manner in which living organisms make use of the feedback loop—or I could say the manner in which the feedback loop enabled living creatures to come into being—that insight is yours alone. That insight by itself should be sufficient to put you down on the pages of the history books as the founder of the science of psychology. I am sure you know that I am not, in that sentence, speaking in hyperbole, but in the straightforward, common meanings of the words. In a decade or two, I think, historians of psychology will be naming the year 1960 (when your two articles appeared in _Perceptual and Motor Skills_) as the beginning of the modern era. Maybe the historians will call it the Great Divide. The period before 1960 will be treated much as historians of chemistry treat the period before Lavoisier brought quantification to that science.

. . . .

You have bestowed thought, time, paper, and computer screens, not to speak of hospitality, on everyone who has evinced the slightest interest in PCT. You have understood the internal upheavals suffered by those of us who try to comprehend this strange new world—our intellectual foot-dragging and our anguished obsequies muttered at the graves of our long-cherished beliefs. You have been patient with misunderstanding, persevering in the face of disdain, forbearing of invective, and modest under praise.

In all of this, you have been aided immeasurably by the intelligence, stamina, and love of Mary.

I owe you, for your help to me, a great debt. You have given me a way, after all these years, of laying hold of a system concept, a psychology, that is more than a grab-bag and a tallying. You have given me a way to set down thoughts that will come to more than a mere rearrangement of what every other psychologist would say. To join you and your other followers in the effort to make PCT available to others is, for me, here in my last years, a joy, a privilege, and a comfort.

Thanks, brother.

Letters and enclosures

Page no	*Letter from Phil*	*Letter from Bill and Mary*	*Enclosure (pages)*	*Note*	
3	850723_Phil				
9		850729_Bill			
18	850808_Phil				
19	850826_Phil				
20	850909_Phil				
—			850909_CognitiveSimilarity 17 p	See website	
42		850914_Bill			
—			850914_Byte articles 54 p	See website	*
46	850923_Phil				
48		850925_Bill			
51	850930_Phil				
53	851007_Phil				
54	851009_Phil				
59		851018_Bill			
62	851104_Phil				
66			851104_Emergents 6 p	In book	
72		851108_Bill			
76		851109_Bill			
78	851206_Phil				
82	860106_Phil				
83		860108_Bill			
85	860108_Phil				
87	860108_Phil_diagram				
88		860118_Bill			
521			Marken_Farewell 6 p	In book p. 521	
92	860125_Phil				
100		860131_Bill			
—			860206_SpiritOD 18 p	See website	
—			860206_Handout 40 p	See website	
104	860208_Phil				
104		860213_Bill			
105	860217_Phil				
106		860221_Bill			
109	860306_Phil				
110	860404_Phil				
122		860412_Bill			
124		860420_Mary			
125	860421_Phil				
127	860422_Phil				
129		860503_Bill			

* 328 pages of enclosures are located at www.livingcontrolsystems.com, listed on this volume's web page. 31 pages are included in the book itself.

Page no	*Letter from Phil*	*Letter from Bill and Mary*	*Enclosure (pages)*	*Note*
133		860518_Bill		
135			860518_Ford_Feelings 2 p	In book
137	860519_Phil			
138	860603_Phil			
139	860611_Phil			
141		860617_Bill		
145	860618_McGrath_Phil			
151	860620_Phil			
155	860625_Phil			
155	860627_Phil			
156			860626_Phil_Survey 4 p	In book
160	860630_Phil			
161		860701_Bill		
165		860708_Bill		
167	860718_Phil			
167		860725_Bill		
168	860724_Phil_McGrath			
174	860728_Phil			
176		860730_Bill		
178			860730_Bill_vita 4 p	In book
182			860800_Exh2-4 4p	In book
186	860805_Phil			
188	860806_Phil			
191	860809_Slater_Phil			
197	860811_Tom			
198		860818_Bill		
199	860902_Phil			
203		860906_Bill		
206	860915_Phil			
—			860901_Phil to Hart 22 p	See website
—			860915_AsIf 28 p	See website
—			860915_HartBrainSchool 4p	See website
208	860917_Phil_Slater			
216		860918_Bill		
—			860918_NaturalKinds 5 p	See website
219	860922_Phil			
—			860722_Generalizing 48 p	See website
222	860924_Phil			
223		860928_Bill		
228	861002_Phil			
231		861018_Bill		
238	861106_Phil			
—			861203_Answers 24 p	See website
245	861206_Phil_Williams			
252	861211_Phil			
255			861211_Traginology 5 p	In book
260	861215_Phil			

Page no	Letter from Phil	Letter from Bill and Mary	Enclosure (pages)	Note
261	861218_Phil_AmPsy			
267		861230_Bill		
271	870105_Phil			
273	870106_Phil			
277		870109_Bill		
280	870122_Phil			
283	870123_Phil			
284		870131_Bill		
286	870217_Phil			
289	870221_Phil			
290		870228_Bill		
293	870303_Rick_Phil			
295	870309_Phil			
298	870310_Phil_Rick			
306	870314_Phil_Slater			
307	870320_Phil_Rick			
310		870328_Bill		
316	870402_Phil			
321	870704_Phil			
324	870708_Phil			
325		870719_Bill		
—			870724_gen_ch4 8 p	See website
328	870724_Phil			
329		870730_Bill		
335	870823_Phil			
—			870823_SimultaneousCausation 12 p	See website
336		870826_Bill		
338	870901_Phil			
339	870914_Phil			
340	870916_Phil			
341		870917_Bill		
343	870922_Phil			
345		870926_Bill		
353		871001_Bill		
358	871015_Phil			
359	871106_Phil			
364		871204_Bill		
392		880123_Bill		
393	880124_Phil			
394	880131_Phil			
394	880000_Phil_Mary			
395		880206_Bill		
397	880217_Phil			
398		880416_Mary_LordHanges		
408	880424_Phil_Mary			
410	881013_Phil			
413	881118_Phil			

Page no	Letter from Phil	Letter from Bill and Mary	Enclosure (pages)	Note
414	881120_Phil_Slater			
417		881121_Bill		
419	881126_Phil			
420		881201_Bill		
422		890218_Bill		
424		890223_Bill_Shepard		
428	890617_Phil			
436	890624_Phil			
437		890704_Bill		
438	890720_Phil			
441		890725_Bill		
444	890800_Phil			
448		890809_Bill		
451	890817_Phil			
452			890914_ReviewsABC 5 p	In book
457		900114_Bill_Gillian		
462	900316_Phil			
464		900320_Bill		
469	910925_Phil_Suls			
472	911112_Phil_Bourbon			
474	920203_Phil_Bourbon			
478	920616_Phil_Mitchell			
—			920700_MakingExplaining 24 p	See website
480	920910_Phil			
—			920910_RunkelMcgrath 10 p	See website
481	920924_Phil_Judd			
484	930408_Bill			
486		930817_Mary		
487		940306_Bill		
489	940418_Phil			
—			940418_Probabilistic 5 p	See website
—			940418_Replication 9 p	See website
490		940422_Bill		
492	941016_Phil			
493		941024_Bill		
495		950714_Bill		
495	950901_Phil	950901_Bill		
496	991013_Phil			
499		991014_Bill		
500	070608_Claire			
501	070611_Dag			
501	070613_Bill			
503			OrphanedPage 1 p	In book
504			Kindergarten 1 p	In book
505			Commandments 1 p	In book

Page number reference Quantitative Analysis of Purposive Systems

In his very first letter, Phil Runkel references pages in *Quantitative Analysis...*
You may have the reprint in *Living Control Systems; Selected papers of William T. Powers.*
This table provides a conversion.

Quantitative Analysis... Psychological Review 85, pages 417-435.	*Quantitative Analysis...* reprint in Living Control Systems, pages 129–165.
1. ... I made that deduction from some sentences below the middle of the second column on page 427	1. ... I made that deduction from some sentences in the middle of page 150
2. ... page 422, first column, 3-5/8" from the top edge of the page: "The output quantity will be related to many other external quantities .... "	2. ... 2/3 down page 138: "The output quantity will be related to many other external quantities .... "
3. ... the caption to Figure 3 on page 423	3. ... the caption to Figure 3 on page 140
4. ... page 425, a little below the middle of the second column you write: The engineering model would show	4. ... near the end of page 145 you write: The engineering model would show ....
5. ... 426, lines 7 and 8 (not counting the page heading): "... may reflect only properties of the local environment."	5. ... page 146, middle: "... may reflect only properties of the local environment."
6. ... the end of the second of the paragraphs beginning on page 426, you say that the "open-loop explanation contradicts itself."	6. ... on the second line on page 147, you say that the "open-loop explanation contradicts itself."
7. ... on page 426. In the paragraph beginning near the top of the second column, you say, "... we are not seeing the function *f* that describes the bird ...."	7. ... in the middle of page 147, you say, "... we are not seeing the function *f* that describes the bird ...."
8. ... beginning on page 427: "A Time-State Analysis with Dynamic Constraints."	8. ... middle of page 149: "A Time-State Analysis with Dynamic Constraints."
... I did not study with great care the first column on page 428, though I gave it more attention than mere scanning.	... I did not study with great care your derivation on pages 150–151, though I gave it more attention than mere scanning.
... the ninth line of the second column on page 428, "with the dynamic constraint, ...	... the fifth line from the bottom of page 151, "with the dynamic constraint, ...
9. ... On page 434, at the beginning of the first paragraph that begins in the second column, you say, "The natural tendency ...	9. ... On page 164, in the first new paragraph, you say, "The natural tendency ...

Page number reference
Behavior: The Control of Perception (B:CP)

In several letters, Phil Runkel references pages in *Behavior: The Control of Perception.* Phil had the 1973 edition. You may have the slightly revised 2005 paperback edition. This table provides an approximate page number conversion.

B:CP Chapter title	1973 Page	2005 Page	Δ +/–
The Secret of the Great Machine	N/A	v	
The Dilemmas of Behaviorism	1	1	
	5	5	
	9	10	+1
Models and Generalizations	10	11	+1
	14	15	+1
	18	20	+2
Premises	19	21	+2
	23	25	+2
	31	30	-1
	35	34	-1
	40	40	
Feedback and Behavior	41	41	
	51	51	
	55	55	
The Control-System Unit of Organization	57	57	
	68	69	+1
A Hierarchy of Control Systems	70	70	
	77	77	
	80	80	
First-Order Control Systems: Intensity Control	82	81	-1
	89	88	-1
	98	97	-1
Second-Order Control Systems: Sensation Control or Vector Control	99	98	-1
	107	106	-1
	113	113	
Third-Order Control Systems: Configuration Control	115	114	-1
	123	122	-1
	128	127	-1

B:CP Chapter title	1973 Page	2005 Page	Δ +/–
Fourth-Order Control Systems: Control of Transitions	129	129	
	135	136	+1
Fifth-Order Control Systems: Control of Sequence	137	137	
	146	147	+1
The Brain's Model	147	148	+1
	152	154	+2
Higher Levels	154	155	+1
	164	165	+1
	175	176	+1
Learning	177	179	+2
	189	190	+1
	197	199	+2
	203	205	+2
Memory	205	207	+2
	211	213	+2
	220	222	+2
	229	231	+2
Experimental Methods	231	233	+2
	241	243	+2
Emotion	N/A	252	
Conflict and Control	250	262	+12
	266	278	+12
Appendix: Control System Operation and Stability	273	285	+12
Glossary	283	296	+13
Bibliography	N/A	301	
Index	289	303	+14
End Papers	N/A	312	

Part I

About these letters

When *People as Living Things* was nearing completion in August 2002, Phil Runkel sent me an e-mail:

> "I have a file of my paper-mail correspondence with Wm Powers that started in 1985. I have no more use for it. Do you want it?"

Of course I did. Phil sent a one-inch stack of letters. I sent a CD with a note to Bill and Mary Powers:

> The collection Phil sent me is not complete. Phil did not consistently keep copies.
>
> I hope that you have saved Phil's originals and kept duplicates of Bill's letters, but even without all of Phil's letters, the collection is a wonderful PCT tutorial, written in a most exquisite, loving way by two eloquent gentlemen as they explore the subject from many angles.
>
> There are autobiographical sketches here and there; opinions on the field and players in it.
>
> I would like to put together a volume with these letters. I think this would make a wonderful resource of great educational and historical value. How about it?
>
> Mary, do you have Phil's letters? Any missing letters from Bill?

Bill replied a few days later:

> Dag, I have received the CD-ROM and have spent several hours reminiscing through that old correspondence with Phil. It seems as if it happened in a different world, but only yesterday. Phil truly brought out ideas I had only halfway considered, and made me think carefully where I had been careless. I have come to think of him as Brother Phil.
>
> Mary is collecting all the materials that we have relating to these conversations, and will be shipping them to you pretty soon. Aside, perhaps, from referring to some cyberneticians as "dilettantes," I don't think I have said anything too damaging about anyone.

The history of Perceptual Control Theory and the scientific revolution it brings to the life sciences will be thoroughly studied in the future. Fortunately, Mary Powers was a librarian and kept archives in order. She sent a package with this note:

> Here are letters from Phil to Bill. Some have replies attached. There are also letters from and to Phil, to and from other people. Phil sent Bill copies to show what he was up to. There are also some draft papers of Bill's, with commentaries from Phil on sticky notes. This is all rather disorderly. I decided not to weed some of it out, leaving the choice of what to use and what to exclude up to you. I hope this is useful to you.

Mary sent a four-inch stack of originals and copies. The combined record appears rather complete.

This volume contains all letters supplied by Phil and Mary. Some enclosures are included among the letters. Where Phil sent copies of published articles I list references. Other enclosures are of minor interest, or very long. These are listed and posted at the website.

More pieces of this particular puzzle will no doubt surface in the future as scholars have an opportunity to scour Bill Powers' archives for additional insight into the history of PCT.

In this volume, letters are presented in original form, typically reduced to 95% of original size. Footnotes and other comments by the editor are visually set apart as regular print text.

Enjoy!

UNIVERSITY OF OREGON

July 23, 1985

Ans. July 29

Dr. William T. Powers
1138 Whitfield Road
Northbrook, IL 60062

Dear Dr. Powers:

I hope this letter reaches you. Some years have passed since your article "Quantitative analysis of purposive systems" was published in the Psychological Review in 1978. I was captivated by it when it first came out, but I have only recently got round to studying it with care. I am still captivated by it.

Some interpretations of your article are deep, such as the matter of what we allow ourselves to learn from experiments, especially highly controlled experiments. Other interpretations are simple (or so they seem to me just now), such as the common observation that workers on the assembly line can be more influenced by possible accusations from their fellow workers of rate busting than they are by exhortations to increase production, since the sensory input of accusations from fellow workers can usually be kept close to the desired rate (the reference standard is typically zero) without jeopardizing the receipt of wages--another desired input. Or the interpretation that many students pay more attention to signals that a good grade is forthcoming than to signals that their understanding of subject matter is increasing. Am I right about those simple interpretations?

I am still studying your article, but there are several points where I need help.

1. Am I right that when you write "open loop" you mean a feedback circuit with a break in it so that no feedback gets through? I made that deduction from some sentences below the middle of the second column on page 427.

2. On page 422, first column, 3-5/8" from the top edge of the page: "The output quantity will be related to many other external quantities...." Here I tried to fill in an example for myself, but I had no confidence. Can you give me an example or two?

3. In the caption to Figure 3 on page 423, I am not sure about the antecedent of these and this. When you write "these variables" and "This is not a model," are you referring to Figure 1 or Figure 3?

4. On page 425, a little below the middle of the second column, you write:

Graduate Studies and Administrator Certification • Information and Field Services • Center for Educational Policy and Management
(503) 686-5171 (503) 686-3409 (503) 686-5173
DIVISION OF EDUCATIONAL POLICY AND MANAGEMENT • COLLEGE OF EDUCATION • EUGENE, OR 97403-1215

Dr. William T. Powers
July 23, 1985
Page 2

> The engineering model would show a reference input to the system, the effect of which would be to adjust the setting of q_i* and also to affect indirectly the objective consequence. As mentioned, no such input from the outside exists in natural N systems (in none of them, at any rate, that I have investigated).

I agree that no reference signals go directly into natural N systems. But do we not try to alter reference signals indirectly--and often succeed? Do we not bring influence to try to alter a habit so that a new q_i* remains after the outside influence is removed? Isn't that what we call teaching or learning? Is it an example when a parent tells a child to look both ways before crossing the street? Or when a person learns the temperature at which the spouse prefers his or her coffee? Or when a neophte photographer learns to fill the frame with the object being photographed?

5. Page 426, lines 7 and 8 (not counting the page heading): "... may reflect only properties of the local environment." Here again I tried to fill in an example and was not confident. Is one example the example given lower in the column of the light rays from bug to bird and the retina of the bird?

6. At the end of the second of the paragraphs beginning on page 426, you say that the "open-loop explanation contradicts itself." It seems to me that the resolution of the apparent contradiction is in your phrase four lines earlier: "As indeed it very nearly does." The retina is so sensitive to spatial changes of excitation that only very small displacements are required for the head to turn a little. The head does not "exactly compensate" (seven lines higher), but almost exactly does so. OK?

7. Now I get to a question that is crucial if I am to feel that I have not missed the crux of your article. It is about that bird on page 426. In the paragraph beginning near the top of the second column, you say, "... we are not seeing the function f that describes the bird...." Where is the boundary between the bird and the environment? You say, "... we are seeing the function g that describes the physics of the feedback effects." But the retina is inside the bird. And the head turns because of neural and motor activity inside the bird. The head would not turn if there were no bird there with a head.

I have no trouble with the mathematics--I think. I can trace the argument from Equation 7a--I think. My trouble is in understanding what you say about what the head-turning tells us. It might help me if you would give an example or two of what you mean at the top of the column when you say, "The organism function f, on the other hand, may be both nonlinear and variable over time." What kind of behavior would be an example of that? And it might help if you would given an example that is "... well known to control engineers and to those who work with analog computers."

In this letter, Phil references pages in *Quantitative Analysis Quantitative Analysis of Purposive Systems,* as published in Psychological Review 85, pages 417-435. For equivalent page references that apply to the reprint in *Living Control Systems; Selected papers of William T. Powers,* pages 129–165, see table on page xlvi.

Dr. William T. Powers
July 23, 1985
Page 3

Or it might help me most if you could give an example of a sort of psychological experiment in which the investigators typically fall into the behavioral illusion you describe. I enclose a couple of articles by William Dember. * The 1965 article describes some experiments with rats. Do those experiments suffer from your illusion? How?

Dember proposes a very interesting kind of interior reference standard--and a <u>movable</u> one--beginning in the lower half of the second column on page 157 of his 1965 article.

I know I am asking a lot of work from you. I'll be grateful for any answers, long or short, that you can give to my questions.

8. And I don't think I have properly grasped your section beginning on page 427: "A Time-State Analysis with Dynamic Constraints." I guess I <u>do</u> think of feedback effects "as if they occurred separately, after one response and before the next...." I did not study with great care the first column on page 428, though I gave it more attention than mere scanning. My ability to read mathematics is not what it once was, and the introduction of the "opt," "crit," and "ss" values makes my head swim. I might be able to figure it out with a couple more hours of scrutiny. I guess my trouble starts with the rationale for "allowing the output to change only a fraction of the way...." Anyway, when I got to the sentence starting at the ninth line of the second column on page 428, "with the dynamic constraint, the discrete analysis shows that behavior follows the same laws of negative feedback whether the feedback effects are instantaneous or delayed," I said to myself something like, "Yes, given that sequence of mathematical reasoning, that certainly seems to be where we come out." But I also said to myself something like, "And so--? What does he want me to do or not do?"

I don't think I have pin-pointed my trouble very well. But I hope you can say something to me about this. Anything would probably help. I think this question, such as it is, is also crucial to my understanding.

9. On page 434, at the beginning of the first paragraph that begins in the second column, you say, "The natural tendency of any human being is to deal with the unfamiliar by first trying to see it as the nearest familiar thing." Naturally, I agree. So did Krech and Crutchfield (<u>Theory and Problems of Social Psychology</u>. McGraw Hill, 1948) in the restatement on their page 98 of their Proposition III: "... a change introduced into the psychological field will be absorbed in such a way as to produce the smallest effect on a strong structure." Though I like your words better.

Is the manner of dealing with the unfamiliar one of trying to match an interior signal (the familiar thing) with incoming feedback by attending selectively to features of the feedback image that match the reference, rejecting the rest? I suppose my interpretation of your sentence goes beyond your experiments, in which the feedback is unambiguously unidimensional--has only one "feature." Doesn't understanding the unfamiliar thing require building a new

* Dember, William N.: The new look in motivation. American Scientist, 1965, 53: 409-427 (reprint) Motivation and the Cognitive Revolution. American Psychologist, March 1974, pp 161-168.

Dr. William T. Powers
July 23, 1985
Page 4

image or "reference signal"? What am I doing as I try to understand your articles? I grant that my images and sentences here are very vague compared to the inputs and outputs in your experiments. I am trying to apply your principles beyond such precise and simple situations.

My wife has Alzheimer's disease--at least our physician says that's the best label he can put on it. Although she forgets most current events within a very few mintues, I think most of her long-term memories are still there. I think they get out of order. You might say that her memory of the addresses of her memories is erratic. So she often doesn't know whether one memory comes before another. In particular, she may think that her memory of getting ready to go on a trip (a trip that actually occurred 20 years ago) is a memory of just a minute ago. But (this is my hypothesis) her sensory input at the present gets into contradictions with that old memory that she is trying to respond to this minute, with the result that the old memory loses detail. She cannot tell me when the train leaves, from what station, whether she has a ticket, where her destination is, or who might be expecting her to arrive at the other end, even though her urge to respond to the out-of-place memory by getting dressed and leaving the house is very strong. I often drive her around in the car until she forgets the urge. I admit I am no neurologist. My hypothesis seems to fit every occasion--but then, so does astrology. Anyway, I do the best I can.

When my wife has a memory of visitors being in the house, but when, in fact, there are no visitors in the house, she will often see objects as people. She will point to a blanket lying over a chair and ask, "Who is that?" And her interpretation of the blanket does not change until I pick it up and shake it out.

So what is happening to my wife's reference image of "person in the chair" in comparison to the sensory input? Where, as I shake out the blanket, does here selective perception of the blanket snap over from one set of features to another?

(I don't want you to think I am this intellectual about my wife's affliction. It is a terrible experience for both of us, and we do a lot of crying. But I can't help having intellectual speculations about it.)

10. Is the attached diagram correct?

I note that you wrote a book entitled, Behavior: The Control of Perception in 1973. Am I going to have to read that before I can understand your article?

I am setting out to write a book on life in organizations. It will be a sort of list and explanation of what you need to know to live a half-way decent life as a member of an organization--what you need to know about individuals, dyads, groups, interfaces of groups, and organizations. Acting to control input is of course one of the vital things to know about individuals--and the other levels of human system also. So I want very much not to make a botch of how I talk about the process.

Just the sheer idea of paying attention to control of input is more than worth the time I have spent on your article, but I would like very much to have a more sophisticated understanding than that. I would like to have some confidence that I can avoid the pitfalls you describe. I hope you can help me.

Sincerely yours,

Philip J. Runkel

Philip J. Runkel
Professor of Education and Psychology

PJR:dvc
Enclosure

P.S. How come I don't hear my psychological colleagues taking about controlling input? I don't talk with them much, but they still seem to be talking about responses. And I don't think I've read <u>anything</u> in the psychological literature in years that cites your paper or book or even distinguishes between controlling input and output.

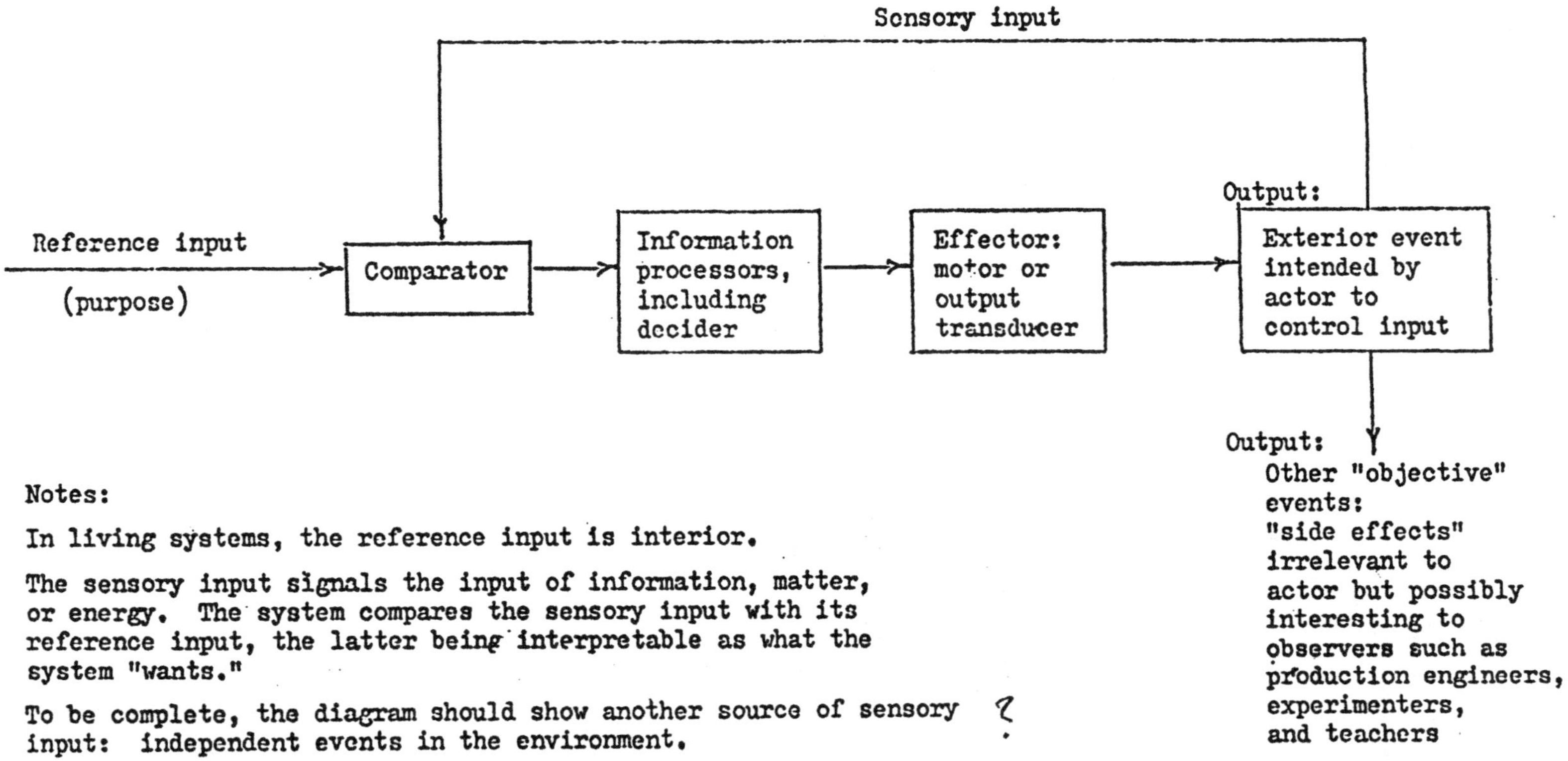

Notes:

In living systems, the reference input is interior.

The sensory input signals the input of information, matter, or energy. The system compares the sensory input with its reference input, the latter being interpretable as what the system "wants."

To be complete, the diagram should show another source of sensory input: independent events in the environment. ?

This diagram is my interpretation of Powers's text, especially that at bottom right of page 419.

July 29, 1985

Dear Dr. Runkel,

Your letter implies a pleasant project, which I will undertake immediately -- that is, now. More or less in the sequence of your letter:

1. Yes, "open loop" is a bit of engineering self-contradiction which actually means straight-line, but as you suspected it implies a broken loop. In artificial systems, breaking the feedback loop is a technique used for measuring properties of a system without the confusing presence of feedback effects. Human beings can only be measured open-loop for a fraction of a second; they immediately find out how to close the loop in a different way and regain control.

2. An output quantity might be something like a muscle force. A muscle force is "related to many other external quantities" such as limb angular acceleration, skin pressure, body velocity, positions of objects, and so on. If the control system in question is sensing and controlling joint angle, then all those other effects are irrelevant to its operation. Only the effect of the output force on joint angle plays a part in THIS control loop. The other effects of output are side-effects.

3. The variables shown in the figure are all observable externally to the intact organism ("these"). I'm referring to Fig. 3. A "model of the organism" would require taking apart the function f to show how its form is accomplished in "wetware." While I've spent a lot of time on such models, I avoided them in this article to show how feedback effects and control processes can be studied even by those who eschew such models.

4. External reference inputs: I am a very literal-minded modeler: even when I don't know what I'm talking about, I am trying to talk about a real brain and how it really works. So if there are circumstances suggesting that "Telling a child not to cross the street" effectively sets a reference-signal in that child, I want to ask, "How?" Literally, all one can do is send sound-waves to that child's ears. It is the child's brain that must detect them, make phonemes of them, make words and sentences of the phonemes, and convert the words and sentences into the kinds of reference signals appropriate for working in control systems that control visual and kinesthetic relationships. And before any of that will happen, the child has to set the higher-level reference signal establishing the goal of doing what he or she was told to do: as any parent knows that reference signal is not available for outside manipulation. The hierarchy of control acts to satisfy its own reference-signals. Of course these come to include goals for the kind of person one wants to be, and they normally entail kindness, altruism, and so on. Also, at the higher levels, we often actively look for suggestions from others -- but we must still understand them, agree to them, and figure

out how to turn them into specific reference signals.

5. The "properties of the local environment" means the property that converts head angle into a position of the image on the retina -- the laws of geometry and optics in the physical world, not in the nervous system. The feedback running from (the muscles that turn the head) to (the position of the retinal images) is shaped by these basic physical properties. By modifying that feedback path (say, by partially and nonlinearly stabilizing the retinal image through some clever optical-electronic gadget), one could arrange for the feedback path to have different properties. Then the relationship between bug movement and head movement would be different even though the path from retina to muscles remained exactly the same. The image on the retina would still be stabilized even though the head did not now track the bug. Back to this point in 7.

6. Yes, you voice a common and correct objection by saying that the retina is sensitive to small changes, and so could provide enough information to account for the head movements. The key words are "could", meaning "conceivably could," and "changes." If the head/eyes moved by an angle proportional to the integral of the first derivative of retinal position, then IF the integration were perfect, IF the retinal response to the first derivitave were precisely quantitative and linear, and IF the muscles could respond to the ensuing neural signals with complete precision and linearity, it is possible that the compensatory explanation might work. Actually it would work in other situations, granted these impossible IFs, but not in the example of the bird. The reason is that the movement of the image on the retina is <u>simultaneously</u> a function of the bug's movement and the head's movement. Pursuit tracking does not involve saccades; it is continuous. The time taken for a disturbance of position to propagate through the nervous system and muscles and back to an effect on the retinal image is a matter of a few milliseconds; during those few milliseconds, neither the position of the image nor the muscle tensions can alter significantly. For all practical purposes the time-delay is zero.

Actually, all you have to do is express the retinal image position as a joint function of head and bug angle, and at the same time express the angular acceleration of the head as a function of the retinal deviation from center, and you have modelled a control system. Your comment concerns the internal function connecting retinal position error to muscle tension. But to analyse the whole loop correctly you must solve that relationship simultaneously with the one between muscle tension and retinal position, in the external part of the loop. That's the part that the SR theorists always leave out. They could have had control theory 100 years ago if they hadn't seen so sure about cause and effect.

7. I consider the behaving system to be the nervous system, and everything else, including the muscles and the body they operate, and effects on the outside world, the environment. The

input boundary consists of all the sensory receptors; the output boundary, all the motor nerve-endings. This division allows me to treat all levels of control alike, with the feedback loop always being completed by a path from output to input through the environment.

Muscles fatigue; the relationship of an increment of muscle force to an increment in neural frequency in the driving neurones depends strongly on the absolute frequency; because muscles are attached in strange ways to bones, the effect of a given muscle tension on physical variables like applied force or acceleration varies with the configuration of the body. Similar considerations hold at the input side, for sensory receptors are neither completely precise in their responses nor linear in their responses. So the organism-function, which expresses output as a function of input, can vary a great deal. Nevertheless, we observe that the overall relationship of behavior to external events is a good deal more repeatable than we know the nervous systems and muscles to be. The only way we can explain this is through control theory. In fact the only way we can appreciate just how precise behavior really is is through control theory (see the experiments at the end of the article). The variability of behavior is primarily a product of a bad model.

```
             |\
    i1     - | \ +
A -rrrrr-----|  \ --------- B
   R     |   | /      |
      i2|    |/       |
         |            |
         |     | |    |
          -----| |----
               | | C
```

Engineering example, well known to control engineers and to those who work with analogue computers. The triangle is an "operational amplifier," which amplifies its input (commonly) from 100,000 to ten million times, and also inverts it: a very small input voltage produces an output voltage of opposite sign and much larger amplitude, . The feedback effect is negative, and holds the input voltage very nearly at zero. Here's one way the analysis can go.

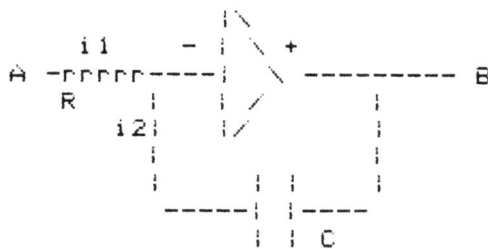

The differential equations:

1: In the current path from A through R and C to B, we have (with A and B being voltages, and e being the voltage at the input of the amplifier):

Physics: voltage across a capacitor is the time-integral of current flowing through it: voltage is int(i/C) with capacitance C in Farads. Voltage across resistor R ohms with current i amps flowing through it is iR. The output voltage of the inverting amplifier is B = -Ge where -G is the amplification factor of the amplifier and e is the voltage at the input connection.
Voltages in series add.

1. i1 =i2 = i (no current flows into the amplifier)
2. A + iR + int(i/C) = B (series voltages add up)
3. e = A + iR (amplifier input voltage)
4. B = -Ge (the amplifier equation)

(Note: the reference signal for this system is zero).

Now substitute:

4 into 3:

6. B = -G(A + iR) = -GA - GiR

solve for i and iR:

7: i = -(B + GA)/GR
iR = -(B + GA)/G

result into 2 with cancellations
(the constant R moved out of integral):

8. A - (B+GA)/G + (1/RC)int[-(B+GA)/G)] = B

(B + GA)/G is B/G + A, so

9. A - B/G - A + (1/RC)int(-B/G - A) = B, or

-B/G + (1/RC)int(-B/G - A) = B.

Suppose the maximum value that A or B can physically attain is plus or minus 10 volts. If G = 100000, then B/G can be, at most, 0.0001 volt, or 0.01% of the full-scale voltage. We can ignore B/G in that case without creating more than 0.01% error in our prediction of the result. By setting B/G = 0.000, we have

10. B = -(1/RC)int(A)

within 0.0001 volt of exactness over the range of A and B from -10 to +10 volts.

The amplifier input is like a sensory receptor: its voltage (the actual "stimulus") depends both on the voltage applied at A and on the feedback current through the capacitor due to changes in voltage B, which modifies the effect of voltage A on the amplifier. Even though the amplifier simply multiplies its input by -G to produce its output, the relationship between the more remote input A and the output B is quite different from that: B is the time integral of A, and the gain G doesn't even appear in the final approximate -- but nevertheless precise -- expression. The overall relationship is almost completely determined by the external resistor and capacitor, given only that the gain G is large enough and negative.

The voltage across an inductance (a coil) is proportional to the rate of change of current through it. Since current is proportional to the input voltage A applied to the resistor R, if an inductor were substituted for the capacitor we would find that for an inductance of L henries,

$$B = (L/R)\,dA/dt.$$

The output would be the first derivative of the input, with the same degree of precision as before.

The amplifier itself, of course, would be exactly the same device. Yet the apparent input-output relationship from A to B is totally different. This is what I meant by saying that the external feedback path has far more effect on the overall apparent input-output relationship than do the properties of the "behaving device" (a fact well known to control engineers and those who do analogue computing) . A set of operational amplifiers, together with a collection of feedback and series input components, constitutes the toolkit known as an analogue computer. Networks of these devices can simulate most physical devices and very rapidly solve complex simultaneous equations -- faster, in fact, than large digital computers can, and with respectable accuracy. And the brain contains circuitry quite capable of behaving in the same way. Not only is negative feedback important in overall behavior, but it is probably at the heart of the brain's inner computations as well. The brain is not a digital computer, but an analogue computer, at least at most of its lower levels.

You can see why I did not want to do more than allude to this subject in the article.

Behavioral illusion: Take a Skinner box, and an easy schedule on which the rat can feed itself indefinitely. The assumption is that reinforcement increases the probability* of behavior. The reinforcement is the rate of presentation of food, and the behavior is the rate of bar pressing. Now the experimenter introduces a change: the schedule is altered so that now twice as many bar-presses are needed to get each food pellet. The rat soon doubles its rate of bar-pressing. Conclusion: "the change of schedule controlled the change in behavior."

*frequency

Now, at random intervals, start throwing extra food pellets into the cup, at various known rates. The rate of bar-pressing will decrease in a closely-correlated way. Conclusion: "Non-contingent reinforcement reduces behavior."

But the rat is actually a control system, controlling the rate of pellet delivery and ingestion by varying its actions. It is trying to maintain the delivery rate at a level related to a long-term internal reference level for food input. If we change the apparatus so that the same behavior produces less food, the rat increases its rate of pressing and maintains the food input the same. If we add pellets, increasing the food input, there will be less error and thus less behavior, once again maintaining the input at essentially the same level. If we force-feed the rat (obesity experiments) the rat will stop pressing the bar altogether, and starve itself until its average food input comes back to what it intends it to be [sic]. Of course the real input variable is inside the rat; we can see only the food input, and must express the control action in terms of controlling it. But we're still closer to the truth than the behaviorists are. Under extreme deprivation, by the way, these relationships do not remain the same. In fact the control system breaks down and we see only pathological behavior. That is what is studied, for the most part, in Skinner boxes.

8. Dynamics. Move your hand slowly across your visual field, like a Chinaman doing Tai Chi in the park. You are slowly altering your reference-signals determining hand position in three axes, and your control systems are maintaining the perceived position in a very close match with the changing reference position. You are controlling the perception of your hand in space (including how the arm feels), resisting gravity, friction, and (if any occur) disturbances of other kinds.

The time required for a disturbance of position to propagate through your nervous system and back out to the muscles is on the order of 50 milliseconds (the longer delays called "reaction time" appear only when there are sudden large changes of conditions, and involve higher-level phenomena). But this does not mean that during the 50 milliseconds nothing is happening; in fact there is a continuous flow of changes in neural signals, the effects reaching the muscles 50 milliseconds later but still being smooth and continuous, involving hundreds of impulses per second. There is a delay between input and output, but there is no alternation. All processes, input and output, proceed continuously and overlap in time.

If you analyze this control system properly, using differential equations and taking the time-delay into account, you will come up with a model that works essentially as the real hand does, including resistance to disturbances. You will be able to see the effect of increasing and decreasing the sensitivity of the system to errors. You will also find that if the model is to be stable, it must contain some sort of filter that eliminates the effect of the time-delay. This filter might be something as simple as the mass of the arm, which prevents any movement from being instantaneous.

. Once you have a stable model, you can investigate the effect of the time-delay by varying it. What you will find is that if the time-delay is reduced to zero, there will be no interesting effect on the behavior of the model. The very conditions needed to achieve stability in the presence of the time-delay make the system work as if the time-delay were zero. The filtering simply slows the system until no significant change can take place during one such unit of delay.

"Allowing the output to change only a fraction of the way" is just my trick for inserting the needed dynamic filtering. We compute the error, and using the error compute the next value that the output should have. But if we then used that value of output for the next iteration, we would find that the next error would be of the opposite sign and very much larger, and the system would rapidly run away. So, we might decide to reduce sensitivity to error -- but then the model would control so weakly as to be useless. The solution is to build that filter into the model, so we can keep a high sensitivity to error, but smooth out and slow down the response so the system can remain stable. Of course "slow" is a relative term. Compared with a stable straight-through system or a system without filtering that is stable because of low sensitivity, this system is the fastest of all. That is because we can set the gain very high and select the optimum filtering.

The division of behavior and stimuli into "events" is an artifact: both perception and behavior are normally smooth and continuous.

9. I have a little different idea of how perception works -- not original, but not very popular, either. I think we construct perceptions out of lower-level information (at many levels) so that each <u>kind</u> of perception is the output of a device that continually receives inputs and transforms them according to some perceptual computation into an output signal. Thus one such perceiving function reports only one kind of perception, the amount of signal indicating how much of it seems present. The difference between kinds of perception is the difference between different computing devices. There are some cross-connections between devices (the rabbit-vase) to complicate matters, but I'm not ready to deal with that.

We perceive the new as something familiar simply because we do **not** yet have a perceptual device organized for perceiving in terms of the new kind of perception. Only the old ones respond, however weakly. To the man who has seen only axes, a hammer looks like a strange and inferior sort of bent axe. It takes a while to construct a hammer-recognizer, so we can see it as a damned good hammer. As you may guess, I have personal acquaintance with this difficulty.

* * * *

10. Yes, the diagram is correct. And yes, you are going to have to read my book. Most University libraries have it, I think: I know that Willamette* does (and also contains a few fans).
*I'm enough of an Oregonian to know that the accent is on "a".

The asterisks bring part of your letter to here. No, I am not put off by your intellectualizing your wife's plight. I understand the double purpose behind it: to gain a measure of relief for a little while, and to grope toward an understanding of what is happening. You needn't apologize for either purpose.

It's clear to me that control theory has little to say -- except in a general reorienting sort of way -- about higher mental functions. It does, however, encourage us to take subjective reports seriously, which psychology has scandalously failed to do for most of this century. When you have a model that works, even if only a partial one, that very fact encourages you to examine other problems with more hope of understanding them, if not of gaining control over them. It lets you think about topics that have been taboo in science.

I have been encouraged by some successes in modelling behavior. As a result I have given a lot of thought to the problem of awareness, since my model clearly has no place for it, yet awareness (just as clearly) exists. I have come to regard it as a phenomenon utterly distinct from perception, from memory, and even from thought. To be aware is to observe. The objects of observation consist of what we call the external world (really signals in our brains, the external world remaining hidden, though suggestive of its presence), AND ALSO of what we call our internal worlds: thoughts, emotions, and everything else mental. But the Observer is none of those things: the operands are not the operator. For the Observer, the brain is a set of meter readings that represent to it the world of sensation, and also all the operations on that world that produce form, relationship, sequence, logic, and principles. All these, the observer observes.

Your wife's brain is malfunctioning. I would be very surprised if her awareness were, although of course I have no idea what awareness is. Part of her mind is upset because other parts are not working right. If that were not so, why would it bother her to forget, to make mistakes? But part of her, like part of you and part of me, is not upset, but only observes, and acts. I think that communication among people is, in the final analysis, a way for one observer to know that another is there. Once you have discovered the signs of the other observer, the rest becomes of much less importance in comparison, although it is never unimportant.

My ideas on this subject are not well-formed, nor have they hardened into convictions. There is no great comfort in them, either, because we all hate to be approaching the end of life. But if I am talking about a real phenomenon, one we might be able to learn more about, then there is certainly more to life than meets the eye. Even before we understand, we can get a glimmer of the real mystery of life, a mystery that religion has only visited in its outskirts (and in doing so, has offended science into ignoring anything that even sounds like religion, a thalamic response if I ever saw one).

You and your wife each know that the other is there. That is part of the pain, but without it there would be nothing, and never would have been anything.

Your final comment was, I hope, intended to be wry. I've been sort of wondering, too, when they'd start talking about controlling input. For about 30 years.

Regards,

Bill Powers

William T. Powers
1138 Whitfield Rd.
Northbrook, IL 60062

August 8, 1985

Dr. William T. Powers
1138 Whitfield Road
Northbrook, IL 60062

Dear Bill Powers:

Thanks very much for your sympathetic and instructive letter of 29 July. I shall study it carefully soon.

In the meantime, I went to the library and got your 1973 book. I read chapter 16 on experimental methods. Wow! I did the rubber-band experiment with a friend. Wow!

More later.

Thanks again,

Phil R

Philip J. Runkel
Professor of Education
and Psychology

PJR:dr

August 26, 1985

Dr. William T. Powers
1138 Whitfield Road
Northbrook, IL 60062

Dear Dr. Powers:

I have now read the first six chapters of the library's copy of your 1973 book. I continue enthralled. I have ordered a copy of my own.

Some years ago I read an article by J. G. Miller on the information-transmission capabilities of different levels of living system. When his 1978 Living Systems came out, I read his chapter 5, which reviewed a series of such studies. He gives no summary table. Partly from numbers in the text and partly by reading the scales on his graphs as best I could, I made the summary table enclosed.

I thought those numbers were interesting. The higher levels of living system take in and put out fewer bits per second than the lower levels at their maximum outputs, but they are much more efficient than the lower levels (O/I ratio). But I never knew what to do with that interesting information, I didn't know what other interesting behavior to tie it to, until I read pages 52-54 and 75 of your book.

I know you deal almost entirely with the single organism in your book, but I'm a social psychologist trying to think about groups and organizations, and I don't mind extending to groups and organizations your principle of the higher-order reference signal taking longer to operate while altering the lower-order reference signals to suit. I know you'll worry about my making over-extended analogies. I know, however, that reference signals are not set someplace in the air among the individuals in a group, but remain in the individuals. Some months from now (not years, I hope), I'll send you what I write. I cannot say with precision yet how Miller's numbers fit with your principles. I'll think it through when I get to writing that section. But there is obviously a correspondence there someplace.

On to chapter 7!

Sincerely,

Phil Runkel

Philip J. Runkel
Professor of Education
and Psychology

PJR:dr
Enc.

September 9, 1985

Dr. William T. Powers
1138 Whitfield Road
Northbrook, IL 60062

Dear Dr. Powers:

I have finished reading your book--I mean I have got to the end of it. I went back and studied some sections several times. I know I'll be studying it more from time to time. In my file of "to read," I found a note about your book that I had put there four years ago. I wish I had bought the book then. Well, that's the way it goes.

Here are various thoughts that came to me as I read. First are a couple that don't apply to any particular page.

Here and there you call your reasoning reductionist. I don't think that's a good description. The S-R people were reductionist, because they had only one order of system. The most complex action resulted merely from a sufficiently lengthy string of conditioned responses. There was no second-order reference signal. In your description of the orders of control system, you label the orders with the emergents that characterize them. Sensation does not arise, in your model, from a heap of first-order systems. It arises from particular combinations of inputs to second-order comparators--combinations from various sites and in various assortments of intensities. From that patterning of inputs from first-order systems, we get the emergent of sensation at the second-order system (as interpreted in awareness higher up, I guess I should say). If an author points out emergents, I don't call him a reductionist.

In these letters, Phil Runkel references page numbers in *Behavior: The Control of Perception,* 1973 edition. See the table on page xlvii for equivalent page numbers with reference to the 2005 paperback edition.

Phil Runkel made notes to himself and underlined passages on the original he received and, sometimes, on his copy of what he had sent. Some notes are in light red pencil, others typed on what are now called Post-It Notes, small, separate pieces of paper taped on top of the text they refer to. Such notes have been designated Note (A), (B), (C), and moved to empty space or a separate page. Otherwise, they would obscure the text. Scanning contrast has been adjusted to keep the text clean while at the same time capturing Phil's notes.

Dr. William T. Powers
September 9, 1985
Page 2

*

Next. If action stops when the error signal is zero, why don't we come to a dead stop more often? Or at all? I've never known anyone, certainly not me, who ever came to a dead stop. Even when I am loafing (which I practice now and then), I keep noting sensory inputs. When I am asleep, I dream like crazy. Whould you say that there are always so many inputs happening that many error signals can be zero for quite a while, but there are always so many others pushing the error signals off zero that we are always active--even in sleep? Well, I am willing to grant that the control of bodily functions (breathing, heartbeat, moving feces along in the intestine, etc.) requires constant (at that time-scale) reaction to disturbances. But aside from that sort of thing, why don't I see people more often just sitting with glazed eyes? Why don't they do it for an hour or two (between meals) instead of a few seconds or minutes? Do people in affluent but primitive cultures do it more often--because of the lack of telephones ringing, clocked jobs to go to, people passing by the office door, etc.? Because of control by error signal, would you call your whole model an equilibrating system?

See (2) in attached letter from P.— His letter of 14 Sept 85

Page 52, bottom. "A control system that is too rapid in its response to disturbance will be unstable." Seems to me there is an analogy, or an extension, to the two-person system. Imagine that A sets out to teach B how to do something. But B is over-eager and tries to anticipate what A is about to tell him, starting to act before A finishes a step of instruction. We would hear A crying, "No, not that one, this one! Wait, you've got the wrong end! Yeah, but you did it too soon! Oh hell, now we've got to start all over again!"

Page 98. I have always (well, for decades, anyway) taken it as a rule applying everywhere that knowledge is the knowledge of differences. But your "leading question" No. 5 on page 98 makes me wonder. Certainly I can be

* For this letter, Phil made notes to himself on the copy he kept, referring to similar notes on the reply from Bill of September 14. This is a scan of the original, which is sharper. Phil's notes have been entered in a script font. It is readily apparent how his notes helped Phil digest the reply.

Dr. William T. Powers
September 9, 1985
Page 3

aware of a particular level of brightness uniform over the visual field. But maybe the rule still holds in the first-order control system: the error signal is giving me the knowledge of difference--<u>this</u> level versus what it <u>might</u> be. Would you say that?

But then I wonder about adaptation. If I experience the same uniform brightness for half an hour, will I still be able to experience it as a particular level of brightness, or will I be unable to tell whether it is low or high? I suppose that is a factual matter. What's the answer? See (3)

Page 165, trying to decide whether to postulate a separate level for language. Another disadvantage in doing so, it seems to me, is that it would be difficult to imagine how evolution could have built a "layer" of feedback systems just for that. Once we have systems for scanning the environment and for imagining dangers that might lurk in the next cluster of bushes, it is easy to image how evolution could build systems for inquisitiveness--for getting more information about what is going on hereabouts. But for language? I think we probably invented it with our capacity for making images of images. And how come you didn't mention Korzybski? See (4) See (7)

Page 171, middle. We do now have chess programs that play at master's level, though I forget whether they yet play at grand master's or world champion level. See (6)

Pages 171-173: 9th order systems. Would understanding the operation of a clock and studying the operation of General Motors both be examples of the 9th order? And my question reminds me of Boulding's list of nested systems. In case you don't know his list, I enclose it. His list is not at all vital to your model, but you might find that some of his ideas are compatible with your thinking. See (8) See (9)

Dr. William T. Powers
September 9, 1985
Page 4

I suppose closure operates in forming a perception of a system? People often get mixed up in reporting, including reporting what happens in their group or organization. They confuse what they saw with what they infer: "But it must have happened that way!" I think that the historians who wrote on clay tablets or who painted hieroglyphics on walls probably put in a lot of inferences: Considering the awesome, god-like person our king is, the battle must have gone that way. *See (10)*

Page 180. I have enclosed a paper headed "Cognitive Similarity." If you find that you want to read it, here is a question about preference orders. Suppose the objects in my diagram are foods varying in water content. Suppose the organism usually prefers crackers to melons. But it is getting thirsty; an intrinsic variable is not matching well the reference signal. So the organism alters its vector to lie more parallel to the dimension of water content. Is that 8th order or is it reorganization? *Oblique answer* * *See (11)*

Page 195. I am charmed by your frequent informalities. A nice one, for example, appears on page 195: "'Nicely' isn't the same as 'correctly,' of course, but we can't have everything." And page 171: "Eventually I will have to give in and speak of learning, the kind that is related to 'reinforcement,' but that time is not quite yet."

Page 200, last paragraph: ". . . only systems in the conscious mode are subject either to volitional disturbance or reorganization." It seems to me your previous exposition said that the reorganizing system could go about its business without consciousness. You surely implied that, at least, as long as you were letting the reorganizing system try out random alterations. So I don't think your sentence can mean ". . . subject either (1) to volitional disturbance *See (12)*

* 850909_CognitiveSimilarity.pdf —enclosure at this volume's web page.

Dr. William T. Powers
September 9, 1985
Page 5

or (2) to reorganization." I think you mean ". . . subject both to volitional disturbance and reorganization." Is that correct? Your example in the next sentence seems to support my interpretation, since it deals only with volition being conscious, and you gave no example arguing for reorganization always to be conscious.

On page 201, at the end of the section, I suppose you mean that what awareness and consciousness are for is to keep the whole shooting-match in good working order. OK?

Page 219, remembering the girl in the red dress. That is exactly what my wife does. She goes around the house looking for the girl in the red dress. And if she "remembers" that there are other people in the house, she will often "see" people where there are actually chairs, large plants, coats lying on chairs, and so on. When I ask her to touch the "person," she sometimes must actually feel the spot where she "saw" the person before her present sensory experience crowds out the out-of-place memory.

Page 233, last paragraph about hunting for hypotheses about the controlled quantity. Your examples of the spot of light and the target on the screen are always very clear. Your remarks about extensions of The Test to more complex situations, especially social situations, are all pretty vague (if I remember right). So I thought it was about time I tried to imagine an application of my own. Below is a story I made up to see whether I could imagine The Test in a complex situation. I'll be grateful to hear any comments you have about the behavior of the supervisor in the story.

THE CASE OF THE TALKATIVE TOILER

or

THE CASE OF THE LOQUACIOUS LABORER

See (13)

Suppose a worker is hired and stationed on an assembly line. The worker is told that it is against the rules to talk with other workers while on the job. Nevertheless, the supervisor finds that the worker frequently engages in shouted conversations (it is a noisy place) with workers at nearby stations. The supervisor doesn't like that. Here are some hypotheses the supervisor might entertain:

1. The worker will gabble if you let him; he won't if you don't.
2. He is new on the job, and he wants reassurance from other workers that he is doing things right.
3. Since he is new in town, he is trying to strike up a few friendships.
4. He chafes at rules, and he has happened to pick on the rule against talking to be the one to violate.
5. He seeks camaraderie--he wants to feel himself to be an accepted member of the work group.

To test one of those hypotheses, we need to choose a "quantity" the worker might be controlling, and then find a way to alter that quantity through means that operate outside the worker. If we succeed in altering the quantity, our hypothesis will be wrong; the worker will have been found not to be controlling that quantity. If the worker brings the quantity back to its former level despite our disturbance, then we will have found the controlled quantity--or at least one of them.

Dr. William T. Powers
September 9, 1985
Page 7

Suppose we try Hypothesis 1 in the standard manner; that is, we ignore feedback theory. The supervisor tells the new worker to stop his gabbling. After a few days, his shouted conversations come back to the frequency they were at before. That doesn't tell us much. It tells us there might be a quantity associated with his talking that he is trying to control, but we suspected that before, since he was going against the rule. It doesn't give us a clue about whether any of the other hypotheses might be a better bet. The supervisor decides to drop Hypothesis 1 and try Hypothesis 2.

If we are going to try to alter a quantity, we must have a measure of it before we begin so that we can tell whether the quantity has changed. For Hypothesis 2, we have no measure of how much the new worker is talking <u>about his job</u>. You might think the supervisor should simply go to the new worker and ask him whether he is talking to the other workers about how to do the job right. Or maybe the supervisor shouldn't even mention the talking. Maybe the supervisor should simply ask the new worker whether he is getting enough feedback about how he is doing the job. If the worker says he's pretty sure he's doing the job all right, then the supervisor could give up Hypothesis 2 and go on to another. If the worker says he wants more feedback on how well he's doing, then the supervisor would know his hypothesis is correct. He could arrange, for example, for another worker to stand by the new worker for a while to answer his questions. After that, if the worker stopped his shouted conversations, the supervisor would know he was right.

Going to the new worker in that manner, however, has at least two drawbacks. First, people often do not know what quantities they are controlling. The new worker may indeed be asking questions about the work, but may think he is simply carrying on "friendly conversation." Or he may know he is asking questions about the work, but is not doing so because he wants the information, but merely to open conversation with his fellows. And he may be unaware of the particular quantity he is controlling when he feels the urge to open conversation with his fellows.

Second, if the supervisor opens the conversation about the talking or the feedback, the supervisor's sally produces environmental happenings that might contain another quantity the worker wants to control. Even if the worker were conscious of the quantity his conversations were controlling, in his shouted conversations, even if he were right about it, he might also want to control some feature of his relationship with the supervisor. He might want to keep the knowledge the supervisor has about him to a minimum. Or he might want to maximize the degree to which the supervisor thinks he is gung ho. The action of the supervisor in opening the topic might cause a disturbance in one of those controlled quantities, and the worker would act to restore his desired relationship to the supervisor, not to act in connection with his behavior at his station.

In brief, by going to talk to the worker, the supervisor would be trying to get a measure of the presumed controlled quantity by getting it through the verbal behavior of the worker. But that verbal behavior could be controlled by a higher-order system that was getting perceptual signals both from the worker's memory of his behavior at station and from his conversation with the supervisor. The statements the worker makes might have little connection to what he "needed" at his work station.

So the supervisor decides not to talk to the new worker, but to try to get a measure of the worker's work-related conversation elsewhere. The thought

Dr. William T. Powers
September 9, 1985
Page 9

of planting a microphone at the new worker's station flits through his mind, but he does not want to violate the worker's civil rights. He decides to try to get the information from the workers on either side of the new worker. The supervisor goes to the workers on either side of our troublesome worker and asks them what that worker talks about to them. If he is talking about the job, the supervisor reasons, surely the other workers wouldn't think the worker needs to be protected from the supervisor knowing that he wants to do his job well.

"You guys didn't talk on the job before the new man came," the supervisor says, "and I guess you just want to be decent to him, not just ignore him, so I can see why you answer him. I guess there's something the new man wants to talk about even if he has to shout. So I'm wondering if there's something he needs that he doesn't want to tell me about. What does he talk about?"

"Why don't you ask him?" the workers say.

"Well," the supervisor says, "he hasn't come to me about anything that's bothering him, so if something is bothering him, he must think it's something that wouldn't go over very well with me. So he probably wouldn't tell me if I asked. And anyway, if there's something you think you shouldn't tell me, I won't push on you to tell me. You just tell me what you think it's OK to tell me, and if I don't get a clue, well, that's that." All the supervisor really cares about knowing is how much the new worker talks about the job. And he thinks the other workers will be willing to tell him that.

The workers say that the new worker talks about various things--the town, baseball, his job, the company, lots of things.

The supervisor asks how much the new worker talks about the job.

Well, quite a bit, says one worker. Some, the other says.

Does he ask questions about how to do his job?

Well, yeah, one worker says. That's part of it, the other says.

Does he talk about the job as much as half the time?

No, they say.

A third of the time?

Well, maybe, one says. I guess, says the other.

The supervisor wishes it had been all of the time or none of the time. But of course the new worker might be trying to match more than one reference standard in his conversations with his fellows. So the supervisor decides to try to disturb the amount of conversing the new worker can do, and to cause the disturbance by acting only on the environment, not by acting through the new worker himself. He asks the other workers not to respond to anything the new worker says for several days. Luckily, they agree. When the new worker buttonholes his co-workers at lunchtime, they say they have nothing against him, but they thought they'd better go back to obeying the rule.

the efforts of the new worker to get a reply from his co-workers on either side decrease rapidly during the first hour or two of the day on which his co-workers stop replying. He tries again once or twice in the afternoon. He gets no reply.

That afternoon, the new worker is late getting back from the coffee room after the break. During the ensuing days, the supervisor observes that the new worker is frequently late getting back after breaks. The new worker is also sometimes a few minutes late getting onto the line in the morning; he is talking with others who are leaving the earlier shift. The supervisor also notices that the new worker often does not walk right out of the plant at the end of the shift. He often waits at the time clock until he finds one or two others who are going his way; then he walks away with them. In sum, the new worker's conversations with others have not decreased; he has apparently transferred his conversations from the line to the coffee room and to the beginning and end of the shift.

So there is probably something in his talking with others that he is acting to maintain. But what? It might indeed be getting information about how to do his job right. The supervisor decides, however, that surely enough time has gone by for the new worker to have picked up anything he needs to know from other workers. After all, it's a pretty simple job. Hypothesis 2 has decreased in credibility as the time has gone on. The supervisor decides to drop that hypothesis.

How about Hypothesis 3? Has the new worker been trying to find friends? After a few more days, the supervisor decides to ask the new worker about that. Surely he ought to know whether he has found friends, and maybe he'll be willing to say so.

The new worker turns out to have no reluctance. Yes, he has found several new friends; he's had a couple over to the house, and they've invited him and his wife to return. So that's not it. That feature of the environment has changed, but the new worker's tardiness after breaks and in the morning has continued. The supervisor crosses out Hypothesis 3.

How about Hypothesis 4, wanting to break rules or defy authority? The supervisor decides to change the environment by changing the rule for the new worker.

"You've probably been wanting to get acquainted with people around here," he says to the new worker. "I know it takes time to get to know the ropes, find out how you're doing, and all that. I guess I'm kind of late with this idea, but I tell you what I'm going to do. I'm going to give you ten minutes in the morning, and ten extra minutes after break, so you can have some time to talk with the other guys. You can tell me when you're ready to go back to the regular rule."

The new worker says, gee, thanks.

The new worker uses his extra ten minutes, or most of them, but he does not violate his new special rule, nor does he violate any other rule. The amount of his talking with others does not seem to decline; perhaps it rises slightly within the ten-minute grace. In brief, his talking with others seems to stay more or less the same as it was. Since the new worker did not act to violate the new rule or some other, the supervisor crosses out Hypothesis 4. He is left with Hypothesis 5.

The supervisor now needs to alter the environment in a way that will change the opportunities the new worker has for camaraderie. If the amount of talking the new worker does with others changes, then the supervisor will have to cross off Hypothesis 5 also and start all over again. If the new worker's amount of talking does not change, if the new worker finds some way of continuing that amount, then probably the supervisor will have found what the new worker needs.

But what to do? How can he <u>decrease</u> the opportunities below what they are already and still allow some way for the worker to find friendliness? Some people say that you should not expect to satisfy all your needs at work. If you need camaraderie, you should find it after working hours. But if the worker has an internal standard for camaraderie <u>at work</u>, that idea doesn't help.

Maybe the supervisor could transfer the new worker to a job off in the corner of the lot where he would encounter no one but a foreman all day long. But if the new worker's need for companionship were strong enough, he'd simply walk off that job to find someone to talk to. Then the supervisor would surely be forced into "disciplinary action," and he knows that punishment rarely gets you the behavior you want. Anyway, why should he arrange things so that the worker ends getting punished for something he, the supervisor, did? That's not ethical.

The supervisor decides that all he can do is to increase the new worker's opportunities for companionship, within the rules, and see whether the new man's communication with others stays about the same.

Luckily, the company has another division in which workers are organized into teams of four and five. Within the teams, workers are allowed to talk all they want. In fact, they are expected to confer about the day-to-day problems that come up and find solutions for those that can be solved within the operations of the team. There is a great deal of interdependence within each team, and the teams show a good deal of self-reliance and comradeship. The supervisor describes the teams to the new worker and asks whether he would like to transfer to one that has an opening. The worker eagerly says yes.

After a few weeks, the supervisor checks with the team leader. How has the new worker fitted in? The team leader says he's OK. During the first week, he seemed to want to talk to everyone, and he talked about more kinds of things than the rest of the team typically does, but since then his communication has settled into the pattern of the rest.

Did he talk much about how to do his job? Well, yes, especially during the first few days, but not more than any new man does. Now he talks about the work of the team as a team, the way the rest of them do. (So that lets out Hypothesis 2.)

How about any tardiness? No trouble about that, the team leader says. He's always on time. During the first week, especially, the team leader says, the new man often cut his breaks short. He doesn't do that as much any more. He does it the way the rest of us do--when there's some time pressure.

Does he hang around before or after the shift? No more than the rest of us, the team leader says.

And he doesn't seem to bother people with more conversation than they want? Oh, no, the team leader says.

And he is doing his work OK? Sure, the team leader says, we're glad to have him.

It is difficult to compare the amounts of comradeship the new worker was getting in his job on the line with the amount he is now getting in the team. On the line, however, he was clearly acting <u>against</u> the "disturbance" of the rules In the team, he seems to have settled into a stable pattern of comradely behavior and does not seem to be acting against anything. the supervisor believes he has found the new worker's controlled quantity.

That's the end of the story. It sounds like a happy ending. All's well that ends well.

But there are some weaknesses in the story as an application of The Test.

First, my hypotheses are rather arbitrary. The plant may be located in Iowa, and the new worker might have been keeping his voice in shape for the

Dr. William T. Powers
September 9, 1985
Page 15

upcoming hog-calling contest. But there might have been other possibilities more likely than that. The supervisor's search might have been much longer than in my story.

Second, the workers on either side might not have been as cooperative as they were in my story.

Third, the job might not have been the kind where the new worker could use tardiness as a way of having for time to talk with others; the movement of the line might have forced him to get back on time or quit. If he stayed on the job, his yearning for camaraderie might have taken a form of action not visible to the supervisor.

Fourth, the new worker might not yet have found friends. That would have complicated the supervisor's detective work.

Fifth, the supervisor's tactic of allowing the new worker ten extra minutes might have had side effects he wouldn't want. Other workers might have complained about the special treatment being given the new worker. Or some of them might have thought they, too, could get ten extra minutes by breaking the rule about talking. Or the new worker might have refused the favor, thinking he would be resented by the other workers; that would <u>increase</u> the "error" between the amount of camaraderie he was getting and the amount he wanted.

Sixth, the company might not have contained a division with the teams in it. What would the supervisor have done in that case? I couldn't think of anything. that's the reason I invented the division with teams in it.

Finally, the supervisor's superiors might not have condoned the time he took and the actions he took to correct the "simple" matter of a worker talking too much on the job.

Despite my effort, in other words, to make my story reasonably realistic, it may not be so. It may be that in most instances in most plants in the United States with assembly lines, a supervisor would be very lucky to be able to apply The Test even as sloppily as my supervisor did.

But the big thing wrong with the story is that throughout, the supervisor wants arbitrarily to control the worker. The whole plant, the assembly line, the very posts and beams of the buildings, are built on the supposition that some people have to control other people.

If the supervisor did not believe that it was his job arbitarily to control the workers, what could he do? He could confer with the new worker. He could say, "Here we are within these fences. We've agreed that in exchange for our wages, we will limit our behavior in certain ways. But we have our individual limits, too, and the company's limits seems to be exceeding your limits. What can we do?"

That won't bring an immediate solution. The usual norms are all against that procedure. The worker will immediately be suspicious. If he is not suspicious, he will probably think the supervisor is a well meaning bumbler who won't have his job very long anyway.

But suppose the company is one--some now do exist--in which a fair level of trust has been built up among the employees, where there is a lot of self-management on the shop floor, a lot of conferring in groups about improving working conditions, and so on. (I have heard of one plant with three rotating shifts, with two shifts doing immediately productive work while the third shift does nothing but talk about how things can be improved and try out improvements!) Then the new worker might try out some problem-solving behavior with the supervisor.

Dr. William T. Powers
September 9, 1985
Page 17

Even then, the effort might falter through the worker's unawareness of his own reference standards. The supervisor might try some verbal exploration: "Do you think you might like it if you were in a job where X happened? What about Y? What about Z? if you were in a job that was so good that you jumped out of bed in the morning eager to get to work, what would it be like? If you were in that kind of job, what might you find yourself telling your wife, when you got home, about how the day's work went?" The idea would be to find clues about reference standards that might be controlling the worker's behavior.

In a group where there is good trust in one another's intentions (these guys won't knowingly do anything to hurt me), that kind of exploration is better done in the group. Members can report to the person his behavior they are actually seeing. That enables the person to see behavior on his part that he was unaware of. And members can make guesses about conditions or behavior the person would feel good about. The person can accept or reject the guesses according to whether they "feel right." Members can offer help or trades. "How would you feel about your doing this and my doing that? Would you be able to promise to do this if I'd promise to do that?"

That kind of process is a groping one, but it often works. It is not nearly as precise as finding the quantity a person is controlling when his is controlling a spot of light on a screen. But it has the advantage of mutual helpfulness. The person comes to see that he can control the relevant part of his environment through agreements with the others who are a part of that environment. It fits the requirement of letting the person tailor the solution to his own reference standards, not to someone else's.

Then the supervisor could try one or more of the proposed solutions, watching to see whether the behavior of all the members of the group will stabilize.

Actually, what happens is ultrastability, not simple stability. As things change, the group returns to finding new stabilities. After a while, they come to understand that continual experimentation is a way of life.

It's not easy to bring a group to the point where they are capable of continuous mutual problem solving. But I don't think it burns up more energy than continual rewarding and punishing, continually patching up a bureaucratic treat-everyone-alike kind of organizing that fits no one well, abondoning buildings to build new ones to try a new organizing experiment (to be impressed on the workers by the designers of the new scheme for controlling their behavior), and so on.

The big difficulty, as I said before, is in trying to set up an island of mutual adaptation in the group in the midst of an ocean of control by others. The old bad norms keep seeping into the new good ones. But you have to start someplace, and starts are indeed being made. Even trials that fail are often worth making, because they put ideas into some people's heads about what is possible. Some people, of course, say, Oh, that was just pie in the sky. Others say, By golly, maybe it will work next time.

There. That's <u>my</u> end to the story.

Page 256 or thereabouts. Does it fit with your model to say that a person can adopt a reference standard (my term for reference signal or reference level) that is dynamic and complex--such as "I'm OK as long as things are going along so-an-so"? For example, a program reeling off as expected, a principle being maintained, a system-concept being maintained? Surely you must be saying that, but it would help me check my understanding to hear you say <u>yes</u>. *Yes* Seems to me this property is necessary to explain how people can carry on a long string of activities such as working at a job or answering a question like, "How did things go today?" *See (2)*

Dr. William T. Powers
September 9, 1985
Page 19

Pages 262 ff. I am very glad to find you coming to the conclusion you do. As my years have gone by, I have found it more and more difficult to believe that anyone is going to find a way to control the behavior of others--that is, bring behavior into a stable pattern--by setting up environmental conditions or by verbal persuasion. The reason, I came at last to see, is that humans construct their own environments. They do that in two ways. First, the environment of any individual is the one he (or she) perceives, not the one the controller perceives. Second, humans operate on their environments. They change them; they alter them from what the controller has set up. Until I read your writings, however, my notions had been about as vague as what I have just said. I could not say much about how individuals go about doing those things. I dc have some convictions from my experience as an organizational consultant and from my reading that most people do form certain sorts of high-order internal standards such as wanting confidence in being able to control the part of the environment nearby in space and time (I won't explain that vague phrase here), wanting to acquire information and understanding beyond one's present point (often called curiosity), and a few more, all of them fitting nicely, I think, into your speculations. I have come to see, too, both as teacher and consultant, the time it takes for reorganization of the higher-order systems, especially when the environment contains other creatures all with their own higher-order systems. One plants seeds.

Many people nowadays are coming to your conclusion--through routes other than yours, most of the routes not nearly as testable as yours. I enclose * a paper by my friend Roger Harrison. You will, I think, find his conclusions See (14) sympathetic to yours; you may find interesting the high-order reference signals he sees around him.

* Roger Harrison, Strategies for a new age
Human Resource Management, Fall 1983, Vol. 22, Number 3, Pp 209-235

Page 264, third line from bottom: ". . . the two methods of control we have just discussed. . . ." Do you mean (1) control by disturbance and (2) control by deception? Or (1) control by disturbance and (2) control by altering perceptions? By the way, those methods correspond to the old argument among social psychologists about whether changing attitudes changes behavior or whether it is the other way round. I got distracted by that argument for some time before I realized that human interaction is a dynamic flow with continuous reciprocal adjustments--adjustments to anticipations, really, the way we move to catch a ball--not an alternation with quiescence between discrete acts or states.

Page 265, end of first paragraph starting there. Here there might be a weakness in your model. People do sometimes harm themselves in the service of their ideologies. Martyrs. Give me liberty or give me death. People who go on hunger strikes do sometimes die. Suttee. Not to speak of soldiers in war who can tell their muscles to move themselves into a hail of bullets or a field where shells are bursting on every side.

Page 266 ff. There is another feature of reward and punishment that you don't mention explicitly, though it may be implied. Rewards "work" only in connection with very specific acts. (That's Skinner's principle of shaping.) If you are a foreman in an industrial plant and you want the workers to carry out certain acts, you must be sure that the workers connect your rewards with those particular acts. Otherwise they'll go off doing the things they erroneously thought that you were rewarding. And then, if you stop the rewards, the workers will interpret the cessation as meaning that you don't care any more, or even as punishment. So you have to go on giving those rewards, maybe even increasing them. It's a very troublesome technique. It requires constant attention to small acts from the reward-giver. No wonder Taylor had to carry his analyses of tasks into such small motions.

Dr. William T. Powers
September 9, 1985
Page 21

Punishment, of course, is even more troublesome. It scatters behavior. Leading by the carrot is fairly smooth, even if you do have to be there all the time with the carrot. Prodding is much more troublesome. The person keeps going off at odd angles. You have to prod first from this side, then that side, and the moment you stop, the prodded creature is immediately out of control. Look at the astonishing amount of effort the shepherd's dog puts out in trying to keep the sheep from straying.

Page 268, near the end of the second paragraph beginning there: ". . . they cannot live up to that principle." Judging from the next paragraph, I guess what you mean by "live up" to it is that they cannot themselves knuckle under to it. Is that right? I think it is right. No place is the inability more obvious than in schools. Yet we go right on believing that rewards and withholding rewards will somehow, sometime, someplace work.

End of book. Wow! I'm with you. Sign me up. Where do I send my membership fee? I don't suppose people who would be appalled by your last few pages would read very far in the book anyway. If they did, I suppose you'd get comments like: This fellow is an anarchist! And, I know we can't really control people, but what else can we do but try? And, if we don't control people, they won't work in armaments factories, and the Russians will take us over! This fellow is a Communist! And, if everybody did what they wanted to do, nothing would get done! And so on. See (1)

Well, as you say, high-order changes can require a long time. I hope we have it. I agree with your implication that feedback theory, inner control-systems theory, goes against the current culture of Western civilization. Yet I think there are signs here and there that there is at least a small movement against the grain. Do you know: L. S. Stavrianos, <u>The Promise of the Coming Dark Age</u>, San Francisco: W. H. Freeman, 1976? See (15)

Well, as you can see, I am very grateful to you for writing the book. And for all the work and thought you did before writing it. And of course all you can do is hope that someone pays attention. I've been talking to people around here about controlling input, though too vaguely, since I read your article in the Psychological Review. Now I'll be able to talk better. And I intend to be writing about it. I hope you will be willing, between your other activities, to help me think. I hope you will be publishing more. If you have something since 1978 that doesn't require calculus to read, please tell me.

Again and again and again, thanks thanks thanks.

Sincerely yours,

Philip J. Runkel
Professor of Education
and Psychology

PJR:dr
encl.

P.S. There is one feature of brain functioning that I wish you had put into your model: the switching effect of emotion. Maybe the rheostat is a better analogy. Anyway, I mean MacLean's hypothesis. I enclose some writing of his: "A Triune Concept of Brain and Behaviour."

See (5) and my reply to P.

Sept. 14, 1985

Dear Philip,

Sorry, I have to turn down the promotion. No "Dr." However, "Bill" suits me just as well.

All I can say is, welcome aboard. Consider yourself signed up. Actually, there is something to sign up for. About five years ago I decided that the proper place for control theory to achieve recognition is inside cybernetics, despite the fact that cyberneticists are uniformly ignorant on the subject and tend toward being wordy dilettantes. So I attended a meeting, gave a paper, and made some friends (never saying "dilettante"). The next year I showed up again, this time with Tom Bourbon, a psychologist at Stephen F. Austin University and a strong convert; he gave a paper and I did a one-hour presentation. The following year (1983 by now) eight of us showed up at the meeting in Los Altos: Tom Bourbon, Rick Marken, Dick Robertson (psychologists), Bill Benzon (linguist), Francis Jeffery (mathematician), Ed Ford (management consultant), Mary (my wife and a non-practicing psychologist) and I. We held three full afternoon sessions and met a lot more people. Then in the Fall of 85, our group appeared at the ASC meeting in Philadelphia, 17 strong (I won't list them all). We had paper sessions every day, plus ongoing computer demonstrations of several kinds. Each year we acquired a few more converts from the ranks of cybernetics -- not yet ready to sign on, but definitely getting there. (1)

This year -- Sept. 19 to 24 in fact -- the first genuine meeting of the Control Theory Group is about to take place (the regular ASC meeting was postponed or cancelled, don't know which, yet). The roster is about 23 people. One attendee will be Barry Clement, an old-time cyberneticist, who will be bringing an invitation for a member of our group to serve as an officer of the ASC. Furthermore, I've been asked to chair a session, in the spring of 86, at a Gordon Conference on cybernetics. I am finally getting the impression that this exponential curve has begun to depart perceptibly from the x-axis. That is, I hope it's exponential and not bell-shaped.

I hemmed and hawed about asking you to the meeting, and finally didn't because I figured you'd have to stay with your wife. Now I'm sorry -- I shouldn't have decided for you. If you can get on an airplane and be there, please do -- you have at least one whole day to arrange the trip! At any rate, you'll be on our mailing list from now on, and will be apprised of all meetings, and will receive our extremely sporadic newsletter.

In fact I'm going to try to call you right now. If you can figure out this temporal paradox, please wait while I talk to you.

... so we continue to correspond. You'll start getting

things from us once in a while, although none of us is yet in a position to devote much time to dealing with the framework.

Skimming through your letter:

If action stops when the error signal is zero --- who sez? (2) If a fourth-order reference signal says "perceive the bicycle moving at 20 miles per hour," can you stop pedaling when it is going 20 miles per hour? Somewhere in there, I think, I point out that we have to think of control in terms of maintaining a constant perceptual signal, but at the higher levels a constant perceptual signal can be created only by continuing action. In any control system, equilibrium is that condition at which just enough error exists to drive just as much output as needed to keep the perception just as near to the reference signal as it is. Equilibrium does not have to mean a static (third-order or lower) condition. A steadily-moving second-hand is steady, yet moves. Constant perception, changing configurations.

Trust your own perceptions. The perceived world is continuous, not discrete. Psychologists have got in the habit of thinking that between changes there is nothing going on. Wrong. There isn't any such time as "between perceptions." Of course change does exist, and we can both perceive and control it. I even devoted one whole level to it. Out of 9. But it's not all there is, as you can easily verify.

Adaptation is an interesting problem: it's as if all (3) perceptions are measured relative to some moving average that changes very slowly. Land's theory of color vision assumes that very explicitly. I don't have that phenomenon in my model. Want to put it in?

I ended up not postulating a separate level for language, (4) but I did end up inserting the level of category control between relationships and programs, so now there are 10 levels. I think we can perceive in categories with or without labeling them. I entirely agree with you that language per se is too recent a phenomenon to have merited a special function of its own in the brain. But what is the function which, when we use it for linguistic purposes, permits language to work? And what does it look like when it is being used for other purposes?

Speaking of programs, MacLean is wrong: a program is not a (5) fixed sequence. A sequence is a sequence. A program introduces choice points: if this condition has resulted, start up that sequence, otherwise start up this one. And I am just about to move the sequence level so it lies just under programs and above categories -- tinker, tinker. I think this will make the boundary between the analogue and the digital worlds of perception occur in a much more sensible place: between relationships and categories.

The chess programs, although they play at a high point- 6

level, still tend to use brute-force algorithms that human players do not use, and depend on sheer speed instead of principles. As far as I can tell; I'm no programming genius.

I did think of Korzybski, a major intellectual inspiration of my high-school days. But when my book was written, Korzybski was considered by critics of the book manuscript to be a cousin of Velikovsky, so I gave up the reference. But I have known that the map is not the territory since I was 15. How else do you think I could have realized that the perception is not the reality? And how can you argue with people who think the map IS the territory, like Gibson and his current defenders? 7

9th order: depends on what you study about them. You can study a drop of water as a system concept, depending on what you make of wee beasties. 8

Nested systems. Funny, Boulding knows of my work, I think, but I've never heard a peep from him. Boulding's hierarchy is more like mine than many others. Most others take one generating principle such as size or complexity, and make a hierarchy by applying it over and over to bigger and bigger units. Others seem to make hierarchies by different ways of drawing circles around groups of objects, always in the same plane. In my hierarchy, there is a right-angle turn every time you go up a level. Boulding's is like that, too. 9

Closure, the way I treated it, can happen at any level, since I propose that it's the filling in of missing lower-order perceptions through the imagination connection. 10

Cognitive Similarity. No wonder you find my ideas so easy to understand. You seem to have worked out exactly the same principle I used for dealing with second-order perceptions: the same idea applies at any level, doesn't it? It hadn't occurred to me that these projections could actually alter the ordering of perceptions. I think you have the germ of a special theory of relativity of perception. Say more. Publish! 11

Awareness and consciousness. Just feeling my way, trying to find a place for phenomena that have to be accounted for. Reorganization probably doesn't have to be conscious, but when it is, doesn't it work a lot better? 12

If my stomach is empty and I therefore reorganize my reading habits, I will probably starve. The problem is how to get reorganization to work on the part of the hierarchy where something is going wrong, and not reorganize what doesn't need fixing. Remember, reorganization has to work before we have any cognitive systems or any other kinds of smarts. It occurred to me that awareness always seems drawn to trouble-spots, so maybe that's part of what it is for: to focus reorganization where it is needed.

My concept of consciousness probably makes my conjectures

harder to understand. I don't equate it with verbalizing, thinking, reasoning, and so on. Those are all acquired processes, things the brain learns to do. We don't learn to be conscious: we ARE conscious, from the start. We are aware. It's consciousness that experiments, pushes the buttons to see what will happen, pushes them again if the result was worse. In silence.

As I said on the phone, The Test can be tricky to apply in higher-level situations, but it isn't impossible. It's just a way of testing for the presence of control. The nicest thing about it is that it doesn't automatically come up with the answer you want: you have to sweat it out, not knowing if you're going to find a controlled variable or not. Contrast that with other theories that just call everything a "response" without any attempt to show that that is the right model. Actually, judging from your hypothetical example, you should be pretty good at applying the test in real situations. What is your wife controlling for? ("Controlling-for" was spontaneously invented by some students learning control theory. They obviously got it). (13)

Applying The Test in a complication situation is complicated, for precisely the reasons you pointed out: there are many hypotheses that must be experimentally eliminated. I should think that to find out what that worker was really controlling for would take a good year of full-time effort. Of course then you'd know something worth knowing.

Page 256 -- see remarks at top of page 2. Yes. (2)

Yes, one plants seeds. But I have found that you can't stop with one seed, or depend on someone else to water it.

Your friend Roger Harrison sounds like a neat person. He speaketh control theory from the intuition, like most people who have avoided being contaminated by received wisdom. I value people who can think clearly at those levels: I can't. Higher-level applications of control theory will really have to come from people like you and Harrison. (14)

I'll obtain Stavrianos. But I warn you, I'm a very concrete thinker, and like most words to have real meanings. I always approach the works of generalists with the anticipation of pain. Horribly unfair of me, of course. (15)

I'm sending a copy of an old paper. 1980 * Nothing of interest since 1978. Probably nothing of interest until next year. I'm too busy now trying to get rich by working on a program for scheduling nurses (while working at a full-time job) to pay any attention to generating new ideas in control theory. God, I wish I could spend all my time doing my real work! But that's now how life works, and I'm used to it.

(squeeze) Regards,

Bill

Bill Powers

* BYTE articles —enclosures at this volume's web page.

Ans Sept 25

23 September 85

Dear Bill:

Thanks for your letter of 14 September.

Here is my understanding of your present ordering of the orders (please make up your mind whether you are calling them orders or levels) of control systems and their conscious experiences. Please let me know if I've made an error.

1st. Intensity.

2nd. Quality of sensation.

3rd. Configuration, position, perception of invariants.

The three above deal with momentary events, not with time-sequences. Any necessary memory, if you want to call it memory, makes use of "lower" neural networks, not the memory functions of the cortex.

4th. Transition, change, tracking, control of movement and other changes of configuration, sensation, or intensity. Time emerges.

5th. Relationships.

6th. Categories.

7th. Sequences, episodes, routines.

How low down in the evolutionary scale would you find the above? Surely reptiles have all those orders of system. Do worms?

8th. Program control, rationality, language, TOTE. And what I call achievement: working your way to a goal that requires more choice-points than running off a sequence.

9th. Principles, strategy, program "writing," heuristics, values in the sense of what one puts consummatory goodness on. Picking out intermittent evidence. Averaging or otherwise "composing" instances. Rats do that averaging when subjected to irregular reinforcement, don't they?

10th. System concepts, perceiving organized entities. Does a worm recognize another worm as a worm? I doubt it. But butterflies, lizards, birds, rats, and humans all recognize conspecifics. Are butterflies doing it merely by sensation--chemicals, without the kind of thing you have in mind for 10th order? I'm supposing you are not equipping butterflies with 10th-order functioning.

And overall: Reorganization.

In response to your remark that MacLean was mixed up about program and sequence, I scanned through his three lectures I sent you and did not find the word "program." Maybe you picked that up from Hart. But I think Hart means what you mean by sequence--a linear sequence that runs off unless a higher-order system interrupts it in the middle.

I have a book on the brain by J.Z. Young in which he uses <u>program</u> in that same way. People without experience in writing programs for computers, or maybe even using them, are likely to use the word analogously to a program for a concert--a series of events expected to go off in the preplanned way.

Sorry your readers of the manuscript were so prejudiced against Korzybski. I've met people like that, too. Some of those same people use some of Korzybski's ideas without knowing where they came from. People are always throwing the baby out with the bath. (I know I do it, too. I even stop reading a book or a paper if the writing style irritates me. One nice thing about old age is that you can be crotchety and most people let you get away with it.)

*

Thanks for sending me the 1980 paper. I am half way through it, and it is just what I needed. You are very good at sending me just what I need.

Thanks again for wanting me to attend the meeting at the ASC. And thanks for putting me on the mailing list.

Despite your lack of time for the things that attract you most, I hope you are well and happy.

Sincerely

Phil

PJR

* BYTE articles —enclosures at this volume's web page.

Sept. 25, 1985

Dear Phil,

The order of each order is exactly the same as the level of each level. However, I have not yet determined whether the level of the orders is the order of the levels. More research is needed. I had originally chosen "order" so as to avoid confusion with "reference level;" that was fine until I realized that the number of derivatives in the system equations determines the order of differential equation involved, having nothing to do with position in the hierarchy, at which point my terms became completely arbitrary and I began getting sloppy. Let the context decide. Anyway, what business does a mere Oregonian have telling an Illinosian he has made a mistake? If the Illinoisian could recognize a mistake, he'd live in Oregon. Seek your own salvation with diligence.

The list is not only correct and up to date, but nicely gathered together from a number of different sources. You put it so clearly. Why does it take me so much longer?

You're right, the reference to programs was in Hart. He does mean a sequence, not a program. So do most modelers who use the term. I have Young's Model of the Brain -- is that the one you mean? Doesn't have much to do with us, does it?

Since you're aware of Korzybski, you might try applying his famous dicta about maps and territories, words and objects, to that list of perceptual levels. The exercise consists of trying to see what the names point to, none of the referents being words. The reason it took me so long to sort out even these few levels (orders) is that I was looking for the territory, having realized some time ago that the brain does not control just the map. These descriptions purport to identify aspects of the real outside world that are really put there by our own perceptual processes. It seems to me that the Artificial Intelligence people, and many others, think the brain deals only with words. I think otherwise.

The first annual meeting of the Control Theory Group was a smashing success. Twenty-two people were there for the peak two days, only a few less for the other two. They represented clinical, experimental, and pedagogical psychology, economics, sociology, management, engineering, and even piano teaching (an old friend who has used control theory to speed up piano teaching by a factor of four). They came from the four corners of the country. There was one nut, but a reasonably nice one. We talked for four days from seven in the morning until midnight, and according to several surveys, every person had extended discussions with every other person. There were no disciplinary barriers at all: we all speak the same language. Astonishing.

Barry Clemson of the ASC was invited and came. He was urging us please not to pull out of the ASC, please even come in and take it over. Nobody wants to take it over, especially not me, and we might just remain independent. Cybernetics seems awfully old-fashioned to us now, although the name can't be beat.

There will definitely be a newsletter, and in the next one we'll poll the membership to find out what level of support seems reasonable. In connection with talk about that, I brought up your name, evaluated your understanding of control theory, mentioned your confining circumstances, and asked if anyone would object if I asked you if there was any level of participation you would be interested in, working from home. Nobody objected, so I'm asking. Think it over.

I am indeed well and happy. At the age of 59 I am seeing, in this control theory group, what I have dreamed for thirty years of seeing, and there is little more I could ask. It's under way, and if I die it will not be lost. I hope that vicariously you can share my feelings. Aside from being a hell of a good scientific meeting, which people said was the best they had ever attended, it was a very joyful affair. The relief of doing something real, at last! If you want to see a REALLY happy man, I can show you a sociologist who now has a theory that works for him.

Lest you think that I am tempted to become a guru, or that if tempted the control theory group would give me the opportunity to succumb, I enclose the enclosed paper, which I wrote with an eye to publication in the new cybernetics magazine Cybernetic, and took to the meeting for criticism. I got criticism. I was told to burn it. Of course they were right, as you will see, but it was fun to write (sometimes my resentment is thinly veiled) and is worth showing to a friend. Do you advise me to stay out of politics? They did.

I hope all is well with you, meaning that you are finding equanimity if not happiness, and perhaps discovering challenges that make tomorrow actually an attractive proposition. Hard or easy, life goes on. It might as well be worth while. Tell your wife that you have some new friends. If she can understand, it might help.

Best regards,

Bill

Bill

PS... forgot one subject, phylogeny.

You are, of course, bringing up a subject that absolutely demands research with control theory in mind. What variables can animals of different species, different evolutionary, uh, levels control? Are they the same as ours, or do we simply project our perceptions onto them, being unable to conceive of any other way of ordering experience than our own? Does a salamander perceive configurations? An Alpha Centaurian on a different evolutionary track? Sometimes I think about Carl Sagan's platinum plaque on that Surveyor spacecraft, and laugh.

The Test for the Controlled Variable can be used to ask questions of an animal without using any words. Ditto for babies and children. The first biologist, ethologist, or developmental psychologist who catches on to this theory is going to find virgin territory waiting, as well as rejection slips from granting agencies. I'm not much good at research like that, and don't know the literature well enough to build a bridge, but I'm content to leave that to others with better credentials and study habits. Who do you know?

WTP

30 Sept 85

Dear Bill:

It was a shock to me, when I read your book, to find that you were putting maintenance of the "intrinsic" state at the top of your hierarchy of control instead of at the bottom. Mpst of us think of mainaining bolidy states (food, safety, etc.) as being "basic"—that is, at the bottom, or before everything else.

But I chose not to say anything at the time. I wanted to let the idea soak and see how I would adapt to it.

And I wondered why you didn't put a number on it—10, or now 11. Though I vaguely remember that you said someplace that reorganization might act at any or all orders of control.

My chief trouble was not the top-or-bottom imagery, but speed. Every higher order of control acts more slowly. And if you think of reorganization as "higher" than everything else, then it should act most slowly. But often it acts most quickly, as at the hot stove. (MacLean says that threat switches control into the limbic system, permitting much faster action than the neocortex permits.) See over

And threat pushes imagery—that is, let's say middle-to-high-order perception—into simpler forms. We fall back quickly to either-or. Either you are with me or against me. Give me a simple yes or no. Don't just stand there, do something. I'm good, you're bad.

(I'm writing this at home, and my copy of your book is at the office, so I'm writing from memory. Maybe after I look in the book again, I'll have to apologize for writing this letter. Oh, well.)

If we think of reorganization as possibly occurring at any level, then we have to think of learning of the reorganizing sort as happening at any level (order of control). Do you remember "inner tennis"? Is that an example? How about optical illusions that flip back and forth? (I remember the eestatic shock I had when I read Lobachevsky. But I suppose that would be system-order reorganization. But maybe not. What was "intrinsic"? Unless you postulate consistent images as intrinsic.

But I guess I do want to think of reorganization as occurring at any order of control. Is that all right?

Phil

Well, I went to the office and looked in the book. I see on page 185 that I have to wait for new synaptic connections to grow. Cancel the hot stove and the either-or.

I guess the first and second-order control systems take care of the hot stove. I guess the reorganizer works more as I said in an earlier letter: to keep the whole shebang in good working order over the longer run. I guess when grandfather says to the youth: Yes, I used to think that way, too--he has been undergoing reorganization.

And on page 186 it says that behavior has <u>indirect</u> effects on the reorganizer.

Page 187 was very helpful.

Well, I'll go on reading, but I'll mail this now.

7 Oct 85

Dear Bill:

I'll answer your recent letter later on. The fall term has started, and I am teaching two classes. So I have only snatches of time for other things.

But here is a note about something I thought I'd better think about. I sometimes use J.G. Miller's theory of living systems in my thinking, and I thought I had better look for correspondences between his subsystems and your orders of control systems. Here is what I think I see.

I think his input transducer and internal transducer correspond to your first-order control system: Intensity.

I think all his other subsystems occur throughout your orders of control.

I think his decoder and encoder functions occur in your comparators.

His channel and net are simply that: all those neurons going every which way.

His output transducers must be at the nerve endings in muscles, glands, or other comparators.

His decider, I think, corresponds to your error signals.

Since, in your model, you take all reference signals to be retrieved recordings of past perceptual signals, Miller's memory runs throughout your orders.

I think his associator corresponds to your control of lower orders by higher orders.

By "learning," I think Miller includes all three of the types you mention, but maybe he thonks more often of the first two, not reorganization.

I know you have other things to do. I don't know that you want to bother yourself with small matters like this. But it is nice to have someone to write to about these things otherthan myself. I could just write myself a note, but it feels as if I'm doing something more "real" if I write to you and keep a copy for myself.

Phil

9 October 85

Dear Bill:

Now to your letter of 25 September.

It was euphoric. It gave me delight to read it. I am glad you had such a good time, such a confirming, invigorating, promising time at the meeting. That's the way life ought to be every week.

I suppose that you are indeed referring to the same book by J.Z. Young that I have. Mine is Programs of the Brain, Oxford Univ Press, 1978. It has chapters in it with titles like Living and choosing; Learning, remembering, and forgetting; Seeing; Fearing, hating, and fighting; Knowing and thinking; Believing and worshipping. It may not have much for you, but for me it has some good stuff on emotions.

Thank you for your kind thoughts about my condition. I'm managing. It's not easy. I always have a lump of anxiety someplace in my body. But I do get something done every day, for Margaret and for other people--and for myself.

I was impressed (favorably) with your report that you had evidence that everyone at the meeting had some useful talk with everyone else. That's not easy among 22 people, even over several days. It is evidence to me that something you did as a bunch encouraged that roaming, that the people wanted to expadd their horizons, and that they had the energy of enthusiasm.

There is always one nut.

I'll be glad to receive the newsletter. I'll tell you what I tell everyone these days: I'll be glad to be on your list, but I don't promise to do anything at all.

I am spending all my "spare" time getting ready to write a book on behavior in organizations. I bought a computer a couple of years ago so I could use it to organize my notes--and to write more easily. (I don't write to you on it, because so far I have written all my letters to you at the office.) I am also a co-author of a book in progress on research methods. I sent a copy of your article to my co-authors; I have had no reply about it.

Thanks for enclosing the paper that your critics said you should burn. I don't think you should burn it; just take out the occasional inflamatory sentence. I enjoyed reading it. I like to read stuff with emotion in it. I get very tired of academic prose written as is the author was bored with the topic.

I wish you had written more someplace on speed of response of the various orders of control. Response obviously takes time. But why can't the neural impulses rip through the ten orders and back again in ten seconds instead of ten years? I suppose the curve of time versus orders is exponential, or at least convex downward. But how can I guess at how much time must go by at a given order of control while I am floundering around hunting for a way to make my perception match my reference standard?

And what is happening to my attention and my emotion while I am floundering? Some mismatches stir strong emotion immediately. Some don't. You left emotion out of your book completely.

I know that sometimes I get immersed in a problem (maybe at the program or principle level) and my mind keeps dwelling on it throughout the day, day after day, even when I am doing other things. You know that writing a book is like that. A little step forward brings glee. Frustration brings anger.

The exercise you posed for me, of finding the extensional referents (Korzybski's term) for the names of the perceptual levels, is a dilly. I am often uncertain whether I am dealing with words or things. I keep nagging other people about it: "But what would you see or touch? What happenings would send light to your eyes, beat at your eardrums, press on your hands?" But it is easier to ask those questions than to answer them yourself.

Anyway, here is what I cam up with after some hours of floundering:

"Intensity" points to your own experiences sush as those described on page 97.

"Sensation" points to your own experiences such as those with lemonade.

"Configuration" to those such as on page 125 ff.

"Transitions" to those such as on pp. 131 and 133.

"Relationships" to those such as on page 155, your own analysis of Bruner, Goodman, and Austin, etc.

"Categories" to experiences with categorizing, a very common human act, including the "either-or," Korzybski's bane. But I can't refer to your own account here, since this level is not in the book.

"Sequence" to those such as at the bottom of page 139 and on page 140.

"Program" to those such as on pages 160 and 167.

"Principles" to those such as on pages 169-171.

"System" to those such as on page 172.

That's the best I can do. I feel like a child trying to guess what is in the teacher's mind. That's not your fault; that's my own doing. You issued a brotherly challenge. So I put aside my childish feelings and tried to work at the problem instead of trying to psych out the teacher.

Before I arrived at the list above, I wrote down a lot of other gunk. So as not to "waste" my struggles, I'll write them down for you here. I'll be glad for you to point out where I went off the track, either in trying to grasp the problem you set or in some other way.

First list

Intensity. Perceptual signals received in the reorganizing system from first-order control systems. The perceptual signals sent "up" from the first-order system consist of a representation of, or analog of, or correspondence with, the frequency, amplitude, or duration of the firing of cells in the sensory nerve endings. More figuratively, "intensity" points to the reorganizing system's perception of the perceptual signals from the sensory organs to the first-order systems.

Sensation. Perceptual signals received in the reorganizing system from second-order control systems, from which the perceptual signals sent "up" consist of a weighting (an input function) of perceotual signals from first-order systems. Figuratively, the reorganizing system's perception of the perceptual signals going from the first-order to the second-order systems.

Configuration. Perceptual signals received in the reorganizing system from the third-order (and lower?) control systems, from which the perceptual signals sent up consist of an analog or an input function of perceptual signals from second-order systems. (But I am uncertain whether the signals sent up might come from both first and second-order systems.) Figuratively, the perception of the . . .

but what should I say here? Here are some possibilities. I don't know which to pick:

(a) Figuratively, the perception of perceptual signals from an input function of second-order systems to the third-order systems.

(b) the perception of perceptual signals from second and first-order systems to an input function of the third-order systems.

(c) the perception of perceptual signals from third and lower-order systems.

Because I am uncertain, I will hereafter arbitrarily use form (a).

Transitions. Perceptual signals in the reorganizing system from an input function to fourth-order systems, the input function at the fourth order being composed of perceptual signals from third-order systems. Figurately, the perception of perceptual signals from an input function of third-order systems to fourth-order systems.

And so on.

Then I tried another tack.

Second list

"Intensity" points to frequency, amplitude, or duration of neural firings, with a conscious experience of (1) unregistered, (2) noticed and usable, or (3) painful.

"Sensation" points to a combination of intensities from various sense organs, with a conscious experience of a state of affairs in some part of the body, including sense organs.

"Configuration" points to combinations (or juxtapositions) of sensations, with a conscious experience of, say, the forearm being bent in such-and-such a way from the upper arm, or a visual edge appearing between an area at the left and an area at the right. An experience of perceptually separable "objects."

"Transitions" points to a contrast between configurations over time, with a conscious experience of change and motion.

"Relationships" points to co-occurrence or co-variation of configurations or transitions: two people walking together, the bottle filling up as the water goes in the top, one person shouting and another turning toward the first, the appearance of death notices within a number of black-bordered rectangles, the smell of gasoline at the sign of the shell. A conscious experience of association or going together.

"Categories" points to configurations of configurations, transitions, or relationships: clusters (configurations) of smooth things (configurations), bumpy things, and sharp things. A flock (configuration) of crows flying (transition) from one cornfield to another versus a flock flying in the other direction. A series of trucks carrying carrots versus a series of cars carrying people. A conscious experience of some similarities or associations being different from others.

I had trouble with "categories." It seems to me it would be easier to go from order to order if categories came before relationships. But maybe you don't mean what I think you mean by categories.

"Sequence" points to transitions in relationships or categories: picking up the water bottle, putting it under the spigot, filling it, and putting it aside. Picking up in a cafeteria first a salad, then an entre, then a vegetable, then a dessert (though maybe that's a program). A conscious experience of departing from one relationship or category and entering another along time's arrow.

"Program" points to transition from one sequence to another. A conscious experience of departing from one sequence and entering another.

"Principles" points to categories of relationships within sequences and programs. Is it all right (category of things fitting and proper) for whites to marry blacks (sequence or program-branch-point)? Does it strengthen command of the center (category) to move knight to queen's bishop's third (program-branch-point)? Will it bring better coordination for peace (category of relationship) for the US and the USSR to threaten each other more ferociously (program-branch-point)? A conscious experience (here there seems to me a discontinuity) of making a choice under uncertainty with the hope that the transitions you have assembled into a sequence or program are a better bet than other assemblies you might have made. Here there seems to be an imagination of future events. How did that creep in? And I don't like having to insert the concepts (excuse the word) of uncertainty, hope, and better bet.

"System" points to configurations, transitions, relationships, categories, and sequences of intensities, sensations, configurations, transitions, relationships, categories, sequences, programs, and principles that are perceived to cohere--they have an overall configuration. A conscious experience of thingness even though containing all those other nine kinds of perception.

After that rumination, it dawned on me that I ought to get personal. After all, you were the person hunting for the thing beyond the word.

Well?

Your admirer,

Phil

Oct 18, 1985

Dear Phil,

I'm burned out on scheduling nurses (that supposed money-making project that is soaking up all my spare time), and your letters beckon. Never let it be said that I take the path of most resistance.

I have a copy of J. G. Miller's giant book, sent to me by D. T. Campbell with a note saying "I can't figure this out, can you?" My overall impression is that form has displaced substance. The building blocks -- the "decider" for example -- don't seem to be serious proposals about real brain structures. Also, Miller's hierarchy seems based primarily on size. You will probably have noticed that in mine, there is no organizing principle that, applied over and over, will carry you from one level to the next. Every level introduces a new dimension of experience, and is not simply a larger grouping of lower-level entities. No matter how you juggle configurations, you will never come up with transitions, and so forth. I like to think that my model is a literal proposal about how the real brain really works (that's quite aside from its correctness). Miller's doesn't impress me as being of that nature (also disregarding correctness).

Also, Miller is too enamoured of generalizations, and fails to look for counterexamples that might cause trouble. For example, his "ingestor" is defined as whatever "brings matter-energy across the system boundary from the environment." I presume that this is supposed to include sensory receptors at some appropriate level of organization. But that is not what sensory receptors do: no matter enters, and no energy flows along a nerve fiber, since signals are carried by a travelling wave of breakdown of the membrane potential. Energy flow, in fact, goes at right angles to the direction of propagation, and runs from the chemical soup outside the fiber, through pores in the fiber's wall, and into it. After the breakdown wave passes, the energy is pumped back out again. The energy dissipated at a synapse comes from the fluids immediately surrounding the synapse, not from the distant source of the nerve impulse. This state of affairs, while perfectly well known, is always ignored by generalizers, especially if they think that information theory has something to do with perception. And of course they all ignore even more difficult problems, such as the "cold-receptor," which responds to a net OUTFLOW of heat-energy from the receptor, or a "producing" muscle which actually absorbs energy from a load if the load is being lowered (there is, in fact, about a 40% recovery of energy during the lowering part of a lifting-lowering cycle -- synthesis rather than metabolism of ATP).

Just one more. I don't think that "decision-making" has much to do with behavior. Once in a while we have to make a decision because of a conflict, but the rest of the time we simply try to keep errors small. The trouble with putting a decision-maker into every system is that you really have to stretch the notion to make sense of it in most contexts. When an archer aims an arrow at a target, is the aiming done by a decision-making process? Does the archer decide between aiming 1.2534 degrees upwind of the bullseye and 1.2535 degrees, or even 1.6 degrees? Of course by postulating alternatives you can make any process seem to involve a decision among several possible choices, and claim that all the possible alternatives but the final one chosen were rejected by some suitable, and hypothetical, decision-maker. But when the possible alternatives amount to an infinity of choices, does that explanation seem plausible? I think there are far simpler, and far more believable, ways to explain most behaviors.

My book by Young is "A Model of the Brain," not the one you have.

To put this response-time stuff in perspective: I did some experiments years ago which seemed to show that detecting a relationship-disturbance takes about half a second, and that to be detected lower-order disturbances took about 0.1-0.2 sec less for each level, down to the minimum of about 0.05 seconds for first-order "reflexes." When I see a programming error message on my computer screen, it usually takes me about 1.5 seconds to realize what a simple error was and start fixing it (Turbo Pascal displays the source code with the cursor located at the point of the error, which starts the timing to which I refer). When something occurs that disturbs my self-concept, I estimate that it takes perhaps two or three seconds to start reacting (if I am willing to recognize the disturbance -- otherwise we're talking latencies of months, years, or a lifetime). So using a time-scale in which experience is sampled, say, 20 times a minute, we can suppose that all the systems at all the levels are acting simultaneously. The reorganizing system, of course, runs much more slowly. I think.

This "teacher" (I much prefer "brother") didn't mean to put you in an uncomfortable position, but I don't blame you for having trouble seeing what was in my mind. I hope you will be intrigued to hear that your first two lists did, indeed, miss the point, and that your final one (the first one you mentioned) is the closest. The terms I use for the levels are not supposed to refer to descriptions or definitions, but to experiences that are contemplated without any words accompanying them. When you look at the world around you and can finally turn off the words, you will be in a position to say whether or not that world appears to you as it does to me -- in terms of certain recognizeable classes of perceptions. This is not a matter of theory (not even control theory), but of fact. If you see the aspect of the world to which I refer by the word "configuration," and if you find some quality of experience that seems to remain the same over all kinds of sensory experiences, and if the term "configuration" seems to suit this sort of experience, then you know what I am talking about. When you experience visual or auditory or other configurations you don't experience neural signals, weightings, levels, combinations, or invariances: you perceive what you perceive, which is none of those things, but is what I try to point at with the word "configuration." (On paper, that word is a configuration). Never mind how obvious this is: it's apparently been too obvious to be understood for what it is. We don't experience the world: we experience PERCEPTIONS.

I usually discourage people from slavishly trying to fit their thinking to the levels of perception I have described. The reason is only partly because I know the levels are uncorroborated. A more important reason is that I know how hard it is to get out of the clutches of language, and I don't want to lead people into playing word-games, starting with my words and running in circles through the dictionary, trying to see what I mean by looking up more words in real dictionaries or in the verbal networks in their heads. I'm not talking about words, but about perceptions. Perceptions unaccompanied by descriptions, conclusions, observations, or implications.

Behavior is the process of controlling perceptions. Only some perceptions are words, not very many. Controlling "real" things is controlling perceptions. If you can grasp what I am going on about, you will someday have the same insight that I had long ago, and that my wife had about ten years ago, appropriately enough while stepping into a bathtub: "It's all input!" Once you see that clearly, personally, you will see why control theory is the ONLY theory of behavior that makes sense.

This is the short-cut to understanding my work. But how can I get that across to a scientific community that spends ninety-nine percent of its time believing that the word IS the object, and the map is ever so much more real than the territory?

Regards,

Bill

Bill

4 Nov 85

Dear Bill:

I'm sorry to hear that D.T. Campbell "can't figure out" J.G. Miller.

I thought I was being verbose in some of my letters to you and some enclosures. I thought I was being burdensome. But it never occurred to me to send you a thousand-page book with the query, "Can you figure this out?"

Please don't throw out the baby with the bath. Also don't expect Miller to write the book you would have written. And similar familiar caveats.

I have been reading through Miller's chapter 4 on cross-level hypotheses to see whether I have fogotten anything I don't want to forget. As I read, I was often reminded of your ideas. If you read that chapter, you will find a lot of hypotheses with which you agree and maybe some empirical support that you didn't know about. You will also find a few hypotheses at which you will cry "Bosh!" But that's all right.

One thing that annoys me about Miller's citations is that he cites experiments with college sophomores (at the group and organization levels) with the same respect that he cites studies of "real" groups and organizations. As far as I am concerned, that's just plain error.

Anyway, here is a minor example from Miller's page 109 with which you must agree:

> Hyp. 5.4.3-4: Decentralization of decision making in general increases the speed and accuracy of decisions that reduce local strains.

You can paraphrase that as:

> Comparators at lower orders of control can set off corrective action faster than comparators at higher orders.

Or in your words on your page 53:

> ... the higher in the heirarchy one looks at behavior, the longer becomes the averaging time and the longer a disturbance may act without being corrected.

By "decider," Miller simply means making a choice between one possible action and another, bringing about one possibility or another. Your deciders are your comparators. Miller is not trying to find structures in the brain; he is trying only to find functions that have to be carried out by living systems of every sort everywhere. He goes to considerable trouble to make it plain that by subsystem he means whatever structures and processes may be involved in particular functions.

I admit that the concept (excuse the word) gets messy at times. He has to bring in the ideas of dispersed function, partipotential systems, and included structures.

Although he talks on page 39 about purpose and goal almost the way you talk about reference signals, at other places in the book he talks about purpose and goal as if they were add-ons, almost embarrassments. That's an impression with which I am left, but I won't take the trouble to hunt for examples for you.

But life is complex and multiform, and who am I to say that Miller should have been neater? I have a book by Alfred Kuhn, "The Logic of Social Systems," that purports to be very neat and logical, and it does help my thinking, but I think Kuhn succeeds in being neat and logical only by ignoring a lot of detailed complexities that Miller meets head-on.

Your hierarchy resides within the individual organism. Miller's is expressly among living systems. So you shouldn't expect much in the way of analogy, isomophism, homology, homomorphism, or whatever the right word is.

I don't think Miller has any "organizing principle that, applied over and over, will carry you from one level to the next," as you say. His cross-level hypotheses don't "carry you from one level to the next." His subsystems don't do that. That is, neither the cross-level hypotheses nor the subsystems tell you that you are at one level or another, nor do they tell you what to expect at the next upper or lower level. Quite the opposite. You have similar functions at various levels in your model, too--the comparator, for example.

Miller gets from one level to another by aggregating a lot of systems at one level into a system at the next higher level, BUT not just by aggregating. He makes it clear that they must become associated in a systemic way--with interdependencies of the sort specified by his subsystems.

And he does make some assertions about the new directions that come about as you go upward through the levels--his emergents. Starting at the level of the organism, I have copied off his emergents and added a few notes of my own in a list that I made for my own reference. I enclose a copy.

When Miller talks about "matter-energy," I think he is wanting to be meticulous about their underlying unity. He doesn't want to spend time, over and over, reminding the reader about the convertibility of the one to the other. In most instances of his use of the ~~hypehation~~ hyphenation, I think he has in mind something like taking in food to get energy.

When Miller talks about "information," he always makes it clear that information is carried on "markers," and that the markers are often patterns of energy--as in your statement that "signals are carried by a travelling wave of breakdown of the membrane potential."

So when Miller talks about the ingestor, he is thinking of a mouth through which "matter-energy" like food gets into the system. When he talks about the input transducer, he is, I hope, thinking about energy impinging upon, affecting, stimulating,energy-sensitive neural organs that then send "information" (patterns in electrical fields, I suppose) inward.

But I have to admit that he does write as if "information" exists outside the organism and is "transduced" for use inside--converted from one form to another. He does not write as if the organism makes information out of the impinging energies. I looked on his pages 379-380 to see if I could find a sentence that would mollify you, and I couldn't find one.

As to your aiming archer, I don't know how Miller might reply, but my guess is that he would say, "Sure, decision making is sometimes a smooth flow. But you have to get somehow from pointing off the target to pointing at it, and I include that in decision making." That's my guess.

I'm going to keep decision making in my vocabulary, because it is a convenient way to communicate with people about processes in groups. But I'll also keep in mind your "smooth action" kind of behavior.

Well, that's enough about Miller.

Thanks for your information about response time. I couldn't understand one sentence: "Using a time-scale in which experience is sampled, say, 20 times a minute, we can suppose that all the systems at all the levels are acting simultaneously." I can understand that systems at all levels are simultaneously busy with their own business, but I cannot connect that sentence to increasing time lags as you go up the levels.

I use my computer only for word-processing and sorting. I have no present use for computing (numbers), and I can't foresee that I ever will have. I don't analyze data much any more, and when I do I get a graduate student to do it on one of the university's computers. But I keep thinking that I might someday, for the fun of it, learn how to do programming on my computer (Kaypro II), and I keep articles about the various programming languages. I've seen good recommendations of Turbo Pascal. Do you think I could learn it in a year?

I knew it, I knew it. Sooner or later you would get to Zen. You remember Korzybski's exercise of dropping the pencil as soon as you hear words in your head. Sometimes in our human-relations training work we try to get people to listen to themselves and include exercises such as staring at the peel of an orange, trying not to think about it, but just to see it. Or lie on a hillside on a sunny day and let the configurations flow into you. I still use exercises like those several times a week, but briefly, not as long or as often as I ought.

Well, I'm glad I was moving in the right direction, anyway.

It's all right to "discourage people from slavishly trying to fit their thinking to the levels...." But you should allow them to do so until they get disturbed doing it. Harrison wrote an insightful article about learning in which he talks about how to change from one dimension to another. If you are accustomed to dealing with people by dominating them and you want to change to be able to show affection, Harrison says, you will not do well by trying to go directly from the one to the other. You should first practice being submissive for a while. That will enable you to understand the full dimension of dominance-submission and understand that affection does not lie somewhere on that, but is indeed in a different direction from the one you have been using. Then you can try moving in that other direction.

So when we meet a new idea that runs off in a direction strange to us, it is useful to "go to extremes" for a while.

Word games. I often have students who say that if you want to do research on X, you should be sure to define X carefully, so you should look in the dictionaries and read all the books in which the authorities have used the word you think stands for your X, and so on. Then I have to tell them to look inside themselves, not inside the dictionary.

I read an article recently, otherwise very good, in wich the authors spent three pages on the etymology of their terms. Ugh.

I think I'll go home and step into the bathtub.

Yours,

Phil

EMERGENTS

I

Individual human

(1) Emotion in humans and other mammals--in comparison to organ or cell or many other species of organisms.

(2) Gamma-coded language; see J.G. Miller, pp. 404, 439 ff. Miller distinguishes three levels of language. Alpha-coded: Pheromones or other odorous substances like those in urine are received by sense of smell or taste. Alpha codes deal very fleetingly with images, if at all. Very little cortical memory is used, if any. The concepts of image and cortical memory are trivial at this level.

Beta-coded: Signals received by sight, hearing, or touch, such as signals for danger or harmlessness. Beta codes require images in cortical memory, but not images of images--not consciousness.

Gamma-coded: Symbols recognized as such--that is, images of images.

II

Dyad

(1) "The ability to extend over a much larger spatial region than a single organism can.... Markedly increased physical separateness and more autonomous mobility of components in physical space." Those two sentences are Miller's emergents (b) and (e) on his page 575.

(2) "The possibility of shifting a subsystem process from one component to another" (Miller, p. 575). This possibility is a coin with two sides. (a) One person (component) can substitute for another, making for flexibility. In an individual organism, in contrast, the heart cannot substitute for the intestine. (b) Certain subsystem processes can be assigned more or less permanently to one person; that is, roles can be established, as between the male and the female in a marriage. Organs within an individual also have permanent roles (functions), but an individual within a dyad or group can take on several subsystem functions, and those functions can quickly be reassigned.

(3) "The ability to perform motor activities and to make artifacts that would be beyond the capacity of a single organism" (Miller, p. 575). For example,

each member of the dyad can pull one end of a large fishing net through shallow water or between boats. Two or more persons, too, can create concepts, proposals, plans, and so on that no individual member could have produced from his or her own head.

(4) "The sharing of a single component by multiple [dyads or] groups" (Miller, p. 575). A husband can play chess with a guest and also intermittently talk to his wife.

(5) "The capability of creating and implementing an implicit or explicit charter for a new [dyad,] group, or higher-level system" (Miller, p. 575). "Charter" means an agreement about the ways the new system will behave.

(6) "The ability of components to be integrated and to coordinate and control one another by symbolic languages" (Miller, p. 575).

Though the skill of symbolic language resides in the individual, and though individuals can talk with themselves (and thus "coordinate" one of their actions with another), the components of individuals (such as heart and intestines) cannot coordinate through symbolic language. The components of a dyad can do so.

The coordination in a dyad (or group) gives rise to skill that can be exercised only by dyads and higher levels of system, not by individuals. Most interpersonal communication exhibits that kind of reciprocal skill, whether high or low.

(6a) Symbolic language enables the two persons in a dyad to exchange images of past, present, and future. One can influence the other to move in a direction to match the first person's image of where the other person might be in the future. Person A can describe to B a possible future action of B and connect with that image in B's mind the likelihood that A will be pleased or displeased with the action. Or whether B is likely to be pleased or displeased. "If you do it that way, you are likely to be here after quitting time." Or whether some other person C is likely to be pleased or displeased. "The boss wants them kept in alphabetical order."

(6b) Symbolic language enables the two persons to

compare their images and thus get an indication of the satisfactoriness or reliability of their images--social reality. They can use paraphrasing or impression checking.

III **Group**

(1) "Longer duration of survival" (Miller , p. 574). A basketball team can remain the "same" team even though its members come and go. But a marriage is not the same marriage if the husband leaves and another takes his place. And of course no marriage remains at all if the husband leaves and no other is substituted.

If one person leaves a dyad, the "group" vanishes. If one person leaves a group of (say) six, the remaining five can usually make do. The three-person group is the limiting case. Threatening to leave a dyad is catastrophic. Threatening to leave a triad is awful, but the departure still leaves a minimal group. Threatening to quit larger and larger groups becomes less and less a threat to the functioning of the group.

(2) The coordination through symbolic language has emergent features in groups of three or more persons.

In a dyad, each person can talk only to one other person at a time. In a group of three or more, one person can talk to more than one at the same time. All the others can hear simultaneously what is said to others. All members have the same input (more or less, allowing for differences in reception) with which to deal. Communication does not have to be relayed from one dyad to another.

Groups of three persons or more exhibit high or low levels of skill in communicating within the group. New kinds of communicative skill emerge beyond the kinds characterizing the dyad.

In a group of three or more members, one person can observe the communication between two others. In a dyad, a member observing the communication within the dyad must perforce observe himself or herself. But in a triad or larger group, the observer can be "outside" the dyad. The observer can become, temporarily, a separate system interfacing with the dyad. Thus is born third-party helping and consulting.

In a triad, a reliability check of broader scope can be made. That is, person B can not only check whether B's paraphrase or impression check matches what A thought he or she meant, but also whether it matches what the observer C thought A meant. The awareness of both A and B of their own meanings and interpretations can sometimes be broadened by such an interchange.

Furthermore, members of a dyad do indeed observe their own interaction as well as their separate acts. In a triad, A and B can ask C for C's image of the dynamics of the communication between A and B during the last few minutes. A and B can get a check from C about their effectiveness not only as individuals but as an interacting duo.

In a group of four, still further reliability checking can be done. For example, does C's understanding of the interaction between A and B agree with D's?

That kind of sequence can be extended to larger and larger scope. For example, in a group of five, persons D and E can compare their images of the third-party helping within the triad A-B-C. Some groups have the skill and the norms that enable them to make the kind of reliability checks I have been describing.

Even in groups without much awareness of this kind of communicative dynamic, reliability checking nevertheless goes on among large proportions of the people involved and makes use of some complex combinations. For example, after a meeting of the whole faculty of a high school, the mathematics faculty and the science faculty might discuss, separately or in a joint meeting, their understanding of what went on at the meeting of the whole. At the next meeting of the whole, members of the mathematics faculty and of the science faculty might portray their images to the assembly.

Because there are more and different combinations of persons or subgroups acting as third-party helpers as groups get larger, one might claim that new possibilities are emergent with each larger size of group. I think, however, that the important discontinuity appears between the dyad and the triad. The possibility of one member of the group

> standing apart from the remainder to act temporarily as an interfacing system occurs first within the triad. I will consider the more complicated combinations in larger groups to be a matter of degree.

Most of the other emergents at level II can also be considered to be matters of degree in groups of three and larger. In the comments below, the numbers correspond to those used under level II.

(1) With three or more members, the group can spread over or "occupy" more and more space. (2) With more members, it is easier to include more kinds of ability among the members, and substitutions can be made, on the average, with less likelihood of having to accept inferior skill. Similarly, in larger groups, roles and divisions of labor can be more elaborate and detailed. (3) Larger groups can perform tasks that are more complicated and require more matter-energy than smaller groups. (But one should also note that nowadays immense machines and immense supplies of energy can be commanded by one finger. That adds fateful complications to groups and organizations.) (4) When a group is larger, more of its members can simultaneously be members of other groups. Finally, (5) a group of any ~~six~~ size can create a charter for a new group or higher-level system.

IV

Group-to-group interface

Representatives emerge when the interfacing groups are sufficiently large that communication in a meeting of all members of the two or more groups would be inefficient or when one or more of the groups cannot spare the matter-energy to act as committees of the whole.

When a group is small, the entire group can represent itself when interfacing with another. The mathematics faculty can meet with the science faculty. When groups are large, individuals or subgroups must take the roles of representatives--they must be the interfaces.

The communicative dynamics of interfacing groups, as their numbers increase from two, are similar to the dynamics of communication among individuals as they grow in number. That is, representatives of two groups cannot easily stand outside their roles as representatives to watch their own interaction. But representatives of a third group can act as third-party helpers to the representatives of two other groups. And so on.

The skill of communicating as a representative is **different from that of communicating as a member.**

V and above

Hierarchy appears--that is, two or more levels of decision making.

V: Organization.
VI: Organizational interfaces.
VII: Society.

No further discontinuities appear at level V or above. Aside from layered decision making, the capabilities of organizations are different in magnitude from those of groups or interfacing groups, but not in kind. Organizations can organize people over larger geographic areas, build bigger buildings, and so on. Perhaps pluralism, maintaining several subcultures, or the like are easier in organizations and societies than in smaller systems.

No doubt, as primitive organizations and societies evolved into the modern sort, new forms of social organizing emerged. Here, however, I use the word emergent to ask what we see today in the U.S. as we look from one level to another. What discontinuities, not matters of degree, do we see at a higher level that cannot occur at a lower?

New skills and techniques come into play at level V and above, though they seem to me extensions or expansions of skills and techniques at lower levels.

Nov. 8, 1985

Dear Phil,

My argument with Miller is similar to the argument I have with most theoreticians in psychology, flavored to an extent I am in no position to assess by my own professional jealousy. In my defense, I try to be honest and keep a fine strainer over the drain, but what I find after the last gurgle is usually just a wad of hair.

Miller, like many others, says things with which I can agree. But that isn't enough for me. Before they came to understand what I am about, even strong supporters used to send me reams of useful material showing that so-and-so back in 1937 (e.g., Tolman) stuck his neck out and insisted that behavior is, e.g., purposive. I would write back and say thanks, but I would also explain that thousands of people have had the feeling that behavior is purposive, and have said so, and I can't possibly acknowledge them all. Nor am I inclined to: if all I had to say was that I, too, think behavior is purposive I might as well have stuck to engineering. So my friends caught on, and I no longer get such materials unless the author also offers an explanation of what a purpose is and some attempt to say how purpose works, from which the conclusion follows irresistably. Needless to say, I don't get much of that stuff any more.

It's easy to make proposals to the effect that this or that phenomenon exists or occurs. Most "theories" in the life sciences do no more than that. To me, however, such proposals are just the start of a theoretical effort: the real question is not what happens, but HOW IT WORKS. Anybody can guess about properties of behavior, and find both data and other people to agree with the guess (given a friendly interpretation in both cases). But to find an explanation that not only fits the data but is internally consistent, rigorously defined, non-statistical, and plausible in terms of what we know about the physical capabilities of an organism -- that is the real problem. That's the only problem I consider worth the effort to solve. I don't care if other people agree or disagree. That's a side-issue to me. All I want is a model of behavior that I can't poke holes through, a model I can test, a model that doesn't depend on my faith in it or on unspoken assumptions. I am my own worst critic: I put questions to my own efforts that few others even know how to ask. This is not because I'm smart, but because I KNOW SOMETHING THEY DON'T KNOW: control theory.

Behind essentially every theory of behavior I have ever seen, Miller's included, is a basic assumption about the nature of behavior. It's expressed under various names: stimulus-response, input-output, antecedent-consequent, dependent variable - independent variable, and so on. The assumption is that behavior results from influences acting on organisms. This is the only model of a behaving system that most life scientists understand. It underlies EVERYTHING they say. Let me quote Miller, p. 448:

> "Some individuals are stronger, larger, healthier, more talented, better educated, or more disposed toward a certain activity than others. [Who could argue with that?]. Consequently, within the range of species norms for different processes, individual organisms differ in their characteristic input-output relationships."

Aside from the fact that the "consequently" could just as well go with the first sentence (moved to be the second one), this quote shows how the old input-output model is almost invisibly taken for granted. My first reaction to sayings like this is not to the substance, but to the assumption: who says organisms have any characteristic input-output relationships in the first place? I can prove, in fact, that they don't (all you have to do is consider the role of reference signals -- or just look at behavior). This results in my losing interest in whatever conclusions follow.

Miller, of course, knows a little about control processes, but like most others who do, he relegates them to homeostatic systems; p.448, title of section 5.2: "Adjustment processes among subsystems or components, used in maintaining variables in steady states." The idea of controlling through varying a reference signal has never occurred to him, or if it has, he hasn't seen what it means.

Looking higher on page 448 I see " ... when different messages arrive at the two eyes or ears simultaneously, a number of factors influence a person's ability to respond appropriately to them...". The embedding paragraph isn't even about S-R theory -- that's assumed without defense. It's concerned with information theory, and the peculiar idea that "messages" are always clamoring to get into the brain, which has to filter out what it can use to avoid being overwhelmed. The tricky term "appropriately" isn't explored at all -- just lucky for the organism, I guess.

And so it goes, sentence after sentence, paragraph after paragraph, page after page, book after book. The life sciences are in the grip of a wrong model of behavior, a model that has never been tested, a model that is based on blind faith in a few basic assumptions that aren't even recognized as being testable theoretical assumptions. I don't care how many guesses agree with my conclusions if the basis for them is simply wrong, or worse, non-existent. That doesn't make me right, of course, but why pursue what we know is wrong?

In school, I was always the guy who raised his hand during the introductory lecture. If I can't swallow the basis for an argument, I just can't see any point in hearing the whole tedious thing worked out. I am as certain as I can be that Miller's fundamental assumptions about the very nature of organisms are false to fact. I'm willing to stipulate that his logic is impeccable -- but so what? Garbage in, garbage out. Sorry.

I'm sure this testy essay hasn't convinced you of the vacuity of Miller's book, but we'll get back to that sort of thing, without doubt. If I know you, you'll call my bluff.

Response time. What I meant was that if you view behavior on a scale where the least unit of time is about three seconds, the inherent lags at the various levels will become invisible, since the longest ones are, I think, about two seconds. The lags are probably different, and slower as you go upward because of stability considerations, but even the longest ones really don't play much of a part in behavior (they had better not, if the whole system is to be stable). Most of the slowness that shows up is probably processing time, and can't be attributed to delays that are inherent in the system. It takes longer to multiply 9967 by 37 in your head than it does to multiply 17 by 11, but that has no bearing on signal-transmission times or irreducible reaction times. When I was trying to identify levels using reaction-time experiments, I always tried to find the simplest possible example of a perception fitting the proposed description of a level, used subjects who were practiced to saturation, and looked for the minimum reaction times. For task A to qualify as being of lower level than task B, <u>all</u> the reaction times for A had to be shorter than the <u>minimum</u> reaction time for B. Needless to say, it took a long time to find even four levels that qualified. When you're trying to discover real facts as opposed to statistical ones, most of your experiments fail (note gauntlet lying on ground).

Your letter to a friend about words answered most of my questions about what you know on this subject: everything I know. I suspect that you don't hold my suspicions against me, however.

The lesson with the orange peel is the kind of Zen I meant. I have been doing the same thing with transitions, relationships, categories, sequences, programs, principles, and systems concepts, too: every kind of perception I could catch myself perceiving. Sorry to say that this did not end in Satori, although it has certainly changed my views on "knowledge." Lest you conclude that I have some exceptional facility (or delusions) in these regards , I should mention that the rate of production of new levels has been about one every three years. Also, I periodically shake off the theoretical fumes, look around, and ask myself, "WHAT levels?" I'm constantly having to reconstruct the theory from scratch. I have a horror of falling into self-delusion, having experienced plenty of it in the past. But who can say for sure?

Computer:

Kaypro II is good, and there is a Turbo Pascal for it. Yes, you can learn to program in Turbo in a year: you'll be writing decent programs in three months, and useful ones in six months.

In a year you'll be fluent. Turbo Pascal is everything they claim for it: bug-free (version 3.0), and by my measure, 30 times as fast to compile as the nearest competitor -- which is downright unbelievable. The only times I use any other language now are when (a) there is no version of Turbo for the computer I'm using, or (b) I have to get absolute maximum speed and so must program in assembly language. Or (see enclosed), it's necessary to assume the user knows nothing but BASIC. The enclosed, by the way, reflects an editor's style more than mine: I could not keep that cute character out and still publish. The grammar and other verbal twitches, I hope you notice, are not all mine either.

Borland will soon come out with a Turbo Modula-2 (purportedly better than Pascal) and a Turbo C (which seems to produce more efficient coding, hence faster running of programs). The C language is in some respects nicer than Pascal, but I gave up using it when Turbo Pascal appeared, because the error-correcting cycle was ten times as long as it was with TP. Once you learn any of these these languages, the others will come easily, since they're quite similar. I'd suggest starting with Turbo Pascal 3.0, all in all, just because it's a mature product, while the others will go through debugging for many months after their release.

Why learn to program? The best immediate reason is to learn control theory through simulations. Making a model run on a computer, I've always maintained, keeps the theoretician honest. It's easy to SAY that a model will behave like the organism you're trying to explain, but DEMONSTRATING that it does is a completely different proposition. As near as I can tell it's just not done in psychology except sometimes at the AI level. The model won't behave at all unless you make your assumptions specific and quantitative, and when it does behave the chances are that it will do something quite different from what you expected. I wish Pribram had tried, 25 years ago, to make that TOTE model work on a computer: he never would have been party to its publication if he had. That damned model has done immense harm to my efforts: it has given people the impression of understanding control theory while utterly misleading them. Ah, well.

When you get your copy of Turbo, also get the Turbo Tutor.

Do you have a modem? If you did, I could occasionally send you some programs to compile on your machine. Otherwise I'd have to send you a printout, since the Kaypro can't read my disks.

I'm running down, and must get back to the other computer, which is vainly trying to find a schedule for some simulated nurses. It needs me. Also I'm at the end of a page.

Best,

Bill

Nov. 9. 1985

Dear Phil

Becker's article arrived too late to be included in the last letter, but it gives me the chance to expound further on what my theory is and isn't about.

I think, tentatively, that human beings have the neural equipment to perceive in roughly ten identifiable classes to which I have given the names (November. 1985) intensities, sensations, configurations, transitions, relationships, categories, sequences, programs, principles, and system concepts. I claim that this is what we inherit. I claim, further, that there is no particular example of any of these levels of perception that is a built-in aspect of human experience: all the specific examples are constructed by the individual out of his own experience, which of course includes trying to make sense of other people's ideas, through language.

Becker is only one example of thousands. He is transmitting Rank's proposal that one particular principle, "dominant immortality-ideology," represents Man's "deepest innate hunger." I doubt it, just as I doubt that Christianity or Islam or Capitalism or Communism represents Man's highest goal. I doubt that any particular example of a system concept that anyone could give -- and people are always coming up with proposals, aren't they? -- represents any more than a cultural/experiential accident, something that seems to work for someone. We adopt specific perceptions among the classes we are capable of constructing simply because doing so seems like a good idea at the time. The only constant factor is that we construct perceptions in these ten classes and no others. I have not been trying to create a taxonomy of perception. I have been trying to understand human perception in a way that is so general that it no longer is attached to any specific examples -- because it covers all specific examples.

Life scientists have been trying to discover the secrets of the brain's workings by examining particular examples of things the brain does. They compete with each other to find the example that is the most general, the deepest, the highest, the most generic, the most dominant. But this is an empty exercise, because they are missing the obvious: the fact that we do perceive in terms of system concepts and the other levels. What they are trying to do is just like studying the output of a computer to determine the most general kind of program that runs on it, not realizing that an infinity of different programs can run on it, as long as they all use the facilities the computer really possesses. They mistake the products and activities of the brain for the functions of the brain that create those products and activities.

Sure, it's quite possible that some people consider immortality the highest goal, and it's perfectly possible that the way they think about this amounts to what I would call a system concept. The same holds for people who think that power is the highest goal, or love, or belonging, or God, or the Superego. There are unlimited examples of system concepts. But once we understand that people perceive at this level, haven't we understood the main thing? Once we see that system concepts are at the top of the hierarchy, and account for all other goals at lower levels, shouldn't we begin studying system concepts with an eye to finding ones that more people can live with, in less conflict with themselves and each other? As long as we focus on particular system concepts, all we can do is fight over whose is really right. When we see that system concepts are INVENTED, we can begin developing them consciously, for higher purposes. And by understanding their relationship to all other perceptions, we can learn to deal effectively with conflicts.

As you can see, I am speaking as if from a level above that of system concepts. I am quite sure that there is a law of awareness: one is never aware OF the level he is aware FROM. So it is now: I can't characterize what I am doing now. I can only do it. When I become able to grasp the point of view from which system concepts look like means rather than ends, I will add it to the model -- but then, of course, I will have moved again. Perhaps it isn't possible for me to move again. Too bad, then. Someone else will have to do it.

Perhaps when I put it this way, my obstinate rejection of most received wisdom will not appear so arbitrary.

Best

Bill

6 December 85

Dear Bill:

Here is a letter from a friend.

I don't know whether you can make head or tail of "On Powers: First Meander." I am going to write to her and ask her to translate her highfalutin language for me. But I am sending it to you as it is in case there is anything in it that amuses you. It's also an excuse for me to say hello.

Carol Slater and I were graduate students togetherin social psychology at Univ of Mich, 1952-55. In recent years, she has got interested in philosophy of science and is pursuing a second doctorate in it. As she admits in her letter, she has got so she talks like those people, and a lot of it goes over my head.

I will answer your good letter of some time ago before long. I have been full of papers from students and various other things. I still have a couple of fat dissertations to read.

Phil

December 1, 1985

rubber bands

Dear Phil,

Thank you for the nifty new equipment. I think I am getting the hang of what Powers is up to. When things settle down, I shall try to read the book from the beginning. I'm still having trouble identifying his target(s)--who, in 1973, would have seriously disagreed with him? Or am I missing something? Enclosed is a meander starting with the page on lemonade and ending with the problem of "dissolving" the substantive domain. I think there is a connection.

Just in case you have been knitting up attributions, let me mention that the long latency of response is due to having to take a prelim and undergo a bit of surgery (and compose lectures for Intro.) and not to your editorial comments. Re-reading that paper, I am more than ever struck by my linguistic susceptibility. Rorty writes a fruitcake-y style and I can see, in retrospect, how much I had picked it up by the end of the semster. I shall try to stick to polyunsaturated authors. (At the moment, I am slogging my way through Spinoza and Heaven only knows what that will do. If you catch me going on about eternal essences or infinite perfections, let me know. How can I write 750 words on someone who worries about generating finite modes from infinite attributes????) On the other hand, I have been putting together lectures about the contrast between our everyday heuristics and the deliverances of cognitive high tech (aka, science) for Intro., and repackaging Ross and Nisbett and Tversky and Kahneman for beginners, and it has been fun. It is amazing how much yardage you can get from the framing statement that our everyday heuristics let us down because they avoid disconfirmation whilst science is a matter of looking for bad news. By the next time around, I ought to have it in really nice shape. As you might have guessed, there has been a big pause in my blouse-making activity (although I did acquire a chunk of coffee-coloured silky stuff, which waits patiently in the attic). Confronted with house guests over Thanksgiving (Joseph & cie), I reconstructed a feather pillow hastily and can see why that's not much done any more.

I shall try to drop a note to Joe McGrath and let him know that I am grateful for his hospitable response to my ~~kibbutzing~~ kibitzing. My friends in philosophy tell me that committing philosophy in public is a good way to lose friends. I'm lucky to have a charitable audience.

As ever and ever your friend,

Carol

On Powers: First Meander

Text: "[A] 'philosophical fact' . . . emerges from this theory: perceptual signals depend on physical events, but what they represent does not necessarily have any physical significance. . ..[A]n organized approach to physics which takes the arbitrariness of human perceptions into account <u>at all levels of observation</u> would seem to me a most powerful way of deepening our physical concepts of reality."

Good. The connection between a physical event and our qualitative experience of it is contingent, not necessary. The energy changes which (happen to) make us feel "heat" could have made us have a visual sensation of "white", says Paul Churchland (<u>Scientific Realism and the Plasticity of Mind</u>). To the extent that our report "heat now here" is incorrigible--that is, to the extent that it reports a particular quality of experience--it is irrelevant to any knowledge claim about the world outside. Usually, when we use words like "hot" or "white", however, we are not just using them in this innocent way. Rather, they are being deployed as theory terms in an informal theory and take their meaning from a network of connections with other, related terms ("cool", "illumination", "normal viewing conditions" and so on). As theory terms, they are potential constituents of knowledge claims, but the claims in which they figure are, as theoretical, corrigible. Churchland says, and I agree, that no term which can figure in a knowledge claim (or be relevant to one) is ever purely observational. Thus, the everyday "thing" and "process" terms of talk about "persons", "organizations", "groups" and so on, are, in fact, theory terms every bit as much as our more self-consciously conceptual vocabulary. Some empiricists would like the world of ordinary things to be somehow given; others would like sense-data (or its equivalent) to be similarly "below" theory. I agree with Powers and Churchland that nothing we say meaningfully is nontheoretical--it is just a matter of how explicit we are going to be about the theory. (Putnam, however, says, against this view, that we cannot really call something a "theory" if there is no alternative view anywhere in sight, any more than we can call everything "unreal". The use of terms, on this view, requires contrastive cases. I am sympathetic but unpersuaded.) I think we may have a problem in using the term 'theory' in this way if we think that theoretical entities or properties arc, ipso facto, less real than whatever else we have in mind--the "phenomena" or "things" which we once thought grounded theory, perhaps. I think there is, in the wings, still the ghost of Russell's 'logical constructs' and 'fictions' (cf. Quine's 'posits'.) But those of us who grew up on Sherif and the reality of groups should be able to fight off this ontological snobbism. It does not "dissolve" the substantive domain to consider its entities as (yet another sort) of theoretical constructs unless you think that there is something else to talk about which is somehow more real than theoretical constructs. Insofar as one function of this category is to legitimate the interests of applied social science, there is, I suppose, the difficulty that calling something 'informal' or 'folk' theory is going to be heard as a put-down. This is certainly the case if your criterion for what is real is the goodness of the theory in which it appears: witches aren't real and microbes are because witch-theory is a washout and microbiology works wonders. One might argue, however, that for many purposes, sorting things into 'classrooms' and 'teams' and 'offices' is exactly what we want because those are the categories on which important norms operate. Our everyday sociology may be an ineliminable theory about how the world works because we enact it. It is thus a theory in good standing and its entities are as real as anything else. (Theorists who, like Winch, go in for a <u>verstehen</u> approach, are inviting us to consider the category system of participants not only as one theory but as the <u>only</u> theory worth working with.) I am tempted to suggest that concerns about what we are dissolving and what is real are, au fond, political ones in this case. I recognize that, historically, applied science has been snubbed and that is too bad but I wouldn't bend my philosophy of science out of shape in the hope of remedying this ill. Once you have made the point that we design research to meet human goals and not to capture the reflection of Reality, I think you have done all that can be done to make the world safe for institutional research.

6 Jan 86

Dear Bill:

Happy New Year.

On 6 February, I'll be giving the "keynote" talk at a "conference" here. (It's a sort of small convention.) Enclosed is * a copy of my talk--the version to be read rather than listened to. Also a copy of the handout that goes with it.

Half of my talk consists of your ideas. That part begins on page 8 at "What Humans Are Like." If you want to read it, and if you find any idiocies, please tell me.

In the handout, you have seen page 1 before. Then I have included excerpts from your book as pages 2 through 5. If you don't want your copyright violated, please tell me and I'll remove those pages.

I experimented with an overhead projector to discover whether I could demonstrate the rubber-band experiment to the crowd if another person and I were to do it on the platform of the projector. But I discovered that any kind of spot I could see (to keep the knot over) would be too obvious to the audience. I thought of making a transparency with random dots on it, but it seemed to me the audience would wonder about that, too.

I thought about having some helpers demonstrate it in small groups gathered for ten minutes, but the groups should be small so that everyone could see well, and if we have 100 people at the conference, that would mean I'd have to train maybe 13 to 16 people to do it right, and I didn't want to commit myself to finding and training those people in a month's time, considering my chronic uncertainty about my ability to schedule things.

So I just put the instructions in the handout.

Again, happy new year.

Phil

P.S. I <u>am</u> going to answer your last letter one of these days.

* 860206_Handout.pdf and 860206_SpiritOD.pdf —enclosure at this volume's web page.

Jan. 8, 1986

Dear Phil,

A very good talk -- the best thing about it is the pleasure I get from seeing my ideas grow through sharing. The time is coming when they will no longer be "my" ideas, but just something we all understand. In fact you are already doing with control theory what I could never hope to do, showing specialists in your field how to make it real and comfortable in terms they can grasp and use. Sometimes, watching what others are doing with control theory, I know what artisans have always felt. When Neil Armstrong stepped onto the Moon, someone watching thought to himself, "I designed that boot." Loki thought, "I made that hammer." Imagine how the toolmaker would feel if nobody ever picked up the tools and used them! Needless to say, you don't need to worry about "copyrighted materials." My lawyers will tell you when to worry.

I'm sure this has occurred to you, but in case it hasn't -- you don't have to watch the knot itself over the transparency. Watch it on the screen. There will be some way to judge a reference point, even if it's only the imaginary center of the screen. Have faith. This also demonstrates that control is control of remote consequences, not of outputs. Heck, you could even put your finger into the projector and hold its image on a spot on the screen while someone (slowly) twisted the projector this way and that, but not too far. I've had luck with giving each person a single rubber band as they enter the room, explaining later how to knot them with a neighbor's (and also that if they can't figure out how to knot them, they wouldn't understand the demonstration anyway). One person up front can use BIG rubber bands with you to show how it's done (or you can use the projector idea). The whole demo takes 10 to 15 minutes, counting time for excitement and laughs, and swapping roles. Watching it done isn't anywhere near as illuminating as doing it, as the Bishop said to the actress.

I like your funny complicated friends. It's interesting that Carol Slater should cite Churchland as sharing my ideas: he and I have had a few friendly correspondences, after I wrote him a fan letter about his book, Matter and Consciousness -- it may have helped that his wife, Patricia, recognized me as the author of a book she had admired. Churchland subscribes to what he calls a "network theory of knowledge," (with which I agree), meaning that no isolated perception has any meaning, all meaning arising from relationships (or whatever) among perceptions. When you experience an orange peel long enough, and ask the right questions, you realize that there really isn't any pin-downable difference between its color and its smell. All neural signals are alike. It's all a matter of how intensities ... system concepts act together and depend on one another.

You might relay this to Slater. I think that philosophers, especially epistemologists, have concentrated much too much on passive observation, leaving motor action out of the picture. One way we know that there is a real world is that we must learn what to do with our muscles in order to control our perceptions of it. Making up arbitrary rules works fine as long as we're only observing, since the possible interpretive schemes are endless. But when you try to control a perception, you'll first find whether your actions can even affect it (by itself), and second discover that nature puts requirements on what actions will in fact reliably change it. Somewhere, out there, are rules that make perception depend on action in quantitative as well as qualitative ways. Model-building is an attempt to guess at a structure having rules that are at least similar in overall effect to those that actually obtain when we try to make things happen.

Of course we don't learn in this way what is really out there -- only what COULD be out there. Odd that control theory should bear so directly on philosophy. I've come, through lines of thought like this, to get an eerie sense of how primitive our sciences really are, even physics. The so-called "hard" sciences haven't the least idea of what they're really investigating -- human perceptions, taken for granted. It will not be easy to filter out what is US so we can really begin to suspect the nature of THAT. Doing that is many Scientific Revolutions away.

Slater is right (in citing Putnam, so I guess Putnam is right) that you can't call something a "theory" without an alternative in view. That's why scientists who believe in Scientific Method and the whole Input-Output approach think that they are simply following a basic principle of science. It hasn't occurred to them that the model they are using is a model, a theory. Control theory, since it offers a fundamentally different view of behavior, also turns the assumptions behind Scientific Method into a theory, by suggesting that they could be tested. See if Slater becomes more persuadable with that argument in mind.

Of course this doesn't mean that Scientific Method wasn't theoretical in nature before the contrast became possible, so maybe that is what Slater really means. The appearance of a Necker Cube doesn't seem influenced by theory until it suddenly reverses. Then you realize that the appearance depends on your interpretation. Nothing self-evident seems theory-laden until you accidentally see an alternative -- but the influence of assumptions remains the same before and after being recognized. So be sure to tell Slater that I definitely agree with her, and also disagree. Does your mind try to refute every positive pronouncement you make, too?

Good luck with the talk -- are you a little nervous about introducing control theory out of the blue like that? Never fear. Two or three will understand, at least. That's all we need -- two or three more, each time.

Best

Bill

Bill

Ps: how about an <u>unambiguous</u> address?

Happy 1986
1986
1986 1986 198

8 Jan 86

Dear Bill:

On 4 November (my goodness, that long ago?) I wrote you a letter in defense of J.G. Miller. You replied to that on 8 November. You must have started typing almost before you finished reading my letter! I sent you a note or two later on saying that I would before long answer your letter of 8 November.

So here is the answer. The delay is due mostly to academic duties, but also to my hope that my brain, if I gave it a little time, would find a way out the the bind you put me in. But it didn't.

Your letter shook me up something awful.

In 1972, Joseph E. McGrath (of Psychology at Univ of Illinois) and I published a book on how to think about methods of research in the social sciences. Critics thought it was fresh and brilliant. Academicians still cite it widely, even though it has been out of print for five or six years. It sold very poorly.

Three or four years ago, I wrote to a lot of publishers asking them if they might want to reprint it. One publisher said they'd like to publish a revision. So Joe and I and a third person are just about to put words on paper for the revision.

But, as you point out, the assumption throughout is that you have independent and dependent variables, and when you have found a correlation between input and output, you have learned something about the <u>person acting</u>. I started worrying about that as soon as I got into your book. (I had not worried about it several years ago after reading your article in <u>Psychol Rev</u>, maybe partly because I wasn't then contemplating writing a revision of the Runkel & McGrath, but probably mostly because I didn't understand well enough what you were saying.) I sent copies of your Psychol Rev article to my co-authors. I have had no comment from either of them.

Your letter of 8 November really rubbed my nose in it. Now I am facing myself with the question whether I can make myself write my part of the new book. That makes me very unhappy. I won't bore you with a recitation of the intellectual and emotional ties I have to doing the revision. I'll just repeat that I am unhappy. As I write these words, I can tell that I am beginning to argue myself into withdrawing from the project. But still, I'm not ready to predict which side of me will win the argument.

I am also in the early stages of collecting notes and literature and so on from which to write a book on life (human life) in organizations --what academicians call "organizational behavior." (This is the original reason I bought my computer. I'm coding everything, and I'll use a sorting program to retrieve things by topic.) I have coded lots of stuff in your book so I can use your ideas in mine. Feedback and control of input are now among my codes.

But then you wrote that "it is easy to make proposals to the effect that this or that phenomenon occurs." Well, I knew that. But you make me wonder whether I should be filling pages adding to that kind of literature. Like most writers, I think my proposals are better than the proposals of most other writers. But.... So I am floundering with that knife in my side, too.

Well, I can't think of any better way to help myself just now than to go back to floundering and waiting for my brain to do something clever. So I'll do that.

Thanks for the other comments in your letters.

I'm glad to have your encouraging words about Turbo Pascal 3.0. I hope some day to try it.

Enclosed are a couple of bits in case they are useful to you.

Phil

Is this diagram all right?

For several reasons I won't bother to explain here, I need a diagram laid out somewhat different from any of yours.

The box "Events altering..." means such as lifting the hand holding the glass of water to lead to satisfying thirst or writing a check to the Physicians for Social Responsibility to match a principle.

I have entered here the names of functions taken from your Fig. 5 in the Byte articles.

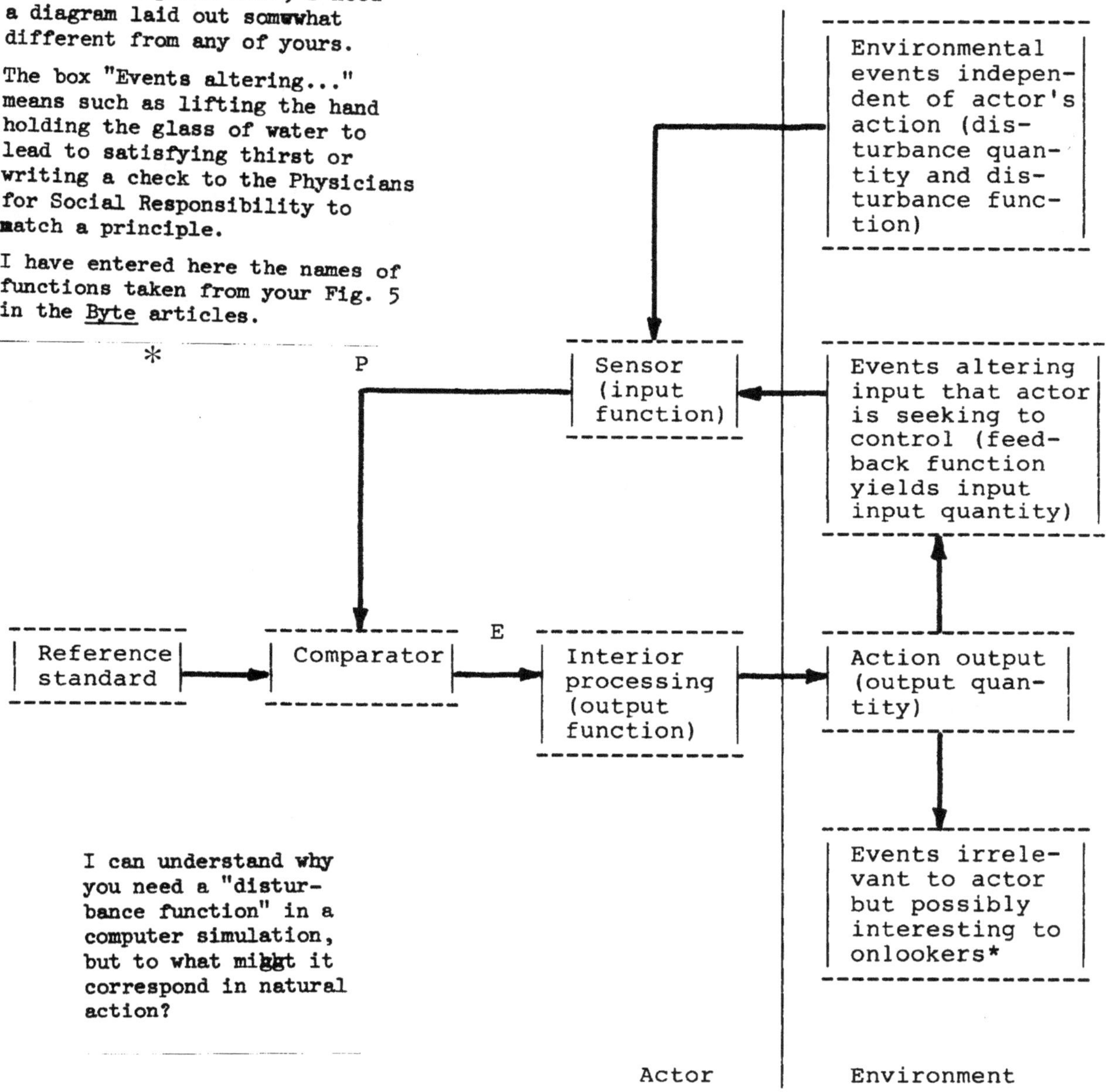

I can understand why you need a "disturbance function" in a computer simulation, but to what might it correspond in natural action?

*Onlookers could include production engineers, experimenters, teachers, audiences, bosses, subordinates, spouses, passersby, and so on.

* This fits with the "1980" paper sent September 14. The BYTE articles were published in 1979. —enclosures at this volume's web page.

Jan. 18, 1986

Dear Phil,

I think that you are now a full-fledged control theorist, even though the shaking-up is not finished. The enclosed talk by Rick Marken may show you that you have company.* Your brain will come up with the right answers -- just remember what you told me about babies and bathwater. We are gaining more than is lost, and not all is lost.

When I fall into doubt, as happens from time to time, I go back to basic principles. I ask myself why I think there is any such thing as a control system. I ask how I know that this idea applies to organisms. I reconstruct the whole thing from scratch. I recommend this exercise to everyone who reaches the inevitable crises, the conflicts between the old and the new that come up again and again. You aren't the first person who has found, to his dismay, that a true understanding of control theory has destructive effects. As Rick says in his talk, revolutions are revolutionary. There's no getting around it. I'm the luckiest one in our little world -- I had no prior career and no prior position in this field, so all I had to give up was a lifetime of prejudices accumulated through informal learning. Perhaps that is all that made it possible to get started.

Naturally, I feel guilty and responsible. I worry lest you accept my every word as if it must be right, just because I have been right about one or two things. I am all too conscious of my well-developed ability to make mistakes, and I am concerned about being too persuasive too quickly. I hope that whatever decisions you make about your writing will remain truly and wholly yours, without defense or apology. My calling is to teach certain ideas, but it is not to pass judgement on what people do with them. If you have good proposals to make, then make them, and don't think of me as hanging over your shoulder to criticize what you do.

* * * * *

The diagram looks perfectly OK, but for one arrow. You have independent environmental events (disturbances) entering the sensor directly, which carries the implication that control depends on the organism's somehow sensing the CAUSE of the disturbance. I would draw the arrow so it runs from the disturbance-box at the top into the Events box below it, via a path outside the organism. Thus the only information the system gets concerns the actual state of the environment, which depends jointly on the organism's actions and on external disturbances.

A simple example. The organism has in mind to pick up a suitcase. The reference perception is a visual-kinesthetic (constructed) experience of the suitcase looking and feeling a certain way in the hand. The actual perception depends, via the input function, on the current position of the suitcase and the

* See page 521

sensations that go with it, in the box just outside the sensor. But gravity (the disturbance) also acts on the suitcase, pulling it downward with a force that depends on the (unknown) contents of the suitcase. The net force on the suitcase consists of the upward pull of the person's muscles (from the action box) minus the downward pull of gravity. The actual position of the suitcase depends on its mass, the net applied force, and any constraints in effect, such as the floor the suitcase rests upon. The person must correct the difference between the desired position of the suitcase and the actual position, without any prior knowledge of the amount of downward force acting on the suitcase. So the position error is simply converted to an upward force, the force being increased until the suitcase begins to move upward, and then being decreased as the final position is approached, until the effort just balances gravity and the error is nearly zero. The person experiences only his own effort and the pressure on the skin of his hand -- he never experiences the downward pull of gravity at all. Of course he calls that sensed effort and pressure "the weight of the suitcase," but he is not sensing the weight of the suitcase, the steady disturbance due to gravity. That physical disturbance is completely invisible and insensible, remaining so throughout.

Another example. A crosswind acting on a moving car gradually increases from 5 miles per hour to 30 miles per hour. The disturbance (top box) is the crosswind, the action is the steering effort, and the sensed position of the car is in the box outside the sensor. The driver "responds to the crosswind" by holding the wheel twisted farther and farther into the wind. But all the driver can sense is the position of the car in its lane: he cannot sense the crosswind independently (as via the arrow you show in your diagram). To keep the car moving straight, all he has to do is compare the actual position of the car with the desired position, and use the wheel to keep the error vanishingly small. The path of the car is the joint effect of the crosswind force and the steering force, so the actual effort produced will appear to be controlled by the crosswind. In reality the crosswind has no effect on the driver's senses: the only input variable that the driver can experience is the position of the car. The output efforts are based strictly on the position error. If the cause of the disturbance were a tilt in the roadbed or a soft tire, there would be no difference in the driver's behavior: the cause is immaterial. The control system simply varies the steering effort in whatever way is necessary to bring the position error toward zero.

Sensing the cause of a disturbance can sometimes -- not often -- improve performance somewhat. If the driver sees a newspaper blowing rapidly across the road on the other side of an underpass, he can use relationship control -- judge the timing and amount of a steering correction to apply as the car comes out into the wind again, and thus roughly compensate for a very roughly estimated effect of the disturbance. This sort of anticipation of the effect of a disturbance is called "feed-forward" in engineering (a much-misused term elsewhere), but it is really feedback control involving a higher level control system (sometimes, the engineer who adjusts the amount of feedforward until the best effect is achieved). It is never enough by itself to produce skilful control. The vestibulo-oculomotor reflex is such a feed-foward system, and you may recall a recent issue of Science News in which someone discovered that the amount of this rough compensation of the eyes for head movement is widely adjustable by some higher system, over a period as short as a fraction of an hour (I forget what fraction, but it wasn't much).

There seems to be a little confusion in your diagram, as both boxes at the upper right seem to contain disturbances, and there is no box corresponding to the controlled external thing itself. "Events altering input" would not seem to be quite the same as "external situation corresponding to sensory experience," which is what I would have put in that box. You include "lifting the hand" as an event altering input, but wouldn't that action belong in the "Action output" box? If you're thinking of the thirst control system, the variable under control (body electrolyte concentration) is inside the physical organism, although outside the nervous system doing the behaving, and the action would then include the ingestion of the water, wouldn't it? When one control action is used as part of the means of achieving another, we can take it for granted and treat it as part of the action, assuming that no serious disturbances can normally disrupt it. The condition sensed as thirst is OUTSIDE the sensor of the thirst-control system, in the environment of the control system although physically inside the organism.

In satisfying a principle, doesn't one normally conceive of the principle as applying in the outside world? Physicians for Social Responsibility are against nuclear proliferation domestic and foreign, and presumably it is that principle one is trying to control by the action of writing a check. Even though it is perceived internally through many stages of input transformation, we act to bring what appears to be the outside world into conformity with this principle. Even though we understand that all perceptions exist inside, when we draw diagrams we are really modelling a hypothetical outside world and assuming that our perceptions have exact counterparts Out There. Even when we speak of "the glass of water," aren't we following this convention? We know that the experienced glass of water as well as the actions we perceive that affect the glass are really neural signals in our own skulls, but since we are also attempting to understand the nature of outside reality, we adopt the convenient fiction that these perceptions are actually outside us just as they appear to be. We don't, in truth, know what is really out there, but this convention represents our attempt to model what is out there. We never deal with JUST control theory -- we are always modelling reality at the same time.

So it would be perfectly appropriate to put the principle of non-proliferation in the box just outside the sensor, and show an internal perceptual signal standing for this external condition.

Writing a check affects the state of this principle in one direction: Reagan and Weinberger affect it the opposite way. We don't need to know what Reagan and Weinberger are doing at any given time: all we need do is observe the net effect. That physics tells us this external condition is really just a congregation of quarks is irrelevant -- that's a different model, though not inconsistent with this one. The action box then contains all the lower-level control systems that give us the means of affecting non-proliferation, so we can use the single-level diagram legitimately.

Dealing with high-level matters is confusing; I'll be the first to admit confusion. The above is just my attempt to get the connections straight. It may not serve your purposes.

Best,

Bill

Bill

Ans 31 Jan

25 Jan 86

Dear Bill:

I'm not surprised that you had a spell of feeling "guilty and responsible" when you read my letter. I hope you have got over it by now. As you know, our bodies give us emotions under certain circumstances. You and me, too. When we <u>don't</u> have emotions at those times, then we should worry.

I notice you shy away from saying anything about emotions within your theory. I hope you will give it a try some day.

Some months ago I read a chapter on emotions that I thought was the best thing I had ever seen on the topic. I must read it again soon, and see whether I can re-interpret it with control theory. I'll write to you about it after I do that.

By now you have the copy of the letter I wrote to my co-authors. I made a couple of errors in my examples, but I don't think they will do any harm at this stage.

I keep forgetting that the comparator is not located inside the sensor, I keep forgetting that what the sensor can sense is limited to the energy or chemicals that impinge on the nerve-endings, I keep forgetting to note whether I am thinking about a conception as an internal reference signal or as a description of some condition I presume is out there is the environment, and I daresay I keep forgetting some of the things I keep forgetting. Anyway, I agree that the arrow was misplaced. I have redirected it. Thanks for your help.

Now please give me some more help.

I have told you that one of my labels is social psychologist. And you saw my brand of social psychology in the OD lecture I sent you. And now here is another aside; I think this is going to be a very meandering letter. In my youth, I had a couple of impressive experiences in organizing people to work productively and happily together. I also had a couple of experiences (and more in later years) of seeing my own patterns of behavior change unexpectedly. In the Canal Zone, about 1947, a friend showed me some writings coming out of the early "experiments" of the National Training Laboratories at Bethel, Maine--the stuff that a little later developed into "sensitivity training." Do those words mean anything to you? Anyway, at an opportune time, my dear generous wife proposed that we quit our jobs in the Canal Zone and that I go to graduate school for a PhD.* So I went to the University of Michigan (1951) and got all imbued, hopped up, excited, indoctrinated, with the best of social psychology at the time and with methodology, experimental method, Coombs's theory of data, and so forth.

* A scan of Phil's doctoral thesis is posted under "About Phil Runkel" —at this volume's web page.

(By the way, Coombs's theory of data has reference signals in it; he calls them "ideals." It could have had feedback loops with no trouble, but neither Coombs nor any of his few followers ever thought of that.)

I remember feeling even during my graduate school days that there were some obvious gaps in what psychologists chose to study, but I don't remember any more what my particular complaints were. As time went on, I got more and more dissatisfied with what I had been taught. I think the idea at the middle of my dissatisfaction was this: that very little of the methodology could be used to do anything in natural life, because by the time all the measurements are taken, the situation you measured has gone by and new things are happening. Underneath that complaint was the idea that action is a flow--that we don't act, wait for a reaction from the environment, assess it, then choose a new act, and so on, but instead we keep altering what we are doing, continuously, to cope with what is happening at the moment and what we expect to be happening in the oncoming moments. Perception and action make a <u>field</u>, not a series of relays. And meaning does not lie in the "stimulus," but in the place you think you are in that field.

But I didn't know how to invent a methodology to deal with that.

I came here from the University of Illinois (Champaign-Urbana) in 1964 to join a research center on education. By 1967, the U.S. Office of Education was urging us to try making some actual alterations in schools. I told my colleagues that the only tradition I knew that held any promise of being able to do that was the tradition from the National Training Laboratories. So we hired Dick Schmuck, who soon became my mentor and friend. And that's how "organizational development" came to the Coll of Ed at the U of O. Even before he was on the payroll, Dick set up a project with a junior high school outside Portland. I went along to watch how he did things. The night before the first training session with the faculty was to begin, Dick and his helpers decided they needed another hand. And that's how I became a consultant in OD.

As a consultant of the OD stripe, helping groups of people draw out their various resources and learn to use them in concert, I saw at last that <u>flow</u>, that continuous emergence, with its plateaus and its sudden leaps, but with its unbreakable continuity. I saw how you cannot teach people "things," but you can give them experiences from which they can draw ways to do things that fit with their own readiness (in your terms, that fit with whatever new pattern of reference signals a reorganization produces inside them).

How, I asked myself, could I have gone 50 years without learning those marvelous things? And how could my culture have been so cruel as to shield me from those marvelous ideas?

And here I am again in that same state.

I often remember the time when, at a tender age, I read Lobachevsky's Theory of Parallels. When I got to the last page, I was overwhelmed with an emotion of sudden freedom, of discovering a vast new universe, of a new power of vision, of invigoration and delight. That's the way learning ought to be, and I am glad it has happened to me several times in my life. Thank you, friend, for enabling it to happen to me once more.

So much for that divagation.

So I want to see how I can talk in control theory about social life, about human interaction. I am putting words on paper about it, and it is a struggle. The question I am using to spur my thoughts is: What's inside and what's outside? So your letter and illustrations are helpful. Here are some questions that came to mind recently during my efforts to write and while I was reading your letter.

1. Where, in any of the feedback diagrams, is the awareness of peripheral events? Take your example of the driver of the automobile. He may be conscious only now and then of guiding the automobile. The rest of the time, he is enjoying the scenery.

I can't help but think we are built to scan the environment continuously. Or, more exactly, to run a continuing check on all the input quantities from the sensorium. Some of those checks result in conscious experience, some not. I was thinking this thought when I got the arrow in the wrong place. I got the disturbance mixed up with the periphery.

I know; you'll tell me we don't "scan" in the sense of taking a sequential inventory; the inputs are all there all the time. All right, they are there all the time. But only some of them become "disturbances." Only some affect the "tasks" we are dealing with at the moment. Some of them get switched through those switches of yours and stop in the memory without producing output signals.

Am I answering my own question?

We do not attend only to one reference signal at a time. And if an input quantity matches a reference signal, the error is zero, and no output signal occurs. I thought for a while I needed still another box on the environmental side of my diagram for those perceptions that had nothing to do (no effect on) with the feedback that was calling for action. But maybe what I need is to remember that the disturbances are potential disturbances. Their disturbance can be zero, in which case the output quantity is zero.

So look at the revisions on the next page and tell me if the words are better.

feedfig.mss

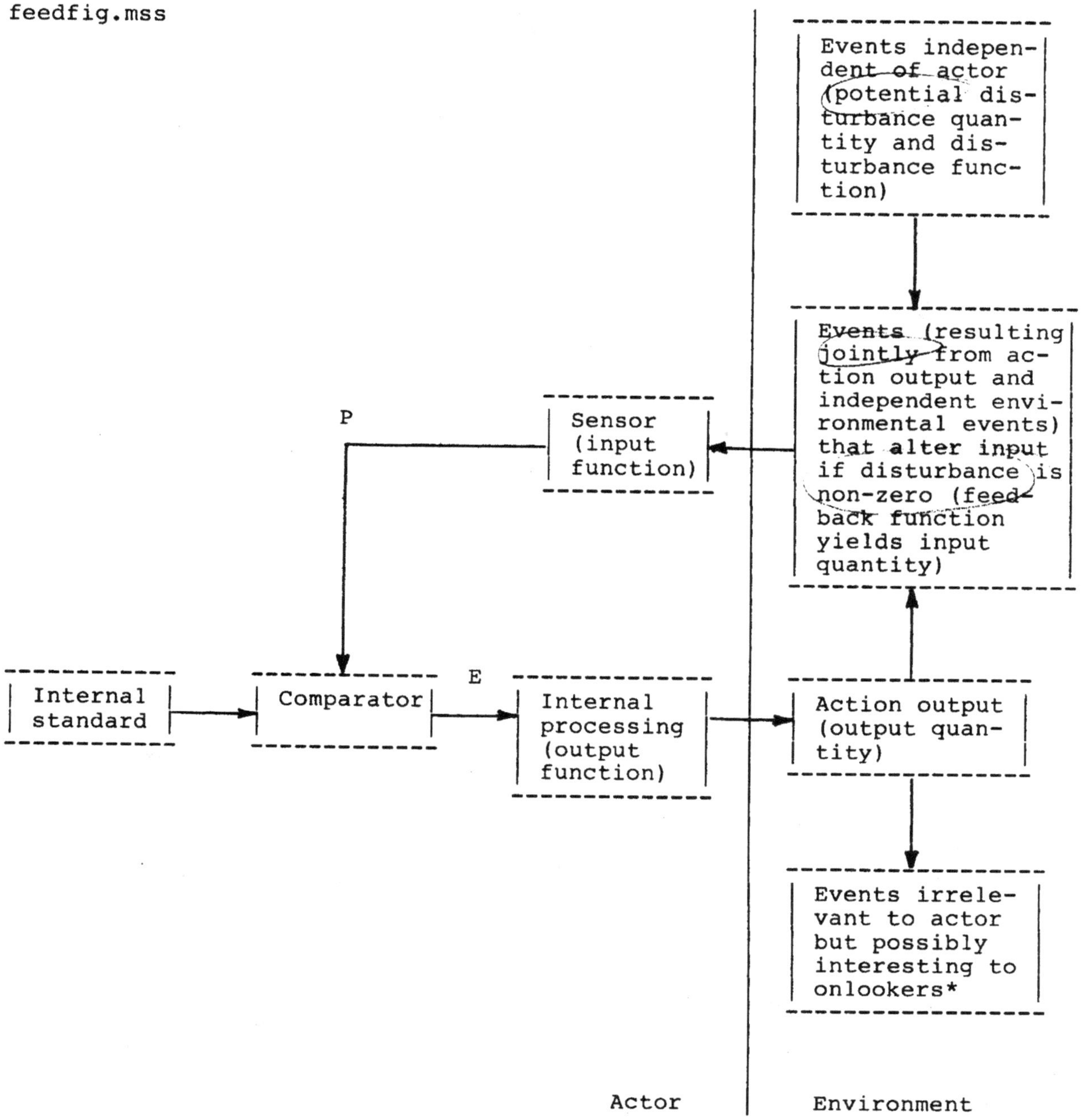

*Onlookers could include production engineers, experimenters, teachers, audiences, bosses, subordinates, spouses, passersby, and so on.

I know, I am trying to pack too much into one diagram. But for now the diagram is only a reminder to myself. I'll simplify it when it goes into a document for a wider audience. I guess.

Have I answered my question? I think so. But a piece of it remains: Why (how) are we conscious of undisturbing inputs?

2. The driver sees clouds come over and some drops of rain hit the windshield. He think the road may become slippery and slows down. I guess that's feedforward. I guess the input quantity occurs at the eyes, gets "interpreted" (excuse the expression) up through the levels, including configuration and transition, and gets combined with various memories at those various levels, and gets controlled eventually at the level of principles.

Where is the disturbance? No exterior force is acting on the foot on the accelerator trying to make it press harder than the driver wants. Someplace there must be a reference signal for speed during rain. For comparing inputs to eyes that get interpreted as judged speed through the scenery, as readings from the speedometer, and those get compared with a resultant of memories of skidding, of trying frantically to compensate with the steering wheel--the disturbance must operate inside with memories. One remembered "image" must disturb another.

"Image" is one of my favorite concepts. I think we "think" because we are built to keep matches between one image and another. I think that is like comparing input and reference signal, but it occurs with the stuff of memory, not with input from outside. We are motivated by "contradictions."

So I think you can get disturbances inside as well as outside.

3. You said that feedforward must act rather imprecisely. But what about catching a ball? I always marvel at the passing and "receiving" in football. And at the way a dog catches something thrown. Animals can do that only because they can calculate where they ought to be when the ball comes down to the level where it can be caught. Isn't that feedforward? And it is very precise.

4. About writing a check to control nuclear proliferation and "observing the net effect." How the hell do I do that? I guess I have to use my eyes and ears to detect millions of input signals that I can interpret as language, and I have to interpret that language as having something to do with nuclear proliferation--how abstract and loosely coupled all that gets! Gorbachev (did I spell it right?) is proposing that all nuclear missiles be done away with in 15 years--the most wonderful proposal that has yet been made. Was he influenced by his physician? Was the physician emboldened by the world-wide encouragement of people who write checks? Where the hell are the input and output signals and the comparators for a process of such

high-level phantasmagorical complexity?

5. Where are images? I am using "image" to mean a sensation, a configuration, ... a principle, a system-concept. I have an image of the path (or alternative paths) from office to home. I have an image of minimizing the intrapersonal conflicts in people working together in a group. And so on.

I try to equate image with reference signal. Let us say I use the image or reference signal to control output signals at lower levels. But where is the image or reference signal that I can put in memory and be conscious of? (I don't suppose you are fond of this kind of question.)

Looking at your Figure 15 on page 107 of the third part of the Byte articles, I conclude that the reference signal to control action in the systems at Level L is really a combination of reference signals from systems at Level L+1. "It" is in effect a vector of reference signals, and the unique vector specifies the unique reference control at Level L. But is there one loop up the line someplace that can set that vector into operation? I'm sure I'm over-simplifying something. *

I think my trouble is that I can't get it out of my head that up at the top, at the level of system-concepts or maybe in the reorganization wiring, there must be one loop labeled ME or BOSS or GOD or something. I know that would render redundant or short out most of the rest of the circuitry. But I have a strong feeling that some notion like that is getting in my way.

But if the guidance, the control, is all in the circuitry, if your model is a model in your sense, then there must be a vector of currents in a vector of reference signals, someplace, that corresponds to ... the "thing" I want to "locate."

I can probably give up yearning to locate the thing if that is your advice.

6. Here is a question about Part 3 of the Byte articles. I can't figure out why the comparators at Level 1 don't all get the same reference signals. Each system at Level 2 puts out the same signal to all the comparators in Level 1. So all comparators at Level 1 get the same combination of signals to convert into reference signals. The only way they can resolve the incoming signals into different resultants is by the action of the M-matrices at Level 1. Well, I read carefully from page 102 through 104, but I still can't understand what sets the entries in the M-matrix. I made some sense out of "destination" and "source," but since each system at Level 2 puts out a single signal, I can't see how the information gets from the S-matrices at Level 2 to the M-matrices at Level 1.

* BYTE articles —enclosures at this volume's web page.

It might be that part of trouble is not knowing enough about matrices. I studied some matrix algebra many years ago (and I could still, if I were not too lazy, pull the right book off my shelf), but I've lost the address in my memory.

Those are all my questions today.

If you haven't already heard it, here is a joke you can probably use: Some one asks a man whether he believes in baptism. "Believe in it!" says the man, "I've seen it done!"

Three of my students are now going round with rubber bands in their pockets. No faculty yet.

You said I am now a "full-fledged control theorist." If you mean I can now fly, then no. But I am a fledgling; I am now able to fall out of the nest without help.

You said not all is lost. Here are some things I think are not lost. (1) Studies of the natural history sort: case studies, accounts of what happened where, in what order. For example, there is archaeological evidence that backs up some of your remarks for D.T. Campbell on the emergence of specialization. (2) Correlational studies where you can ignore the imputations of causation that you don't like. For example, I have a book showing the ways employees of a multinational company answered a questionnaire about values. The central tendencies are very different in different countries. Culture provides us with both physical and social "stuff" to act on in maintaining desired inputs. So studies like that can give clues to types of actions people will choose to oppose disturbances. (3) Studies, even when done under the input-output persuasion, that contain elements of the feedback loop. They can help you speculate about what might have happened if the whole lopp had been investigated. For example, I sent you a couple of articles by Dember about the "pacer stimulus." That's actually a postulated reference signal. It's a postulation, I think, at the level of principle: seeking new information. What's new at one time is old at another time, so the principle is a relative one, maybe a transitional one? Do capacities at lower levels get recycled, so to speak, at higher levels? Are there relational principles, categorical principles, etc.? Anyway, one of the nice things about the Dember experiments is that the rats and children don't have to be starved, you don't have to control a lot of other variables. And the "disturbance" is of a non-obvious sort. That's all I can think of just now.

Thanks very much for sending me the lecture by Marken. It is always reassuring to know that someone else is in a similar boat. I'll write to him.

You said you had been spared some pain by not having been brought up a psychologist. Yes. But I am not in as bad a spot as Marken. I started violating the academic rules about 20 years ago. It wasn't too difficult. Professors inquire very little into what others do in their classrooms. And anyway, I was already then a full professor, and what could they do to me? You have to commit obvious plagiarism or sexual transgressions before you really get punished. And when you do violate the rules, you discover that there are a few others who have also been wanting to do so, and you collect a small circle of admirers. So I teach what I like. And the students mostly like it. And it's pretty difficult for one professor to complain about another if the other's students like what they're getting. I gave up about 20 years ago reciting one study after another. Instead, I taught the students how to <u>do</u> things in their social world. In the light of the academic "standards" of the psychology department here, that was pretty radical and not very respectable. But nobody ever tried to make me stop. A few people even made admiring remarks.

One time I was assigned to teach a course on conflict. I gave three lectures, and then said that was all I knew about conflict and all I thought anybody knew. Then I gave the students some simulations to put them into conflicts and helped them think about the experiences. The next term, a graduate student was assigned the course. I asked for a copy of her syllabus. Sure enough, it was a term full of lectures reciting studies.

Now I'll go back to trying to describe to myself how social life is possible, or can be managed without too much pain, if we are all self-seeking, self-centered, egocentric, selfish control systems. You made a fine start in your paper to Campbell, but there are a number of things I need to add. If you have any more writing in that vein, send it.

Phil

Jan. 31, 1986

Dear Phil,

Oh, yes, I got over it. I secretly control my conscience, so it never goes too far.

Your list of questions is rapidly getting longer than my list of answers; connecting column A to column B doesn't always end up with something at both ends.
some extra answers: yes, yes, no, not yet, and 3.

Emotion. There was a chapter on it in the original book ms, but the editors didn't like it -- too radical. Never got around to publishing on the subject, but here's roughly the outline.

Emotions can be felt. Therefore they are perceptions. But there's more to it. To cut the story short: When one prepares to do something, reference signals are adjusted at all lower levels in the hierarchy. At about order three, the hierarchy splits, one set of reference signals branching into the hypothalamus and thence to the pituitary instead of going through the brainstem to the motor systems. The reference signals reaching the pituitary are converted into chemical reference signals that set the states of all the major organ systems. The effect of changing these chemical reference signals is to alter the biochemical/physical state of the organism. These states are sensed, and become part of experience as sensations, configurations, transitions, sequences, and so on -- of feeling states. Above about order three, they become part of perceptions that also contain proprioceptive and exteroceptive information, so that the experience includes shades of feeling that are an integral part of informational perceptions about the outside world. With our usual verbal muddle, we say we have feeling about perceptions, or that perceptions give us feelings, or that feelings influence perceptions, as if feelings were located in the same place the other kinds of perceptions come from.

The upshot is that emotions are caused by what we want, and particularly by not getting what we want (to fight, to flee, to get help, to undo). They reflect what our bodies are aroused (a lot or a little) to do as backup to the behavioral systems. If you want to know why you are having a given emotion, just ask what you want that you're not experiencing. Of course you have to be actively wanting, not just imagining. So emotions are just part of behaving. The inside part.
You're going to ask about "good" emotions. Who cares?

Never got into "sensitivity training" and that sort of stuff -- whatever was good about it seemed overshadowed by the soppiness of the people who did it. If I'd known you I might have taken a different attitude.

Awareness. The most difficult of all questions. I can answer your questions about it by giving you some words, but since I'm not sure what they mean, it's not likely that when you get through with them we will still be on the same subject.

Here's an interesting fact about awareness. You know that we have perceptual signals in our nervous system at many levels. Some of them, intensity signals, come right out of sensory receptors at the lowest level. Others are constructed after many layers of processing and probably make big networks of signals at the highest levels. And we can be equally aware of ALL these signals (not necessarily simultaneously). Where does this put awareness? It's not anywhere in the hierarchy; it seems to be more like everywhere in the hierarchy.

Here's an image. Awareness is a receiver that exists in a space of higher dimension than our usual three. Thus it can be connected to all points throughout a three-dimensional volume, receiving from any of them. The brain is a three-dimensional interface between awareness and the presumed external reality; the brain is awareness' sensor. Awareness can receive information from perceptual signals, but all the mechanisms that create perceptual signals out of others are in the brain, not in awareness. We can be aware of thoughts, but awareness does not think. We can be aware of principles and system concepts, but awareness has neither principles nor concepts. Awareness simply observes. It is the Observer. What it observes is the world of perceptual signals in the brain, at all levels. The brain's input functions create the world that awareness experiences.

See? Descartes almost had it.

Furthermore -- contrary to much previous thinking -- awareness is never aware of itself. It is only an observer of other things. Among those other things are thoughts, for instance sentences that contain the word "I". An example of such a thought might be, "I am not aware." That sentence can certainly exist. It can certainly lead to confusion if, in our structure of thought, we interpret the word "I" in that sentence to be the thing that is thinking the thought, as Douglas Hofstadter seems to like to do. Awareness, however, is never confused, although one of the phenomena of which it can become aware is a state we label "confusion," which consists of thoughts and feelings in the brain. In the brain's opinion, the brain is confused; awareness can note that state of affairs. Needless to say, awareness has no feelings, either. Feelings are perceptual signals in a brain.

Awareness seems to focus selectively on parts of the whole brain -- for example, it can focus on the level where configurations are perceived. When that happens, the entire world appears to be composed of configurations. But the FACT THAT THIS IS SO is not in awareness: instead, one notices the sensations of which various configurations are composed. At the configuration level, the incoming information consists of sensation signals. It seems that awareness "occupies" this level, and from that point of view is aware of what that level is receiving, the signals from below. The interpretation given those signals becomes the interpretation that is taken for granted, since there is no representation AT that level OF that level. *This cube is red on one side, has sharp edges, looks fuzzy.... what do you mean, "What cube?"*

There are some interesting games that two can play, involving calling attention to the level at which awareness is currently involved, thus forcing a jump to the next level. There is no infinite regress, as logic might insist. This method can become a devastatingly effective psychotherapy.

Awareness is the input function of some unknown and probably very eldritch system, the output of which is called volition. I'm sure this has something to do with reorganization. I'm also sure it has very little in common with this conceptual artefact we call "the brain." I'm also sure that I haven't got anywhere with talking about it in a way that makes any sense. Sure is fascinating, though.

The diagram is now perfect. Get out the mallet and chisel. The REAL actor, of course, is out of the plane of the paper.

You don't HAVE to have disturbances. If the wind blows you around the curve just right, you don't have to steer, so that amount of wind constitutes zero disturbance of the car's path. As for any variable, a possible value is zero, as you say. A zero reference signal says "avoid this perception." Zero is a useful number, as good as any other. When there are literally no disturbances, the perceptions become what you want all by themselves. But how often does that happen? Usually, nature wants to go one way, left to itself, so we have to steer to make it go another way. With respect to our intentions, the failure of dinner to appear spontaneously is a disturbance. If you're going to investigate control systems, you DO need disturbances, because that's how you find out what is being controlled.

You observe the effect of writing the check exactly to the extent that you observe it. You want to observe the REAL effect, right? Too bad. All you can perceive is what you can perceive. You might find that you have to imagine the effect in order to get any satisfaction out of writing the check. We do a lot of that.

About 20 years ago a guy wrote an article in Science showing how a baseball player catches a long fly ball. It turns out that he moves so as to keep the vertical angle of sight to the ball rising at some moderate constant rate, right to the moment of catching it. Forget the distance to the ball -- it's only the angle that matters, isn't it? You control the angle by moving. No prediction is involved.

The input and output signals and reference signals are all inside of people and nowhere else. People are all there is. You can make believe that organizations are control systems, but they aren't. Only the people are. The people, of course, can follow rules that make the organization simulate a control system, which works as long as they continue to follow those rules. Take away the individual people and what have you got? Empty desks and charts on the wall.

Images are in your head. You asked.

Byte article. The fascinating thing about that multi-level simulation was that no one higher-order system could determine the reference signal of any one lower-order system. The lower-order reference signal was always made up of many higher-order output signals. Our lower-order goals don't necessarily have anything ~~understandable~~ obvious do with individual higher goals.

In a general multilevel model, each lower-order system would receive a set of reference signals from some (not all) of the systems of the next higher level. Furthermore, some of the signals are inverted (inhibitory) after they leave a given higher-order system (the M matrix), and the rest are not. Even if systems A and B receive reference signals from the same set of higher-order systems, A might receive them in the pattern +--++-+, while B receives them in the pattern +-+--++. But those are just the choices needed to maintain negative feedback. The higher-order systems also receive the lower-order perceptual signals (copies) and subject them to different input transformations, which is where all the action is as far as determining what gets controlled is concerned. You have to know what these transformations are before you can set up the + and - signs in the M matrix, because the point is to rule out positive feedback. The DESIGNER sets up the M-matrix. Of course you could also devise an automatic internal system that would set it up -- it would try a change of output, and if the result (reflected back into perception, through comparison, and to output again) was a change in the same direction it would invert that entry in the M-matrix. In the Byte article, I just gave a rule that would look at the S-matrix and from that, determine what each entry in the M-matrix had to be to maintain negative feedback around each loop. This was done OUTSIDE the operation of the system.

We are self-seeking, self-centered, egocentric, selfish control systems only if we choose that class of system concepts and principles. While we can never know exactly what another person really is, we can still strive to know, struggle to communicate, wish the best for those dimly-seen Others, and put the totality of an inspiring system above the particular one we call a Self, in our scales of values. The fact that we must set the goals ourselves, and make the judgements ourselves, doesn't mean that we therefore reject responsibility. In fact it means that if we do choose to be responsible, that act is totally voluntary and selfless, because we are free to do otherwise. Only in a world without "oughts" and "shoulds" can there be real altruism and true morality. That's probably the garbled message that comes through, in the obverse, as "original sin." I have a feeling that somebody back there figured all this out, but all we have left of her insights are some slogans.

Merry Mithras.

Bill

Bill

8 Feb 86

Dear Bill:

* The talk on organizational development and the ten levels of control systems went very well. Attention was constant, applause was generous, and one person said she was going to run right over to the library and get the Powers book.

It turned out that I was the center of attention, or mostly so, during the two days. My goodness. Friends with words of love crawled out from every piece of woodwork. I was astonished.

** Among other things, they gave me a fat album full of letters from friends connected with the OD side of my life. I can't resist sending along some of the more effusive ones.

Phil

Feb. 13, 1986

Dear Phil,

What a wonderful retirement present! Thank you for sharing it -- I was as pleased by all those words of praise as if they had been for me, and I didn't even have to say "Aw, shucks." It's no mystery to me why you are so appreciated; I am not astonished, even if you were. There is nothing those people said about you that I haven't seen for myself, in your letters. You're a good, kind, smart, aware man, Phil, and it's a privilege to know you.

Love,

Bill

Bill.

* File 860206_Spirit_OD.pdf 18 pages. Also 860206_Handout.pdf 40 pages—at this volume's web page.
** See *About Phil Runkel*, a link on this volume's web page.

17 Feb 86

Dear Bill:

You said all is not lost, and I put some words on that in a recent letter. Your implied question is: what is not lost?

Well, the obvious just dawned on me.

Any time you have a question about the effects of input on output, then the input-output model will give you what you want.

And sometimes that is a good thing to ask: Despite the varying internal standards people have, what will happen if we put this "input" in their environment? How many of them will act as if it is a disturbance? How many will take one kind of action or another to counteract it?

How many drivers will find a knob on the dashboard more easily than a lever when they want more or less heat? And so on.

Years ago, Robt Blake did some beautiful little experiments on the ways people will find to reduce a disturbance. Here is one:

One of the buildings of the U of Texas at Austin (I think it was) had a main door in the long side and another door at the end. On some days of the week, he put a sign in front of the main door: DO NOT USE THIS DOOR and then counted the proportion of people who went in there anyway and the people who turned away. On other days, he put a sign: DO NOT USE THIS DOOR. USE END DOOR ----- . Naturally, a lot ~~fewer~~ smaller proportion went in the main door when they were given an idea of ~~where they would go~~ another place they could get in. Seems awfully simple, but the design is a good one if you want to explore proportions of people who will accept a suggested way of reducing a disturbance.

Doesn't tell you much about how humans function, but it is often useful in practical affairs.

Phil

Feb. 21, 1986

Dear Phil,

Seems to me, judging from your letter to Rick and your latest to me, that you've reached Stage 2 of the Pilgrim's Progress Toward Control Theory, which is called "So What?"

The foot is poised for the next step, the old road is abandoned, and you're ready to go -- only where is the signpost? For that matter, where are the other roads? How come everything looks just the way it did before?

It finally dawns on one that there aren't any other roads. We're used to travelling down roads that somebody before us made, and the idea of just striking off into the thickets, hacking your way through underbush, fighting off mosquitos, and stubbing your toes, and so on, doesn't sound like something one would do on purpose. Hacking through the underbrush isn't nearly as fast as walking down the road. You can't see where you're going. It's messy and tiring. But some people seem to like it.

Once in a while a sense of wonder comes over me, sort of like the feeling I can remember when, at the age of about eight, I first encountered science fiction and first saw the stars through binoculars, all in the same summer vacation. This is a new world! We know something, you and I and a tiny handful of others, that nobody else has ever known before. We know how behavior works. It really never has been understood before -- in fact, all the life sciences have been way off the track, studying some imagimary creature that never existed, and puzzling over the difficulties of getting anything that looks like data out of real ones. Organisms don't behave because of what happens to them: they behave as part of making the worlds they experience become or stay like the worlds they want.

It isn't easy to think up new experiments that test the facts of control theory -- the existing ideas of research don't help. So you have to start with very simple things. But when you do the simplest things you can think of to test the principles of control, THEY WORK. That's going to be the difference, when we get the hang of it. It won't be necessary to do statistics to see if anything happened. A properly done control-theory experiment always works precisely as you expect it to, within a few percent, and if it doesn't you know you got something wrong. No more "facts" that are true only eighty or ninety percent of the time. Control theory facts are true all the time, of everyone. The rubber-band experiment always works. My little tracking experiments always work. Rick's mind-reading experiment always works. These are very simple and trivial truths we are finding, but as the years go by they will get more complex.

We have to build the base first, just as physics did when the most complicated thing Galileo knew how to do was to run balls down inclined slopes or time pendulums with his pulse.

Before he did that, nobody understood about acceleration and gravity -- NOBODY. Galileo and a lot of others had to go painfully through all the stuff that is now taught as boring simple laboratory exercises to freshmen -- but if they hadn't gone through it, there wouldn't be any physics. The laws of gravity would still be rules of affinity.

We are now exactly where Galileo was. The life sciences have never gone through that development that took place after Galileo. The life sciences still think that events can cause other events, that tendencies mean something, that statistical generalizations are of some use in understanding nature. They think that what happens to organisms makes them behave. Practically everything that is really known about organisms is not "life science" at all -- it's just physics and chemistry done inside organisms. As Rick would say, physics and chemistry have been doing just fine, thank you. But the life sciences are still in the Dark Ages.

Galileo got into a lot of trouble, and so did many of the scientists who tried to follow the new lead. They had trouble with established religion. We're going to have, are having, the same problem: the religion we're fighting is called Science, the brand practiced by biologists, neurologists, behaviorists, sociologists, linguists, and the others. All the others who think that behavior is an effect of prior causes. There is nobody around to hold our hands, help us out when we're puzzled, show us what all this is going to mean, or take our side against all the misunderstanding, criticism, and hostility that will come our way. Nobody is going to give us a million dollars to establish a control theory institute where we can work in peace. We are revolutionaries, like it or not, and we are finding out what that means.

Not all is lost. At least science has taught us patience and honesty, and we've had a few experiences of what it's like to find a solid fact of nature that was never known before. There might be one or two facts that can be salvaged from the past. But as far as I'm concerned, we're starting from scratch. The labor of picking out of the literature observations or findings that have some vague relationship to control theory seems hardly worth the effort to me -- even people who said cogent things didn't know why they were right, and the people who discovered relevant experimental facts never took all the data we need. If you have to do all those experiments over anyway, why not just forget them and do the ones that are interesting now?

Lots of people have known that human beings are purposive. Lots of people have known that we seek goals. But before Galileo, lots of people knew that balls will roll down slopes and that pendulums will swing at regular rates. The problem with what everybody knows is that so much of it is wrong, and until science comes along, nobody knows which part is wrong. Science is supposed to give us better knowledge than the haphazard approach of common sense. I think it does, when it's done right. We're not beating a dead horse: we're changing from a horse to a car.

I try to tell my friends that they must stop being satisfied with data of the kind they're used to. Instead of $p < 0.05$, I say we should settle for nothing less than $p < 0.00001$. Certainty within the limits of measurement. Some of them tell me I am too demanding, that I am throwing away perfectly good knowledge. I say it isn't good knowledge if we can't use it in extended reasoning. If you know four facts about people, each of which is true of 80 percent of them (pretty good for a psychological fact), then if you make a deduction about an individual that depends on all four facts being true at once, the probability of truth of that deduction is 41%: it's probably false. If we want a science of life that is comparable in value to physics and chemistry, we have to start demanding more of our facts. We need facts whose truth value in any given instance is 0.99 or better. Then we will be able to make deductions that might get us somewhere. With control theory we can get that kind of reliability. We're getting it now. Of course right now the only facts we have that are that reliable are pretty simple ones. But so what? We have to start somewhere; namely, where we are. Why, I ask my friends, why go on accumulating facts that will stand by themselves forever, useless?

Well, maybe my friends are right and I am asking too much. I only wish they could see the other side: the tremendous joy of looking at the results and knowing that they are good to 1%. I am absolutely sure that we can keep on doing that if we just refuse to settle for less. Simply the idea that we can get solid data will suggest ways of getting it. But first you have to make that your goal.

Best,

Bill

610 Kingswood Avenue

Eugene OR 97405

6 March 1986

Dear Bill:

I'm thinking of adding a hard disk to my computer. I've been cudgeling my brain about how to make certain things work easily if I do. As with many persons of more renown, the answer came to me in a dream.

In my dream, I discovered that everything would work right if I stuck out my tongue at the machine. That is, everything worked right when I stuck out my tongue--if I was careful to put a colon beside it.

Some day you might be glad to have that bit of information.

Phil

4 April 86

Dear Bill:

I got to wondering, is it really true, how true is it, that social science researchers, and especially psychological researchers, accept the stright-line input-output conception of behavior, trying to predict that event of stimulus A will produce behavior or response B, with maybe a moderating variable or two in between? In other words, D.T. Campbell's famous O_1 X O_2.

Maybe my colleagues are being accused unfairly by W.T. Powers, and on the other hand, maybe Joe McGrath and I, when we wrote our text, were over-estimating the number of people who were familiar with O_1 X O_2 or who followed it, knowingly or not.

So I sent a query to 16 members of the psychology department here, all the members who I thought knew more about me than just a name. I purposely included all the people in the specialties of sensory and physiological psych. After two mailings, I got 11 replies.

I took the simplified diagram that I sent you some time ago, took out the link from output to input, bent the resulting diagram out flat, into a straight line, and changed a few words.

I thought some people would wonder why I would ask for comment on such an obvious matter, so I used the excuse, in my introduction, that I was adding the "sensor" to the customary diagram. As it turned out, one person, Mick Rothbart, head of department, indeed could not understand why there would be a sensor in a psychological diagram, and if there were, why it would come before personality instead of afterward.

Three people said they simply didn't think in terms of an overall plan, model, or design for experimentation. That was a surprise to me.

Nobody said something explicit like, "Whee the hell is your feedback loop through the environment?" So in the Psychol Dept at the Univ of Oregon, you are quite right. One person added an internal loop: Mary R (wife of Mick). One person attached a diagram from a book of his that did include, in a small sub-diagram, a feedback loop through the environment, but in his hand-written notes on my diagram, he didn't seem to feel any contradiction: Norm Sundberg.

I didn't get a reply from the person I thought might know most about feedback loops, Michael Posner.

So there you are.

Phil

24 February 1986

Dear psychological colleague: Pete L

I'd be grateful for five minutes of your critical acumen. Is the following diagram pretty much the standard way we think about connections among variables? I wouldn't ask, except that most diagrams (outside psychophysics) omit the sensor, seeming to take it for granted. Does this diagram suit you all right?

Thanks. — Phil Runkel, DEPM, Educ

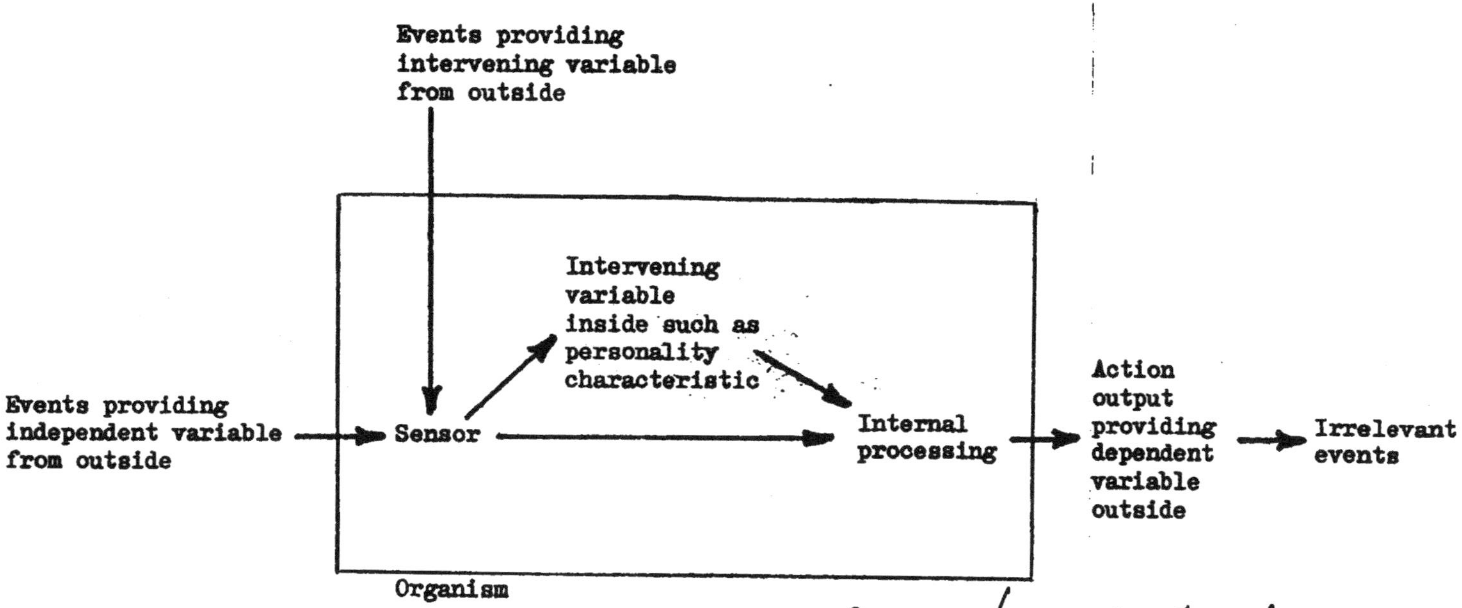

I think the diagram is pretty clear (and standard) but needs clarification in text or caption.

Regards,
Phil

24 February 1986 (second mailing 10 March)

Dear psychological calleague: Jake B —

I'd be grateful for five minutes of your critical acumen. Is the following diagram pretty much the standard way we think about connections among variables? I wouldn't ask, except that most diagrams (outside psychophysics) omit the sensor, seeming to take it for granted. Does this diagram suit you all right?

Thanks. — Phil Runkel, DEPM, Educ

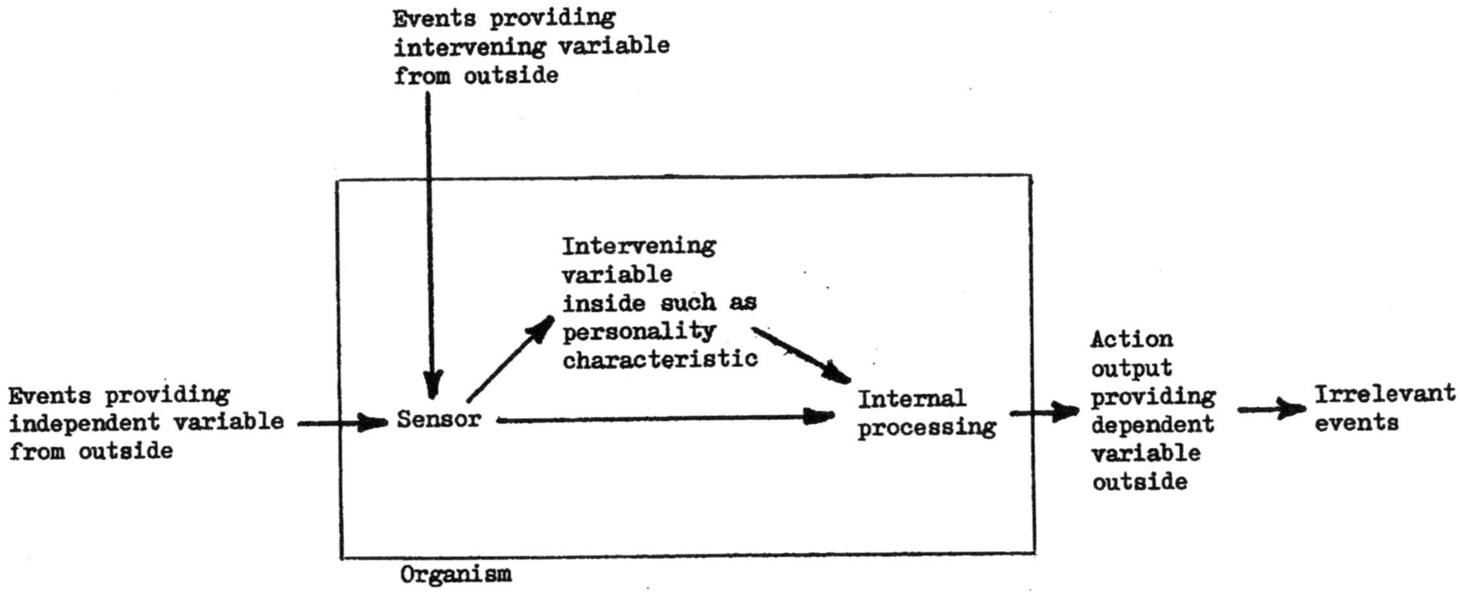

OK by me. I am not certain what intervening variables from outside are.

Jake

24 February 1986 (second mailing 10 March)

Dear psychological calleague: Dick L——

I'd be grateful for five minutes of your critical acumen. Is the following diagram pretty much the standard way we think about connections among variables? I wouldn't ask, except that most diagrams (outside psychophysics) omit the sensor, seeming to take it for granted. Does this diagram suit you all right?

Thanks. — Phil Runkel, DEPM, Educ

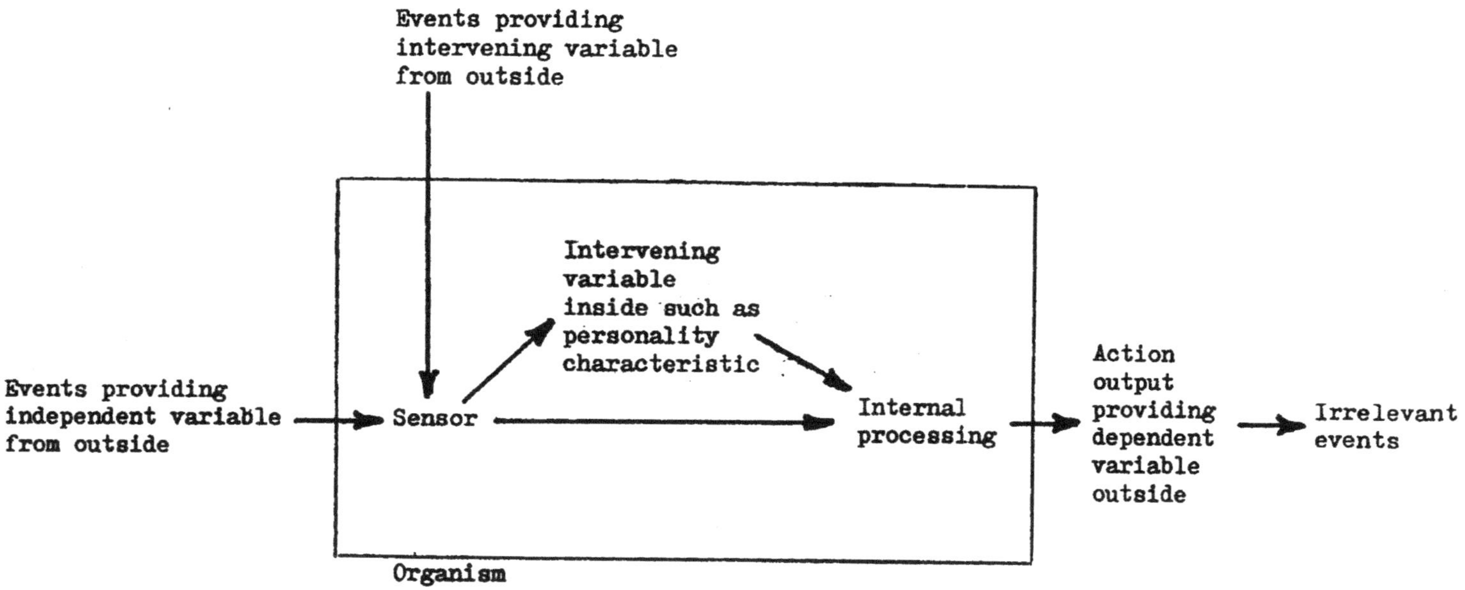

Phil, Sorry for the delay. I never got a first mailing. Anyway, I think that Egon Brunswik's Lens Model shares a similar connectedness.

Dick

3/1/86

24 February 1986 (second mailing 10 March)

Dear psychological colleague: Norm S

I'd be grateful for five minutes of your critical acumen. Is the following diagram pretty much the standard way we think about connections among variables? I wouldn't ask, except that most diagrams (outside psychophysics) omit the sensor, seeming to take it for granted. Does this diagram suit you all right?

Thanks. — Phil Runkel, DEPM, Educ

Phil— Looks very good. Enclosed is a more complex diagram I've used, which is of some relevance. —Norm

—And personal greetings. I hope all goes well!

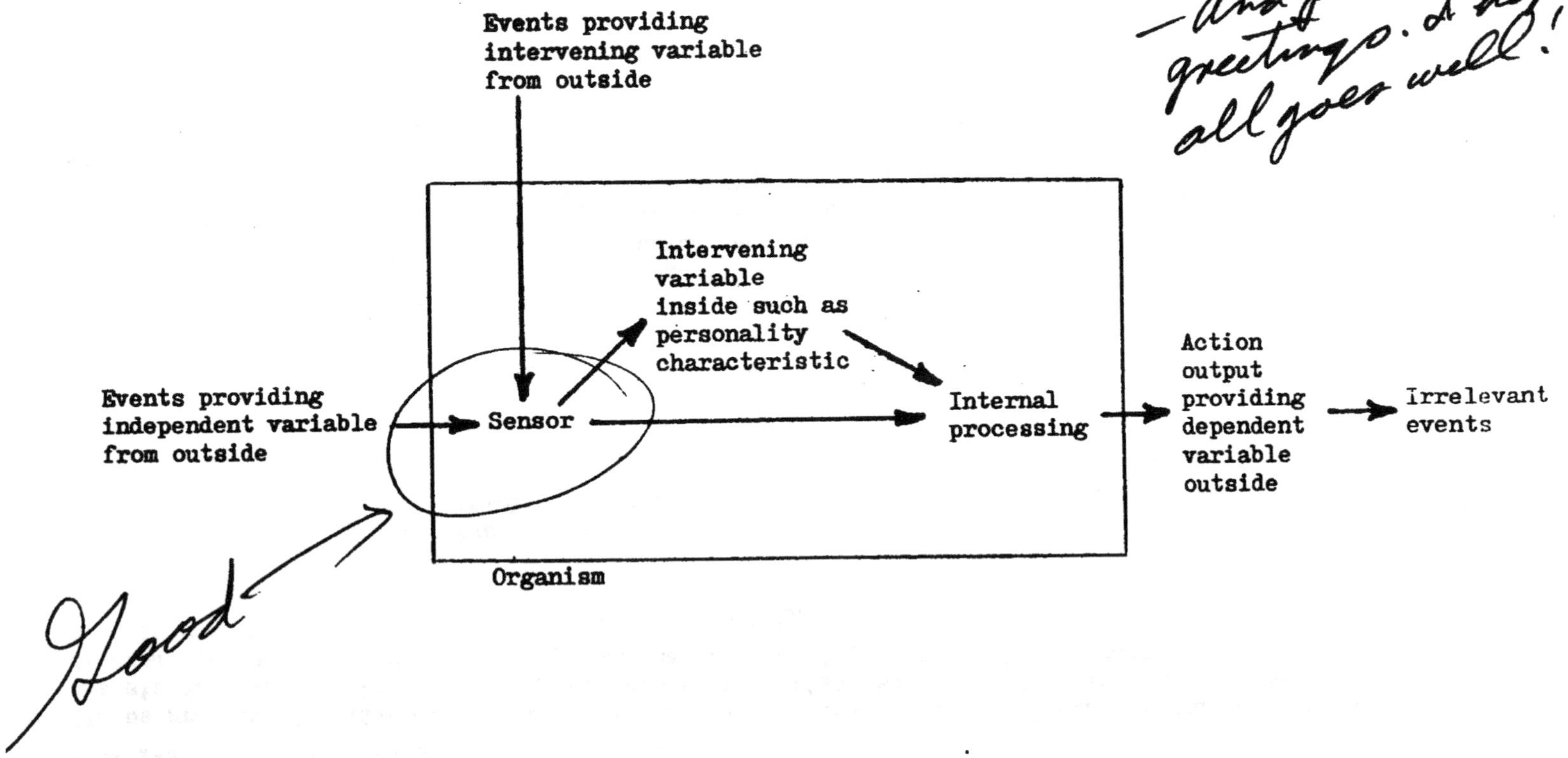

Levels of Interpretation
(from Sundberg, Tyler, and Taplin, 1973, p.143)

and

Proposal for Model of Internal and External Interaction
(From Sundberg, 1977, p.284
Assessment of Persons

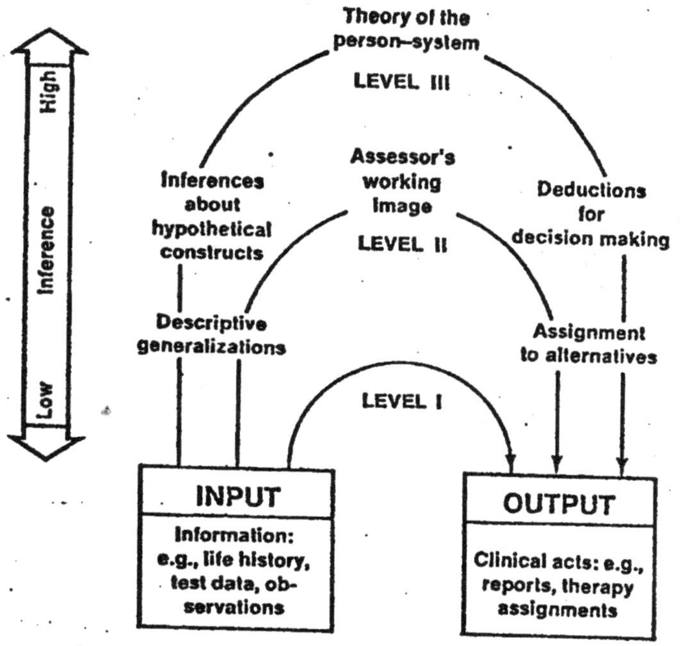

Figure A. Levels of interpretation.

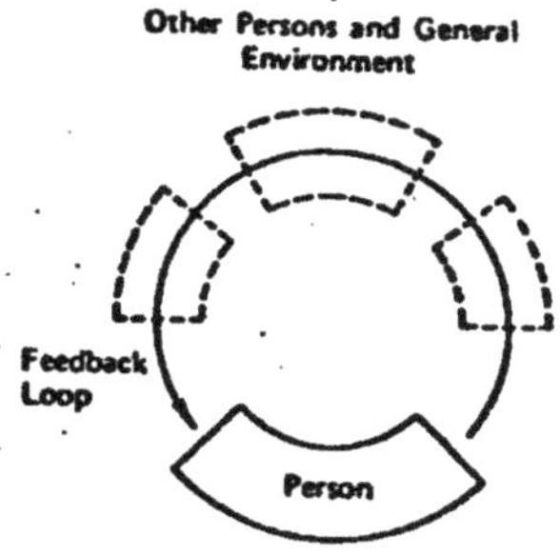

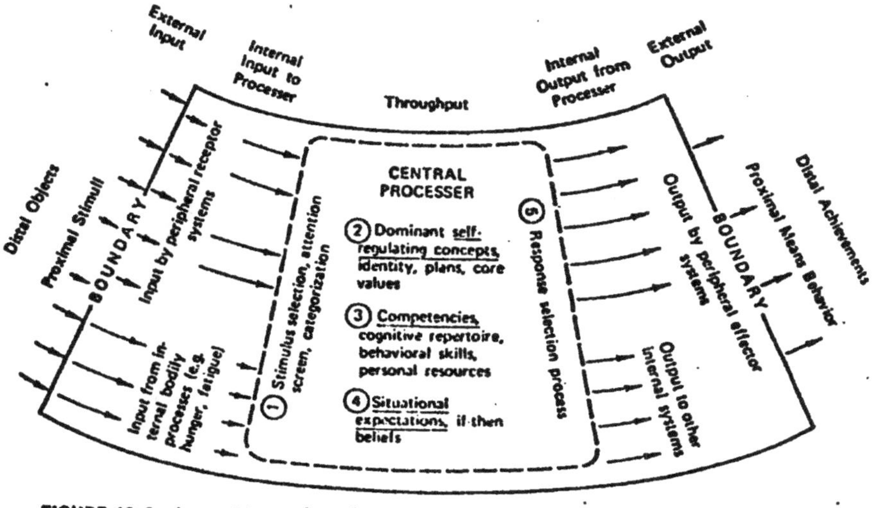

Figure B

FIGURE 12-2 A cognitive-*ecological* view of the personal sector in the continuous flow of interaction with the environment.

24 February 1986

Dear psychological calleague: Carolin K—

I'd be grateful for five minutes of your critical acumen. Is the following diagram pretty much the standard way we think about connections among variables? I wouldn't ask, except that most diagrams (outside psychophysics) omit the sensor, seeming to take it for granted. Does this diagram suit you all right?

Thanks. — Phil Runkel, DEPM, Educ

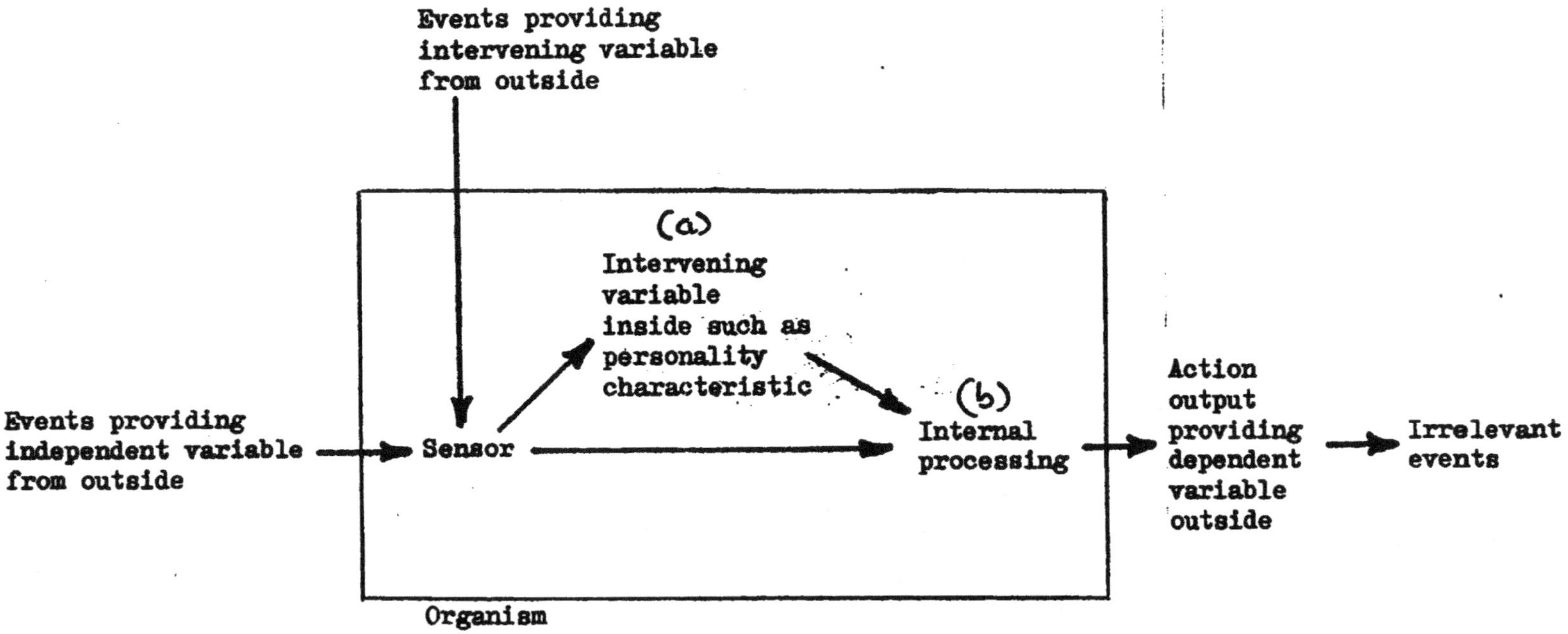

Looks fine to me except that I view the intervening variable (a) (mood, set, openness, etc.) as part of the internal processing (b)

24 February 1986

Mary R

Dear psychological colleague:

I'd be grateful for five minutes of your critical acumen. Is the following diagram pretty much the standard way we think about connections among variables? I wouldn't ask, except that most diagrams (outside psychophysics) omit the sensor, seeming to take it for granted. Does this diagram suit you all right?

Thanks. — Phil Runkel, DEPM, Educ

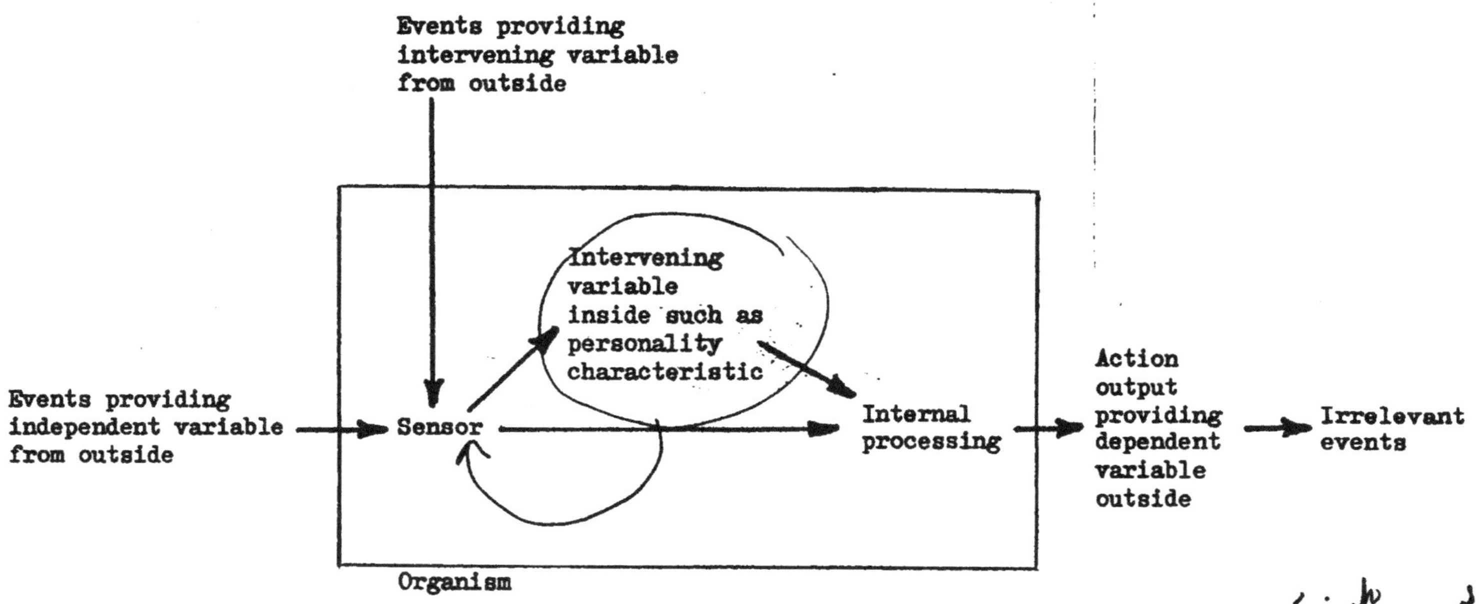

Phil — I would just add another arrow here. I tend to think of internal events in terms of brain systems now, and an activated ARAS influences sensory processes.

Mary

24 February 1986

Dear psychological calleague: Marv G-L

I'd be grateful for five minutes of your critical acumen. Is the following diagram pretty much the standard way we think about connections among variables? I wouldn't ask, except that most diagrams (outside psychophysics) omit the sensor, seeming to take it for granted. Does this diagram suit you all right?

Thanks. — Phil Runkel, DEPM, Educ

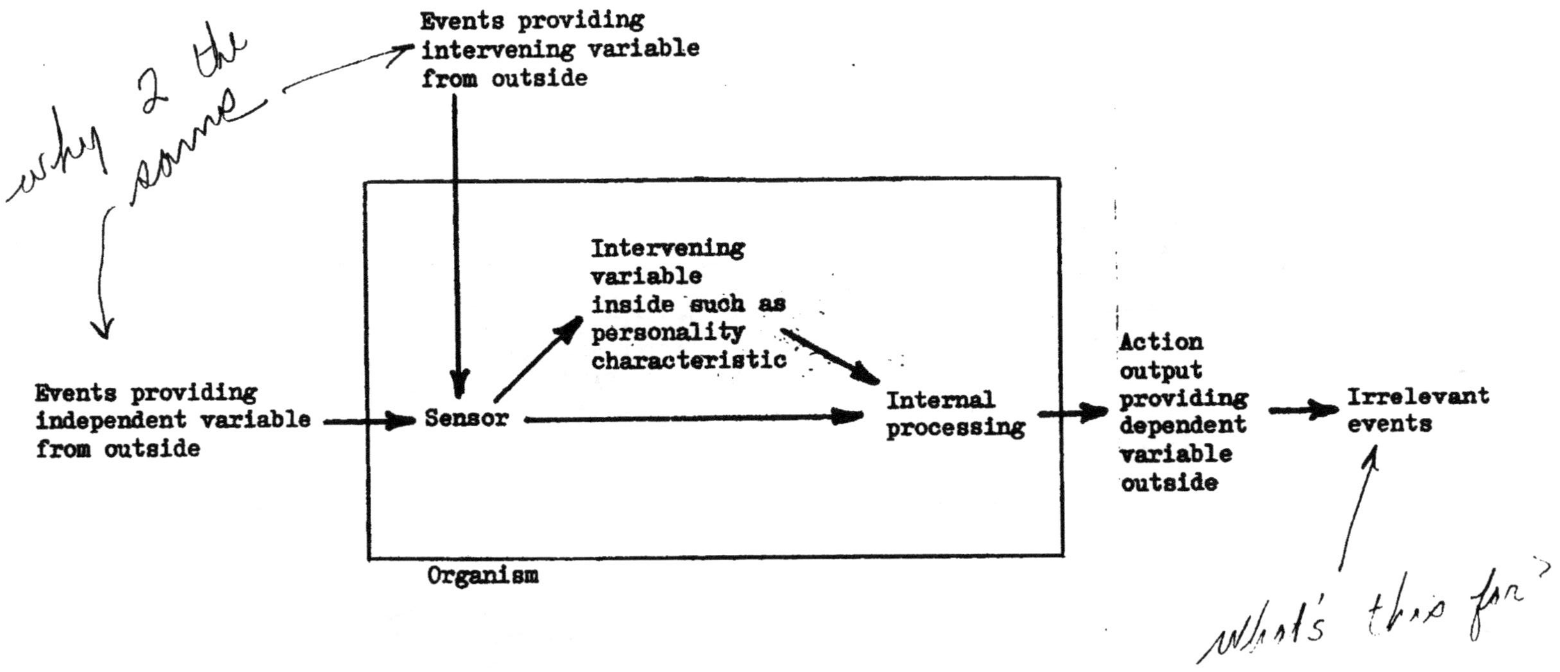

M. Gordon-Lichty

Dear Phil,

Thanks for sharing with me the interesting article and excerpts from letters from W. T. Powers. I have to admit that the "quasi-static analysis" was beyond my comprehension but I can grasp the importance of this revolutionary way of re-thinking scientific psychology. And still keep the concept of purposive behavior!

I will be interested in seeing what your poll of psychologists re including the "sensor" in the diagrm reveals.

Do keep in touch! Carolin

Carolin Keutzer

2/26/86

UNIVERSITY OF OREGON

March 13, 1986

Phil Runkel
DEPM
Education
Campus

Dear Phil:

Sorry I didn't respond to your diagram earlier, but I was puzzled both by your question and the diagram.

I still don't understand it. A "sensor" within the organism ? What does that mean? Homunculus? Perceptual Apparatus? Why does "personality" come after the sensor (rather than before)? Why are "sensor" and "internal processing" separate constructs? And why is "irrelevant events" the final link in the causal chain? If you like, I could raise even more questions. Sorry, but confusion reigns (in my mind at least).

Best regards,

Mick

Mick Rothbart

4906

24 February 1986 (second mailing 10 March)

Dear psychological colleague: Doug H

I'd be grateful for five minutes of your critical acumen. Is the following diagram pretty much the standard way we think about connections among variables? I wouldn't ask, except that most diagrams (outside psychophysics) omit the sensor, seeming to take it for granted. Does this diagram suit you all right?

Thanks. — Phil Runkel, DEPM, Educ

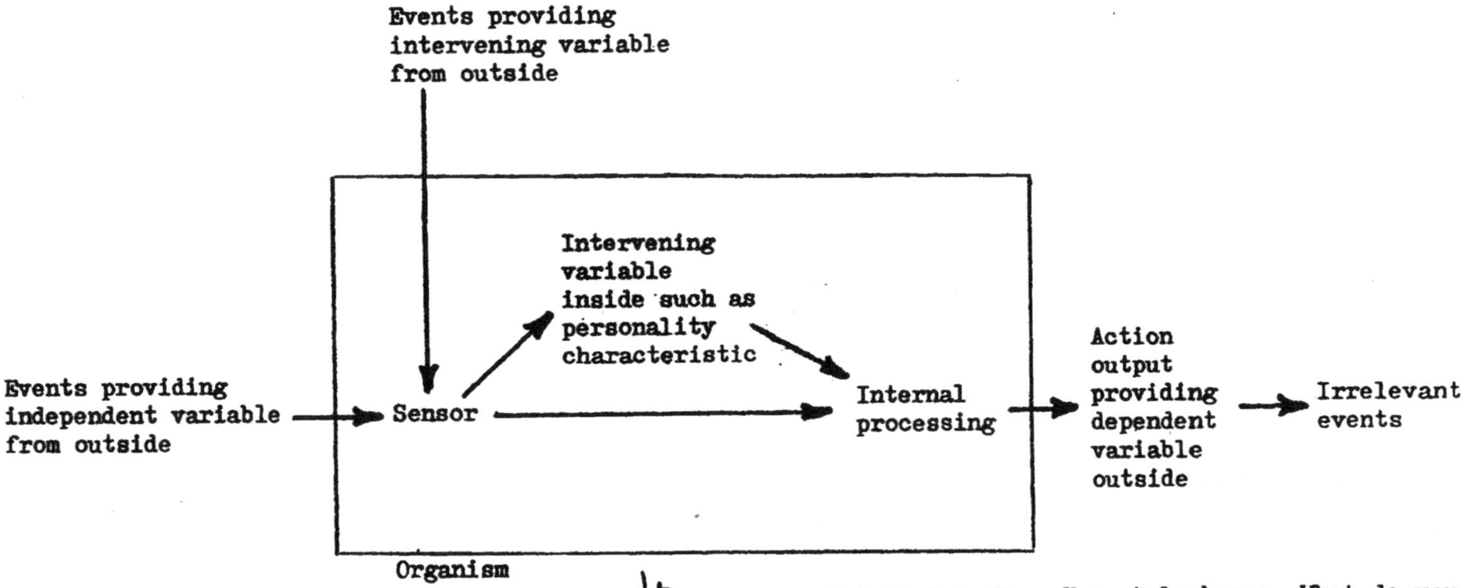

Phil:
Sorry, in 5 minutes I can't make any sense out of this.
Doug

I talked to Doug H on telephone: What do you say to an undergraduate who asks you what this research stuff is about, anyway? Doug: Well, you manipulate things, and you observe the effects.
Runkel: Where do you look for the effects?
Doug: In behavior.

4904

24 February 1986 (second mailing 10 March)

Dear psychological colleague: Barbara G-L

I'd be grateful for five minutes of your critical acumen. Is the following diagram pretty much the standard way we think about connections among variables? I wouldn't ask, except that most diagrams (outside psychophysics) omit the sensor, seeming to take it for granted. Does this diagram suit you all right?

Thanks. — Phil Runkel, DEPM, Educ

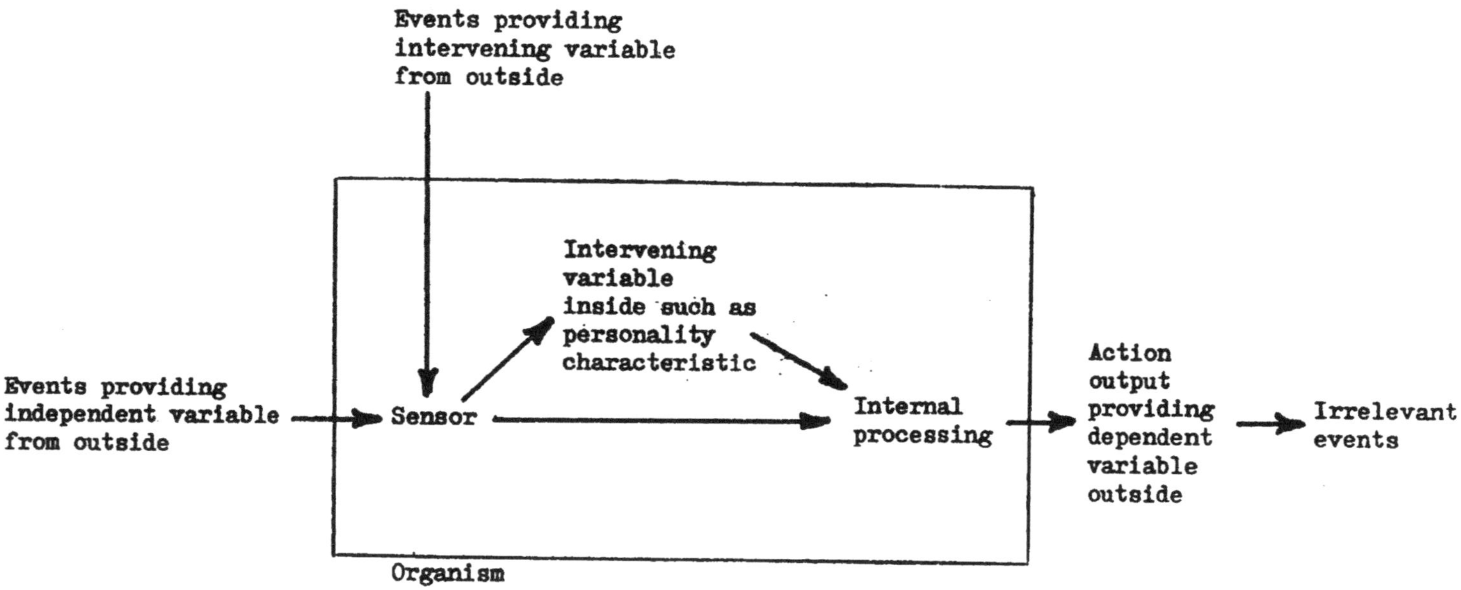

This diagram doesn't mean anything to me. I don't think in these terms

I called Barbara on the phone. How <u>do</u> you think about doing research? Barbara: I think about what the cells in the brain are doing. Me: How do you find out? Barbara: I stimulate them in some way and see what behavior results. Me: What does a cell do when it behaves? Barbara: You get a change in potential. Me: So you get an electrical output? Barbara: Yes. And from a collection of cells, you can often detect an increase in a chemical. Me: Anything else? Barbara: No, that's about it.

Apr. 12, 1986

Dear Phil,

Magnificent. In fact, I hope you will consider this little survey as a pilot project, and turn it into a nationwide survey. This is very important, especially to me: I've been accused again again of beating a dead horse; I've been told that behaviorism is dead; I've been told that nobody believes in this stimulus-response stuff any more, so why do I keep on about it? The problem is obviously that people have just changed what they are calling this basic cause-effect belief, and think they have abandoned the old in favor of the new. This is the main reason why the importance of control theory hasn't been seen: its tenets are so unbelievable that people automatically bend them to sound more like what they "know" is right, and while this is nice of them, it makes my life more difficult. As long as people think they are making progress by renaming SR theory, they will remain unconvinced that control theory is necessary. When will they realize that they have made no progress at all?

I'm working on a paper for Science that I think will have some impact. It's essentially done, but I still have to produce the graphs (real experiments). When all is ready, I'll be sending copies to the whole Control System Group for criticism, and especially for relevant (necessary) references. I have a hard time getting references (interlibrary loan is about the only way, and it's terribly slow). Of course there really aren't any directly pertinent references, but if I don't mention apparently relevant work, editors will find out how ignorant I am and not publish my stuff. I'm sure you or people you know can help.

Your advice on organizing a school was good control theory, well-disguised. Funny how it just sounds like common sense. I think, though, that we have a long way to ge before we really see how control theory is going to influence the way we do things. For example, what about the things that are taught in the school? What about the idea of school itself? In that connection, we really have no theory of teaching at all. All that a teacher can do is make it necessary to learn something, giving rewards or punishments according to whether the student does or doesn't figure out a way to learn. If the student doesn't learn, or learns the wrong thing, there is nothing the teacher knows how to do about it (except whatever comes from a natural gift for teaching, without benefit of science). Here's a homework problem: come back with it solved. How? Sorry. Either you figure it out or you don't. If you don't, all I can say is "try harder." Or I can show you the steps I went through to solve it. But to help you get from not understanding my solution to understanding it, I have nothing. Nothing, nothing, nothing.

As far as I'm concerned, we're starting the whole business of understanding human nature from scratch. Since we've always misinterpreted the very nature of behavior, how could we have come up with any theory of how behavior is learned? People like Skinner don't even make a distinction between acquisition of a new pattern of behavior and the execution of it once it is acquired. He says to the pigeon, if you can walk in a figure-8, you get to eat. So do it. A terrible conceptual jumble.

First we have to understand what behavior is and how it works, after it has been learned. Then we can start looking at the process of acquiring new abilities to control. If we can't accurately characterize the final product, how can we hope to discover the processes that lead to it?

You should be getting another Newsletter early in May, and my new paper, too. I do wish you could attend the conference (Aug 20-24), but I understand why you can't. You seem to be accomplishing a lot from where you are, though.

Best regards,

Bill

Bill

Dear Phil

We are returning all your originals - have xeroxed them, however.

Bill is serious about this being a pilot study - not that the U. of O. can't stand as a microcosm of the world of psychologists, but it would be fascinating to hear from some more people.

Could you drop a note telling us which of your respondents specializes in what? We don't know these guys.

If you can't come to the meeting Aug 20-24, may I stand in your place and make this available to the group?

Also, if anyone at the meeting wants to carry this survey to their own bailiwick - may they?

Bill + I have one more question, which is: what does OXO (or $O_1 \times O_2$) MEAN? Could you either explain or point us at the appropriate (article, book)

4 questions is enough.

Just for the heck of it, here is a photo of the heretofor faceless entity called WTP. He's a lot grayer, though. I had some copies of this made 6 years ago when a picture was wanted to go with some article or other -

Mary P.

21 April 86

Dear Bill:

Do you agree with the following?

It will be more difficult to hit upon good guesses with which to start The Test to the extent that

1. the opportunities for action in the environment of the person being observed are too many, so that the person uses a variety of actions to control the input quantity, and we are distracted in trying to find the common element.

2. the person's opportunities are too few, so that although we can get clues to input quantities immediately being maintained as steps in maintaining a higher-order input quantity, we cannot get clues to that eventual quantity. An example would be a person repairing a machine who visits the stockroom, returns, sits down, and reads a book for the next several hours. In this example, to carry through a program for repairing the machine, the person needed an essential part that was out of stock and could not be obtained before tomorrow, and the person could do nothing until the part arrived.

3. the person's feedback function reaches far into the environment and requires steps that take a long time, so that we cannot spare the time ourselves to track them.

4. the person's internal standard for which we are hunting is high in the hierarchy. At the level of principles, for example, input is maintained by averages and trends over long periods, and it will be difficult for us to separate the positive and negative instances.

5. our own (the observer's) opportunities are too few, so that we cannot find a way to disturb what we think may be the input quantity. We might hypothesize, for example, that the repair person is still wanting to repair the machine even while reading the book. We might want to pull some pieces off the machine to see whether the person would stop us. But we might find a guard at the door telling us we are not authorized to enter.

6. our own internal standards limit the ways we act on the environment to get information. In observing another, we ourselves are acting to maintain one or more input quantities. If it is part of our job, for example, to note when people arrive at work, we can operate at the level of program to maintain the tally, but we are unlikely to learn much about human behavior or much about any one person we observe. The outcome will be similar if, as social scientists, we adhere to a rigid methodological prescription that is an end in itself. That is, we may use a principle that one should carry out one's work in a proper manner, or a system-concept that social life works through individuals carrying on legitimized occupations, but the program of the methodological prescription may be the one the higher standards weight most heavily, so that the routine is very unlikely to be altered. Or we may have a system-concept that the only way you can influence behavior in the environment, whether of people, donkeys, or rocks, is by pulling or pushing the thing in the direction you want it to go, and the more directly the force is applied, the better.

Phil

This must be the picture ☺: 4x6 print glued inside the front cover of Phil's personal, marked-up copy of *Behavior: The Control of Perception*

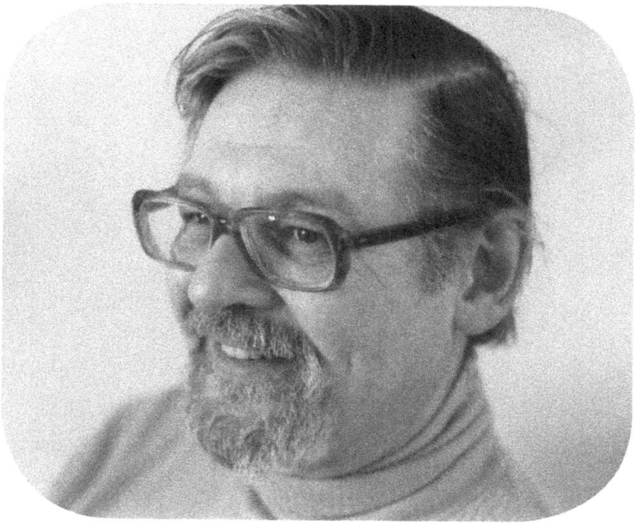

22 April 86

Dear Mary and Bill:

Thanks for the picture. Bill looks absolutely human. I'll dig up a picture of myself one of these days. I don't remember a recent picture of Margaret that looks like her (I mean that looks to <u>me</u> the way she herself looks to me), but I'll hunt.

No, I am not going to make a national survey. I am sure that the Oregon Psych faculty is not representative of U.S. universities. There must be a faculty somewhere in the country where a few people are aware of feedback loops through the environment. But the people here are not, certainly not, by the usual standards of academic psychology, ignoramuses or nincompoops, and to find that pattern of response in this bunch is enough for me.

Of course you can do anything you want with the diagram and with the copies of the responses you made. No copyright, and no original idea, either. But if you or somebody wants to use the diagram, it would be good to change a word. I have indicated it on the enclosed copy. "Moderator" at that place is a more correct word , in the methodologists' lingo, than "intervening."

And to drop off an irrelevançy, you might want to have two output arrows, one each for "action output" and "irrelevant events."

Also enclosed is a list of members of our Psych Dept with a word or two about their specialties. I have checked the ones who replied.

Since Bill is acquainted with Donald T. Campbell, I am astonished to learn that you don't know about O X O. Campbell, as far as I know, is the inventor of that notation. I enclose a copy of what McGrath and I wrote about it (some of what we wrote).

Nice to hear from both of you.

Phil

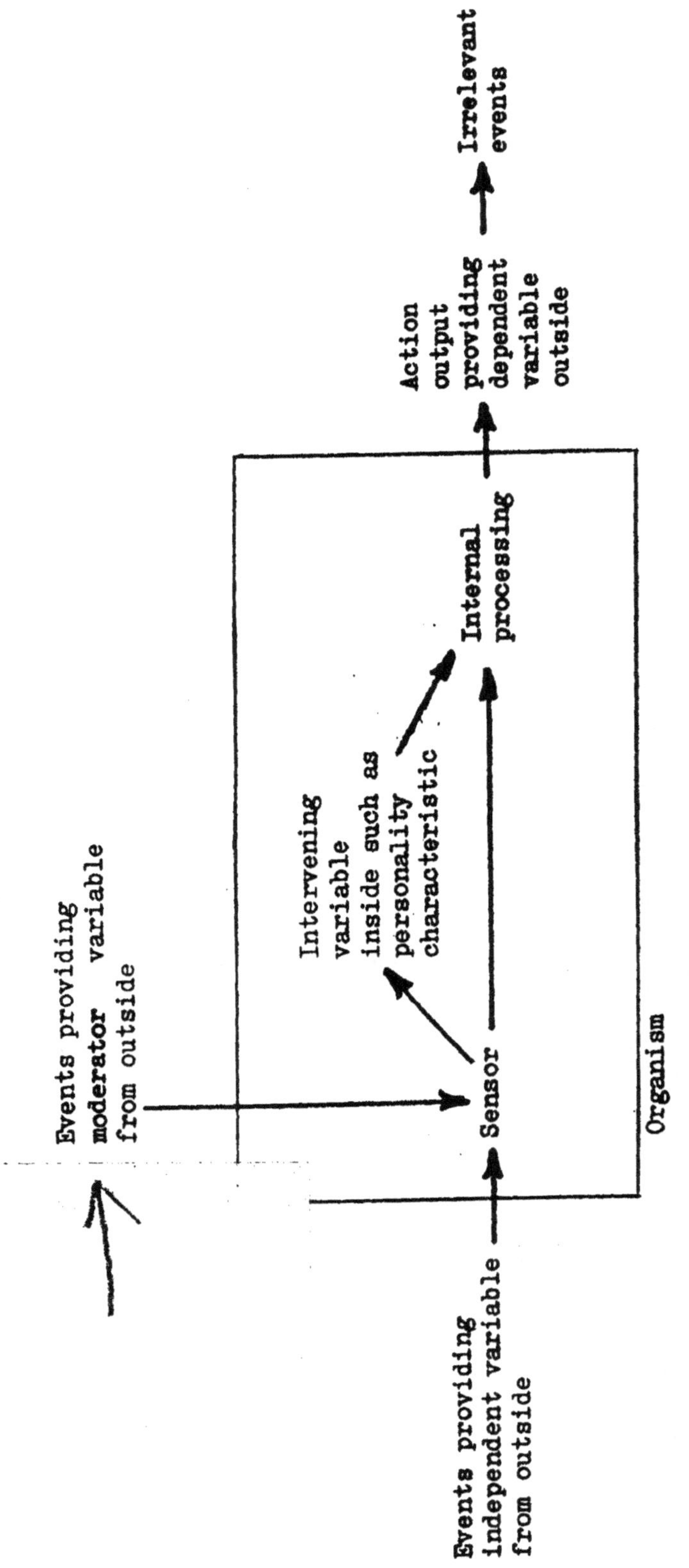
Events providing independent variable from outside
Events providing moderator variable from outside
Sensor
Intervening variable inside such as personality characteristic
Internal processing
Action output providing dependent variable outside
Irrelevant events
Organism

May 3, 1986

Dear Phil,

Yes, I have known Don Campbell for a long time (25 years) and no, I didn't know what O1 X O2 meant. Put that in your pipe and smoke it, to use an old Oregon phrase. My attitude toward statistics has caused me to avoid certain subjects with certain people -- I've read a number of Campbell's papers, but certainly not all of them. Mainly the ones he selected for me, thinking they might be of interest to me. Maybe he saw through my tact and did some avoiding of his own.

There's nothing wrong with that approach -- it does tell you if the treatment had some effect. All it can't tell you is why, and what the specific relationship is. I claim that what you are most likely to discover in this way is a relationship between behaviors and disturbances, but that's a starting point. Once you understand control theory, you can take the next steps, for discovering "relatedness" is just a bare start. The next thing to look for is some variable affected both by the treatment and by the behavior that fails to change as it would if these two influences were each randomly related to the variable. That gives you a hint about the nature of the controlled variable. With that hint you can probably refine the definition of the controlled variable, and choose both a new treatment and a new measure of behavior that are more closely related to the controlled variable. When you are pretty sure of the nature of the controlled variable, you can begin predicting the behavioral changes that will occur under OTHER treatments that disturb the same controlled variable. This will let you pull together a bunch of different-looking treatments and a bunch of different-looking behavioral changes, and show that they are all related to control of one variable. This process will turn aimless collection of statistical facts into useful knowledge.

Your observations about the problems with identifying controlled variables were all agreeable. As I was reading them, however, a thought began to form, and when I was done, it came out like this: "Why do we want to know what people are controlling?" At the moment the thought occurred, I had a much clearer picture of what it meant than I do right now, a week or so later.

To reconstruct, the idea was that conventional approaches to behavior have been organized around trying to predict what people will do under various circumstances, and (to no small extent) trying to figure out how to make them do something else. Since behavior was thought to be the outcome of events happening to organisms, this boiled down to learning the effects of various events on behavior. Why? So that by manipulating events known to have effects, we could produce behaviors more in keeping with what is thought to be desirable.

Ah, the next part just popped up. Because the goal has always been that of controlling people's behavior, the approach to human subjects has always been, in one way or another, surreptitious. Furthermore, simply asking people what made them do things seemed to elicit answers that seemed wrong -- according to the theories that were believed. If you ask a child why he hit his sister, and he says "Because I wanted the toy," this answer is considered inaccurate, because "wanting" isn't an acceptable cause. Subjective reports have got a bad reputation, partly because of reasons like this, partly because of "illusions," and partly because of asking questions that really can't be answered correctly in terms of accepted theories. Also, of course, partly because people aren't aware of a lot that goes on inside them.

As a result of all this, the approach to behavior came to be "objective." That is, you had to find out what made a person behave in a certain way without asking, and when you applied tests, you had to do it without letting the subject know what was really going on -- preferably, without even knowing a test was in progress.

Odd how it comes back in sequence. Next: The objective approach is, of course, difficult to carry out. Furthermore, the results aren't very clear -- you're lucky to find any effect at all. So behavior begins to look very mysterious and murky, and you begin to suspect that the real causes are hidden very deep. You start to look for subtle indications of these hidden causes, and when the indications are fuzzy, you're not surprised: after all, we're delving deep into mysteries lurking under the levels of consciousness. Look at the Freudians: no matter what you say, you meant something else.

So, to wind this up, psychologists of all ilks have got used to searching for subtle clues about the forces underlying the surface manifestations of behavior. And as a result, they have failed to see the BIG IMPORTANT facts that are right there in plain sight. For example, people stand and walk. How do they do that? People utter sounds that other people recognize. How can that be? People go to work every day: why and how? People sharpen pencils, drive cars, do their income taxes, tune radios, go to stores and buy things, tell stories, ski, shoot guns, and whistle. Surely there is something remarkable about all these things! Ordinary life is simply chock-full of behaviors, multiple overlapping unceasing behaviors -- why is it that psychologists have such a hard time finding some behavior to study?

I think the answer is that they take practically everything in their own experiences for granted. If life is just proceeding as usual, nothing is "happening." To make something "happen" so you can study it, you have to set up special conditions, get your instruments ready, and apply the stimulus so you know when to watch and what to watch to catch something happening. The unceasing flow of behavior is ignored. The continuous presence of perception is dismissed -- that's just the World Out There: you have to do something special to get a "perception" out of it.

A long time ago I concluded that everything needed to construct a useful picture of behavior was completely open to observation all the time. The whole trick was to notice the obvious. That's where my "levels" came from, noticing the obvious. I really don't think there is very much that is hidden about human behavior. There's a lot we don't attend to, or perhaps haven't attended to for a long time, but it's all available to inspection when we turn the right way and open up to understanding instead of taking everything for granted.

So, the question was, why try to find out what people are controlling? Now I see that it should have been, "How do we find out what people are controlling?" The answer is, "Look for the big obvious things, not the subtle and hidden things." No matter what people do, and they're always doing something, they're controlling. First just grasp what they're controlling. When you understand that, the next level will become apparent, and so on. It just isn't hard, it doesn't require formal manipulations of data until you finally want to pin down the characteristics of control. Look at the forest before you concentrate on the trees. Look at what's right in front of you instead of trying to peek beneath the surface -- and ignoring the surface. The data we need to understand what people are controlling is lying around in great heaps - why not get those out of the way first?

The Bersheid piece is excellent -- lots of good control theory there. Goes to show that I should have published more and earlier, as they don't cite my model at all. I agree with the Schachter model almost completely. The main thing I have to add, if it's even an addition, is that emotion is simply part of the behavior, the part that results when actions are backed up by changes in physiological reference levels. I've written to you about this, haven't I? If not, let me know and I'll give you my essay on the subject. Emotion is a lousy word, of course, since it implies something separate: it's really what we perceive from inside as we act or prepare to act. The old classifications schemes don't inform us much: joy, grief, fear, and the like come about when we act or would like to act in certain ways, but these words are bound to pass out of use eventually. They don't really refer to "things" as they seem to. At least they ought to be verbs: "grieving" makes a lot more sense than "feeling grief." Of course, "angering" is awkward, even though it's really what is going on. But we'll make much more sense out of emotion when we get used to describing what it is we want: I want my mother not to be dead, I want to be somewhere that bombs aren't falling, I want to bash that bastard in his lying mouth. Of course there are feelings that arise when we want such things, many of them futile preparations for action that will never take place. But the feelings would make no more sense without the goal than the goal would without the feelings. It's all one integrated system, not a bunch of unrelated phenomena stuffed into the same bag. Thought, action, feeling -- different parts of the same process.

A nice Saturday afternoon for running on. I think, though, that I had better go mow some of it.

Best,

Bill

Bill

May 18, 1986

Dear Phil,

Thanks for the useful comments -- in the enclosed you will see that I have complied with many of them. Rather than getting into the literature of statistical analysis, however, I have simply modified my statements to avoid saying wrong things. If I haven't yet avoided doing so completely, let me know. I'm still several revisions away from a finished product. Lots of good, and sometimes severe, comments from the Group.

I must say, this is the first time I've had a paragraph actually physically returned to me on warrantee -- usually the customers just object and (heh heh) let it go at that. I'm not sure you have a legal leg to stand on, but in the interests of PR I will try to explicate.

A "treatment" always involves altering some variables in the subject's environment (that's the only way we can affect anything). As a result, we see a change in the subject's behavior from O1 to O2. Statistical analysis and the various precautions you have described make sure that the effect is real, if unexplained and rather uncertain.

Somewhere in that bathwater is a baby. Under control theory, we assume that the apparent response to X results because X has disturbed something under control by the subject. Since psychological "treatments" normally entail extremely complex operations (however simply represented in words), large numbers of different variables are affected by the treatment. One or more of them, or some function of them, is a disturbance. The disturbance is tending to alter a controlled variable. The change from O1 to O2 also is customarily extremely complex, but among all the changes, some aspect of the change in behavior is acting to stabilize the controlled variable, protecting it from the disturbance or restoring it after a successful disturbance.

Thus starting with O1 X O2, we can begin a search for a controlled variable, a variable affected in one direction by X and in the opposite direction by the change from O1 to O2.

No doubt many possibilities will suggest themselves. OK, you take them one at a time. Suppose you have a variable v1 that you can see would fit the definition. Now you can also see which aspect of the behavioral change would have affected this variable, and which aspect of X (oppositely). So you do a new experiment, redefining both the behavioral measure O and the treatment X so they bear more directly on v1, and do the experiment again. If you're getting warm, you'll find that v1 is indeed stabilized, and the (negative) correlation of X and delta-

O is higher. If that isn't true, you guessed wrong, and you look for v2, and so on through v3...vn, until you either give up and conclude that this isn't control behavior, or you find the definitions of v, X, and O that produce the beautifully precise control relationships we have come to know and love. Like the ones in my paper. What you're looking for is r > 0.95, I suggest. Then you know you've got it right.

Of course this means that the point of the experiment isn't to prove that O or X is important. Almost certainly, the O and X you start with are the wrong ones. Only that little hint of a correlation makes them worth anything at all. O and X are just the initial means of looking for opposition to disturbance, and either one is subject to immediate modification in the interests of finding the controlled variable, which IS the point. Your conception of O and X is bound to change significantly before you're done. What is usually taken as a publishable result, therefore, is considered by control theorists to be no more publishable than a report on sharpening the pencil with which you record your data. (God, I sound like Watson talking about "the behaviorist." Well, at least I now know what he meant. But unlike Watson, I've DONE the analysis of "simple reflexes".).

The chapter on emotion is embedded in the only two copies of my manuscript that exist, a version put together for a 13-week student-sponsored seminar I gave at Northwestern in 197 uh 2. I'll see if I dare take it apart to Xerox the chapter. Don't expect the writing to reflect the 14 years of practice that have intervened. I'm enclosing a copy of Ed Ford's version, gleaned from phone conversions with me and expressed in his own terms. Ed Ford is our link to the world of laymen, I think. This piece is part of his effort to write a book, Freedom from Stress: a basic introduction to control theory. It doesn't exactly represent my views, but it's satisfyingly close.

I'm looking forward to seeing your 200 pages.

Best,

Bill

Bill

Control System Psychology
Feelings
Ed Ford

Bill - thought you might like a copy Ed

Summary of ideas on feelings:

Anytime we set a goal, certain feelings are going to be attached to this goal but the kinds of feelings and the degree of awareness are going to depend on how that goal relates to what we presently perceive, or the extent of the perceptual difference. It is the interaction of the goal with our perceptual system called the perceptual difference that literally determines what the feelings are going to be and the extent of the energy created within the body.

Certain changes take place in the physiological organism when you have set a goal and are preparing to accomplish that goal. All the physiological changes are exactly the same, except that the more energy that you demand as a result of wanting something, the more pronounced the changes become. The feelings that reflect these changes we call by different names even though the physiological changes are precisely the same. We call them fear, anxiety, anger, stress, but they reflect the same changes.

Now if this is true, and experiements show that it is, then the internal changes within us and the actual sensations that result from those internal changes aren't all that different. So what's the difference? The difference obviously isn't in the sensations you feel within your body. It is in your cognitive goal. And people get mixed up between their goals and the sensations from their bodies, the combination of which they call their feelings. It's as if they can't tell the difference when they feel something that's got a thought component in it. They think that the thought is coming from their bodies. They'll say "I feel afraid, I feel upset, I'm stressed, etc." Their interpretation of their feeling has their cognitive goal tied into it, but they don't recognize their control over that cognitive goal. Rather, they blame the feeling over which they don't believe they have control. What would they feel if they could somehow turn off all of the thinking that was going on within them and just pay attention to what specifically are the sensations that are coming from their bodies? They'd find that those sensations are really not specific to any particular goal.

If people would only reflect on their goals, setting aside their feelings and independent of their sensations, and evaluate these goals at all different levels, and try to resolve the problems that lie therein, then the feelings would take care of themselves. Feelings are really a person's view of what's happening within his or her body. You feel all keyed up. That feeling may not be precise but it gives you a pretty good picture of what's going on within you. You get sort of an overview of your present physiologic state and that's what you call your feeling state.

The feeling state also has within it a hint at what your goal might be. For example, I feel afraid may indicate your belief you might not be able to handle something that might do harm to you. I feel upset may indicate your belief that what people are doing around you isn't to your liking. Maybe it's something else.

Now what that feeling state is telling you (if you are indeed feeling all keyed up) is that your heart rate has increased, your breathing has become deeper, adrenaline has entered your blood stream, the blood has pooled toward the center of the body, the blood vessels in the periphery of the body have constricted, etc. etc. You are ready to do something. When the behavioral side of you does something, ultimately you use up all that energy you've generated through having created a goal, namely something you wanted, and the perceptual difference that has resulted. Once the energy is used up, then you calm down. It takes time to recover. If you haven't satisfied the goal, you get ready again.

Unfortunately, not having satisfied the goal, you'll get back into the same physical state, the same feeling will return. Why? You haven't satisfied the goal. The key in all this is to examin the goal, work out a plan to satisfy the goal, and then the feelings won't return.

Again, feeling is sort of a general term that includes goal. Sensation sort of takes the cognitive part out of it and leaves you with how does your chest feel, how does your stomach feel, and so on. The cognitive part of it, which is what you want, your goal, is the key to resolving the disturbing feelings. Dealing with feelings independent of the cognitive goal is like clipping the wires on a fire alarm that is sounding it's horn rather than checking to see what it wants, namely, someone to put out the fire.

19 May 86

Dear Bill:

I have come to the end of the writing I was doing about how to find persons, groups, and organizations. I picked up my computer from the repair shop today, and if it really fixed, I'll be copying typewritten copy into the computer, editing, rewriting, and so on for, I suppose, a week or two. I am going to hire some extra help to take care of Margaret so I can spend some solid hours at the computer. I hope to have the document to you in ample time for you to read it before you go to the Control Systems Group. I'm not sure why I think there is a connection. But I don't spend much time wondering why my mind makes connections between things. I usually just let it go its own way.

In the meantime, I am writing to answer several of your earlier letters.

Yours of 31 January: You wrote a long passage about awareness. To me it sounds mystical, not like you at all. I'm not going to think about that right now. I'll let it soak a while.

Same letter, thanks for the information on how to catch a ball without using feedforward.

Thanks for the clarifications of the Byte articles. It's pretty complicated, but I don't think beyond me. I'll think about that later on, too.

Well, it turns out that those are the only things that need mention in the letters of yours I had in a pile to answer. I've commented on other things in other letters I have written, and you'll find comments on still other things in the long document I am about to put into the computer.

Here is a sentence from Hofstadter that you have probably already come across: You are under the control of this sentence, because you will go on reading until you have come to the end of it.

And here is a question I have been meaning to ask for some time. It is about memory and the four modes. What is the circuitry for remembering what seems to me must be an error signal: a yearning, a threat, a lack of something wanted, a time when you were very cold, and so on? I don't find any path in the diagram on page 221 of your book that would permit remembering an error signal. Maybe we remember two perceptions, one used somehow as a reference signal, and we send upward a relationship between the two. That is, one perception would be what it is like to be neither too cold nor too warm, and the other would be the feeling of too cold at the remembered time. Or what?

Phil R.

3 June 86

Dear Bill:

Glad to have your letter dated 18 May, postmarked 29 May; arrived today.

I'll read your revision of "Purposive Behavior" after a while. I'mm too busy with my own writing just now.

Sorry, but Ed Ford's piece does nothing for me. I already know that point of view (Schachter), and Ford's simplified version does not add anything for me.

Sorry, but my memory does not tell me to what you allude in your sentence: ". . . this is the first time I've had a paragraph actually physically returned to me on warrantee. . .". I am, however, glad to have your paragraphs following that. They agree to a T with what I have written in my long document. I'm glad to have your corroboration.

I am now editing my long document and fixing the files in the computer so that the document will look pretty when I send it out for criticism. Dear computer: Don't start a new page <u>here</u>, you dummy!

I sent copies of Marken's delicious paper on random reinforcement to a couple of devotees of reinforcement theory here. I sent their replies to Marken. He wrote me an interesting letter.

You must carry on a Gargantuan correspondence and reading schedule. I admire your industry, and of course I am myself forever grateful to you.

More, MUCH MORE, later.

Phil

11 June 1986

Dear Bill:

I have now read your third version of "Purposive Behavior." It goes along very nicely, I think. I hope Science accepts it. If not, I hope you will find another place for it. When it is published, please send me a reprint. Or at least tell me where it is published so I can go to the library and make a Xerox.

I have only a few places for possible alteration.

On page 2, starting at line 10, it might go a little more smoothly like this:

> . . . we will examine six experimental demonstrations, variations on one theme. The results apply very generally to behavior, but agreement with this claim is not prerequisite to accepting the analysis presented here.

On page 8, line 9: ". . . traditional logic would predict. . .". Yes, going very strictly with the theoretical position that the stimulus produces a particular act, so traditional logic (theory) would predict. But I never find experimenters ignoring common sense (their version of it, of course), no matter to what limited domain their theory tells them to hew. Would the traditional researcher, no matter how hidebound, insist on ignoring the fact that the subject is seeing the cursor movements occurring at only half the instructed amplitude? Maybe it is not your duty to comment in your article on my question. I grant that. But maybe you might add something like: . . . despite the fact that the subject is seeing only half the amplitude he or she was instructed to maintain, because . . . .

Page 10, 10th line under "Discussion, Experiment 1": "common sense." Whose common sense? It is common sense, to use your own example, that the driver stays on the road despite wind gusts, bumps in the road, and threats from other drivers. It is common sense that the attendant on the airplane does not spill the drinks while walking down the aisle during bumpy air. It is common sense that of the door is locked, you hunt for a door that is not locked. And so on. I think most readers who have read to page 10 will have this same comment.

Page 13, 14th line from the bottom: period after "al."

Page 13, 9th line from the bottom: When you write "Figure 13" as a title, it is a common practice to capitalize "figure"--though the Chicago Manual doesn't like it. When you use "figure" merely as a common noun, don't capitalize. I write out all that, because I remember that you did this in the earlier version, too. Maybe you just didn't catch that "F" in the revision. Also line 13 on page 9.

That's all.

I sent Marken's article on "random reinforcement" to a couple of colleagues here who are devotees of reinforcement theory. The gist of one reply seemed to be, to use his word, that it was "just bullshit." The other one gave a more reasoned reply, saying that the actions of the ~~xuxx~~ random movements of the cursor were negatively reinforcing, and that the instructions "set the goal" for the subject.

The reinforcement people nowadays seem to say that to make reinforcement work, you have to find out what is reinforcing for the subject. That seems to be saying that you look to see what condition the subject will work to maintain. That seems to be the same as saying that people have purposes, something I thought reinforcement theorists were not supposed to say. Of course, Hull and his followers, I forget how many years ago, postulated that the organism was motivated by "drives." * I remember vaguely something about maintaining certain conditions in the "tissues." At the time I was reading that stuff, I didn't think of "drive reduction" as a goal. But now it seems to me that they did put purpose into the theory. Tolman did explicitly.

If you are going to take the instructions as providing a goal, then I don't see why you have to bother at all with the idea of reinforcement. You can simply take the movements of the cursor as information that the subject uses to make corrections to maintain input. Am I getting closer to Chicago? Yes. Then keep driving in this direction. Why do I need to say that I am "reinforced" (in any sense other than getting useful information) by the signs telling me that I am getting closer to Chicago? I made this same argument to Marken, but he didn't seem to think it helped much.

Phil R.

* Text reads: postulated that the organism was motivated by "drives."

June 17, 1986

Dear Phil,

Incredible! Marvellous! Heavy!

I got back from the Gordon Research Conference on Cybernetics yesterday, to find your five pounds of warming up for a book, and read it immediately, end to end (with only a little page-flipping at places where I knew what you were going to say). This is exactly what I had hoped to see as a result of my work. It's fascinating to see the bare bones begin to flesh out as you explore the details of every concept in control theory; this is something I could never do, never even aspired to do. Your mastery of control theory is amazing to me: I don't know anyone else except perhaps Marken who has assimilated absolutely everything, made it his own, as you have done.

I especially appreciate your long discussions of The Test. It does seem cumbersome and difficult at first, but as you show what the alternative is the whole procedure begins to look simple and practicable. This is just the sort of thing I wouldn't know how to do. You've expressed regrets over your previous commitment to traditional statistical methods, but if you hadn't become an expert in that field, and a proponent of the approach, you wouldn't now be able to build the bridge found in Inside and Outside. I've always maintained that the proper persons to develop and extend my theory are the people who know how the traditional approaches proceed, and what beliefs they contain; without that kind of knowledge, comparisons can't be made in any useful way. Only the insiders can make this revolution work. You are obviously going to play a major role. This is a great relief to me: now I can get hit by a truck and not worry about it.

What's the developmental schedule on the book?

Letter of the 11th. Suggestions will all be used.

Hmm. I need to make something clearer, obviously. The point of the "half as much" movement of the cursor is not that common sense would accept it after it had happened, but that common sense would not have predicted it on the basis of experiment 1a. Once you see what really happens, of course, you can find an explanation -- but the point was to build a model based on conventional assumptions, and show that it leads to a wrong prediction. Common sense, which assumes along with conventional science that behavior is the final outcome, would predict that any interference with behavior AFTER its generation would simply affect the outcome. The idea that the outcome would remain the same while the means of producing it varied doesn't jibe with our normal conceptions of a chain of causes and effects. So I have to make that transparently clear in the paper, don't I?

In some regards, common sense is aligned with control theory. We don't expect the flight attendant to spill the drinks. In fact, I stopped telling my co-workers, mostly blue-collar, about my theoretical work after enough of them had responded to the basic idea by asking, "But doesn't everyone know that?" Of course we have intentions. Of course we resist disturbances. You have to know a lot more about what science believes to understand that control theory throws a monkey-wrench into the works, or as you say, sand into the gears. Most ordinary people greet a description of what scientific psychology believes with incredulity.

Of course even common sense balks at the notion that we control perceptions, not reality. Control theory isn't quite a total waste of time for the uneducated man. But it's going to be a large job to find the kind of meaning in control theory that can be communicated outside academia in such a way as to have a real effect. We have to remember how few of Us there are, and how many of Them. They, not We, control the world.

The Operant Conditioning people, as you say, have really abandoned their principles when it comes to practical applications. Behavior Mod begins by asking the subject's permission to suggest changes, and there is a lot of emphasis on awareness and goal-setting, none of which makes official sense. Experience with real subjects has simply forced them to acknowledge the realities of control theory, whatever words they use. Tolman, as you say, decided that purposes were real -- but because he had to work within the old model, he came up with some complicated way of showing that this was just a way of talking about certain special SR relationships, after all. What else could he have done?

I agree with your assessment of reinforcement, of course. Basically, all they're saying is that if you're controlling for something, that something is reinforcing. They can't quite bring themselves to say you want it, or need it, or intend to get it.

* Mary found a booklet used at Harvard that discusses the last days of phlogiston theory. As the new theory of matter came into being, with new phenomena being observed left and right, phlogiston theory began to get more and more complex, exfoliating ** all over the place in an attempt to get the observations to come out right. We've probably been seeing the same effect since the late 1940s when behaviorism began to sense a threat from control theory. So there really isn't any way to overcome reinforcement theory from the outside. If you believe in it, you'll find a way to make it right, no matter what the data are, and no matter how baroque the explanations become. Rick Marken is right: correctness won't help much with believers in reinforcement theory.

* *Harvard Case Histories In Experimental Science* Volume I and II (1957)
The Overthrow of The Phlogiston Theory: The Chemical Revolution of 1775-1789. Vol I pp 65-116.
Vol 1 free download: http://www.archive.org/details/harvardcasehisto010924mbp
Vol 2 free download: http://www.archive.org/details/harvardcasehisto007156mbp

The Gordon Conference was an unexpected success for me. I had decided that if the core cyberneticists continued to ignore me, I would opt out of that movement. In fact, one observer of the scene told me that the split ended up about 50-50, far better than the minimum I would have accepted.

The current Guru is Humberto Maturana, who with several sidekicks is pushing the Spencer-Brown approach -- basically the mathematics of categorizing. Heinz von Foerster is God, and while I love him personally (as everyone else does) I find his old-fashioned approach, frozen for the past 20 years, to be a real impediment to cybernetics, particularly mine.

During this meeting I finally understood that Maturana's main contribution has been a new epistemology that is essentially identical to mine. The main practical users of his approach have been the family-systems people (family counsellors); Maturana's concept of "consensual domains," which I call system concepts, together with his understanding that reality is constructed in the nervous system, has turned these therapists away from a strictly objective behaviorist approach into a far more relativistic one. In effect, Maturana's ideas have encouraged people to think in new terms, at a high level in the hierarchy, and in much more detail than I have considered. So I can genuinely find value in his work and in what is being done with it. This helped me find a better attitude toward the competition in cybernetics, which in turn, I am sure, helped me present my own ideas in a more tolerant and more appealing way.

For many years, people have been telling me that Piaget's work has natural connections with mine. I've had trouble seeing the connection, largely because Piaget relied too much on assumptions about objective reality in formulating his "levels," and didn't try to study adult levels of organization. But I was wrong about that, because I was giving my own concepts of what the levels are more weight than they deserved. I learned this at the meeting, too. Three people from the Piaget Institute were there (including Inhelder), invited because it was thought they would come to support the Maturana approach. Instead, they went away enthusiastic about control theory. They have obviously been looking for a model all this time, and have been aware of the lack of one, which factor I had never suspected. My only function at the meeting was to chair one session, but I was able to give an informal talk one afternoon using a working model of an arm that I brought along, and it was at that meeting that the Piagetians got the word about control theory. And where I learned I had misjudged them.

It now seems that my ideas have become a factor in cybernetics. Several articles have been invited for the new publication being organized, including one criticising the lack of awareness of control theory, and another comparing my

** The modified phlogiston theory is discussed on pp. 109–111. See excerpt on following page.

epistemology with that of Maturana. At the start of the conference there were many hostile remarks about the concept of control, all due to the misunderstanding of the nature of control theory (they thought I was talking about ways of controlling people). By the end of the conference that misapprehension had been, I think, pretty well dispelled. This development opens the door for control theory finally to enter cybernetics, and I believe that most of the people in that group are now prepared for this to happen. The old-time leaders are resisting mightily, but I think in vain.

Already I have had a number of occasions to mention that I don't want to take over anything, organize it, or run it. I hope the old-timers stay firmly in control. They deserve the respect, they know the ropes, and they do the work. Spread the word: I am a lousy administrator, and power would instantly go to my head.

It will take me some time to become familiar with your book. In the meantime, may the wind be at your back.

Best,

Bill

Bill

P.S. Perception of ~~reference~~ error signals. Yes, maybe. You tell me. All is possible

From the jacket of *Harvard Case Histories In Experimental Science*, Volume 1:

Edited and with a foreword by James B. Conant, distinguished scientist, teacher and diplomat, these histories were first prepared for students in the humanities and social sciences at Harvard College. They are now offered to a wider public in the belief that a detailed knowledge of a few epoch-making advances in science will provide a key to a better comprehension of the modern world.

Excerpt from page 111:

... The story of the last days of the phlogiston theory is of interest, however, in illustrating a recurring pattern in the history of science. It is often possible by adding a number of new special auxiliary postulates to a conceptual scheme to save the theory at least temporarily. Sometimes, so modified, the conceptual scheme has a long life and is very fruitful; sometimes, as in the case of the phlogiston theory after 1785, so many new assumptions have to be added year by year that the structure collapses. Most of the illustrations of this pattern, it should be pointed out, concern concepts and conceptual schemes of far less breadth than the phlogiston doctrine. They may be ideas that are useful in formulating merely some relatively narrow segment of physics, chemistry, astronomy, or experimental biology. What has just been said applies none the less.

DEPARTMENT OF PSYCHOLOGY
UNIVERSITY OF ILLINOIS
603 E. Daniel St.
CHAMPAIGN, IL 61820

June 18, 1986

To: Phil Runkel (& David Brinberg)

Dear Phil,

Got your large book -- Inside and Outside -- on Monday, 16 June. Could not deal with it until Tuesday. Spent all of Tuesday (yesterday) and today reading it. Then, today, a brief letter from you re my chapter-draft. It indicates, among other things, that you feel a need to have a response re that book before we can proceed much further on planning our joint "theory of method" book. So here goes -- but remember that I just spent two days absorbed in your book and have not had a lot of time to think it all over.

First, it is fascinating -- and somewhat intimidating. Anything 400 pages long with meaningful prose on each page is a dose of ideas to be reckoned with. And if it comes from someone with whom past interactions have been filled with a high proportion of thoughtful and profound ideas, that is even more an omen that what will be in that big book is formidible. And indeed it is.

It is also fascinating, as I said; and intriguing; and in places poetic. And, as always from you, well written (though in this case not particularly well organized at the section level -- more meandering, as I think you intended.)

And, it is powerful -- full of innovative and compelling ideas.

Some of it is about methodology and the doing of research. But a lot of it is about being a consultant to a specific organization. I know they are related; but they are not the same thing.

Some of it is about methodology, but a lot of it is about what groups and organizations (and individual humans) are like. [That is, in VNS terms, it is theory about the substantive domain.] That is also worthwhile, and partly related to the business of methodology and of doing research (and very interesting to me, because I spend some of my time dealing with those content areas as well) -- but it is not the same thing as being about methodology.

As I read it, it has two very important things to say about methodology. First, it largely rejects current methodology -- principally the logic of evidence which underlies it, but also many of the specific tactics and methods by which we carry that out. It says: "Our usual 'components of variance' approach, the Campbellian O_1 X O_2 strategy for getting information from the environment by systematically probing it and observing it, not only does not work well; it does not work at all." It says, further, that many of our ways of getting information -- e.g., questionnaires, rating scales, observations -- and many of our ways of processing that information -- e.g., averaging, difference testing, correlations -- are of little use because they yield little information about the true functioning of human systems.

Second, it offers as an alternative general methodology or logic of method an extension of Powers' control theory, with the key tactic/method by which we carry out that logic (that is, by which we probe and observe) being the Test. It goes beyond that to spend (a) a page or two listing 8 things social scientists could do to help (that part is one of the parts I was least able to feel I understood); and (b) many many pages showing how the O X O approach really doesn't work at individual, group, and organization levels.

Beyond that, it does give me some new ideas for how to look at groups, and organizations, (and individuals too). But it doesn't give me much that helps me in learning how to "probe and observe" (or whatever the control theory equivalent is) in new ways that are compatible with control theory.

In short, it casts out all I know about research strategies, research designs, methods of manipulation, measurement, control of variables, ways to combine and process and reach conclusions with resulting evidence, and the like. But it does little to help me replace all of that stuff with stuff that is more fitting to the new view, control theory.

That may (or may not) be a good thing for someone to do re their own professional program -- to renounce all extant methods, and then start looking for replacements. But it sure as hell seems to me to be a funny way to begin building a book about method.

I too am disturbed about the limitations of our methodology -- all of it. Our logic of method has serious flaws in it. Moreover, all of it gets played out amidst mutually conflicting desiderata; hence it is always not doing a lot of what you know you want to do. I am perhaps as completely disenchanted as you with all of our current methodology. But I am clearly not as ready as you to cast it all out. That is so for three reasons, I think; and let me try to sell you those reasons:

1. First, I certainly don't have anything better to replace it with. I think Powers stuff is extremely interesting, and opens up some whole new ways of looking at human systems, and of understanding our past evidence, theory, and methods. But I do not think Powers' ideas are in themselves our methodological salvation. [Put another way, I am a "methodological ecumenicalist". I am no more ready to buy into a "new religion" such as control theory than I am to adopt one of the currently established ones -- laboratoryism, simulationality, united church of the questionnaire, or whatevver.]

2. One reason I am not ready to buy into a new salvation is because I don't think it is possible. That is, I really believe that <u>our purposes</u> in the scientific enterprise are locked into "dilemmatic fields", sets of mutually conflicting desiderata. I think that constraining fact of life applies to control theory just as it does to our more common methodological (and conceptual) practices. I think we want, all at once, several things; and those things cannot be had all-at-once. Try realism, precision, and generalizability of methods. (And, re concepts and substantive systems, try parsimony, scope, and comprehensiveness of concepts; try system effectiveness, wellbeing, and cost.) There may be other, more descriptive, sets of terms reflecting this dilemmatic nature of the field. The point is, control theory does not circumvent this, it merely tackles the dilemmas from another angle. That's good -- but it does not alter the dilemmatic character of the field.

3. And in spite of all the negative things I have said (and you have said) about our current methodological ways of doing business -- in spite of their <u>inherent</u> limitations as well as the often-poorly-thought-through ways they have been applied -- they are really not all that bad <u>if we do not ask of them what they cannot do</u>. We should ask them only to deliver "reduction of uncertainty" in a probabilistic, contingent, cumulative way. That is what they can do -- and pretty well, if we are careful. But that is all they can do. Instead, we tend to ask them to deliver epistomological certainty about an ontologically murky world. We should not ask them to do that; they cannot.

If I summarize those three points they come out something like this: All our methods are flawed, but I don't want to throw them away because: (1) I don't have anything better to put in their place; (2) they are not all that bad if properly applied; and (3) I believe it is not possible to have a set of methods that do for us all we want done.

But maybe I am wrong; maybe I am merely unable to look at this through the new perspective. Maybe control theory and its extensions can provide a full-fledged substitute methodology for letting us do what we regard as valuable research. If so, fine; but in my present state of limited knowledge I cannot really contribute to a book that presents such a methodology.

That does not mean that I want to write a traditional book about traditional methodology. What I would like us to do is write a book <u>about what methods have been, and what they could be</u>. I have been calling it a "Theory of Method" but that is a bit inaccurate. What would be more accurate would be a "Metatheory of Method".

Are we at a contratemps? You saying you can't partake of a book that upholds the old methodology because you no longer believe it is of any substantial value in trying to understand human behavior; me saying I can't contribute to a book that abandoms that and presents a new methodology, because I don't know very much to say about that new one? Maybe. But before we give up, lets try something really radical, to wit:

Can you envision a book written in two "modes", as we might say in this age of computors-aplenty -- perhaps literally in two type faces or two colors, to emphasize the matter -- along the following lines:

First, there is a book that presents a "theory of method", that tries to tell the methodological story the best damn way we know how -- perhaps following our current 20 chapter outline, or something like it.

Then there is, <u>in the latter half of each chapter</u>, (perhaps literally in a second typeface or color) a <u>counterpoint</u>, or <u>refutation</u>, or <u>alternative version</u> of the chapter's topic, written from the point of view of Powers and control theory.

David and I would produce most of the chapters of the Mode One book. Phil would have to produce most, perhaps all, of the Mode Two <u>counterpoint</u> material.

Obviously, in some places it might be better to have that <u>counterpoint</u> as a full chapter at the end of a "part" rather than as a part of each chapter in that "part". Or, in some spots, it might make more sense to have the <u>counterpoint</u> inserted as the latter part of a section within a chapter. It depends on how topics break down as to their point-counterpoint status.

Of course, we would want our planning to stay flexible enough so that Phil could add in or revise chunks of material in the "mainstream" part, and/or David or I could add in chunks of <u>refutation</u>, either of the mainstream stuff or of the counterpoint (that would be counter-counterpoint).

That sounds to me like an exciting possibility. Do you agree that it is exciting? And do you think we could carry it off?

It would have some advantages if we could. First, it would allow our book to be as radical as we (really, as Phil) could make it; yet it would still touch base with the current methodological forms, hence would deal with topics readers expected to meet. Second, since the new ideas are bound to get developed at an uneven pace, this form would permit us (or Phil) to push the new ideas as far as we can <u>in any given topical area</u> (for example, regarding overall strategy of data collection), but not be obliged to fill it out in full detail in every single topical area. For example, we might have a lot to say about questionnaires, as is now the case in Phil's "Inside and Outside" book, but not necessarily have much "counterpoint" to say about archival measures, or robustness explorations.

Suppose we tried it, and it didn't come off. What are the worst case scenarios, as they say?
1. We might come to intellectual blows and produce no book at all. That would be a shame for the field, no doubt, but we could live with it if we did not let it become ad hominum.
2. We might try the two mode book, and be unable to carry it off in a form suitable for publication. We might then end up with a more traditional book (from the mainstream part) such as we are now starting out to do, or a shorter treatise on control theory and radical methodological ideas (which may be something Phil would like to do anyhow), or both of them as separate entities. That might be quite a big gain for the field and for us. It certainly would be no loss -- and we would have fun and learn lots along the way -- except it would cost us (especially young David) some opportunity costs along the way.
3. We might find, as we go along, that <u>reorganization</u> takes place among our internal standards, as Powers/Runkel might say, and we come to develop an <u>intergrated version of the two approaches</u>, which then becomes a <u>really sensational book</u>. That, of course, would be even better than (though undoubtedly take longer than) the intended two-mode product.

Having a try at such a two mode book has one additional advantage: It would let us go forward, directly and as promptly as we are able, with the "mainstream" portions of the book. We would have a <u>plan</u> for what to do with control theory (and, for that matter, other radical methodological ideas). We could modify that plan (and, indeed, the two mode plan itself) as we go, perhaps discovering integrated solutions, perhaps discovering fatal flaws with the two mode plan, perhaps finding even more richly articulated ways to present the opposition of ideas. But meanwhile the "project" would be going forward.

Please give me some feedback on this, Phil and David. Meanwhile, I will keep puttering away with our current 20 chapter outline as per my prior letter.

Anxious to hear from both of you, Joe McGrath

20 June 86

Dear Bill:

You seem to have bought yourself a new printer. It seems to me to be dot-matrix, but it is as clear as daisy-wheel.

Your letter was mailed 18 June and got to my house today, the 20th. I think that's good service. I believe I told you of an earlier letter that took five days to get to my office. I think sometimes a letter stays a day or two in the campus distribution labyrinth. I don't suppose statistics of this sort are what you wait breathlessly to receive. My mind probably turns this way because of my father. He was a devoted letter-writer, always answering mail the same day he got it. And he fumed at any inconvenience the Post Office caused him. He was outraged that the post office did not deliver mail on holidays.

Well. I sit here with my fingers poised over the keys and a lot of phrases running through my mind that I might use to tell you how happy I am with your reaction to INSIDE AND OUTSIDE. And it takes me a long time actually to hit some keys, and when I do I go at the matter obliquely, because on the one hand I don't want to seem like a dependent baby whose stomach is being tickled, and on the other hand I do want you to know how very highly I prize your opinion. (How's that for an example of conflict?)

I know there are several weak places in the document, and I thought it likely that your letter would tell me I am wrong in this place, confused in that place, missing a better emphasis in another, and so on. But your statements that I have succeeded in "fleshing out" and in contrasting The Test with traditional strategy make me melt with pleasure and gratitude.

I know that you were reading for gist; I am no end flattered that you read it in one day after getting home from a trip. So you were looking for the overall shape of things, not for points to be improved. I hope you will dip into it again one of these days and let me know where the ideas don't flow well.

I sent out pages title-to-24 and 247-249 to 44 friends, colleagues, and some members of the Control Sys Grp. I enclose a copy of the covering letter. So far, 8 people have asked to see the whole document. Four of* them are students of mine, or ex-students.

One person to whom I sent a copy was Heinz von Foerster. I find that communicating with notables brings the same range of response as communicating with non-notables. I mean that both famous people and my students, for example, give me responses ranging from "Gee, thanks!" to "Don't bother me." I enclose a copy of von Foerster's reply. I suppose my disappointment makes me over-interpret his reply, but to me it means something like, "I see you are piddling in my territory. Here are three

* Four of

papers containing THE WORD. Go off to a mountain top and study for five years, and then I may be willing to talk to you."

Von Foerster enclosed three papers, one by him, one by McCulloch, and one by Pask. I enclose a page from von Foerster's. Look at that giddy leap of logic from that little diagram to "here is the origin of ethics."

But I found the paper by Pask highly worth reading. Maybe you have read it, though I do not find it cited in anything I have of yours. It is full of postulations and evidences that match exactly your theory. If you don't know the paper, let me know and I'll send you a copy. As you see from the enclosure, it was written in 1969. There are a couple of pages on which Pask was, as it turns out, overly optimistic about the influence of cybernetics on psychology. But put that aside. If you know the paper, I'd guess it helped you a lot in your own thinking. If you don't know it, I think you'd like to see it.

By chance, there is a section in Pask's paper on the topic I recently raised with you: consciousness of the error signal. Indeed, Pask proposes that consciousness arises from the discrepancy in a higher control system between the perceptual signals from two or more lower systems. I enclose the pertinent pages.

You asked about the "developmental schedule" for my book. I don't have any. The idea of writing it first came to me, maybe six or seven years ago, because I saw many writers on organizational theory saying that organizational theory was in disarray, or sketchy, or contradictory, or that there just wasn't any such thing. I thought, on the contrary, that there were a lot of ideas and evidences that fit together nicely, and I wanted to test my supposition by fitting them together in writing. So I began collecting the paper bearing the ideas I thought would add up to something coherent. I have so far about ten feet of books and three file drawers of articles. Part of the trouble is that the literature keeps accumulating, and where is one to stop and write? But I'm not really bothered by that. My writing will gradually overtake my reading, and when it does finally, I'll stop reading until I get the manuscript completed. But that will be a while from now. I may not get even a tentative outline settled for a couple of years. Who knows--I may die before the thing gets ready for a publisher. If I do, write to Dick Schmuck here. I have willed by professional papers to him.

Mary's booklet on phlogiston theory sounds fascinating. If you want to copy a page or two, I'd like to see it.

Thanks for telling me about the Gordon Conference. I'm glad to hear that you discovered helpful connections with Maturana and Piaget. (I don't know Maturana, and I know Piaget only cursorily.) And I'm delighted that you and your ideas got lots of attention. And it is nice to hear that someone is going to write an article comparing your ideas with Maturana's. In the scholarly world, when someone undertakes to compare you with someone else, it is a sign that you have "arrived." It's like comparing Bacon with Shakespeare.

I'm not surprised to hear that some people thought you were talking
* about "controlling" other people. Though I am not surprised in hindsight, I am always surprised and disappointed, especially when it happens to me, when people cling to their connotations for a word despite the new context in which they are hearing it.

You had a good time at the last CSG meeting, and now you had a good time at the Gordon Conference. Things are looking up! I'm happy for you.

** I wrote you a letter the other day that was typed messily, and I was going to have a secretary pretty it up. But some faculty and all the secretaries have been moving offices, and services are temporarily disrupted, so I am enclosing the letter in its draft form and to hell with it.

Phil

* about "controlling" other people. Though I am not surprised in hindsight,

** That will be the next two pages, treated as part of this letter.

Phyllis:

Low priority.
One copy for me.

--Phil R

Single-spaced throughout

Mr. Wm. T. Powers
1138 Whitfield Road
Northbrook IL 60062

Dear Bill:

More thoughts prompted by the paper "Purposive Behavior."

1. Why does the "subject" follow directions? I grant that traditional researchers rarely answer that question in their research reports, but I think you are more obligated to do so, since you claim subjects are following their own bents and are not slaves to the stimulus.

I think the answer is simple, but it is an answer that unfortunately is likely to raise more questions in the mind of the uninitiated readers than it answers. We do borrow internal standards from one another. We learn early in life that doing so is a convenient way to build feedback functions. So it is reasonably easy to find people at moments when they are not pressed by other disturbances who will (to satisfy whatever internal standard) adopt for a few loose minutes the internal standard of keeping the cursor in the place we request.

2. Page 5, end of the first paragraph under "The Cause-Effect Model." You are saying, "Let's suppose that traditional researchers would take the target as a visual stimulus." And, "Let's suppose they would not take the cursor as one," because the handle causes the cursor to move, and therefore the cursor cannot be a cause of the handle moving. Causation, in the traditional view, can go only one way.

But traditional researchers often do postulate and discuss circular causation--especially the path-analysis people. It is true that their loops are not like your loops. Their loops are embedded in the straight-line scheme of causation like this:

Just leave space here;
I'll draw in the diagram
myself.

* So your reasoning that the researchers' reasoning would be that the handle causing the cursor to move prevents the researchers from taking the cursor as another cause does not seem to me to be

* So your reasoning that

unassailable. It is unassailable for those researchers who eschew any loops along the way, but not for those who admit them.

I think you often, as here, set up a traditional/researcher ,orthodox, unreconstructed who hews to the simplest, unadorned, unmodified cause-effect model. But it is typical of social scientists to think of all kinds of things that might have gone wrong and stick patched-on explanations for them into the simple ~~xxxxx~~ theory.

Note (A) pasted here

3. Page 8, end of paragraph under "Experiment 1b" and second paragraph under "Results and Discussion, 1b." But the subject is still instructed to keep the cursor one cm (OK as is) above the target. If the subject follows that instruction in 1a, why not, by anybody's logic, in 1b? Seems to me the reinforcement people would argue that in experiment 1a, the subject who finds it rewarding to keep the cursor there, and therefore accepts your instructions, will also find it rewarding to keep the cursor there in 1b, and now, in 1b, will "press the lever" twice as far to get that reward.

I'm not very confident, I admit, ~~xxxx~~ of my guesses about what the reinforcement people would say. I have never been able to predict well what they might say.

As you no doubt know, ~~guess~~, I am playing devil's advocate.

I'm still on your side,

Philip J. Runkel
etc.

Note (A)

→ But I also accept your remarks in your letter of 17 June about showing what the traditional bare-bones theory would predict.

25 June 1986

Dear Bill:

How come you never told me
about Carver & Scheier?

I just now got it from
the library.

27 June 1986

Dear Bill:

Well, now I know.

I went into the Carver-Scheier book eagerly and came out wilted.

I read chapters 2, 3, 4, 8, 15, and 18. Overall, I just found it dull. I get sleepy when I plow through recitations of what this and that researcher found. Especially when what they found seems connected tenuously to the topic at hand. I thought C&S often overstrained themselves to find connections.

And in some places, I disagree with their interpretation of control theory. As one example, on pages 157-165, they give a couple of examples of what they call positive feedback. One of the examples is continuing to take actions to convince yourself that you are as unlike a group of people you don't like as possible. To me, that is simply reducing your perception of your similarity to that group to zero--which is negative feedback.

So that's that.

Phil

P.S. They do list a couple of writings of yours that you haven't mentioned to me. If you want to send them, feel free.

Ginsberg
Utica College
of Syracuse University
Burrstone Road • Utica, New York 13502

UNIVERSITY OF OREGON

This group of four pages to Bill about June 26. – editor

June 10, 1986

Pauline Ginsberg
Utica Col of Syracuse University
Behav. Studies, Burrstone Rd.
Utica, NY 13502

Dear Dr. Ginsberg:

Is this a season when you'd like to do some reading?

Ordinarily, I'd recommend Amanda Cross, Rex Stout, Evan Hunter, Arthur Lyons, Gregory McDonald, Donald Westlake, or the like. But this time I'm recommending myself.

About a year ago I read a book by Wm. T. Powers and found many of my previous ideas about doing research on human behavior turned upside down. After stewing a while, I found I had to write out my upside-down ideas. Now I need a few people to tell me whether I should continue with my new thoughts or abandon them.

The enclosure will tell you what I have written about. If it tickles your fancy, let me know and I'll send you the whole document.

If you decide you want to look at the whole thing, then naturally I'll hope you will send me some comments after a while.

Sincerely,

Philip J. Runkel

Philip J. Runkel
Professor of Education
and Psychology

OK, Prof. Runkel, I'm hooked! I would like to see the whole manuscript for 2 reasons. First, a few years ago I read something by Bill Powers, was very interested, wrote to him & was told it would take me 2 yrs. to understand control theory. I haven't had the 2 yrs., but your presentation here simplifies the concepts enough that I have the sense that it would provide the basics on a much shorter time table. Second, my research interest is in bureaucracy & the side-effects of regulation. & I teach group dynamics. Your discussion of groups & organizations promises to be right on track. Thanks! I promise comments. Pauline Ginsberg

UNIVERSITY OF OREGON

THE UNIVERSITY OF CHICAGO
Master Of Arts Program In The Social Sciences
5828 University Ave.
Chicago, IL 60637

June 10, 1986

Dr. Frederick F. Lighthall
The University of Chicago
Department of Education
5835 Kimbark Avenue
Chicago, Illinois 60637

Dear Fred:

Is this a season when you'd like to do some reading?

Ordinarily, I'd recommend Amanda Cross, Rex Stout, Evan Hunter, Arthur Lyons, Gregory McDonald, Donald Westlake, or the like. But this time I'm recommending myself.

About a year ago I read a book by Wm. T. Powers and found many of my previous ideas about doing research on human behavior turned upside down. After stewing a while, I found I had to write out my upside-down ideas. Now I need a few people to tell me whether I should continue with my new thoughts or abandon them.

The enclosure will tell you what I have written about. If it tickles your fancy, let me know and I'll send you the whole document.

If you decide you want to look at the whole thing, then naturally I'll hope you will send me some comments after a while.

Sincerely,

Phil

Philip J. Runkel
Professor of Education
and Psychology

Dear Phil – I'm pleased and on the road, so I'll just write a note on your letter. I, too, have read Powers in earlier versions & heard him speak at APA some years back. I'm looking forward to reading whatever you have made of his ideas. Fred

Okay – send the whole thing!

I certainly will, Phil. I'm still reaping the benefits of your comments on my ms, which I have almost finished re-writing in direct response to your thoroughgoing comments. Thanks again for all that. Fred

Speed Letter®

To: Dr. Phil Runkel
College of Education
University of Oregon
Eugene, Oregon 97403

From: **Karl Weick**
Dept of Management
College of Business Administration
University of Texas
Austin TX 78712

Subject

- No 9 & 10 FOLD

Message: Dear Phil: Thanks for sending the "teaser" about Wm. Powers and Inside/Outside. I have been a Powers fan for years, taught a seminar using his book at cornell, and have never had the time to do what you are doing, namely, thinking with vigor about the implications of what he proposes. I applaud your efforts. I'd like to read what you are doing, but with a jammed up summer, I can't promise how soon or in what form I could respond. If you can still spare a copy, I'd love to work on it amidst other projects. If you can't spare a copy under these conditons, I understandand think you've chosen a superb project. Date ______ Signed ______

I assume you know of Glasser"s effort to explicate Powers in Stations of the Mind. If not, you might browse that book for comparison.

Reply

Thanks very much for thinking of me. The idea that all we can ever know about the world are first order changes in intensity still boggles my mind as a starting place. Engineers who have worked through Powers say he is right on the money! I suspect you are too. Best of everything.

- No 9 FOLD

No 10 FOLD

Date 6-24 Signed Karl Weick
Karl Weick

WilsonJones
GRAYLINE FORM 44-902P 3-PART
c1983 • PRINTED IN U.S.A.

86

RECIPIENT—RETAIN WHITE COPY, RETURN PINK COPY.

Human Systems Testing, Inc.
621 Plainfield Road Suite 203
Willowbrook, Illinois 60521
(312) 654-1454

June 20, 1986

Dr. Philip J. Runkel
Division of Educational Policy & Management *
College of Education
University of Oregon
Eugene, OR 97403-1215

Dear Dr. Runkel,

I read with considerable interest and excitment your preliminary draft of "Inside and Outside." I would certainly enjoy seeing your other chapters and providing you a critique. Powers, Clark and myself did the original writing and research at the Chicago VA Research Hospital from 1957 to about 1962 which Bill fails to mention in his book. Geniuses are like that! I will be happy to supply you with a bibliography and reprints of much of that earlier work. It might save you from re-inventing the wheel.

Dick Robertson, my long-term colleague, has mentioned that he also received your materials. Perhaps you and he can enter into some collaborative agreements since he appears, like you, itching to get into print these days. For the immediate present, I have other irons in the fire which make it impossible for me to do likewise.

Your discussion of the rubber band demonstrations have finally made it clear to me what they are all about. Powers can sometimes lose even his long term admirers. Keep up the good writing! The Control System paradigm is long over due in acheiving its rightful place in psychology.

Best regards,

Robert L McFarland

Robert L. McFarland, PhD.
Director

* Bill comments on this letter in his P.S. to letter of July 1, 1986

30 June 1986

Dear Bill:

I found some parts of the enclosed article by Pribram to be of no interest to me, but I found other parts arousing my imagination, even though some of the details were beyond my comprehension. I have marked in red the passages I am glad to have read.

Another topic: Have you thought of sending an article to <u>Behavioral Science</u>?

Phil

July 1, 1986

Dear Phil,

Nice package -- full of problems for me. Jeez you guys. I'm getting behind on letters again. Yours of the 20th ought to cover it:

Not new printer, new program. Lettrix. Does one line twice, spaces 1/144 inch (on my Epson FX80), does it again. Very slow, though it does look better. This font is Gothic. Be nice or **I'll send you a whole letter in Old English**.

Looks as if you're getting a good response to I/O. McGrath has what looks like a good idea, if this isn't his way of avoiding learning control theory. I've heard this agonizing before: what do I do instead? The answer, of course, is "If there were anyone to tell you, this wouldn't be a revolution." You saw Marken's Farewell Address, didn't you? McGrath raises a real problem that has to be dealt with. It's what the computer people call the "migration path." Anyone who can write some useful material on how to get there from here, answering the question of how one goes on making a living and at least maintaining position, would be performing a vital public service. I don't know the game well enough to write it.

It's going to be a while before I can get into details on your book. Don't anticipate anything bad, however.

von Foerster is an old time guru of cybernetics, who is living on some generalizations and clever ideas he had 20 years ago. Hmm. I'd better be careful about saying things like that. He calls me his friend and I call him my friend, but that is a phenomenon very much on the surface. He is afraid I am taking cybernetics away from him. The Piaget people at the Gordon Conference went back to Switzerland full of enthusiasm for control theory, which was not the von Foerster/Maturana plan. As control theory's star rises, von Foerster's sets. Along with that of the dilettantes, scholastics, and groupies who cluster around it.

The Pask paper is indeed interesting, but as opaque as ever when it comes to the crunch. What does he mean by a "programme with fully specified goals?" And do you remain conscious of the programme when the goals are only 99% specified, but lose it when that last percent is added? If so, why? How is knowledge-of-results feedback different from vanilla feedback? How come man becomes conscious when at least two processes at once are going on? And who says that is true, and why is it true? This is the old cybernetic flim-flam. I know there are some good statements there (i.e., I'd agree with them), but there isn't any MODEL.

If there is consciousness of the error signal (sometimes), how would we know that it is the error signal we are conscious of? I mean, what should we look for in experience, to tell us we are not simply looking at a relationship (difference) between two perceptual signals? There is much to work out but I don't want to just propose a lot of facts manque. Every hypothesis implies an experimental program, doesn't it? While I'm interested in the possibilities of control theory, I'm most interested in the ones I can think of a way to test -- and those are still pretty simple-minded.

On to letter on Purposive Behavior.

To answer the question as to why subjects follow instructions, I'd have to get into the whole hierarchy, wouldn't I? I'm not claiming anything in general: I'm talking about the experiment at hand, leaving surrounding ideas alone. I stipulate that the instructions caused the subject to agree and to try to do the experiment. Given that, you still can't explain what happened in traditional terms. That nullifies the stipulation.

The answer is that the subject, and the subject alone, decided to agree to do the experiment. There is no way for the subject to take in someone else's reference signals even if he wants to. All he can take in are sound waves, which he must then perceive at many levels even to make sentences of them. Then he has to interpret the sentences in terms of an image of what is proposed to happen. Then he has to compare an imagined picture of doing that (from his own memories) with his own goals, to see if error would result. If there are no errors, he has to judge whether acting as a subject fits his concept of the principles, self-concepts, and so on involved in the relationship with the person who is asking. Mary volunteered to be a subject because now she is beginning to feel a personal stake in my work with control theory, which was not always true, and participating now looks different to her. She would have done it before if I had asked, but only because she is my wife: she would have been annoyed at the idea of spending so much time wiggling a handle for purposes she didn't understand.

We can in effect borrow organizations from other people, but doing so is a very active generative process, and what we end up with probably isn't the organization of the other person anyway. We only think it is. What we actually do is make up something we think is like what the other person means, and then claim we have taken on the other person's idea, organization, etc.. That's why I am so grateful to you: when you saw for yourself the relationships to which I was trying to point, you constructed a line of reasoning about them which, when you relayed it back to me through those whispy verbalizations, gave rise to meanings in me that seemed awfully familiar. That's how I know that you, too, have invented control theory. How I know you "understand" me.

Your next point creates something of a dilemma for me. The "traditional" model I set up follows the traditional logic I am trying to refute. The objections you raise are based on someone's claiming to follow the traditional logic, but deviating from it wherever necessary to handle contradictory facts. I'm assuming that the opponent is able to stick to a model and see how it works; you're telling me that he is too slippery to be pinned down that way. That, of course, is Marken's difficulty with the reinforcement-theory people. Theories based on nothing but words can't be wrong if you don't want them to be wrong.

So what do I do? Basically, I am trying to convey, without saying it, the phrase "Put up or shut up." I admit my interpretation could be a straw man, and invite others to substitute a fairer one. But it has to predict what happens. I HAVE sent you the revised version, haven't I? V3? If the path-analysis people have a way of analyzing this series of experiments, let them do it, and show by predicting the behavior even better that my statements are wrong. (I'm not impressed by seeing "loops" in diagrams, unless their implications are systematically worked out). *

Your third point is even harder to deal with, for similar reasons. Maybe I can get this into the paper, but I doubt it.

The subject is instructed to keep the cursor 1 cm above the target. She does so. The question is, HOW does she do it? Experiment 1 proposes one traditional explanation, and shows that a model based on that explanation does in fact work. Now, changing nothing about the instructions, the subject, or the nature of the stimuli, we make a change in the link between the subject's response and the cursor, changing the effect of the response on the cursor. At this point, before doing the experiment, and basing the prediction on the model we have just tested, what is the predicted behavior of the cursor?

The basic prediction has to be that the same response as before will occur. If some other response is predicted -- such as that the cursor will still behave as instructed -- then the original model must have been wrong. And that is what I am trying to show. If the opponent claims that the cursor will still stay one centimeter above the target, then he is rejecting the model we just constructed, because it predicts the wrong result.

OK, thanks to you I have now worked out the right answer before your very eyes, and will put it in the paper.

Best,

Bill

* See DEMO1 and DEMO2 tutorial programs for DOS, available free at www.livingcontrolsystems .com. *Living Control Systems III: The Fact of Control*; by William T. Powers (2008), ISBN 0964712180, features updated, more interactive versions of these simulations for Windows interwoven with an explanation of PCT.

P. S. Re the note from McFarland.

By all means get the reprints from him.

At last year's meeting, he spent an hour telling everyone what an ingrate WTP is, how WTP never gave any credit to McFarland, and how McFarland, being a genuine world-wise scientist, would be a much more appropriate person to lead the Control Theory Group into the Promised Land from here on out. He concluded this recommendation by saying, I quote, "Fuck off, Powers." I figured that he had a lot on his chest, and that there was some justice in his claim that I used to be pretty self-centered. So I said nothing in rebuttal, except that I wasn't planning to lead the group anyway, and if they wanted him to, that was all right with me. I didn't notice any rush.

* His history is a little skewed, however. Clark and I worked on the first parts of the theory together, one night a week, starting in 1952 or early 1953. I worked alone the rest of the time. Clark got himself and me jobs at the VA Research Hospital at the end of 1953, organizing a new Medical Physics department, making it possible for me to devote about half my working time to control theory. McFarland (contacted by Clark) offered his help in conveying the word to the world of psychology in, he says and I agree, about 1957. I departed from this group in 1960, not 1962. One reason was McFarland's continued insistence that I had already made my major contribution in the form of inventing the reference signal, that I was unlikely to make any more, and that from now on he ought to call the shots. Clark seemed to agree with him, so I said "All right, go ahead and develop control theory. Without me." That seemed the best way to settle the question of who had contributed what, and who would contribute what.

The book, I assure you, was written by me from scratch, in the years from 1960 to 1972. The work was started, of course, by myself with the help of Clark and later of McFarland, which I would have acknowledged if I could have transcended the conditions of my leaving by the time I finished the book. I left angry, shocked, and depressed, and took a long time to get over it.

You have a taste of McFarland's style. You see, it isn't that you have an important book in the works, it's that you're itching to get into print these days. If Bob didn't have other irons in the fire, he'd be publishing too, see?

Are you astonished at what he, an important co-founder of this movement, said about the rubber band demonstrations? I'm not.

3

Bill Powers

* For much more detail, see CSGnet archives, posted at pctresources.com. See Haimowoods recollections starting with [From Bill Powers (2002.11.04.0931 MST)], and especially [From Bill Powers (2002.11.05.0854 MST)]

July 8, 1986

Dear Phil,

The second Glasser CT book, I've been told, isn't too hot. I helped Glasser write his first book using control theory. He paid me a percentage, like an honest man. Then he suggested a seminar series on control theory, 30% of take to me ("You can quit your job at the Sun-Times!"). I went to organizing meeting. Found that he wanted $1500 investment, a signed agreement giving him total control, including decision as to how much if any to pay anyone else, and offered me 3%, maybe. When I said I at least wanted to check out the lecturers to make sure they understood control theory, Naomi Glasser got mad, asked why I shouldn't get certified in Reality Therapy, in that case, and that was that. I told the Glassers to stuff it and left. Haven't talked to Bill since. Naomi reminds me of Nancy Reagan.

Bill does not understand that perception isn't "out there." He thinks reorganization is something to be avoided at all costs. His therapy technique, at least in role-played simulations, consists of bullying people into being realistic. That's why it's called Reality Therapy. The reality in question is Bill's. Notice that he calls my book "highly theoretical." That was his pitch: he knows how to explain my ideas to people: I don't.

More general answer: I don't tell you what other people have written about my work when (a) I don't know about it, (b) I've forgotten about it, or (c) I don't think it's worth mentioning. As to papers, there was a long time when I couldn't afford reprints, so I just asked people to do their own Xeroxing. Remember, I've never had any handy institution behind me to pick up the tab for anything. Enclosed is a list of my publications: let me know what you don't have. If I have a copy, I'll send you one.

The letter from Karl Weick reminds me of an old Reader's Digest-type joke. Comedian is persuaded by Chinese friend to do act in Chinese theater. Reluctant, doesn't know if Chinese will get his jokes. Gets on stage, big intro in Chinese, bright lights, huddled forms. Desperately goes through routine. Dead silence. Absolutely nothing. Afterwards complains to friend. Friend's mother is there. Friend says, tell her a joke. Comedian does. Mother crinkles up, big boffo grin, rocks back and forth, doesn't make any noise.

At the Gordon Conference I found out that there's a big group of Family Systems people in the psych department at Texas Tech. They teach three courses in my theory and use it all the time. It never occurred to them that I might like to know about that. Another fellow from Utah said he thought my book should have sold 100,000 copies. Said he was very interested, sort of watching from sidelines. Another one at Univ. of CO taught a course in my theory for several years. Whoever told me didn't remember who it was. It's weird to be a famous person nobody talks to. I suppose they all assume I'm a Big Wheel somewhere, too high and mighty for a pat on the back. On the other hand, it never occurred to me to write to Albert to tell him I liked his relativity stuff.

Working five days this week because of vacation schedules.

How does a non-PhD get a grant to work at home?

Best,
Bill

Bill

I haven't forgotten that I have your photos — will return them soon. You look just like Phil Runkel — we both reacted with recognition (!) Oregon looks beautiful as always — too bad 55 mph makes it so much further away than it used to be —

Mary

18 July 86

Dear Bill:

Will you permit me to send copies of Purposive Behavior, version 3, to a dozen or so colleagues? Please reply soon, because I think I will have finished the paper to which I want to attach it in four or five days.

Phil R

July 25, 1986

Dear Phil,

Certainly, send it out. Usual comments about not citing, etc. until it's published (or rejected). Enclosed is another paper in progress that you might like to see. The last part, the man who believed in phlogiston, would go at the very end: the rest of the writing that remains to be done goes just before that. Don't know what will become of this one: I'm just writing it. Maybe it will turn into the start of The Book. It still needs a lot of work.

Since I got a new ribbon I'm using draft speed for typing.

Best

Bill

24 July 86

Dear Joe:

Here at last is my answer to your letter of 18 June, in which you commented on INSIDE AND OUTSIDE.

As usual, your comments of 18 June are perceptive, demanding, and generous. And I felt satisfied that you understood that I wrote INSIDE AND OUTSIDE primarily for myself, second for you and David and Carol for whatever useful ideas it might have on method, and third for a few other colleagues whom I might ensnare into commenting on it for me. You understood that no place in it was I saying, "This is what the methods book ought to say or look like."

The enclosed GENERALIZING is all about method. Various sections in it deal with most of the opinions you gave in your letter. I will, however, just because I never know when to stop talking, say a few short things here about some of the remarks in your letter.

You said I did not give you much help "in learning how to probe and observe . . . in new ways. . . ." Well, this is one more place at which theory and method are so intertwined that you can't really tell them apart. As we said back in 1972, what you think you learn from observations depends partly on what you have seen and partly on the assumptions with which you open your eyes. Theory tells you what method can tell you. As I thought hard about generalizing, I foundthat I was not throwing out any particular methods, techniques, strategies, settings for research, and the like. I found that I was revising my notions of what you can <u>do</u> with those paraphernalia--what you can confidently learn (generalize to) from one procedure or another. So the answer to your request for new ways is this (you will find it elaborated throughout GENERALIZING): You don't have to learn any new <u>methodological ways</u>. Go ahead, continue using (1) studies on random samples, as the Survey Research Center does and (2) studies on individual humans, as psychophysicists do. But pay careful attention to what those methods can be expected to tell you. One tells you one kind of thing and the other another. But as I just said, what you think the methods can tell you depends on your theory of what humans are like. No methodology can shake loose from that. I hope what I have written in GENERALIZING makes that more clear.

So what you must do to cut out deadwood and bring in sharper tools is adopt a new theory. Sorry. Read GENERALIZING and see if you think I've gone off the deep end. Well, yes, of course I have. That's the wrong metaphor. I mean see if you think I am seeing mirages.

I sympathize with your reluctance to get swept up in a new religion. I can't even accuse you of singing, "That old-time religion is good enough for me," because you say plainly that it is <u>not</u> good enough for you.

All I hope you will do is add one old-time religion to the list you put in your letter. Add O_1 X O_2.

As I say in the last part of GENERALIZING, I suspect, I think it very likely, I'd be surprised if it were not so, that researchers must face dilemmas. But I am not sure where they are anymore. I'd have to go through all those we brought up in our book and scrutinize each one. And I might even think of one we didn't think of then. To illustrate my present perplexity, I dealt with two of the dilemmas in GENERALIZING.

Here is a sentence that should have gone in the paragraph where I spoke about sharper tools: In case you want an example of how much a new theory can sharpen old tools, I am also enclosing a copy of a paper now inpreparation by Powers.

Now to your generous offers about THE BOOK.

First, I agree with your two comments about what's at stake beyond our intellectual compulsion to "neaten things up." First, the book might (just might) help David in his career. So he probably does not want to putter around with it for year after year. (Remember that it was ten years from the time you and I first talked about our book until it got printed?) And IT WON'T bring David admiration from rank-and-pay committees if it is TOO RADICAL. Second, you would like to influence students learning methodology, and you won't influence many of them if the book is too radical, because professors won't choose it as a text. (Not many chose our 1972 book as a text, either, though a lot of them liked to cite it in their writings.)

I don't feel pressed by either of those concerns. I have no "career" anymore. I'll never get another promotion to anything. (But it was nice that in February a batch of colleagues held a conference in my honor, and a group of ex-students named an award after me!) And long ago, even while most people, I suppose, thought I had a "career," I gave up trying to influence people through the establishment. I'm satisfied to influence a coterie.

As far as I can tell, I have two motivations to continue collaborating with you two on this venture. (1) I don't seem to be able to stop writing about things. (2) I feel some obligation to the project, because I am the one who resurrected the 1972 book and enticed you into its revision.

But motivation No. 1 does not drive me into writing what somebody else wants me to write. I can write whatever I damn well please. (I have discovered during the past couple of years that the bank balance stays balanced.)

And motivation No. 2 is not strong, because you have taken over the main initiative and drive, and whatever book is produced is not going to be crippled if I withdraw.

I'd rather not right now say that any one of your proposals is best for me.

I don't think a book of text with rebuttals would sell. I don't think many people in academia would consider it more than a passing curiosity. Even in texts that display "all sides of the question" or "include all important viewpoints," it turns out that all the side of the question lie to the southeast, and all the important viewpoints are variations on a theme. A few books, a very few, have been published that display sharp conflicts or chasms between theories or bodies of data, and they have typically fared poorly. Maybe I am overstating the case. (I often do.) But you get the idea.

Your proposal for an integration appeals to me most. Indeed, I think I have achieved a pretty good integration in GENERALIZING. But an integration, because of the time it would take, may not appeal to David at all. And my feelings won't be hurt if he says so.

Perhaps my pride in GENERALIZING is the pride of a father with a new baby no matter how ugly it looks to everybody else. But I want to wait now until you have read it before we decide where I belong with THE BOOK.

As you may have gathered from previous letters and from the first few pages of INSIDE AND OUTSIDE, I have another book going, too--on life in organizations. And that's big enough to keep me busy all the time for five years. Indeed, because I have come to see how inseparable theory and method are, I was really stirring around my thoughts for both books in both INSIDE AND OUTSIDE and GENERALIZING. The latter, however, is the closer one to THE BOOK.

I am ready to withdraw from THE BOOK at any time that my maverick attitude threatens to stand in the way of purposes you or David have that I have not.

On the other hand, I have demonstrated to myself (writing GENERALIZING) that I can still get fascinated with methodology. And there is no one I'd rather collaborate with than you.

So there I stand on a little hillock, looking about me at the territory and wondering which lay of the land might make the greener pasture. But mainly hoping that we can work out some manner of joint tillage.

Phil

P.S. I will now, or within a day or two, read the draft chapters you sent. But you will not have to wait for a later letter to get comments on what you have written. I am sure that most of what I have written in GENERALIZING is commentary on a lot of what you have written in those chapters.

24 July 86

Dear David:

I understand about remodeling a house. I've done it three times. No apology necessary.

I am glad that you are willing to learn something from individual cases. The enclosed GENERALIZING examines that method under the heading of "Specimens." *

Control theory is not an equilibrium theory. I could go on here to give you arguments that it is not. I will not do so, however, beyond saying that you can have an internal standard for variety, exploration, and activity, too. In the language of the differential calculus, you can not only control the static value of a quantity, you can also control the first, second, third, and so on derivatives.

I won't engage in much argument in this letter, because I do not want you to feel that I am trying to convert you to control theory. Indeed, a few people here have got interested in it, and I have cautioned every one of them that it is dangerous to do so. It can put you beyond the pale. In academia, it is all right to be "innovative" and "creative" as long as you do so without violating the assumptions with which your colleagues are familiar. But if you start saying things that don't fit those frames of reference, you will get from most colleagues one of two reactions:

1. Oh, he's just saying the same thing as _____ but in different words. I don't know why these young whippersnappers think they are doing something when they invent a new vocabulary.

2. He's crazy! If I believed half of what he says, I'd have to give up half of what I've always believed! I'd have to tell people not to read most of what I've written! He has no respect for his elders!

A while back I read an article by Marken. I enclose a copy. To me, his experiment put reinforcement theorists in a terrible bind. So I sent copies to a couple of my colleagues here who are devotees of reinforcement theory to see how they would cope with it. One of them gave response No. 1. He explained Marken's results, however, by appealing to the goal or purpose of the subjects--a concept I thought was verboten in reinforcement theory. The other gave response No. 2. I found his response mostly incomprehensible. His chief point, as far as I could tell, was that the experiment was no good because the subjects <u>couldn't</u> have responded that way. His summary, written in a large scrawl at the end, was "IT'S BULLSHIT!"

* Phil likely enclosed a draft for Chapter 11, Testing specimens in *Casting Nets and Testing Specimens*

Next question. Yes, it seems to me that cognitive consistency is like reducing the error signal. Maybe the way it works is that if one idea easily calls up another one from memory, we don't want the two ideas to be telling us to be doing two different things at the same time. Of course, we are clever about reducing conflict by keeping contradictory ideas in separate compartments. Idea No. 1 applies under these conditions, and idea No. 2 applies under those conditions. I was once on the governing board of a church--an Episcopal cathedral, actually. The janitor was coming to work irregularly and committing other sins. We were discussing what to do--should we fire him? I asked what would be the Christian thing to do. Another member of the board, General Ridgeway, later famous in Korea, said that this was not a matter of religion, but of sound business.

You asked how you can predict that an individual will select one alternative from a set of them. Correct me if I do you an injustice, but I don't think you do try to predict that an individual, a particular individual, will do that. I think you predict that a majority in a certain cell of your design will do that. I talk about predicting proportions in GENERALIZING.

It is easy to answer your question about falsifiability. If the prediction does not work for <u>every</u> individual, then control theory is no good. If it does not predict the individual's behavior at a very high percentage of the data-points, with something like 98 percent of the points being the goal, not just enough of them so that the probability is beyond chance, then control theory is no good. (Given the proviso, of course, that the measurements are proper--a proviso we all, I think, accept for any kind of theory.) Those requirements, obviously, are much more stringent than the requirements given in any methods text I know of.

Your two examples of The Test. I'm not sure I understand No. 1, but if I do, it's OK. As to No. 2, I have to ask for more specificity about your internal standard. Do you insist that every last one of your colleagues respect you? In that case, you would not "ignore" your colleague except temporarily while you were figuring out how to regain his or her respect. Or will a majority of them do? In that case, you might ignore that colleague and go round checking on the respect of the others to be sure you still had a majority. You not only have to look for the direction of the standard, but also the level, just as the "level" in your standing-up example is "vertical."

About "the delivery and impact of the disturbance function on the variable of interest." Here you are mixed up about who is in charge of the "variable of interest." If the event you are guessing is indeed a disturbance changes the person's behavior in the way the laws of physics would predict, if the person does nothing to oppose some feature of it, then you have guessed wrong. You have "a lack of a predicted effect," as you would say, and you would <u>not</u> conclude that you have identified an input quantity. It's the variable of interest to the subject that counts, not some variable you <u>wish</u> the subject cared about. Of course, sometimes you can persuade a subject, for a temporary period, to care about the variable <u>you</u> care about. For a temporary period, I repeat. For example, in the experiments by Powers and Marken that I enclose, they persuaded the subjects to care, for a temporary period, about the pattern of movement of a cursor on a computer screen.

There are answers, or at least comments, to your other questions in GENERALIZING.

I repeat, I am not trying to coax you to start reading up on control theory or to start designing experiments with it. You did me the honor of asking questions, and I am returning the courtesy by giving answers. Indeed, if you decline to read the articles by Powers and Marken or even GENERALIZING, you won't hurt my feelings.

As I said in my letter to Joe (enclosed), I do not know at this point how or whether I can be of use to McGrath-Brinberg-Runkel. I must wait to see whether you and he want to be bothered with me.

I hope your life is now settling down.

Phil

28 July 86

Dear Bill:

Some comments on your last two letters.

You said you "don't want just to propose a lot of facts manqué." You are very good at resisting the urge. I am not as good. I have a hard time resisting the urge to put down my speculations for others to look at. But I'm trying to resist. I didn't resist in INSIDE AND OUTSIDE, because those pages are primarily talking to myself. If I use the ideas for writing for publication, I'll no doubt use only a portion of them.

I was gleeful to see your word <u>manqué</u>. I've seen it in print now and then, but it has never become part of my writing vocabulary. Maybe I still won't use it; it's pretty esoteric. Anyway, it is fun to see a skillful wordsmith at work.

In your explanation of the reason the subject in the experiment adopts the reference signal of keeping the cursor to a certain pattern (an explanation I agree you are not obligated to give the reader), you sound as if you copied from the section on language in INSIDE AND OUTSIDE. Ha!

Whispy is spelled <u>wispy</u>.

In the organizational development business, when we say back to someone in our own words what we think the person means, we call it <u>paraphrasing</u>, which means to us more than the rhetorician means. We mean by it what you described in your paragraph about "borrowing organizations."

About "slippery theorists," I daresay you are right: leave well enough alone.

About changing the scaling between handle and cursor but observing that the subject still keeps the cursor 1 cm above the mark, I'm glad you found a way to clarify the expectation. I hope you can pare it down to somewhat shorter than it took you in your letter.

Those are comments on your letter of 1 July. Now to 8 July.

Glad to have your remarks about Glasser. You have saved me a lot of scanning through his books looking for something I ought to know about. I have talked with a few people who have undergone his workshops or have picked up his ideas from his books. I think his exercises in using verbs to remind oneself that one constructs one's own higher-order perceptions are useful. But I think he overdoes the theme that you always do what you choose to do--that other people or events do not "make" you do something. Strictly, that is true. But it overlooks inner conflict. If you pound that theme into workshoppers too hard, the result is that people feel guilty and incompetent and self-blaming when they encounter difficulties that are especially hard to cope with. A certain ability to forgive oneself is, I think, not only a useful skill, but a virtue. It enables one to forgive others more insightfully. Frequently, one needs to give one's insides some time to figure out how to cope with things. If you just patiently let it happen without reviewing every hour how disappointed you are in yourself, your insides often sort things out very nicely. I'm certainly glad I learned that before Margaret came down with the Alzheimer's, but I have to admit that trying to care for her with understanding and generosity does strain my ability to forgive myself and also my ability to give up my habit of being a teacher. But all in all, I think I am managing pretty well.

It is one of the wonders of human life that academicians talk about science as a communal venture, they practically slaver with gratitude when you show that you have read their stuff carefully and say nice things about it (I'm not exaggerating; I could tell you some pitiful stories), but yet most of them almost never do unto others as they'd like to get done to.

YOU DID NOT enclose a list of your publications. Please send.

Phil

July 30, 1986

Dear Phil,

Glad you and Hugh are getting together. He and Don Campbell had a lot to do with the fact that I published my book. Clark McPhail is a tiny person of modesty and depth. He held the group spellbound last year from behind his beard, describing how he followed the Weathermen about at the 1968 Chicago convention, taking notes on the run.

The meeting this year is now up to 25 people, with three weeks to go. Pretty soon we will have to get organized. Any suggestions?

I looked up whale, whether, wharf, whilom, and whisper: they are still spelled the way they always were. Somehow, however, the spelling of wisp has been changed. Thank you for bringing this plot to my attention before it was too late. I will cunningly pretend to go along.

Regarding the change of scaling: I think I've managed to get the idea across in a few words, by stating that to predict the same cursor behavior -- success in following the instructions -- is the same as rejecting the traditional model, which predicts that the subject will not continue to follow the instructions.

Your remarks about making workshoppers feel guilty reflect my own views of Glasser. Turning theories into rituals and slogans destroys whatever good is in them. Are verbs always better than nouns? Well... I'm not sure "verbs" captures it. Anger is a clearly recognizeable configuration of thought and feeling. It arises from wanting what we want, but what we want isn't simply "to anger." Glasser cites Langer, The Psychology of Control. Mary got it. It isn't about control theory.

I wonder about your letters to your friends Joe and David. You seem to understand what you are asking of them, but are these people like you, in temperament and position in life? How can you convince them you haven't gone over the edge? That they won't regret it in the end? That there is capital-t Truth here to be learned? That being devoted to Truth, hang the consequences, has rewards of a transcendental kind? I offer the bait, and I snap up those who rise to it, but I never tell anyone they will be better off if they follow my way. Quite the opposite. I tell them they will have hard times. And they do. Dammit, this is a revolution. Tell them the truth.

Here, I think, is the real problem. It's easy to tell people that control theory will give them 98% accuracy of prediction. But you have to SHOW THEM HOW. You are going to have to set up the experimental design, and if you can't find someone to carry it out, you will have to do it yourself. I know, you've retired. But you saw this albatross flying by and you said, "Oh, cute, here birdy birdy birdy...". Now it wants to be fed. If I were in your position, I would be gathering some young uncommitted people around me and enlisting their aid. We know the theory is right, but applying it and getting results is going to take the energy of youth, who never know when they've bitten off too much.

Your two books (I'm no longer wondering where you find the time -- now it's how can you type so fast?) are aimed at your brethren, mostly. I can see you warming up, getting the ideas straight, working out the relevance, tidying up loose ends. I'm sure you know you are still aiming at a specialized audience. Of course, that wouldn't be a bad audience to convert. What's your real aim? Maybe it's too soon to ask. I'm long past hovering over you like a mother duck anxious to see that you've learned to swim. I'd better start thinking about keeping up.

OK, finally I've enclosed a vita and a list of publications (not mentioned, seven published science-fiction stories of medium quality). A very miscellaneous life. I don't know if the list is complete. Let me know which of my publications you don't have. If I have any copies, I'll send them to you. Rick Marken probably has more than I do since he's volunteered to be my Boswell.

Mary pointed out to me this morning that as success looms, I seem to be getting more depressed. I suppose I had thought that with success I'd find a way to drop irrelevant pursuits and join in the fun, and am realizing that life is going to go on pretty much as it always has. My life made sense as long as I was assuming that control theory wouldn't have its real impact in my lifetime. Now that premise seems to be changing, and I'm not sure what to do about it. My efforts seem to be scattered; I can't settle down to a Project. Perhaps I'm reorganizing: if so, I wish the pointer would stop spinning pretty soon.

Best,

Bill

Bill

VITA - William T. Powers
as of July 30, 1986

Address: 1138 Whitfield Rd., Northbrook IL 60062
Phones: Home: 312/272-2731 Work: 312/321-2063

Wife: Mary A.
Children: Denison C. (29), Alison M. (27), Barbara K. (25)

Education: BS (Physics), Northwestern Univ. 1950.
1 yr Grad Sch. of Psych, Northwestern (1960,no deg.)

Employment:

1979-present: Systems engineer, technical services Dept., The Chicago Sun-Times. Developed microcomputer system for receiving, formatting, and typesetting stock tables (Marshall Field Award received for this project). Currently working on system for receiving newsprint manifests by wire. In spare time, as for the past 35 years, worked on developing a control-system model of human behavior (see publications list).

1974-1979: Independent consultant in control electronics; writing and research on behavioral model. Principal client, Diffraction Products, Inc., Woodstock Il. Devised control systems for laser-controlled diffraction-grating ruling engine. System currently producing the most precise gratings in the world.

1960-1973: Chief systems engineer, Department of Astronomy, Northwestern University. Designed and built low-light-level television systems for astronomy. Helped design Lindheimer Astronomical Research Center. Designed and built Corralitos Observatory, including building, telescope controls, computer controls, and semi-automated supernova search program. Designed and built automatic all-sky photometer for use on moon (Apollo 18, which never flew). Started part-time while attending graduate school in psychology.

1953-1960: Medical physicist, VA Research Hospital, Chicago, IL. Designed many devices for medical research. Principal item, a curve-tracer for plotting isodose contours in beam of radiation from Cobaslt-60 therapy machine. Also in charge of radiation safety.

1952-1953: Junior medical physicist, Argonne Cancer Research Hospital, Univ. of Chicago.

PUBLICATIONS IN PSYCHOLOGY:

1957

Powers, W.T., Clark, R.K., and McFarland, R.L.; A general feedback theory of human behavior. Counselling Center Discussion Paper III, No. 18, 1957 (University of Chicago).

Powers, W.T., Clark, R.K., and McFarland, R.L.; A general feedback theory of human behavior: a prospectus. American Psychologist 12,p.462, 1957. (Abstract of paper given before APA meeting).

1959

McFarland, R.L., Powers, W.T., and Clark, R.K.; A preliminary report on a clinical rating scale ... derived from a hierarchical feedback model. Newsletter for Cooperative Research in Psychology 1, No. 4, 1959. Baltimore VA Hospital.

1960

Powers. W.T., Clark, R.K., and McFarland, R.L. (1960). A general feedback theory of human behavior. Perceptual and Motor Skills 11, 71-88 (Part 1) and 309-323 (Part 2). 1960.

Both parts reprinted in General Systems V, 63-83, 1960.

Part reprinted in Smith, A. G., Communication and Culture, New York: Holt, Reinehart, and Winston (1966).

1971

Powers, W.T. (1971). A feedback model for behavior: application to a rat experiment. Behavioral Science 16, 558-563.

1973

------ (1973). Feedback: Beyond behaviorism. Science 179, Jan. 26, 351-356.

Baum, W., Reese, H. W., and Powers, W.T.; Feedback and Behaviorism. Exchange of letters in Science, 179,351-356, 1973.

Powers, W. T.;. Behavior: The control of perception. Chicago: Aldine (1973). Now published by Walter de Gruyter.

1974

------ Applied epistemology. In Epistemology and Education, Smock, C., and von Glasersfeld, E. (Eds). Univ. of Georgia Follow-through program. Dept of Psychology, Univ. of Georgia, Athens, GA.

------ Degrees of freedom in social interaction. in Communication and Control in Society, Krippendorf, K. (Ed)., 267-278. New York : Gordon and Breach: 1979. Paper before American Society for Cybernetics, 1974.

----- Some cybernetics and some psychology. Cybernetics Forum, 6, 4-9, Winter, 1974.

Bohannan, P., Powers, W., anbd Schoepfle, M.. Systems conflict in the learning alliance. In Theories for teaching, Stiles, L.J. (Ed). New York: Dodd, Mead (1974).

1975

Powers, W.T. The logic of social systems by A. Kuhn (book review). In Contemporary Sociology, pp. 92-94, March 1975.

1976

------ Feedback theory and performance objectives. Journal of Psycholinguistic Research 5, 285-297, 1976.

------ The cybernetic revolution in psychology. Cybernetics Forum 8, 72-76, Fall-Winter 1976. Paper given before APA in Washington, D.C., Sept., 1976.

------ Reply to Katz' analysis. Cybernetics Forum 8, 143-146, Fall-Winter 1976. Accompanied paper by Katz, S.: The theory of knowledge in Powers' model of the brain.

1978

------ Feedback principles in behavioral organization. In The Psychology of the 20th Century Vol 5: Pawlow und die Folgen, Zeier, H., (Ed.). Zurich: Kindler-Verlag (1978). In German by translator.

------ Quantitative analysis of purposive systems. Psychological Review 85, 417-435, 1979.

------ The nature of robots: Byte Magazine 4.
Part I: Defining Behavior. June, 132-144.
Part II: Simulated Control Systems. July, 134-152.
Part III: A closer look at human behavior. Aug. 94-116.
Part IV: Looking for controlled variables. Sept. 96-112

------ A cybernetic model for research in human development. in Ozer, M. (Ed.); A cybernetic approach to the assessment of children: Toward a more humane use of human beings. Boulder, CO: Westview, (1979), pp. 11-66 .

1980

------ A systems approach to consciousness. in The Psychobiology of Consciousness, Davidson, J. and Davidson, R. (Eds). 217-242. New York: Plenum (1980).

Also publications in astronomy in 1962, 1963 (2), 1964, 1966, 1970, 1972.

In computing:

Weller, W., and Powers, W.; An Editor-Assembler System for 8080/8085 - based computers. Chicago: Northern Technology Books (1978).

Exhibit 2

1

CONTROL THEORY

Philip J. Runkel
August 1986

Control theory is not about controlling other people. It is about controlling your own perceptions. That may sound odd, but read on.

The Neural Net

I begin with a short description, too short, of the human animal as I think it looks to Powers(1973). Or better, I should say that I will describe as best I can the image of the human that has formed in my mind in reaction to the disturbances Powers's book brought to me. To start out in a simple manner, I'll claim there are three key ideas:

1. Humans are purposeful. They do not act at the mercy of stimuli.

2. We act to control input, not output. We act to maintain desired levels of incoming perceptions, not to "master the environment."

3. We maintain desired input by means of feedback loops that run through the environment. The stimulus is not the beginning, and the response is not the end.

Those assertions are not "philosophy". They are easily demonstrable; I won't take space here to argue about them or to offer a lot of evidence. I will take space, however, to describe one exercise taken from Powers (1973, pp. 241-244) with which you can demonstrate those assertions to yourself and others.

The Rubber-Band Experiment

Get two rubber bands three or four inches long. Knot them end to end as shown in figure 1 on the next page. Lasso a friend.

Exhibit 3

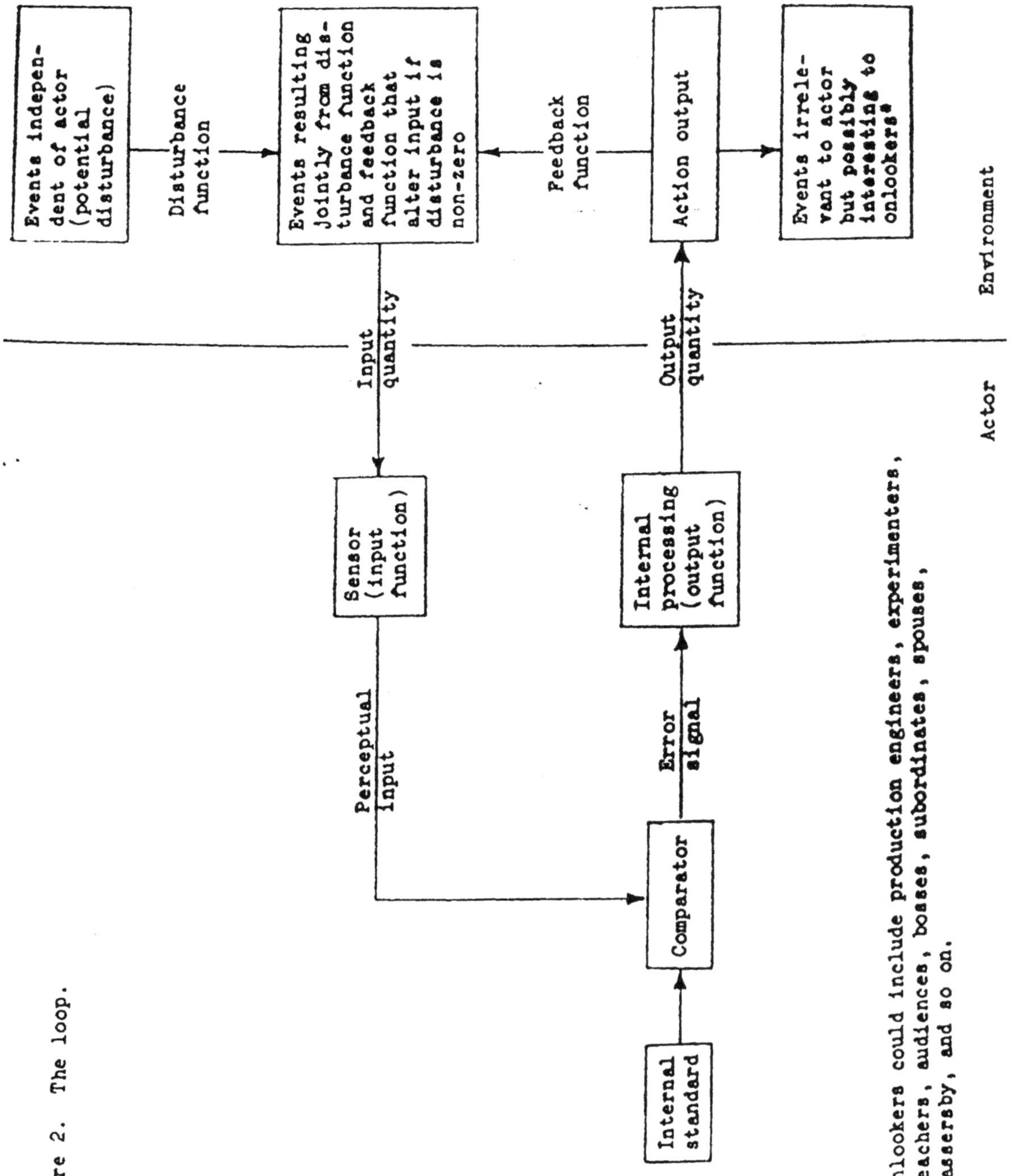

Figure 2. The loop.

* Onlookers could include production engineers, experimenters, teachers, audiences, bosses, subordinates, spouses, passersby, and so on.

Exhibit 4

EdPM 507: Mgt & OD, Runkel, fall 1986. CHARACTERISTICS OF CONTROL THEORY vs "LINEAR" SOCIAL-SCIENCES THEORIES. The following is over-simple, with subtleties omitted, but what do you expect on two pages?

	Control theory	Linear theory
Action consists of:	continuous maintenance of desired perceptual input, a continuous interaction between external disturbance and internal standard. See figure 2 in the Kinko paper.	a series of distinct episodes like S-O-R, each set off by an external event (stimulus). See figures 3 and 4 in the Kinko paper.
The person is motivated by:	a discrepancy between an internal standard for a perceptual input and the incoming actual perception. The person acts to reduce the discrepancy.	a change in an external variable. The person acts to change some other variable. The variable X is what causes Y, not the person.
The researcher wants to discover:	the level of a perceptual input variable that the person wants to hold constant (at zero deviation from the internal standard). Researcher hunts for the perceptual input that has <u>zero</u> correlation (does not vary) with external variables.	an external variable a change in which will cause a change in a specified (pre-chosen) output variable. Researcher hunts for input variable having <u>maximum</u> correlation with output variable.
Researcher expects to be able to predict:	continuous action (though no particular action) to maintain constant level of input. Particular actions will depend on what is available in the environment to serve the person's purposes--a handgun, for example, if you want to stop a person from disturbing your input.	specific action on a particular environmental object or class of objects--for example, hostile acts toward other people, purchases of certain products, memorization of certain strings of words or their equivalent, or change in favorability toward certain things or ideas.

characteristics 2

	Control theory	Linear theory
Researcher finds little or no interest in:	the varieties of acts chosen to oppose disturbances. Researcher wants first to know what is held <u>unchanging</u>.	actions predictable a hundred percent of the time, or patterns that stay the same, such as opening the drugstore for business every morning. Researcher wants first to know what <u>changes</u> with what.
Practical advice:	Find the kinds of events (those affecting the person's desired input levels) the person will act <u>against</u>. Either remove those events or provide environmental resources that will make it easy for the person to counteract the events.	Find the environmental variables changes in which will push the person to the particular acts you want the person to exhibit.
Social psychology:	Other people become both disturbances and resources. Find ways that actions of others can become aids to reaching your own goals, not obstacles. See p. 36 ff. in the Kinko paper.	Other people are stimuli; their actions change the variables that will produce changes in other variables. Find ways you can act that will push people toward the acts you want them to take.

See also items 1, 2, 3, on page 1 of the Kinko paper.

5 August 86

Dear Bill:

Your letters are always a high-point of the week.

Wisp, as in will-o-the-wisp, never was spelled with an h. Not in my lifetime. But I'm glad you came across some other nice words.

No, verbs are not always better than nouns. But to use your example, I think that when Glasser wants you to say "I angered myself about that" instead of saying "You made me angry," he wants to remind you that you "chose" to act that way. I think for many people, at least those who told me about it, that the exercise is a good way to help people relinquish some of their habit, if they have the habit, of feeling as pawns in the events around them, to help people ask themselves, what choices do I have, or might have next time? No exercise works the way you hope with everybody, of course.

Yeah, I got sucked into Langer, too. The advertising certainly made me think it was about control theory. But luckily I didn't lose any money on it; Sage thinks I am one of their authors (whether I will turn out to be, with McGrath and Brinberg, is still undecided), so they sent me a free copy.

I thought I explained very clearly to David that I warn people about the dangers of control theory, and I thought I painted clearly how his colleagues might treat him. And I have just now a letter from McGrath, written after he had had a face-to-face conference with Brinberg, and neither of them is going to leave the beaten track. I instigated this book-in-progress with Joe several years ago, as a revision of our 1972 book, and then after I read the Brinberg and McGrath book I was greatly taken with it, and proposed that we change our original idea and build the new one around Brinberg and McGrath, and Joe thought that was a marvellous idea, and that is what is going to happen. I may turn out to be merely a reader-for-rhetoric rather than an author, and I wouldn't mind doing that. I like to do editing. And it wouldn't be hard work; Joe is a good writer. Joe can see quite clearly how hard it would be for me now to write new prose about the old methodology.

Yes, it would be nice to think up some experiments. But very few people who come to a department of Educational Administration have that turn of mind. And I am just now full of writing that I have to let spill onto paper. And I wouldn't go over to the Psych Dept and try to recruit students. I can't remember any new idea that has come out of that department during the 22 years I have been here. By the way, one of my colleagues most enthusiastic about reading INSIDE AND OUTSIDE and GENERALIZING is a fellow who was once my colleague in the Psych Dept here. He got frustrated and disgusted with the department and is now a vice president at Weyerhaeuser.

I am thinking of making an amalgam of GENERALIZING and the methodological parts of INSIDE AND OUTSIDE and making a book for publication. But I don't want to start until I get more replies from people who asked for copies.

Yes, sure, my two recent documents were written primarily with social scientists in mind, though somewhat consultants, too. You ask what is my "real aim"? I don't know. Are you hinting I should write for the supermarket shelves? I wouldn't know how to do that.

Thanks for the curriculum vitae. I am impressed with your accomplishments. For some years, I was a draftsman working with engineers of various stripes, so I know a little about how to appreciate engineering types.

I am supposing that your more recent writings supersede some of the earlier. But I would like to have copies of:

Bohannan, Powers, and Schoeple in Theories for teaching.

Powers. Review of A. Kuhn.

Powers. A systems approach to consciousness. In The psychobiology of consciousness.

If you don't have a copy of something but you think Marken does, would you please send the extra copy of this page to him and relay my request? Thanks. And if you'd like, I'll be glad to send you back some extra copies of what you send me.

I did not see in your list the 1980 paper "Control-theory Psychology and Social Organizations, etc."

By "General Systems V" do you mean the yearbook of the society? I belonged to that for a while, but later I gave away all the yearbooks I had collected. I also once had a copy of Smith's Communication and Culture. And in neither book did I read your chapter! Thus do diamonds slip through our fingers.

I am sorry you feel depressed. I wish you well. It seems to me obvious that you are reorganizing. You are having your priorities nibbled at. But all I can say is that I wish you well.

Thanks very much for sending me the two recent articles. I have read them with delight, but I'll postpone comment. I'm busy just now with Brinberg and McGrath, and soon I'll have to start planning the two courses I'll teach in the fall.

Phil

6 August 86

Dear Bill:

I said I probably wouldn't reply to your two articles for a while. That shows you how poorly I predict my own behavior.

Both pieces are thoroughly delightful—to me, anyway.

The first paragraph of the criticism of Skinner brings to my mind once more the question of who are the psychologists' psychologists? Skinner's remark, like similar remarks by numerous psychologists, assumes that some people must set up the stimuli to control other people. Who will set up the reinforcements for the people who set up the reinforcements for the rest of us?

Maybe I said the following in some other letter to you. I'll try to remember that I have now said it and not repeat it again. In getting acquainted with a new academician, it is customary to ask, usually in just these words, "What are you interested in?" When a person is being considered for a job, that question means, "Tell us what you would work at if we were to hire you." It seems to me that a reinforcement theprist, and some others, too, ought to answer, "I don't believe in 'interest.' I will do here whatever I get reinforced to do." But I never hear reinforcement theorists talk like that. I rarely hear any other brand of psychologist applying their own theories to themselves. Maybe the clinicians come closest to it. But it is very rare from any sort.

On page 3, paragraph beginning "But this is not...," I need some help. The reinforcement people, when they cannot count on an obvious "reinforcement" like food to a hungry rat, say that you have to fool around a little while and find out what <u>will</u> be reinforcing to the subject. And you say that you have too fool around a while (The Test) and find cur what input the subject <u>is</u> controlling. What's the difference? I could probably think it through for myself, but I'm tired today.

Page 4, line 5: Please do not write "This is because." It raises my hackles.

Turning to "How Behavior Works," I was delighted to have the sketch of history; I didn't know any of it. Well, I knew the names of Weiner and Shannon and Weaver, and I once knew the formula for entropy, but that's all.

I take it that the section on H.S. Black is saying the same thing as pages 6065 in your book.

Page 5, line 3: By "undocumented," I think you mean without citations to publications giving evidence for what you say. But some readers will take "undocumented" to mean without evidence--that no one has produced any evidence.

Page 5, 9th line from bottom: Add "here" to "There are no shades of grey." Or something so that too-attentive readers like me will not stop to wonder whether you have never seen shades of grey.

Page 5, 2nd line from bottom: Save "observations" to mean sensing; use some other word to mean statements or assertions, so that readers don't get confused between the two meanings.

Your sections on quantitative analysis gave me more information (and ammunition) I am glad to have. And you lead up nicely to showing the difference between the output to the muscles (and the feedback function) and the perceptual input.

Page]3, line 2: occurrence.

Page 13, same paragraph: I suppose you have seen the cartoon showing one rat saying to another: "I've got that guy conditioned so that every time I press the lever, he gives me some food." When I first saw that cartoon, it seemed to me a perfectly reasonable interpretation, and not very funny. But the fact that the author of the cartoon thought it was funny, and presumably many readers of it, points up what you say about commitment to phlogiston.

Send me some more on phlogiston. Or tell me where to find the book. Or something. If you give me the full reference for the book and our library doesn't have it, I might be able to get it through interlibrary loan.

Thanks, thanks, thanks.

Phil

REASON IS BECAUSE: *Because* means *for the reason that;* so if you use both *reason* and *because,* you are being redundant. Just drop out *the reason . . . is,* and your sentence will be all right:

> [The reason] they do it [is] because there is no one else to turn to.

Or you can drop *because:*

> The reason they do it is that there is no one else to turn to.

REASON WHY: Perhaps you find you have written: "We wanted to know the reason why the subjects acted as they did." *Reason why* is redundant. Instead of that, write *to know the reason the subjects acted* or *to know the reason that the subjects acted* or *to know why the subjects acted.*

THIS IS BECAUSE: Two troubles arise with *this is because.* One is the likely ambiguity about the antecedent of *this* when the pronoun is meant to refer to the action in a previous sentence, not merely to a noun near its end (see THIS, THAT). The other is the misleading *is.* Almost never does the author who writes *this is because* mean to give a reason for something being or existing (*is*). Almost always, the author wants to say that what happened in the previous sentence did so because so-and-so, or what was asserted in the previous sentence is true because so-and-so, or the like. Instead of "This is because the experimental design is inadequate," one can write:

> This is so because . . .
> That is the case because . . .
> That happens because . . .
> The reason is that . . .
> The reason is the inadequacy of the experimental design.

Here is an actual example:

> Path models rarely include experimental variables, even though they could be exogenous variables in a model. This is because of the problems that polytomies present in analysis.

And a recasting:

> Because of the problems that polytomies pose in analysis, path models rarely include . . .

---

from Philip J. and Margaret Runkel. A guide to usage for writers and students in the social sciences. Totowa, NJ: Rowman and Allanheld (Helix Books), 1984.

Carol Slater,
social psychologist
@ Alma College
in Alma, Michigan

August 9 1986

Dear Phil,

I have just finished GENERALIZING and feel like hugging you, you * are so right about what really matters. But we are too far apart for that sort of critical response, so how about three heartfelt cheers, instead? There just has to be a difference between even the truest of empirical generalizations about higgledy-piggledy collections (even in those cases where we somehow manage to look at every case extant) and the kind of generalization that we take to be the goal of science. We were brought up, of course, not to think about the difference. The story we were told was that the relationship between observation statements and the generalizations they supported was one which could, with cleverness and luck, be captured in the vocabulary of first order logic. In its simplest form, this was something along the lines of

> An observation statement O is evidence for an hypothesis H if and only if O is a logical consequence of H (and some innocent auxiliary hypotheses).

Exactly how this logical relationship was to be characterized became a set of puzzles which kept graduate students in philosophy off the streets at night: some very appealing and strongly intuitive thoughts on the subject turned out to have some distinctly unpalatable logical consequences--the hypothesis that all ravens are black turned out, for example, to be confirmed by sighting a yellow pencil. Since only professional philosophers could take the Raven Paradoxes seriously (I have never been able to get anyone in the science division to sit still through an explanation of how we get into this trouble), the rest of us went blithely on, buying into a doctrine which our elders and betters assured us was the Last Word on Science. Word had got around, of course, that induction was problematic, but most of us thought that switching from universal generalizations to statements of probability would take care of the matter. As you note, it does not. Not only don't we know for sure that the next batch will be like the batch we just netted--it turns out that our predictions are not constrained by the content of our net. Logic cannot tell us what similarity to bet on. The classic formulation of this problem is due to Nelson Goodman, who called it "the new riddle of induction". (If you've already heard this a dozen times, skip rapidly down the page!) Goodman conjures up for us

* Phil likely enclosed a draft for Chapter 11, Testing specimens in *Casting Nets and Testing Specimens.*

a scenario: every emerald we have seen up to this time, t, has been green. Does this constrain us to the generalization that the next emerald we encounter will also be green? Heavens no. Consider that we could just as well describe our observations by saying that every emerald we have examined has been grue, where grue means 'green at any time prior to t and blue thereafter'. If every emerald has been grue (and it has) then why aren't we predicting that the next one that comes along (after t) will be blue? We don't, of course, any more than we think that a yellow pencil counts as evidence for 'All ravens are black.' The point is not that we should switch to grue when we predict but just that from the point of view of the things we have been allowed to consider, grue is on all fours with green. (Strenuous attempts to show that there is something formally different about the two predicates have not been successful.) News about grue made no more impression on non-professionals than did the Raven Paradoxes. Cumulatively, their effect on the professionals has been to bring about the decline and fall of the logical empiricist program. (There is a lovely account of this, told in heartbreaking detail, in Harold I. Brown, Theory, Perception and Commitment; the new riddle of induction is presented in Goodman's Fact, Fiction and Forecast.) To the extent that the official philosophy of science of psychology remains some version of logical empiricism, we are going to be myopic about different sorts of generalizations and numb and vague, as the saying goes, about what we are up to. (Long before he produced the new riddle, Goodman was pointing out that we intuitively sense the difference between a true generalization--'All the people in this room speak Russian'--and a lawlike statement of precisely the same logical form--'All copper wires conduct electricity'. One simple way to display the difference is to haul in a counterfactual claim: we are willing to agree that if this wire (which is not copper) were copper, it too would conduct electricity, but we are not comfortable with the claim that if this person (who is not in the room) were in the room, s/he too would speak Russian. (And attempts to locate a formal, logical difference between these two have also not been successful.) Somehow, the generalizations which we feel entitled to make do not seem to depend exclusively on their logical relationships with observations (or, more properly, observation statements).

You suggest that the place to look for the difference is the status of what we are observing: if we have a specimen of a species, we will be heading toward lawlike statements but if we have a sample of a collection we have no reason to think we are moving in this direction. I think this is lovely, correct, and should be embroidered on everybody's heart. In the circles in which I move (slowly), the going terminology is 'natural kinds' and 'projectible predicates'. Beyond recognizing the distinction you have made, not a lot of progress seems to have been made. There is, in fact, lively debate going on about what endorsement of natural kinds amounts to, whether talking about them commits us to (gasp!) Aristotelean essences and other odd entities, whether theory terms in science are natural kind terms and, as such, follow different rules for establishing reference than do other kinds of terms, and so on. Quine has a very nice essay called "Natural Kinds," in which he attempts to deal with the question of what sorts of similarities matter in science. (His answer is a rather dazzling Pythagoreanism.) Stephen Schwartz has edited a collection called <u>Naming, Necessity and Natural Kinds</u>, which ranges from material accessible to any interested reader to some extremely technical stuff which only a philosopher could tolerate. The introduction, in particular, provides a very helpful overview of some crucial issues.

As psychologists, we have, I think, a particular need to think seriously about the issues you raise--more so than, say, the physicists. For one thing, it is clear that if anything deserves to be called a natural kind it is the theoretical entity featured in a mature science--such sciences give us reason to haul in the term in the first place. It is not that-all clear that <u>we</u> have any natural kinds. Your example of the developmental psychologist observing the child learning to walk or acquiring language is right on: we can imagine natural kinds and projectible predicates (green rather than grue) in developmental psychology just as we can in physiological psychology. But <u>are</u> there natural kinds in the subject area we call personality? The failure of the grand testing programs suggests that if there are any, we're not glomming onto them. Closer to home, you raise the wonderful question of whether there are species of organizations. Wow. I would like to think that there might be, that 'bureaucracy' might turn out to be a natural kind term, but I share your hesitancy. (Is 'loopiness' a projectible predicate, do you suppose?) For a long time, the presumption was that the theoretical terms of any science whatsoever would be logically equivalent to terms that feature in the laws of physics. To the extent that psychological theory terms prove resistant to such reduction (and they bid fair to prove so, most especially in our area), the question of what it means to have different kinds of kinds (so to speak) hangs heavy over our heads. (Once upon a time, it seemed that stubborn refusal to reduce to physics was tantamount to inviting in angels and demons. Nowadays theorists seem more relaxed but the threat of the incorrigibly mental still spooks some folk.) (Jerry Fodor's "Disunity of Science," which is in his Representations and also in Ned Block's Philosophy of Psychology, vol. 1, is a crunchy but persuasive discussion of the improbability of psychological terms yielding to translation into those of physics.)

When Kurt Lewin said that there was nothing as useful as a good theory, he was, I think, betting that the natural kinds of social psychology would line up (or be alignable, at least) with the categories of the lived-in world. If there is no way to get from 'classroom' (which is very unlikely to be a natural kind term) to anything which is such a term, then our theory has no purchase on the world about which we care. The relationship between 'pure' and 'applied' seems to me to be just that between the category system chosen to maximize explanatory power and prediction, on the one hand, and the category system which functions to define and regulate our transactions with each other. It certainly should not become a big status trip. I could not agree more that science is continuous with what we do in our everyday life: indeed, when I teach Personality, I begin with three weeks on our implicit, everyday theory of human action, and then go on to show how professional theories arose from that matrix. The notion that there is some logically (or methodologically) specifiable distinction between science and less lovely sorts of inquiry belongs, I think, to the program which crashed. It is worth pointing out to beginners, nevertheless, that as a cognitive technology, the conventions of science do represent rather a considerable departure from what comes easiest to us in the way of fixation of belief: we do not <u>naturally</u> go looking for disconfirmation, worry about probabilities or bias in our sample of observations, or even constrain our claims by careful counting. (Oh, dear. More alliteration.)

I am impressed, delighted and points west by your ability to see what we were taught not to look at. I also suspect you may have a more sympathetic audience out there than you might imagine. If you think it would be helpful, you could, without distorting your views, make natural and graceful connections with some pretty central issues in philosophy of science and language. Indeed, doing what you are doing, I don't see how you can avoid making such connections. Metamethodology is epistemology for practitioners, no?

I'm not quite sure what THE BOOK is right now. If it is McGrath and Brindberg, I really do not want to go through it. I want to read what you write because I have the deepest respect for your insight and experience: you are the most elegant and serious of us and I care what you think. I really don't care that much what Joe McGrath thinks. It is more trouble than it is worth to try to straighten out the sort of thing he seems apt to say. (I'm sorry if that sounds snobbish. I was like that even before I began doing philosophy.) If something nice comes your way, I'd love to see it. Otherwise, let's leave things as they are.

I'm sorry you are so unhopeful about McKeachie's project. At most, Alma will be a pretest site for developing measures. Perhaps the process will encourage people to think more about what they are doing and what they might be able to do, even if the research itself is not particulary edifying. (I never got to comment on your fascinating observations about possibility as an aspect of what we study. They are provocative and elusive--I'd love to hear more.) We do very little in the way of systematic consideration of what we're up to, and perhaps having an occasion to talk about it will at least bring some new thoughts to the surface. Since we're not being X'd (hardly even O'd), I doubt that any harm can come of it.

It must be odd to get very different readings of your books. (That is, at any rate, what I gathered must be going on from your replies to your readers.) Not intellectually surprising--you of all people would anticipate just such constructive and interpretive variation--but odd, nonetheless. Rather the way I feel when I see what a roomful of students has made of a reasonably complex lecture. I hope that the work you have done has brought you further than you were and that there will be more to come. It has been nice to have summer time to read and think about your ideas. But then, it is always nice to spend time with you.

love,

Carol

STEPHEN F. AUSTIN STATE UNIVERSITY
NACOGDOCHES, TEXAS 75962
11 August 1986

DEPARTMENT OF PSYCHOLOGY

Professor Philip J. Runkel
Division of Educational Policy and Management
College of Education
University of Oregon
Eugene, Oregon 97403-1215

Dear Dr. Runkel:

The selections you recommended for summer reading, back in June, looked attractive. I was especially interested in the final item on the list: the sampler from your manuscript. I want to take you up on your offer to send the whole thing. In exchange, I will send the comments you requested.

My first reactions are that you have a refreshing approach to the subject of control theory and that you obviously understand mamny of the basic principles. I could make some rather minor remarks about details, but I will save them until after I read trhe entire manuscript. Some of the details you add to the basic diagram of a control loop resemble the additions I make in handouts for my psychology students.

Do you plan to attend the meting of the control theory group in Wisconsin, August 20-24? If you do, perhaps we could exchange the manuscript and some remarks, for I will be there.

Sincerely,

Tom Bourbon

Tom Bourbon, Ph. D.
Professor of Psychology
(409)569-4402

Aug. 18, 1986

Dear Phil,

Two days to the Meeting: no time to finish anything, too late to get anything started, so I'm feeling very much on hold. I'm giving a talk and showing a few computer programs; otherwise I just plan to attend. The talk is enclosed. It should last eight minutes. I'm Bored, let's have some Action.

Thanks for Tom's letter (he's an especially good friend). Carol's letter, strangely enough, was easier to read the second time through. I suppose it takes time for unfamiliar trains of thought to establish tracks in my head. I have a comment for her if you'd like to pass it on with or without embellishments.

How do you tell who speaks Russian? By listening. Discovering that someone speaks Russian implies nothing about where that person is. How do you tell if a metal is copper? By seeing if it is malleable, ductile, and conducts electricity. If it does not satisfy all three criteria it is not copper.

Knowing that a person speaks Russian implies nothing about that person's position, because the criterion for being a Russian-speaker is not position, but utterances. Knowing that a metal is copper, however, does imply that it will conduct electricity, because one of the criteria for a metal's being copper is that it conduct electricity: it cannot be copper and not conduct electricity.

Generalization in the sense Carol uses is an attempt to make these implications work backward. The problem is in a word that is implicit but is left out: "necessarily." If a person not in this room were in this room, that person would not necessarily speak Russian. But if this wire, which is not a conductor, were made of copper, it would necessarily conduct electricity. When we make that word explicit, the two original proposition are of the same logical form but have different initial truth-values:

All the people in this room necessarily speak Russian.

All copper wires necessarily conduct electricity.

The truth value of the first proposition is FALSE under any normal premises. The truth-value of the second is by definition TRUE.

When you extract the principle from that, which I am too lazy to do right now, I believe it will constitute the logical formal difference that Carol mentioned has not been successfully established. Next.

Regards Bill

2 Sept 86

Dear Bill:

I was glad to hear via the telephone that you were happy with the meeting of the CSG. I am glad you are getting lots of admiration. Couldn't happen to a more deserving person.

Please don't practice your Olde English typeface on me.

Carol Slater's grue and people in the room speaking Russian failed to tie me in the knots of paradox. But I am still embarrassed by the cawing of the black ravens. Can you find fault with the black ravens?

I enjoyed your review of A, Kuhn. I'm glad to have it. I found a lot in Kuhn that I liked. Indeed, for a while I was intending to use his detector, selector, and effector things as a simplification suitable for use as codes in my literature-retrieval scheme. But when I actually coded some bits of literature, the selector pocket got very full in a hurry, and the other two sat there starving. Then I read Powers. Actually, I've read so far only about half of Kuhn. I'll at least scan the rest of it one of these days.

But I am grateful to you for your remarks about what he says on his page 31. I would not have been able to pick out the lack of an "active system" there. I went back and read the page again, and it was obvious to me why I would have missed it (if I had had control theory in my head when I first read that page). The page lies under the heading "Propositions about Acting Systems." And his description of equilibrium in a water tank there is OK. At such a point, my thought, typically, is "All right, no doubt he'll tell me in a little while why he told me about the water tank." He doesn't say on page 31: "And this is the way humans function." Indeed, the following chapter is entitled "Human System." Maybe he would tell me there where the water tank fitted in. If I had had control theory in my head then, would I have been able to leaf backwards after a while and trace what I did not like on later pages to the omission of active control on page 31? No one will ever know, will one?

I am enclosing some copies of the review. Anything else you'd like copied?

Phil

RICHARD J. ROBERTSON, Ph. D. & Assoc. Ltd.
Clinical Psychology

(Hyde Park)
5712 Harper Ave.
Chicago, Il. 60637
(312) 643 8686

(Loop)
30 N. Michigan, Suite 429
Chicago, Il. 60602
(312) 782 5989

August 19, 1986

Prof Phil Runkel
DEPM, College of Education
University of Oregon
Eugene OR 97403

Dear Phil (If I may presume),

I have been greatly enjoying reading your book and wanted to send off this quick note before I leave for the CTP conference in dear old Haimowoods. No, your seeds haven't fallen on dead soil, but I have been frantically publishing my book (on Kinko's captive audience "professor publishing" plan, not a commercial publisher, dammit,) to use with my 100 level course right after Labor Day.

I have especially liked your discussion in the section on "partitioning variance" and would love to have it available next time I teach experimental psych. In fact, I had been thinking I wouldn't teach it anymore, but with your material (with some of Bill's and Rick Marken's) it could be a whole new ball game.

I am sorry you can't make it to the conf. I would enjoy meeting you and hashing out ideas with you. I was very sorry to hear about your wife. I wish there might be some hopes, but it would be polyanna to say anything very cheerful about Alzheimer's.

If I can get to the post office before I have to dash I'll send a couple of the things I threatened to drop on you. Otherwise, I will do it the first thing after labor day.

Best Wishes,

Dick Robertson

UNIVERSITY OF SOUTH CAROLINA
COLUMBIA, S. C. 29208

DEPARTMENT OF SOCIOLOGY

20 August 1986

Professor Philip J. Runkel
Division of Education Policy and Management
College of Education
University of Oregon
Eugene, OR 97403-1215

Dear Dr. Runkel:

I did not receive your letter and manuscript until I returned from a trip in July but waited until my "desk was cleared" to read it several days ago. I wish now that I had read it sooner.

I have also found that using the rubber-band experiements an excellent introduction to Powers. I use my "experiments" in classes. Your writing describes very accurately my experiences with my students. It was a pleasure to read your work.

I have read several attempts to present Powers but yours is the best. It is clear and very easy to understand. I can find nothing with which I disagree (which is unusual for me). It was a pleasure to read.

I would very much like to receive the entire manuscript. I promise to give you all of the comments - good and bad - that I can generate.

I suspect that you know about the Control Systems Group and its newsletter. The group would be very interested in your manuscript.

Thanks again for sending me the manuscript and I look forward with great anticipation receiving the rest of it.

Best regards,

Charles W. Tucker

UNIVERSITY AT BUFFALO
STATE UNIVERSITY OF NEW YORK

Office of the Dean
Faculty of Educational Studies
367 Baldy Hall
Buffalo, New York 14260
(716) 636-2491

August 15, 1986

Dr. Philip J. Runkel
Division of Educational Policy & Management
College of Education
University of Oregon
Eugene, Oregon 97403-1215

Dear Dr. Runkel:

I've just had time to read your "introduction." Please send me the rest of the manuscript. It is extremely well-written and I know I will learn a lot. I don't know when I will get to the larger work, but I will eventually.

Just two comments on the introduction. First, I tried to get at some features of reorganization in my book, Dilemma of Enquiry and Learning. Unfortunately, I wasn't comfortable enough with Powers to call it reorganization. So, instead, it's cast in a different tradition-- more Piagetian. However, I think I am talking about reogranization. Second, as I have worked with Powers, I have always found what we usually call "perception" the hardest to explain to other people. I urge them to abandon the notion that perception is passive, that it is intimately connected as input functions to loops, and is essentially active. Nevertheless, I don't know how to explain very well the situation of simply opening one's eyes and seeing one's bedroom. Any suggestions?

Finally, I quickly read through your comments on my testing papers. I need to digest them a bit more, but I will respond in due course--probably along with the response to the manuscript.

Thanks again for brightening my routine administrative day. Yes, I know Tom Hastings very well. He is a special friend and extremely close to my wife, Carol Hodges, formerly Carol Wardrop. Hope to see you next year.

Sincerely,

Hugh G. Petrie

Hugh G. Petrie
Dean

HGP:er

Sept. 6, 1986

Dear Phil,

OK, no more Olde English. I am full up with admiration, and am back at work in the real world. Next project, a chapter for a book that Plenum is publishing. I'll send the Mss when the experiment has been done. The nice thing about control theory experiments, at least at my elementary level, is that you write them up first, then do them. They always work. The other "V3" paper languishes, for some reason. Something is telling me it isn't what I want to publish, in spite of all the work, mine and others', that went into it. I don't know what will happen to it. I committed to the Plenum project early this year, so have to do it. Not that I mind: it's short and simple.

You keep asking questions that elicit my answering-response. Here's an assortment.

I think the key to the Raven problem is to be found in the reluctance of "anyone in the science division" to sit still for an explanation. Despite what they think on the soft end of the campus, the hard sciences just don't use generalizations, induction, and so on. They make models: if the underlying reality contained such and such entities with such and such properties (very precisely stated), then we would observe so and so, which is precisely what we do observe (if not, change the model until this statement is true: a control process). Mercury does not have a density of 13 grams per cubic centimeter because it has always had that density before: it has that density because of the way mercury molecules, which have a known (although imaginary) size and weight, pack together in the liquid form. Given the model, mercury couldn't have any other density. Generalizations in hard science don't apply "most of the time." They apply ALL of the time, or they aren't accepted as generalizations. In the soft sciences, "generalization" is a pun. Generally, the attraction between two pieces of matter is proportional to the product of the masses and inversely proportional to the square of their separation. That means <u>generally</u>: everywhere in the universe, all the time. "Generally," ravens are black, except for the albinos, the gray ones, and others that might well show up, for all we know. The two words have the same sound and spelling, but different meanings.

Another tack: The Raven Paradox arises from the branches of science that use statistics and abstract reasoning to find out about nature. An implication has the form, "it is not the case that H is true and O is false," which admits of only one false condition out of the four possible. In logic, if something is not false it must be true: no other value of the variables is allowed. "H implies O" is false if and only if H is true and O is false. If O is not false (a non-black raven is not seen), then (if I grasp the nature of this supposed paradox) the implication must be true, no matter what is observed instead of a raven. Thus observing anything but a non-black raven -- say, a yellow pencil -- leaves the implication "true," confirming the hypothesis.

There are two problems I can see. The first is a confusion between the identity of a variable and the value or state of the variable. The identity of the variable, above, is "raven." The proposition concerns the state of a raven, which is implied to be either black or non-black. By observing the raven, we can determine its state: we will observe, in the idealized world of logic, that it is either black or not black. In that case, the hypothesis is either confirmed or disconfirmed. But suppose we observe a yellow pencil. Now the variable is "pencil," and its state is "yellow" or, I presume, "not yellow." But "pencil" is not the variable involved in the original hypothesis, so its state is irrelevant. The hypothesis concerns ravens, not pencils. So if we have not observed a raven, we can't determine its state, can we? In that case we can't finish computing the value of the implication, H -> O. We have to leave the value of O open, by just writing the name (identity) of the variable, without giving it a value. This is perfectly legitimate in logic. What paradox?

The second problem is more interesting to me. Physical observations are stated in terms of the real number scale, not the binary scale. The physicist does not predict that there will be either some temperature or no temperature: he predicts that the temperature will be 29 degrees centigrade. If it's actually 28.8, the physicist doesn't say, shucks, it wasn't 29. He says "not bad, less than one percent error." The analogue world versus the digital world. Ravens do not come in two flavors: their color lies on a continuum.

You will notice that the hard sciences have done a lot better with their subject-matter than the soft ones. The soft scientists attribute the difference to the excessive difficulty in working with living systems. I think the problem is their method. When your only model is "If something happens n times, it is likely to happen n+1 times," you don't have much to work with. "Similarity" is not a property of nature: it is an observer's opinion, based mainly on the habit of categorizing and aided by the fact that perception has limits of discrimination. If you look closely enough at any two things, similarities disappear and variables become continuous. The raven paradox, if there really is one, is caused by categorizing. When you draw an arbitrary dividing line through nature, you get categories. You give the categories names. At the next level, logic, something is either "name" or "not name." But the world being observed is not limited in that way. Categories have no force in nature. Our acts of categorization are not what makes real things be related as they are. Since categories are arbitrary, propositions relating names of categories can be made true or false just by shifting the boundaries of the categories a little. How may white feathers can a raven have and still be called a "black" raven? As many as you please. How many graphite claws must a raven have to be called a "pencil?" One? Do I detect a lingering whiff of Scholasticism?

The reason that Kuhn's water tank isn't a negative feedback system is probably not as self-evident as I made it out to be. In fact the water level would be resistant to disturbance: scoop some water out and the water level will rise again, add a dollop extra and the water level will fall again. There are lots of similar systems: pendulums, magnetic compasses, a marble in a bowl, a buffered chemical solution. Why aren't they control systems? Basically, because the equilibrium condition can be exactly calculated from the sum of all disturbances acting on the variable of interest. The equilibrium water level is the level at which the disturbance filling the tank and the disturbance emptying the tank become equal: the emptying disturbance changes with water level, while the filling disturbance is constant. Therefore equilibrium will be reached without any need for control. Imagine, however, that we now enlarge the hole in the bottom of the tank. Now less water pressure will be required to bring the outflow up to equality with the inflow, so the water level will drop to a lower equilibrium position. If this were a control system, the water level would remain the same. If you want to see a control system like this in action, look into the tank of your favorite old-fashioned toilet. The float detects water level; over a very small range it varies the inflow from zero to maximum. Even if the stopper leaks, the water level will rise until the float just turns off the inflow. So this system controls water level despite all possible kinds of disturbances, within reason. But keep the mop handy.

Thanks for the extra copies of the review. There seems to be considerable enthusiasm among the control-theory group for your book(s). This pleases me, as one of my rewards for being guru of this group comes from my role as match-maker. I had my 60th birthday party at the meeting (a highly successful surprise), and somehow this makes it more important to see that control theory be handed off to people who can develop it on their own.

Best,

Bill

15 Sept 86

Dear Bill:

Your analysis of the raven paradox by making ravens a variable and pencils another sounds good, but the philosophers of science, Hempel and Carnap and Quine and so on, pose the paradox this way: all ravens are black is logically equivalent to all non-black things are non-ravens. I suppose two variables are implicit there: raven-or-non and black-or-not.

I suppose the paradox is interesting because we do act that way in the social sciences. We hypothesize that children from poor families will do less well in school that children from affluent families. All poor children (ravens) are poor-in-school (black). Then we examine the school performance of an affluent child and find it good, and cry, "That fits!" That is, the good-in-school child (non-black thing) is affluent (non-raven). Indeed, we make "control groups" by comparing ravens with non-ravens.

I think what makes the paradox not very interesting to me is that in social science, we never have the sharp boundaries of two-valued logic and Venn diagrams. Our ravens are never all black--not all our ravens are black--and some of our non-ravens turn out to be black. But we keep wanting, nevertheless, to conclude that it is in the nature of ravens to be black, even though the facts in front of our eyes are that it is in the nature of ravens to be frequently black and sometimes something else.

I thought your first tack was very good: that physicists don't use the idea of generalization. I'm going to ponder on that.

Suppose you found one woman who did not operate by control systems. Then you would have to conclude either that not all humans operate by control systems or that the woman was not human. A traditional social scientist, faced with all those other humans who do operate with control systems, would conclude: "Humans tend strongly to operate with control systems." That's the way social scientists talk. Can't you just see me tending strongly to operate with control systems?

Now, leaving the paradox for the moment, I have a question about your example of predicting 29 degrees C and getting 28.8. You said that was less than one percent "error." Seems to me you are talking about how close to your prediction your observation comes. But where is the base of your percentage? You seem to be talking about what engineers call significant figures. You were two points off in the third significant figure. But compare how close you were with how close an observation of 100° or -100° would have been. How do you figure percentage on those? I don't see how you can go by significant figures. What if you predicted 1° and got 0.8°?

Logically, I can accept your statement that "similarity" is a construction in our minds, not a characteristic of nature. But as a theory about reality, it troubles me. Take your own example of mercury. Time and time again, when you and I and a million other people (so we report to one another) perceive some qualities of mercury (including a visual boundary), we also see all the other qualities we have seen before. Isn't that enough evidence to conclude that for all practical purposes we can safely act as if the is a category of mercury and non-mercury out there? I'll grant that we can knock out an electron and have something else, but I don't think that weakens my question.

It doesn't trouble me a bit to admit that I myself do not exist inside my skin, but only in interaction with things and people. Other people knock "electrons" out of my personality or stick some in every other week. But I can't make myself think that may about mercury.

Happy birthday. I'm 69.

You say some members of the CSG are pleased with my "book(s)"--plural. That's nice, but the plural is wrong. You are the only member of CSG who has seen my "Generalizing." Indeed, the only other member to whom I sent a few pages of "Generalizing" as advertising or invitation or come-on is Marken. He has not requested the document. I sent the invitation mostly to people whom I know are methodological experts. God help me.

Aside from you, I sent "Generalizing" only to four people without sending the invitation first. I have heard from two of them. Carol Slater was almost ecstatic. Joe McGrath said it was full of interesting ideas, and he'd have to read it again. You know what that means.

Enclosed is some correspondence with Leslie Hart, some pages from Vaihinger's "Philosophy of 'as if'," and your Conant book with an extra copy of it. Thanks. *

I don't know whether you know about Vaihinger. I first heard about him in Korzybski's book, but never looked it up until this week. It's a good thing, too. I wouldn't have understood it in 1947. Anyway, it fascinates me how so much of what he says could have been the beginnings of control theory. He set down the core of his ideas in 1867. The English translation was published in 1925.

As usual, I send things only with the risk that you will read them, not with a demand.

Well, I must go correct a course outline in which I made an error.

Phil

* 860901_Phil_Hart.pdf
860915_HartBrainSchool.pdf 860915_AsIf.pdf —enclosures at this volume's web page.

17 September 86

Dear Carol:

I've already said thank you for sending me the excerpts from Schwartz (Ed.), but I'll say it again. Thank you. I've enjoyed watching those three philosophers flounder about with the question of how they can keep words from getting in the way of their vision when they are looking for reality.

Well, they are never going to succeed fully in doing that, of course, because they will always want to check with one another, and words are the only way we have of doing that.

That is not wholly true. You and I can make supper together, quite without words or even gestures, and if after a while I see you eating out of the same dish I am eating out of, I will conclude that we must have experienced at least some parts of the world similarly enough as makes no practical difference. But philosophers will never be satisfied with such low-level agreement. It would be fun, though, if a committee of them would design a wholly non-linguistic cooperative social activity to test in what degree of detail evidences could be generated of behavioral agreement on the perception of natural kinds. And then carry it out.

It seems to me that your three philosophers depose and testify as follows:

> Well, you good ol' scientists you, especially you good ol' physical scientists, you have solved for us philosophers the old problem of essences and natural kinds. Of what things really are. Of the difference between something that's really something, on which we will put a label that's just a label having no implications about things that might have some of the same properties--between that sort of something on the one hand and, on the other hand, something that is a collection of odds and ends for which we put a label that is an invitation to others to let us talk about those odds and ends in one chunk--a sort of stipulative definition, you might say.
>
> If you good ol' scientists say that anything that's H_2O is going to behave like water, that's good enough for us. And if you say that a certain mouse, despite its similarities with other mice, is not really a mouse but is a marsupial, that's good enough for us. Granted that now and then a puzzle such as the duck-billed platypus will show up, and it will be a while before you agree whether it is a natural kind, but we trust you to let us know when you get it figured out.

That seems to me a frail argument, though not one that wholly unravels when you pick at it. I think your three friends omitted two matters that they should have added: reliability and system. Let me give you some thoughts (I'm giving them to myself at the same time), and I'll gradually get to those two matters. It is possible, of course, that your three friends did take those two matters into account and expected me to know they were without having to be told.

What lies behind "water is H_2O"? Some experts make a lot of statements. They fill up a page or two explaining that there are some things called electrons and atoms and elements and molecules and compounds and so on, and there are two particular arrangements bearing the labels H and O. Nobody has ever seen those things, but the experts claim that when you see water, you are seeing some evidences of what they are talking about. If you want to see other evidences, they say, you can build some apparatus in such-and-such a manner, being sure you do it exactly right, standing on the right points of the pentagram as you do it, and then you will see still further evidences of the sort they describe.

Note(A) pasted here

If you see something that looks to you like water but doesn't behave the way they say water should behave, then you are not seeing real water--what they say water is. That's the comfortable argument experts have used since time immemorial: "X behaves so-and-so." "Oh, does it? Well, here is some X that doesn't." "But that's not really X. It's X only if it behaves the way I say it does. If your prayer wasn't answered, the reason is that you were not praying properly."

Note(B) pasted here

The way you tell whether this stuff is X is by whether it behaves so-and-so. So you must use the label X only when you are looking at stuff that behaves so-and-so. Anything that behaves that way is X. In brief, anything that behaves so-and-so behaves so-and-so. The label is irrelevant.

That may sound at first hearing trivial and almost circular. But it is not so. Its usefulness depends on being able to recognize unambiguously so-and-so behavior. That kind of recognition is of course the stuff of science. That's what I meant earlier about reliability. It means not only that you have seen those specifiable (recognizable) things behaving that (recognizable) way in the past, but also that you can put those things in the specifiable (recognizable) conditions again, and you will see the same behavior again, and you do it, and you do.

You will notice that I talk a lot more about behavior than your philosopher friends do. To me, all things and all properties of them, the "nature" of them, are evidenced in interaction or behavior. It is "behavior" when H and O combine into H_2O. It is the "behavior" of certain pigments when they absorb certain wave-lengths of light and reflect others and we see red. Just as it is our behavior to pull up the word "red" when those wave-lengths impinge on our retinas.

Is schizophrenia a natural kind? Schizos behave so-and-so. If it behaves so-and-so, it's a schizo. Same argument.

What's the difference? It seems to me the difference is reliability. Do I have O in that bottle? Every time someone makes that claim, we can do some things that everybody (everybody who is anybody) agrees will tell us whether that stuff is O, and we can all agree that we are seeing the same evidences. We can point at things, use words that we all agree are telling us what to point at, and so on, and agree that yes, we are all getting perceptions of the same sort and of the sort that tells us whether that stuff is behaving the way the specifications for O tell us it ought to behave. So it's O, the stuff that behaves so-and-so. We agreed that when we saw that behavior, we would all cry, "O!"

I agree that's about as close as we can ever come to "knowing" whether you and I are both dealing with O.

Why don't we agree that Carol is a schizo? Because her interaction with other things (and people) is not as reliable as H's interaction with O. She has a cluster of properties, and some of those properties are included in the specification of schizophrenia, but some are not. And conversely. We can't agree on whether we have her in a bottle, not to speak of whether we are seeing the same (reliably recognizable) interactions. We do agree, unfortunately, that we do not always see her behaving so-and-so when we put her into interaction with Hal.

So schizo is nominal or analytic or attributive, not rigid or referential.

I think the argument your friends make will hold up better—or at least be more convincing to people like me—if they add reliability to it. If they add the requirement of <u>always</u>. They sometimes seem to imply always, as I think Schwartz does in the middle of his page 36. But no place in the three articles did I find it explicit.

I don't understand Schwartz on pages 38-39. He says at the end of the first paragraph on page 39, "The new theory is led into this error because of the failure to clarify what the referent of a natural kind term is." But it seems to me that the earlier argument was that it is not the job of the philosopher to do that, but of the scientist.

When Schwartz says "what the referent of a natural kind term <u>is</u>," it seems to me that he must mean all the time, every time. It seems to me that it is then easy for philosophers to avoid the error that Schwartz fears. Namely, if scientists have not described a referent that reliably appears every time (like the evidences of O), then the thing is not a natural kind.

The trouble with "bachelor" is not that you can imagine other worlds in which there are unmarried males. The trouble is that none of us can find a behavior such that when you put a bachelor into interaction with others (or certain specified natural-kind others), you get that behavior every time, and when you put married males into interaction with others, you fail every time to get it.

It is not good enough, of course, to say that you can tell bachelors from married males by watching to see whether they have at one time behaved the way people behave in a marriage ceremony. It is not enough to define a natural kind by whether the thing has behaved just once in a specified way. If we discovered some males who every spring went through courting and marrying ceremonies, like bower birds, and other males who never did, then we could nominate "marrieds" and "bachelors" to be natural kinds.

(Nowadays we can make just one test of the stuff in the bottle and declare whether it is O, but we can do that only because we are told that is the way it worked in every previous trial. I suppose this is the kind of parenthesis your three philosophers find it unnecessary to insert.)

Your three friends seem (if I am not reading too hurriedly) to define a natural-kind term as one refering to a kind for which scientists can give an explanation of "how it works." If so, it seems to me they have left out an important specification: system.

Take Schwartz's example of "pencil." He says that the word "pencil" is nominal-or-analytic-or-attributive, not rigid-or-referential. I certainly agree. But suppose I define pencil as any substance that will leave some of itself on paper when dragged across the paper with a pressure of one ounce per square inch. If you wish, I can even specify the minimum amount of residue in weight or number of molecules per square inch. Then physical scientists could describe very accurately how that would work, and they would be right every time. Should I then accept the class of anything that rubs off on paper (to those specifications) to constitute a natural kind? Not me.

Natural kind, it seems to me, must have thingness. It must be some sort of "system," as we say nowadays. It must have interdependence of some sort among its parts and lack of interdependence between its parts and non-parts.

Granted that the boundary is not always sharp. The earth travels within the "atmosphere" of the sun. The behavior of the earth is interdependent with that of the sun. Is the earth therefore not a separate system? Most people would say that it is a separate or identifiable system even while being a subsystem of the solar system. For me, it is sufficient to say that one thing is independent of another if there is a sharp declining gradient of interdependence as we go from "inside" to "outside." That's what physicists seem to accept for the "existence" of an electron in an atom.

Anyway, I claim that the class of rubs-off-on-paper is not a system, a thing, and that's what keeps it from being a natural kind. When we talk of what O does in interaction with other elements, we don't mean just a class of things that share a common property, no matter how accurately scientists can specify the appearance of that property in every single case. We mean that we can also specify the system, and the system itself is a part of the explanation of how it works. Namely the O-molecule.

You can put graphite into the scale of hardness. But when you do that, you don't make rub-off-onto-paper into a natural kind. Rather, you make it into a behavior of graphite and paper.

A sensation or a mental event is not a natural kind, because it is not a system. An electrical current in a wire is not a natural kind. The wire is a natural kind, and the current is a behavior of it. A biological neural net is a natural kind. The sensation or the mental event is its behavior. A leaf turning toward the light is not a natural kind; the plant is.

On page 41, Schwartz implies that he will accept a "process"—such as a dream—as a natural kind. If your friends are going to accept processes as natural kinds, I don't want anything to do with them. As far as I am concerned, natural kinds must be tangible, delimitable, bounded things—even though at the limit, every part of the universe is interdependent with every other part.

I realize that I am rushing in where philosophers fear to tread. That's called learning by trial and error.

Now I have a nice puzzle for you.

Remember those puzzles called "What's wrong with this picture?"? Well, what's wrong with this proposal:

how is not a process?

> If you take a natural kind and make it self-reflexive, you get a new natural kind.
>
> More formally:
>
> A new natural-kind Y appears when at least one natural-kind X becomes a part of system Y in such a way that circular causation sets in—that is, in such a way that the behavior of X affects the behavior of Y and the behavior of Y affects the behavior of X.

When that happens, I think you will also observe a focal system (Y) and an environment. But that's an aside.

You can get a new natural kind, I think, by bringing two natural kinds into closer interdependence. For example, here are a bar of iron and a bar of copper. When you heat them, they get larger, but they retain their proportions—their shapes. But now rivet them together. When you heat that assembly, the two bars bend—a behavior that does not occur when the bars are separate.

Or when you bring H and O together so that they share an electron or two, then you get the new behavior of water.

Those are not examples of reflexivity. I mention them merely as introduction. Or as a way of saying that natural kinds are distinguished by behavior or interaction or "how things work." And reflexivity is a special sort of how things work--a sort that you always find in living creatures.

Now I'll turn to the analogy of the thermostat. As a sensing device, some thermostats have in them a couple of strips of different metals riveted together. In my high-school days, if my memory is right, that device was called a thermocouple. But maybe my memory is wrong. I looked up the word in the dictionary (AHD), and it says "...two dissimilar metals joined so that a potential difference generated between the points of contact is a measure of the temperature difference between the points." Well, be that as it may, for the purposes of this letter, I am going to call those two joined strips of metal a thermocouple. And again, for purposes of this letter, I will claim that a thermocouple is a natural kind.

If you heat a thermocouple more and more, you just get more and more bending in the same direction until the metals melt.

But if you hook up the thermocouple to a switch, and run wires to the mechanisms in the furnace, and enclose the whole apparatus within the walls of a house, and arrange the "on" and "off" switching so that you get a negative feedback loop, then the thermocouple no longer behaves that way. It bends in one direction for a while, then in the other, and never melts.

So I guess I am claiming that a heated house is a natural kind. Think of that.

It may be that bringing two natural kinds into close interdependence does not always produce a new natural kind. For example, a tennis ball and a tennis player can have strong interdependence, but I don't think they make a new natural kind. Maybe only because the interdependence is temporary? Maybe if we found someone who was <u>always</u> in strong interdependence with a tennis ball, we would have a new natural kind? I don't know. I am only saying that I am not ready to claim that it always happens.

But I think that when you add a negative feedback loop to a natural kind, you invariably get a new natural kind. When a glob of chemicals starts acting to affect its own chemistry, lo! we have a new natural kind--a living creature.

Also, negative feedback loops always have in them a "reference signal" or "bias" or "preferred setting." In the case of the thermostat, the preference is set by the Deus ex machina--to stretch the term only slightly. I don't know how the preference comes about in living creatures, but it does somehow come about. Anyway, all negative feedback loops thereby have purposes.

Most of the remarks in the last three paragraphs are asides. My main point, the one I offer you to puzzle over, is that when you add a negative feedback loop to one or more natural kinds, you get a new natural kind. The main idea feels right to me, but I won't be surprised if you think of an example that ruins my "always."

Here are more asides.

Part of human reflexivity is that we use language to talk about language. We have words for words, symbols of symbols, images of images. I think that on that grand day when all the sciences are unified (whatever that can mean), the philosophers, linguists, logicians, and artists will still be in business under their old shingles. They are the people who will still be talking about what people are talking about.

What distinguishes humans from other animals? That question, though vague, is still a good one. Some people are claiming that we can no longer hold that humans have language but other creatures do not. I agree. But are there any other creatures who have language about language? Washoe, I know, has a sign for herself. But I don't think she has a sign for that sign.

At one time many people held that the danger in the fruit of the tree of knowledge was knowledge about the self. (Suddenly Adam and Eve realized that they were NAKED!) But I think (Washoe is only a small part of the evidence) that creatures started behaving as if in reference to the self long, long before hominids appeared. Apparently self does not require symbols of symbols.

If it turns out that only humans can learn symbols for symbols, that would make a fairly sharp distinction. And it would give the artificial-intelligence people something to play with.

Now. What does my idea mean for the "reality" of human groups? I think it means just what I said about "loopy groups" in "Inside and Outside." I think we can dispose of the question of "always?" easily. Just as your three philosophers use the principle of "if we put H and O together, then ..." or "when we do," similarly when everyone in the group is acting to maintain the shared principle of cooperation, then.... So the new natural kind "loopy group" does not exist at every random moment, but it does when the people are put together in such a way that the negative feedback loop is operating.

Is there a sharp demarcation between a collection of humans and a loopy group? Do we get a surprise of the sort we get when we put together H and O? Well, we certainly get that kind of breath-taking contrast between a "mere" collection and a "minimal loopy group" as I defined it. But do we get such a contrast between ten people maintaining cooperation and nine of ten people maintaining it? I don't know. Do you? One of the difficulties in attacking that question empirically is the matter of how long you should observe before claiming you have an answer. The new behavior of H and O in H_2O appears almost instantaneously. To observe the new behavior in the thermostat, you should allow some number of minutes. How long should you observe loopy and almost-loopy groups?

So there.

You can see that the articles you sent set me to thinking. Or ruminating.

Thank you very much.

Love,

Phil

In a recent issue of Science News, I saw two pictures. One was a model (a lot of balls with sticks among them) of what atoms on the surface of a piece of metal ought to look like. The other was a picture of what some researchers got from a scanning tunneling electron microscope, or some string of words like that. The author of the article wanted me to be impressed by the match between the model and what the microscope showed. Well, I could match the microscopic view to grandmother's patchwork quilt better than to the model. Reminds me of Feyerabend telling about the frustrations Galileo had when he asked people to look through his telescope. Most of them simply could not see what he saw. Note(A)

Note(B)

I have made a couple of snide remarks in the two paragraphs above. I didn't do it to say that scientists or your philosopher friends are incompetent. I did it mainly to remind myself of the pitfalls we can encounter.

Sept. 18, 1986

Dear Phil,

The Xerox Kid is at it again. But interesting. Your friend Les Hart makes the same mistake a lot of others make. Behaviorism is dead, they say. Pooey. Anyone who thinks behavior is output is still an S-R theorist, correct? I do agree that Hart and I would have few arguments over substance. We have both noticed the same phenomena, which are probably more important than theories.

* Quine is an elegant writer, but if he really represents philosophers of science I wonder if this stuff is worth bothering with. It seems to me he is just playing with words -- he seems to have trouble telling a noun from an adjective, so maybe he needs to play more, not less. If we're talking about the noun, raven, then as long as that category remains fixed we can discuss the presence or absence of attributes of ravens. What we say about non-ravens has exactly nothing to do with the attributes of ravens, has it? Maybe these people need to think in levels (logical types?). The category "raven" is made of its shape, its color, its appendages, its mating habits, its cry ("Seldom!") and a number of other aspects: when enough are present, we say we are looking at a raven. But is there some essential raven-ness aside from this collection of attributes, casting a shadow on the wall? One is permitted to doubt. I have a secret feeling -- supported, suprisingly, by some biologists I have met -- that there really isn't any such thing as a raven. The species form a continuum, which for our own reasons we mark off into categories within which differences are agreed to make no difference. Our mutual friend Korzybski would probably have supported this view.

As to grue, how can you categorize anything that will not be observed until tomorrow? This procedure falls within the rules of verbal games, but not within the rules of observation. And when tomorrow comes and you pick up the emerald, hasn't the definition of "grue" correspondingly changed? When we get to "tomorrow," what does "tomorrow" mean? I suppose you could substitute a date. But jeez, this sounds like material for lawyers, not scientists.

Quine give me one strong impression: he thinks words "have" meanings. Here is the word "emerald." This word really means something real, you know, an <u>emerald</u>. Out There. Objective.

The critical word that "has" a meaning is "similarity." Quine is looking for some primitive inbuilt concept that describes what different objects have in common, their similarity. I notice you circled "really" in connection with his discussion of "what it means really for a to be more similar to b than to c."

* 860918_NaturalKinds.pdf —enclosure at this volume's web page.
Carol Slater mentioned this 860809. Now Bill has received it.

Consider: is a Buick similar to a horse? Of course -- if you're a small creature trying to get across a road. They're both big and dangerous, and thus more similar to each other than to an ant. Similarity depends on the dimensions in which you're perceiving, but more important it depends on your capacity to perceive distinctions. At the first level, every perception of the same intensity is the same as every other perception of the same intensity. To an animal capable of perceiving only light and dark, objects fall into only two categories: light and dark. Light objects are similar to light objects, and so on. So our own perceptions are what create similarities: our own failures to see differences.

Re your remarks on mercury. I think we have to work very hard to create categories like that. You're right: the whole point is to create perceptions that are reliable and repeatable, so we can use them with confidence. I think that's what the physical sciences are about. But I still don't think that we have to believe that there is something Out There that corresponds to the experiences we call mercury. Or better, I don't think we NEED to know if there is -- that the quest for objectivity just gets in the way of science.

Base of the percentage: you got me. Of course 28.8 degrees centigrade is only 6.6 parts in ten thousand different from 29 -- on the only temperature scale that is meaningful for such statements, Kelvin, where zero has a physical meaning. In our accepted model, of course.

If I found one woman who did not operate by control systems I would have to say that there is something wrong with the idea that all organisms are control systems. It would be like finding a piece of matter than fell up: we'd have to rethink everything. Explaining this anomaly would take precedence over everything else, because, you see, control theory is supposed to be <u>general</u>. No exceptions allowed, or it's back to the drawing board, which isn't necessarily bad. I'm sure you appreciate exactly what I'm saying.

Back to similarity. If similarity is something objective, then we must ask what objective effects it has, by itself. With no organisms present, the similarity of the shape of a cloud to the shape of a locomotive has zero physical significance. The best we can say is that similarity provides organisms with something to consider similar. But I think the best way to deal with this word is to say it describes the limits of our ability or inclination to notice differences.

<u>Generalizing</u> is strong medicine, I suppose. Not having been into statistical treatments of any importance, I probably can't appreciate the radical nature of this work (not fully -- I

do get the idea). You are no doubt beginning to feel the same thing I feel: the weight of centuries behind current beliefs. People who struggle to learn what is known become defenders of the faith even without knowing it -- even while thinking of themselves as radical and progressive. I concluded some time ago that behaviorists aren't stubborn: they literally can't help twisting control theory to their own purposes. One has to abandon all committments to science as it is in order to do something new, or like me, never develop any strong committment in the first place. I really had it easy! Behaviorists, and most other scientists, are caught in a web of assumptions, and their sense of belonging to a club doesn't help them escape it.

I've skimmed through Vaihinger, and I see what you mean. I see myself in this man, trying to confront the phenomena of existence directly, without authority, trying to say simple true things. I think people did that more in the 19th Century than they do now. Now everyone wants to generate something complex, something impressive. The simple ideas are for dummies, like me. Stu Umbleby (of the ASC) had a word for people like you and me that I like better than "dummy," though: "self-authorizing" persons.

I have an ambition which will probably never be realized: to survey the last 350 years of science as if we knew that control theory was what people had been looking for all along. Vaihinger would qualify as a near miss, as would many others. That phlogiston pamphlet did a little of what I would strive for -- showing how a simple change of interpretation would have made all the difference. I'm afraid it would take a better scholar than I am to do it, though.

My life is still in a pretty strange state. It's getting too schizophrenic. Days of sleeping and reading and sleeping. Not good. Something must be done. One of our group, Charles Tucker, has offered to try to get me a professorship at the University of South Carolina. My initial reaction was panic -- I don't know how to be a professor, I don't want to live in South Carolina, won't they think I'm too old, do they realize I don't have a PhD? And underneath, do I really have that much to teach anyone? But I'm getting my courage up to call him back and say let's talk about it. I think Mary wants to do it. I think I do. But my confidence seems to be at a low ebb. How about some advice, you being so much older and wiser?

Best,

Bill

22 Sept 86

Dear Bill:

What's schizophremic about sleeping and reading?

I forget the name of the man, but for many years Harvard had a president whose degree was the Bachelor. And I'm not talking about the 1700s or 1800s. I think his tenure included the early 1960s. We have here a fair number of faculty with only the Master's. For all I know we have one with the Bachelor's.

So strong in the academic mind is the measuring-stick of the degree is that it is difficult to talk politely about people without the Doctorate. The current way academicians deal with it is to speak of the degree of a person who is not enrolled as a student as his or her "terminal" degree. I don't use the term myself; it sounds so hopeless and fatal.

Anyway, I'd say that your chances of being appointed a professor depend mostly on the department. If the department Tucker has in mind has 3 or 4 people acquainted with control theory who are willing to take your side with the Dean, if others in the department think individual behavior is important (or at least how the individual "works" is important), and if two or three people in the department who are influential (for whatever reason) think your presence will help them with their own work (no matter how--give them good ideas for research, reduce their teaching load, help them influence other colleagues, whatever), then you have a pretty good chance.

I note from the membership list of CSG that Tucker is in sociology. I'd be surprised to find a sociology department that would meet the criteria I have just listed. Maybe he has in mind a cognitive science group in psychology. Maybe an ergonmics group in the engineering school. Well, you will know.

You don't have to "know how to be a professor." As long as they are polite in faculty meetings, professors have a good deal of autonomy.

Oh, by the way, check on the current financial state of the university. If they are scrabbling, and if they want to offer you a position to be renewed annually, I wouldn't take it unless they bought me an annuity right away. And also by the way, I think it will be especially unlikely that they would offer you a tenured position. But they might. As I say, it all depends on the perceived needs (that means what the people think their own reference signals are like) of a fair number of people in the department.

Anyway, it won't cost you anything but time to look into the matter. They will pay your travel costs and per diem to discuss the possibilities.

If they don't offer you the job, then so what? If they do, then you don't have to worry about those panicky thoughts. They will be answered by the people's action.

And check on their mandatory retirement age. (No doubt Tucker has already told you.) It would hardly be worth making the change for five years, I should think, unless you just want to have the experience. It usually takes a few years to begin to draw the students you want. Students come to you partly because of the recommendations of other students and partly because of the recommendations of faculty advisers. It takes both processes a while to develop. Of course, if the department meets the criteria I mentioned, other faculty would begin sending students to you right away, and that would help a lot.

I wouldn't want to live in a Carolina either. But I suppose there are some nice people who do.

Thanks for your further remarks about philosophizing. When reading philosophy, especially the modern ones, I have your same trouble--when are they talking strictly within the world of formal logic, not about observables? They don't, of course, always know themselves. And sometimes they even know they don't know.

Well, I've had a reply to "Generalizing" from one of my expert colleagues. He said sorry, he was not shocked. He also said that what I said needs to be said--that I should publish the ideas. He also predicted that Guru Cronbach would not be shocked either. (Cronbach is having surgery for cataracts, and told me he wouldn't read it until after that was done.)

Well, this first respondent (Hastings) is one of the more savvy and less hidebound people on my list. I will wait to see what some others say. And maybe I'll send out more copies if I don't get another response in a month or so.

Oh, "belonging to a club" is indeed an extremely powerful influence on academicians. How can they tell whether they are meeting their internal standard for "doing good work" or "being a scientist"? If you are studying gravity or the genetics of corn, you can get pretty good feedback with your own eyes. But in the social sciences, you know that no matter how good you think your experiment is, some people are going to say, "Yeah, but ...." So the main path of feedback for most, most of the time, is indirect--whether you get papers accepted for publication, whether you are invited to give talks, etc. Those things depend on staying in the middle--saying the "right things." Most of the direct verbal feedback social scientists get is adverse criticism. They are always ready to tell colleagues what faults they find in a research report (and no research report is without faults) and only rarely ready to tell colleagues what they like about the work. That's the custom--which I violate right and left, always telling my colleagues first what I like about their work. They are always grateful, and sometimes pitiably so. I don't write many book reviews, but when I do I sometimes have them rejected by the editor for the reason that they are not "critical" enough.

But as I said earlier, academicians will forgive a good deal of idiosyncracy and cantankerousness if only you are polite in faculty meetings.

I'm glad you found Vaihinger interesting.

Phil

PS: WOULD YOU PLEASE review my illustration in "Generalizing" about removing the jacket, running from bottom of page 23 to bottom of page 27 and then pp. 29-31? And tell me whether there is something in it that makes it a bad illustration? Or can you think of a better illustration? It's not as clean and neat as I'd like. I have not yet worked hard to think of a better one. If you could do this within a month or so, or even two, I'd be grateful. *

* 860722_Generalizing. The date and version of the enclosure listed here is uncertain. The content does not seem to match the remarks here. More likely, Phil enclosed another draft for Chapter 11, Testing specimens in *Casting Nets and Testing Specimens.*

24 September 86

Dear Bill:

* I hope the enclosed has some nourishment in it for you.

It is too laconic for my poor sophistication. Even if it were more detailed in its explanation, I might have a hard time following it.

It says plainly in the abstract that it is a model "at the level of the individual" and that it uses "equations linking two multiple-loop feedback systems." But that's the plainest statement I can find in the whole article. Elsewhere, I can't find what the loops run through, and I can't find what were the reference signals used. Maybe you can.

I couldn't read some of the symbols on the graphs. I enlarged the original by 20% to make this copy, and I still need a magnifying glass to read some of it. I think editors insult readers when they use type so small that only 20-20 18-year-olds can read it--if then. The implication is that the reader won't want to read those words and numbers anyway.

Well, enough complaint. I hope there is something in it that will do you some good.

Phil

* Computer simulation of Freud's counterwill theory: Extension to elementary social behavior. Denker MW, Achenbach KE, Keller DM. Behavioral Science. 1986 Apr;31(2):103-41. http://www3.interscience.wiley.com/journal/114041265/abstract

Sept. 28, 1986

Dear Phil,

Thanks for the long and encouraging letter. I suppose for all my innovative bold theorizing, I'm still basically pretty timid in the real world. Being polite at faculty meetings would once have taxed my forbearance, but the rough edges have been rounded off and I can suffer idiots gracefully, most of the time (but see below). Also I have gradually learned' that some academicians really do know something, and that this is not necessarily apparent right away. Other people are timid, too. It's probably just another pipe-dream, so I'm not revising my life-style just yet.

That paper you sent me, although it does drop me a reference-crumb, is disgusting. If there is any virtue in it, it's that Denker et. al. are presenting a model that at least does run. That's the first step toward honest modeling. Most models are simply proposals about the internal organization of some system. There's no proof, however, that the model drawn on the paper would actually behave in the same way as the system being modelled: the idea of running a model is confined to a very few people outside engineering. When you commit your hypotheses to specific functional representations and simulate the consequences on the computer (or otherwise), at least you find out whether your model behaves at all like what you had hoped.

The Denker paper, as you noted, is pretty laconic: a block diagram would have helped the poor reader to understand the relationships among those countermnemonic symbols. I tried to follow the relationships for a while, but the conventions are apparently pretty arbitrary: sometimes multiplication symbols are used and sometimes they are left out -- i.e.,

DTAIN - (F) X (AFF.K) + DC.K

versus

(1/PTQ.K)(OBHG.K - BHQ.K)(WQS) + (1/PTQ.K)(OBHZ.K - BHZ.K)(WQO).

Beyond the fact that there is no way to understand the model from the stated equations (where are the "rate equations?" I hate being referred to a different paper for vital information), the whole model is untestable against reality. Terms like "normal level of affect" and "doubling time for antithetic ideas" and "nervous energy" are devoid of meaning: linking them to quantitative equations is preposterous, insane. These people haven't the vaguest idea of what it is to be a scientist. They're masturbating with the help of a computer. I would have a hard time being polite to them in a faculty meeting. Yuk.

The philosophy-of-science stuff is like chipped tooth you

can't keep from probing with your sore tongue. Your letter to Carol of Sept. 17, of course, is the chipped tooth. I give tongue as follows:

What underlies "water is H2O?" I claim, two things: a model and an experience. The experience of water consists of all the perceivable attributes that we categorize as water: the feel, look, and behavior of the perceptions. In other words, all the perceptions at levels below "categories." These are direct experiences of the products of our own perceptual functions, and there is no science to them. There's probably a lot of variability from one person to another, too -- does water look inviting or scary, smooth or rough, transparent or colored, delicious or dull?

The concept of H2O, on the other hand, concerns a model, something imagined. When we experience water, we do not experience H2O: we imagine that what we see is composed of hydrogen and oxygen in the ratio 2:1. There is no direct experiential way to verify this. Instead, we have to refer to the rules of the model, the laws of chemistry and physics. These physical-science models have been refined until we can state with considerable confidence: "if water were composed of H2O, and if all the associated laws and relationships held, then water would behave as we experience it to behave." It would still, however, look, taste, feel, and behave like water and not like H2O.

In fact we check the imagined world of chemistry by operating on it, and seeing if the experiencable consequences predicted by the rules of the model do in fact occur. We never see the model itself working except in imagination. What we really know is established like this:

1. Perform some act that can be perceived.

2. Imagine the consequences of this act behind the scenes or on a scale inaccessible to the senses, and predict at one or more points some consequence that can be experienced.

3. Compare what was actually experienced with what step 2 predicted would be experienced.

4. If there is a discrepancy, correct the model.

5. Go to step 1.

This process will cease, of course, only when no known discrepancies remain.

So clearly the statement "water is H2O" is incorrect. "Water" refers to a set of subjective experiences. "H2O" refers to a model. The only correct interpretation of the phrase is "water, as we experience it, behaves as if it were made of H2O".

It seems clearer and clearer to me that the difficulties being encountered by philosophers of science come directly from the assumption, conscious or otherwise, that there is really some objective thing to which the word "water" refers. When you take that tack, you have a very hard time with other aspects of reality that seem just as objective, but are obviously not. A is nicer than B. A is more expensive than B. A is a pretty shade of purple. A is moving smoothly across the television screen. B is caused by A. The probability of B depends on the frequency of A. A is orderly whereas B is chaotic. B is affected by A but not on purpose. A is real, but B is an illusion.

Obviously problems like these invite the thinker to divide the objects of experience into different groups, groups that are "real" and groups that are "subjective." I notice that the examples used -- water, ravens, emeralds -- tend to be recognizeable objects. What about the sensations they are made of? Edges, curves, shades, colors, corners? What about the transitions, relationships, categories, sequences, principles, and system concepts they exemplify? Is the philosophy of science stuck at the third order of perception? I think that to understand (and largely dismiss) questions like the ones you've been citing, you have to consider all the levels as being subjective. Then, it seems to me, there's no problem.

What are they ASSUMING?

Best,

Bill

* Commentary on "coat" example

pp. 23 -- 27, 29 -- 31.

The problem with the example, as I'm sure you have realized, is that the "levels" aren't clearly hierarchical. You don't have to remove your coat in order to concentrate. Also you could be totally absorbed in a task and remote your coat without realizing you're doing it. The role of consciousness isn't clear to me in any case, and that includes this one.

Also you have to ask whether you're giving an example of a hierarchical relationship or a conflict. You seem to treat being absorbed and removing your coat as mutually exclusive. That makes it a conflict if you have to do both at once.

I'm not sure that "absorption" is a controlled variable anyway. Can you deliberately become absorbed? It would seem to me that this would be like deliberately ignoring that white elephant in the corner. If you know what you're not perceiving, you're perceiving it.

A try at an example: Higher-order task, keep comfortable, un-sweaty. Means: take off coat if it gets too hot OR (if it's cooler outside) go outside OR open a window OR turn on a fan OR turn on the air-conditioning OR postpone heavy lifting. The real environment is rich in means for accomplishing any given end. Since we generally pursue multiple goals, we can usually find an action that accomplishes more than one goal at a time. If it's too hot outside to walk to the store, you can have left-overs for dinner, or get the grocery store to deliver (sure) or drive in your air-conditioned car.

When you look only at behavior without considering the goal, you find that sometimes a person takes off his coat, sometimes opens a window, sometimes turns on a fan, sometimes goes outdoors. All these different behaviors! To predict any one of them would be hard, because other goals are also in play, and change from time to time. It's difficult, too, because there are so many possible disturbances of the controlled variable. Anything that makes the person warmer results in one of the behaviors that opposes that effect or corrects the error. Which one is used depends on what other goals hjave to be satisfied at the same time. So it looks as if there is a causal relationship, but a weak one. The sun shines in the window so you take off your coat. Whoever is cooking puts a turkey in the oven and you turn on a fan. Your friend hates drafts and closes the window, and you go outside. I'm sure you can elaborate on this idea. If you know what the controlled variable is, all these cause-effect relationships boil down to controlling just one variable. And because this variable -- the level of comfort, the skin temperature, whatever -- is stabilized by the behavior, it

* 860722_Generalizing. The date and version of the enclosure listed here is uncertain. The content does not seem to match the remarks here. More likely, Phil enclosed another draft for Chapter 11, Testing specimens in *Casting Nets and Testing Specimens.*

doesn't change much -- and so it is ruled out because of contributing nothing to the variance.

I don't know if this is what you're looking for, but it seems simpler to me than introducing an iffy concept like absorption. There is plenty of opportunity for bringing out spurious interactions among variables, low correlations, and the general confusion that arises from accepting cause-effect relationships at face value without knowing why they appear to hold. Statistics doesn't tell you why anything happens. You could perhaps use a variable for which reference-levels vary widely among people or with one person's circumstances -- how much salt to put in the soup, or what color a car should be, or how long one's hair should be, or how many people in one room are too many. Then you get the smearing effect from averaging over populations, ruining the data. But your examples are better than mine.

2 October 86

Dear Bill:

Tell me if you want me to refrain from sending you stuff like the enclosed having a high proportion of dross. The title attracted me. I found your name on the first page. Then, scanning the article rapidly, I found no indication that the author had used any of your specific ideas. I will read the artcle eventually, just to be sure I am not missing a good idea even if it is not yours.

Maybe some day I will find a chance to persuade a student to look in a Citation Index for Powers.

Some authors cite a string of other authors just to say, apparently, "All those guys used this word that way, so you shouldn't complain if I do, too." I suppose that's what Ms. Ashford was doing.

Phil.

CITING: It is customary in scholarly writing to cite the work of other authors. Some writers may do so just to show how many books they have read, but there is also a more serious reason: to tell readers where they can get more information on the topic you are writing about. You might cite another author to show readers that you are not the only one holding an opinion you have stated—that someone else agrees with you. You might cite a source of empirical data to back up a factual statement you have made. You might cite a review of literature that can round out the readers' appreciation of the scope of a topic. You might point readers to a more detailed discussion of a topic you merely touch on. You might point readers to a mathematical derivation of a formula you use, and so on.

Many writers, unfortunately, leave us guessing. Some readers will want to pursue further information of a certain kind, but they cannot decide whether to pursue it until they know the kind of information that is in the article or book the author has cited. Here is an actual example of the kind of citation that infuriates us:

> She drew on their respect to maintain control, sometimes in directive ways, sometimes in ways that drew out and developed the controls from within (Redl and Wineman, 1952) the children.

Why should we get Redl and Wineman off the library shelf? What information will we find in the book? Do Redl and Wineman tell how to draw out and develop controls from within? Do they give the theory about controls from within? Do they contrast directive ways and drawing-out ways? Did they invent the phrase "controls from within"—is the author of the sentence merely giving credit to Redl and Wineman? Or do they perhaps tell more about that particular teacher? Or about similar teachers? We may be interested in one of those questions and not others. We cannot judge whether to go to the trouble of getting the book from the library unless we know to what kind of further information the author is pointing us.

No reader is going to dig up every reference cited on the chance that it might be interesting. If you hope your readers will care enough about your subject that they will want to read more about it, do them the courtesy of telling them the kind of information they can find. Here are some ways you might do it:

> Redl and Wineman (1952, pp. 263—75) describe in detail the method of drawing out controls from within.
>
> . . . (the phrase is from Redl and Wineman, 1952, p. 17).
>
> For another example, see Redl and Wineman (1952, pp. 78—84).
>
> Redl and Wineman (1952) review the literature.
>
> (Redl and Wineman, 1952, Chapter 7, give empirical data.)
>
> Redl and Wineman (1952, p. 19) make the same point.

Notice that we included page numbers in our examples. It is maddening to be told that there is a valuable piece of information somewhere in a 600-page book.

Give in to your self-regard. Take it for granted that at least some of your readers will be captivated by what you write and will *want* to know more about it. Then be considerate of them: tell them the kind of information they will find and the page they will find it on.

Finally, some citations are superfluous. It seems to us unnecessary to give credit for an assertion that readers of social science have long taken as true or for words that might be spoken by any of us, scholar or not, any day of the week. Here are three examples:

> Individuals vary in their degree of openness to learning and using new experiences (Alderfer, 1976; Rokeach, 1960).
>
> Aristotle was one of the greatest of the ancient philosophers (Jones, 1978).
>
> In the words of Mark Smith and Claude Johns (1968), "Where is policy formulated and who makes it?"

---

from Philip J. and Margaret Runkel. <u>A guide to usage for writers and students in the social sciences</u>. Totowa, NJ: Rowan and Allenheld (Helix Books), 1984.

After reading and hearing a lot about generalizing during my thirty-some years in the social science business, and uttering a lot of verbiage about it myself, I took time off the other day and went to the science library on our campus to find out how physicists, chemists, and biologists think about generalizing.

The university's subject catalog has no entry for generalization. Under the main heading of Chemistry, it has no subheading for experimental or methodology. The same is true for the main heading of Physics. None of Biological Abstracts, Chemical Abstracts, or Physics Abstracts had any subject heading for generalization.

The subject index for 1985 for Biological Abstracts had headings for experiment, method, methods, and methodology (or variations of those words), but only two or three among the hundreds of entries seemed even faintly likely to touch on what I think of as generalization. Biological Abstracts seems to be written entirely by computer from key words entered into it. The authors who put down "experiment" or "experimental" as a key word, almost all of them, were clearly not writing about experimental design as the main topic. They were simply wanting readers to know that they did an experiment. Almost all those who put down "method" or "methodology" seemed to want to convey the idea of "here is how I went about it." The two or three entries that I thought might conceivably touch on my topic seemed nevertheless so unpromising that I didn't bother to look them up.

The subject index for 1985 for Chemical Abstracts had no headings for experiment or method. The subject index for 1984 for Physics Absracts didn't have those headings, either.

The science library's entire collection of books on science as a general field of inquiry, philosophy of science, methodology, and research as method occupies about six feet of shelf space.

Eleven titles looked to me as if they might contain something about generalization. Only one turned out to have the term "generalization" in its index. One book seemed to have slipped into the science library by mistake; it was a book on experimental design in psychology: incomplete block designs, Latin squares, and so on. It was not the one with "generalization" in its index. Four other books, heavily oriented toward biology (plant breeding, for example) were also like that one. Naturally, they treated the matter of generalizing from a sample, even if the term was not in their indexes. The logic was the same as the logic we find in methods books in psychology and sociology--as is not surprising, given the history of the field of inferential statistics in social science.

Three books seemed to contain non-technical essays on idiosyncratically chosen topics the authors wanted to get off their chests. I'm not saying they might not be worth reading; I'm saying only that they didn't have the organized comprehensiveness that one comes to expect from a text on method in the social sciences.

Oct. 18, 1986

Dear Phil,

Your research on generalization was very interesting. My comments on its use in the hard sciences were made on the basis of general impressions and experience, but not from having searched the literature. I guess I wasn't too far off the track.

Your little project got me to thinking about the subject again, and once again asking myself why I feel that things are done so differently (as your last paragraph comments) in the two divisions of science. It's not easy to put one's finger on such impressions. On the surface, the life sciences seem VERY scientific, with all the trappings of experimentation, objective analysis, cautious advancement of hypotheses, and so on. But why does science work so well in physics and chemistry, and so poorly in psychology? Maybe, I'm thinking, it's something like this:

When we approach any natural phenomenon for the first time, we can't do anything but look for some sort of order in it. Let's call this the "rule-learning" phase. We poke the buttons and see what happens. Basically we're trying to figure out the hidden connections. Without any sort of theory, we can eventually learn how to make some events occur: we just keep a notebook of the results of pushing buttons, and label everything so we don't lose track. There isn't any question of understanding what is happening, however; we're just trying to find regularities, regardless of why they exist. Every science has to start like this. I think I said all this in my book, but now I seem to understand it better.

When we have found some regularities, we can start working on theories. Here, I think, is where the two approaches diverge. The basic question that follows finding a regularity is, "Why does this regularity appear?" There are two directions in which we can search for the answer: one leads to workable answers and the other leads to delusion.

The workable-answers approach goes like this. We look at the button and we look at the event it reliably causes, and we imagine that between the button and the event is a regular universe containing details we can't see (at the moment). Suppose the event is cessation of the buzz of an alarm-clock and the button is the "snooze" button. We find that pushing the snooze button causes the buzzer to quit.

Ah, we say, the button acts on whatever is causing the buzz and renders it inoperative. In our minds we sketch in a little diagram of what is inside the clock. See Fig. A. Inside the box, we imagine, is a buzz maker, among all the other works. There must be a connection from the button to the buzz maker, and when the button is pressed, this connection causes the buzz maker to stop working. We verify this model by pressing the button, and sure enough the sound stops just the way it did before. The model works. Just as we are congratulating ourselves, the sound comes on again. Oops.

We have to modify the model by putting in a timer. The button starts the timer, and while the timer is running it disables the buzz maker. Fig. B. By experimenting we find that the timer runs for 5 minutes after each press of the button. Trying it one more time, we find that the sound goes off and never comes on again. Drat.

Through continuing experiment we find that the button actuates the timer exactly three times before its effect becomes permanent. Back to the drawing board: Fig. C. Now a counter also responds to the button, and when the count reaches three, turns off the buzz maker permanently. But the next time we set the alarm, we find that the snooze button works again, so we have to add a counter-resetter -- and so on, the model getting more elaborate with each new phenomenon we find.

During all of this, we haven't opened the box. But we have put entities into the box having properties that would make what we can see happening happen as it does happen. By the time we have done every conceivable experiment on this box, we have a complete functional diagram of its innards. In fact we have a design which, it it were actually implemented inside that box, would work exactly as the box works. What is actually in the box, of course, might be differently arranged (Fig. D.), but at least we have one possible arrangement that would work. When the model does EVERYTHING we see the box doing, we say we understand the box.

That's the way physics works, if for buzz-makers and timers and counters we substitute electrons, fields, charges, masses, atoms, and so on. This approach works mainly because we demand that the model behave EXACTLY as the real thing behaves under all circumstances.

Now the other approach, the one that doesn't actually work.

We observe that pressing the button stops the buzz. Since we have seen other kinds of buttons and other kinds of sound-generators, we venture the rule, "Buttons suppress buzzes." Then we remember that there are other things beside buttons that can affect buzzes, so we expand this generalization to "Actuators suppress buzzes." One more bright idea leads to "Actuators suppress auditory stimuli."

Checking out this rule, we try many boxes of many kinds. We

find that the rule applies some of the time but not the rest of the time. So the rule becomes "Some actuators suppress auditory stimuli." If we find that 51% of actuators have this effect, we can go back the the original version, actuators suppress auditory stimuli, $p < 0.05$.

But we can't help noticing that the population of actuators is bimodally distributed: one group suppresses auditory stimuli, while another group activates auditory stimuli. Comes the dawn! We do not have just actuators: we have suppressors and activators. Now the truth is becoming clear: suppressor-actuators suppress auditory stimuli, activator-actuators activate auditory stimuli. $p < 0.01$.

Now the research question expands. What makes the difference between actuators which generally look similar to each other, such that some of them suppress and some of them activate? Clearly, some circumstances are activation-facilitating, while others are suppression-facilitating. We notice that on alarm clocks, the right-hand button permits buzzes to occur at certain times, while the left-hand button turns the sound off, most of the time. Clearly, activation/suppression is position-dependent. Extending the research to other areas we find that alarm-clock-shaped objects exhibit this position-dependence, while radio-shaped objects do not. We now can classify objects according to whether the facilitating effects of their actuators are position-dependent or not. What do we do with alarm-clocks that have built-in radios? Well, obviously position-dependence is a dominant trait.

I'm sure you are familiar with this latter way of pursuing truth. You could probably come up with plenty of real examples, where I have to make them up. I think the key to this approach is in its verbal and taxonomic character. The specific phenomenon that first brought the matter to attention is abandoned almost immediately. The first step is to substitute class-names for the specific terms initially used: to look for general categories of which the items in question are only one example. The search for rules is then transferred to relationships among whole categories. Naming and renaming play prominent roles. Basically, we're looking for general ways of stating the rules such that our statement always holds true. At no point to we ask why it is that any one of the rules applies. We're just looking for more and more general ways of describing phenomena.

What is the object of this way of pursuing knowledge? I think it must be to arrive at general rules relating general categories of phenomena, the level of generality being so high that the rules are seen to apply very widely and nearly all the time. But I think this process is inherently endless. There is always a different way of categorizing. And the more general the rule that is found, the less we are able to apply it to any given situation to predict what, in fact, is going to happen. Sure, in circumstances that facilitate position-dependent implementor polarity, time-sequenced event distributions show contravariant stimulus constellations, $p < 0.0001$, but when I press <u>this</u> button on <u>this</u> object, what will happen?

This approach is bound to fail, because what creates the relationship between button and buzz is not the categories into which we can put the phenomena, but the works inside the box.

As biologists sneer at "psychologizing," so do psychologists sneer at "neurologizing." But neurologizing gives us a way to check out models of what is in the box, and checking out models is the very essence of what makes models useful in the physical sciences. A model is put together initially to account for some observed relationship. It's designed to be adequate, in that the behavior of the model necessarily creates, out of the properties we give it, relationships like those we observe. But once those properties are in place, we can look them over from other points of view and see what ELSE they imply that might be observable. An electron is attracted to a positive pole: from its time of flight we can deduce its mass. The electron is also deflected by a magnetic field. From the radius of curvature and the velocity, we can again deduce its mass. Naturally, we demand that the mass found by these two different means be the same. This is what I mean by checking out a model. All observable consequences of the model's details must be verified by observation, and they must remain internally consistent.

In physics it isn't against the rules to take the box apart. In fact, all models are created with the idea that when we learn to take the box apart, we will find all the elements of the model. We're very careful to propose parts of the model that in principle could be observed. There is no rule in this game that says we have to guess what is inside without looking inside, so we model defensively. If we aren't sure how something might be accomplished inside the real box, we just draw a block and state what it must do. We might have a very good idea of what a block has to do by way of causing one variable to depend on another, even though we are unsure about which possible way of doing it is really in there. Without knowing anything about photoelectricity, we can represent a retina by a block labeled "light intensity to neural signal converter." We can draw a "timer" box without knowing whether the timer is electronic or mechanical, or whether timing is done by a little man with a wristwatch.

So the physicist demands that his models not only match their behavior to real behavior, but that everything we can find out about the parts of the model by any means at all check out with experimentation and remain internally consistent. It isn't considered good form to propose models in which most of the parts are in principle unobservable directly or indirectly. Nor is it considered good form to let a model go public while there are still observations of any kind that contradict what the model implies. Such observations indicate that the model isn't finished yet.

What I'm getting at, I guess, is that there is no mystery behind the success of physical models, or behind the failure of models -- "intervening variables" -- in the life sciences. It's just a matter of where you set your standards for acceptance of a model. If a physicist is baffled by failure of his model to predict correctly in just one important situation, he doesn't say "oh, well, it works most of the time," and publish it anyway. Not my ideal physicist at least. He says "Oh, shit!" and goes back to work. If life scientists demanded that their models work with a high degree of precision in all known circumstances and take into account all known facts, they, too, would generate highly successful models -- or, quite properly, admit ignorance. You don't get anywhere by insisting that a model MUST work and that if it doesn't the data must be wrong, and you don't get anywhere by lowering your standards to let models go when they still don't work all the time. But that is exactly what has happened in the life sciences.

Most generalizers I have met object to models. I think they object, without knowing it, to BAD models, models that don't work, because in their fields they have seen nothing else. Unfortunately, when a good model comes along that does work, it doesn't impress the generalizers, because they simply don't expect models to work, to add anything to what observation tells them.

I'm running down on this subject -- probably should have quit a page or so ago. Let's see what these remarks stir in your imagination.

Let me know what the citation search brings up. I did one about four years ago (Social Science Citation Index appears to be the right one) and got 113. Of course most of them are probably like the one in Ashford that you sent me: meaningless. Somebody ought to tell Ashford, by the way, that you don't have to look so hard to find some feedback. She makes it sound as if it only happens on special occasions and takes a lot of effort to get. On page 471 she does the inevitable: "... even if a person receives positive feedback from a target ...". Gaaa. Pretentious blather. I'll kept it in the "horrible example" file.

If this letter has been a little disconnected and run-on-ish, it's because it's the 11th of this weekend. I've stopped pretty-printing because it takes four times as long. What am I going to do if even more people start writing to me? I know, be brief.

The reason I got behind is the enclosed paper -- I can't even remember if I sent you a copy already. If I did you have two. This is to be a chapter in Analysis of dynamic psychological systems, Levin and Fitzgerald (of dept of psych, Michigan State Univ), to be published by Plenum, maybe late next year. It ought to amuse you.

Best regards,

Bill Powers
1138 Whitfield Rd.
Northbrook, IL 60062

(I call up that signoff as a file -- not being formal. Hurry hurry hurry).

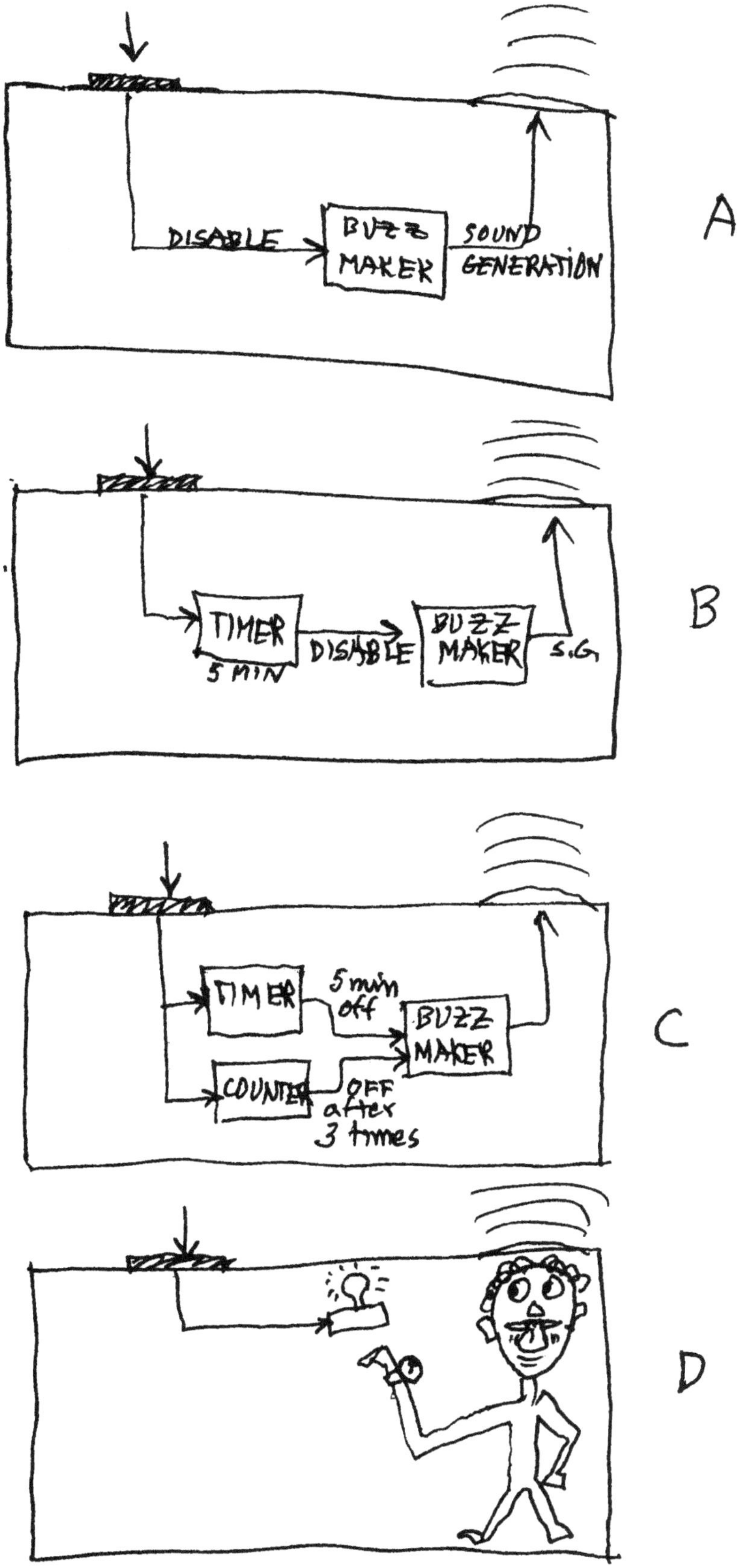
DISABLE
BUZZ MAKER
SOUND GENERATION
A
TIMER
5 MIN
DISABLE
BUZZ MAKER
S.G.
B
TIMER
5 min off
BUZZ MAKER
COUNTER
OFF after 3 times
C
D

Bill:

I've been occupied with demands from students, flat tires, irascible washing machines, and the like. I'll write a proper letter before long.

--Phil R

MEMO TO: Members of class in Management and OD

FROM: Phil Runkel

DATE: November 6, 1986

SUBJECT: What's happening here?

I'd like to tell you about something I see going on that you may not have noticed. Or maybe you have. But I feel urged to tell you what it looks like to me. You can tell me whether you agree or disagree.

I see something happening that always happens when enough people have the attitudes that OD espouses and when there is room in the rules for people to act on those attitudes: (1) people are shaping the way they carry out their duties (as students, in this case) more to serve their own purposes (to maintain the perceptual inputs they want) and less to carry out the purposes of a boss (the professor, in this case) and (2) to do that, they are drawing on the resources of people around them, not merely on their individual resources in one-to-one relation with the boss.

One panel called in Judy Small to add a perspective. All panels have made good use of relevant resources of their members. Some members are practiced group consultants themselves. (I can illustrate points with my own experience, but the experience of other consultants is just as valid as mine.) All panels have designed activities that have enabled the experience of class members to bring out features of the lesson to be learned. (By lesson to be learned, I mean connecting ideas to your own experience in a way that enables you to ponder your experience in a more fruitful way.)

I have done my best to listen carefully to what people say during the sessions. It seems to me that you have spent very little time quoting this authority and that authority. You have spent very little time wondering what might be the "right" answer. You have spent a lot of time saying things like "What this brings to <u>my</u> mind is that when <u>I</u> am doing so-and-so, <u>I</u> . . ." That is what I mean by "fruitful."

All panels have chosen topics (subtopics) from the assigned readings that they found most useful to <u>them</u> to explore with you, or possibly most useful to <u>you</u>, or both. (They are better judges of what might be useful to you than I am.) Those topics, I think, are more likely to have been topics that will lead panel members into further thinking than topics I might have assigned from <u>my</u> interests. I think, too, that the topics have "connected" with other members of the class, because the current experience of panel members is, on the whole, closer to your current experience than is mine.

Look, too, at what is happening with the panels scheduled for 3 and 10 December. The panel for 3 December has decided (or so my second-hand information tells me) to summarize what will have gone on by then. Obviously, they will be drawing on what a lot of you have done. Maybe they will do some interviewing--I don't know.

The panel for 10 December is contracting with an outside group to put on a demonstration. The outside group has been formed recently from within the College and is looking for ways to encourage the use of OD practices in the College. Our class will be one of its first "clients."

I haven't put much "structure" on any of the sessions. What I have said, in essence, is "Hey, you four people, here is a range of topics. Pick a subtopic and do something with it." To the panels for 3 and 10 December, I have said only, "Hey, you four people, do something."

Some professors would say I have abdicated my responsibility. Indeed, over the years, a few professors have said that to me explicitly. My view is that you know better than I what you are ready to learn about--what particular kind of idea you can seize upon and make your own. My view, too, is that you are capable of teaching yourselves. Maybe undergraduates in mathematics might be pretty poor at doing so, but surely graduate students in education, with years of experience in pedagogy, ought to be able to teach themselves. And it turns out you can, doesn't it? And of course you can; we all do it, every day. But you are more self-aware about it than people in other fields.

Notice the kind of magic that springs up when enough people think in the OD way. College classrooms are typically about as isolated as goups can get. Even when a professor invites an outside speaker to appear, the speaker has to ask a lot of questions about what will be relevant and then try to fit into what the professor has planned. The College OD Group, in contrast, is entering into a sort of partnership with the panel for 10 December. It will be a joint venture.

And how did it come about? I was sitting in a meeting of the OD Group. They told me they were wanting to try out an idea for something they might do at the OD Conference in February. They asked me whether my class might be an opportunity for them. I said that one or two of the panels might be looking for resources. So one of the OD Group talked to the panels.

I didn't sit in the meeting of the OD Group because someone had asigned me to do so. Dick Schmuck told me they were meeting. I thought to myself, "Ha! I might find resources there!" Not just for the OD class, but for me personally. Not professionally, primarily--I don't need professional buttressing here at the end of my career--but personally. I thought I might find in the OD Group resources to help me live the kind of life I like to live. Resources with which I can maintain the kind of perceptual input about how I can work with other people that matches my internal standards for who I am and what use I can be in the world.

That's the way OD works. When a gap of some sort comes about (in this case the gap I left in 3 and 10 December), it doesn't just sit there until some administrator discovers it and orders something done about it. People see it as an opportunity to satisfy their own needs. I saw it as an opportunity. So did the OD group. So did the 10 December panel. People feel free to act on their needs when they feel welcome. I did not feel I was "going out of channels" or "abdicating my responsibility." The OD Group did not feel they were "interfering with a professor's teaching." The panel did not worry whether they were "carrying out their assignment" or doing what would bring an "A." We all thought it likely we could work out something of mutual advantage. We were all following Powers' dictum of finding a way of maintaining our own desired inputs in a way that would not prevent others from maintaining theirs. In OD language, we found the win-win deal.

We found the win-win deal, notice, because I left the gap, because I did not dictate every action and every minute of what you should do. And because I did not think that chaos would result, that things would be "out of control." And why did I think and feel the way I did? Because I had confidence, as a devoted OD person does, that the panel, whoever the members were, would have resources to fill the gap, that other resources would be available in the environment of the panel (even though I did not know at the time what those resources might turn out to be), and that the panel would be able to reach out for resources or seize them when they appeared. And because, too, I felt trust. I trusted you to use your time in class for something other than friendly chatting (that would not have satisfied my purposes). I trusted you to welcome my way of offering learning opportunities to you--and to alter it if it didn't suit you. I trusted the OD Group to welcome me personally and to welcome what I might offer to their pursuit of their own purposes. I trusted, in brief, your readiness to seek ways to maintain your input of useful experience without preventing me from maintaining mine.

That's the kind of magic that OD brings about. You do not have to do detailed planning three months in advance. You know that other people will join you in going where you want to go. Once there is general agreement on a rough direction ("This course should have something to do with how OD applies to me"), everybody will help everybody. And you know that if you leave some free space for individual and group action, resources will appear along the way. You don't always know what those resources will be, but you know some will appear, and they do.

Notice that I have not said that the magic will occur at any time an administrator decides that OD would be nice. I have put in qualifications: "When enough people have the attitudes. . . ." "When you leave some free space. . . ." "When you have trust, when you feel welcome. . . ."

I've been talking about OD. Now let me complete this communique with some talk about pedagogy.

The usual view about transmitting knowledge is just that--that knowledge can be "transmitted." It can't. The usual view is that various things "contribute" to learning, as if someone learning something were a kind of "product" or "package" into which various people dump a "contribution." It's not. The usual view is that the teacher tells or "presents", the students put up various impediments of poor previous "preparation" or lack of listening skill, and the result is "learning." That's not the way to think about it.

People make use of what is going on around them. We act constantly, every minute, to use what is available in the environment to maintain the perceptions of input that our internal standards call for. Students make use of a lecture to the extent, and in the ways, that it can serve to maintain their desired inputs. A few people in a class will be so thirsty for new knowledge that they will pay attention indiscriminately to everything the professor says. But not many--and probably not during very many lectures, either. Most people will be hoping against hope for some idea that ties in with some possible way they can act to their profit, and they will be thinking their own thoughts, maybe about what to have for supper, the rest of the time.

What are the ways the students can use information to their own profit? The professor knows very little about that. In particular, the professor knows next to nothing about the connections, or lack of them, that will occur to this student and that during this lecture on this day.

Presumably you come into class--an elective class, anyway--feeling some gap, some descrepancy between the way things are for you and the way you would like them to be. Using Powers' word, some condition in your life or some series of events is a "disturbance" to you; it is not leading in the direction you want to go. You hope you will find a way--a "feedback function"--in what goes on in the class that will enable you to close that gap. If you do succeed in building that feedback function, that is what you "learn." You discover actions you can take (if only mouthing right answers on a test) that can bring you closer to perceiving the inputs you want.

You build the loop, the feedback function. The teacher does not; the teacher cannot. I am not saying, as professors often do, that you must do half the work. I am saying that you will inevitably, willy nilly, do all the work. All the teacher can do is make opportunity by putting something in your environment that you might not readily come across in your own searches for ways to build effective feedback functions for yourself. If the professor is clever, what the professor puts in front of you will be rich in good materials for feedback functions--for your own personal feedback functions, not just for those of the "average" person in the class. In educational lingo, this is called "meeting the needs of the students"--but I do not mean the needs the teacher thinks students ought to have. I mean the needs you do have. You do not always know the needs you have, but the chances are that you know them better than the teacher does.

I hope you are finding ways to make use of some of the ideas and skills you are encountering in this course.

THIS IS NOT THE WAY IT WORKS

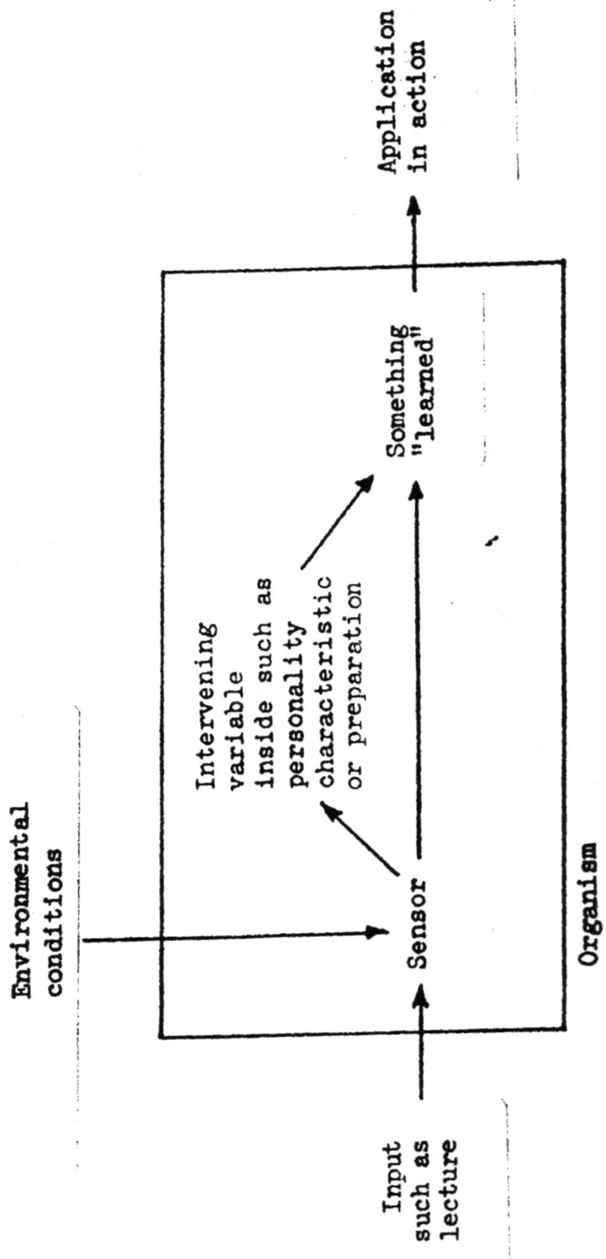

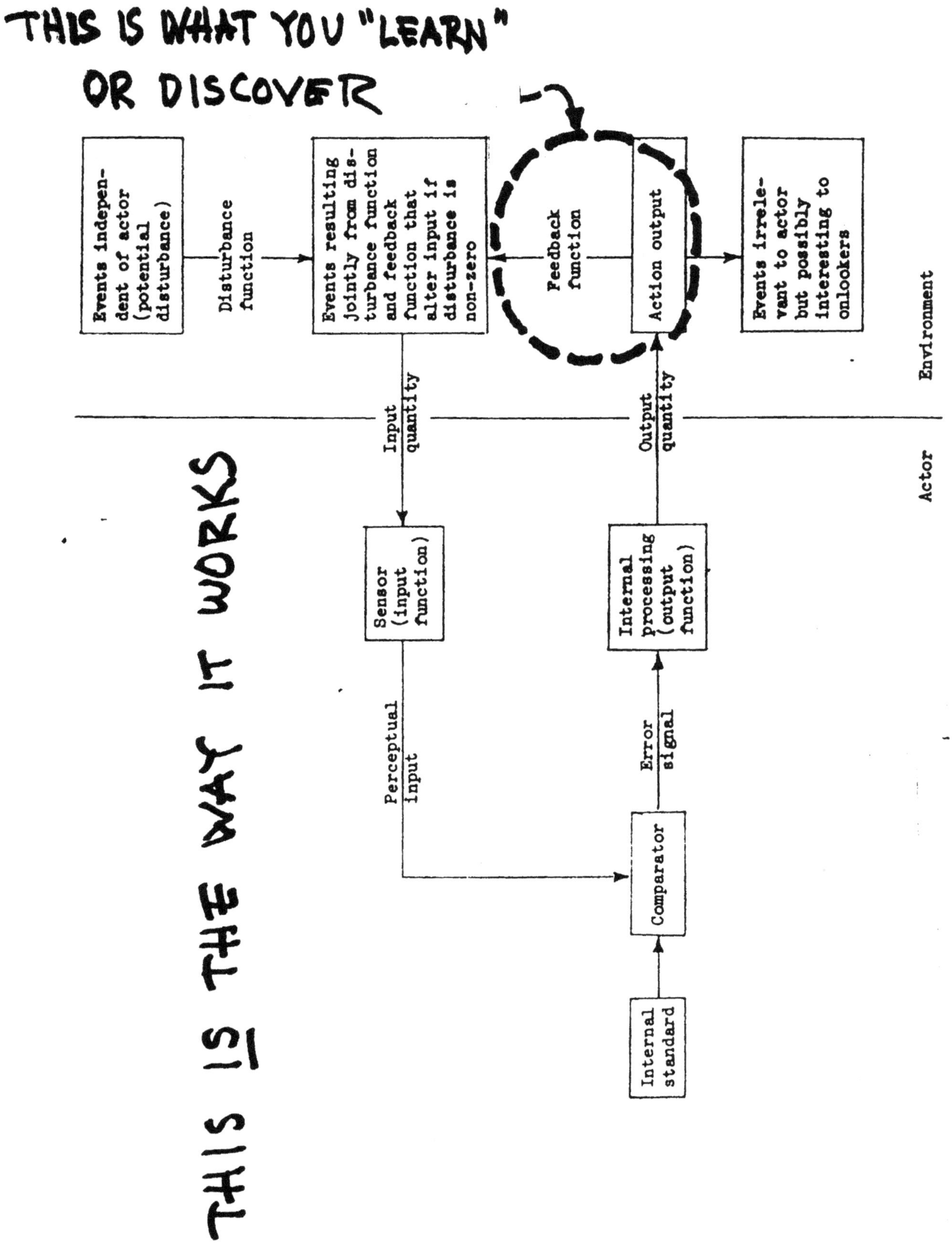
THIS IS WHAT YOU "LEARN" OR DISCOVER
Events independent of actor (potential disturbance)
Disturbance function
Events resulting jointly from disturbance function and feedback function that alter input if disturbance is non-zero
Feedback function
Action output
Events irrelevant to actor but possibly interesting to onlookers
Environment
Input quantity
Output quantity
Actor
Sensor (input function)
Internal processing (output function)
Perceptual input
Error signal
Comparator
Internal standard
THIS IS THE WAY IT WORKS

The lecture is something out there in the environment. You take some action in regard to it--listening to some parts and not others, asking questions, trying out one of the ides, or whatever. You take that action with an eye to counteracting some condition or series of events (the "disturbance") that is not what you want to be happening to you. The interaction of your action and the disturbance yields the "input quantity." Inside yourself, you compare that input with what your internal standard wants and keep acting until the input matches the standard--and keep on acting to maintain the match.

The first thing you learn is what you can do with the lecture--the discovery of what features of it (if any) you can use for your purposes. There are, of course, some other things you "learn." (The word "learn" seems to take on a new meaning here, doesn't it?) Some other things happen, too, of course. You put into memory some things you did about the lecture in case those things will come in handy later. You put the memory of those things together with other memories and construct an altered view of the potential resources in the environment. If what you did brings some surprising benefits, you might even alter the conceptions you have of what the world out there is like or of useful principles for dealing with it. Those additional happenings are messy to try to draw into the diagram, so I didn't try.

But the first thing that happens is the discovery of whether your action alters the input in the direction you want. If you have no reason to act much, if all you feel urged to do is to memorize a few sentences, then not much happens inside you, and even less happens outside you. If, however, you can use what happens inside you to try out further action on your environment, then you don't have to resent having paid out your tuition.

UNIVERSITY OF OREGON 6 December 1986

Dr. William D. Williams
1850 Norwood
Boulder CO 80302

Dear Bill:

I feel an urge to help you with your paper. I also feel an urge not to make an ass of myself. I don't know anything about economics, so I'll have to limit my comments to ideas about how to write a paper so that editors will not throw it down in frustration and readers will be led on to the next page. Even so, misinterpreting the meaning of something you have written, I may propose a change in the writing that won't do.

The only book on economics I have read all the way through (since a sophomore course in 1936) is Exit, Voice, and Loyalty by Hischman. I liked that.

You seem to be saying either that (a) control theory can replace equilibrium or maximization theories or (b) that it can supplement them in a way that explains certain embarrassments such as the Giffen effect while leaving them general, without having to claim that they operate only under restricted cricumstances.

In either case, it seems to me the reader needs to have or be given:

1. Knowledge of equilibrium and maximization.
2. The key or relevant postulates of control theory.
3. The key ideas of the Giffen effect.
4. The logic of the way traditional theory deals with the Giffen effect, in the form: The postulates of _____ and ______ give us the conclusion that _______, a conclusion that is unsatisfactory because _________.
5. The logic of the way control theory deal with the Giffen effect, in the form: The postulates of _____ and ______ give us the conclusion that ______, a conclusion that is consistent both with the Giffen effect and ________.

You may take it for granted that readers of the journal you want to publish in will already know all about equilibrium and maximization. If it is safe to take that for granted, you are lucky. In my field, I find that I cannot take it for granted that readers will know the "right" meaning of such a basic concept as norm. And I find that psychologists hold a wide variety of ideas about what the basic postulates may be in such a staple sub-field as reinforcement theory. So I always take a sentence or two to tell what I mean by my basic terms. If I cannot do it well enough in a sentence or two, I do the best I can in a sentence or two and give a reference to a book where the reader can get more.

As to "2" above, I did not find anyplace in your paper an explanation of control theory. Since you explicitly say that economists who have so far written about control theory have done so inadequately, it seems to me that you must say at least a little about what you mean by control theory. It is not enough to mention Powers's book. Readers of your paper will be justifiably outraged if you say, in effect, "You won't have even a glimmer of what I am talking about unless you go read a whole book."

Exhibit 1, enclosed, is a sketch of how I would rewrite your paper if I were writing it--which I couldn't actually do, of course, because I don't know enough. Section II in that sketch is the place where I would explain control theory. I know you will want to say as little as possible about control theory, because the paper is long enough already, and explaining control theory is not your chief purpose.

Exhibits 2, 3, and 4 show ways I have used to convey at least some of the key ideas--the flavor, so to speak--of control theory to people who haven't the least glimmering. Probably you will want to say a little more than any one of those say, and you will surely want to choose features of the theory more relevant to the Giffen effect than are the features I chose. Despite that, I hope the examples will be of some use to you.

As to "3" above, I think I understood your pages 9-15 pretty well. You can judge whether I did by what I wrote in section I of Exhibit 1. No place in your paper did I find a sentence telling succinctly what the Giffen effect is about. I tried to write such a sentence as the first sentence in section I of Exhibit 1.

And no place did I find a sentence telling why the Giffen effect is an embarrassment to economists. I found lots of sentences telling what economists have done to try to rid themselves of the embarrassment, but no sentence telling why they care to go to the trouble to do so. I found several sentences asserting that the Giffen effect was an embarrassment, but none (none, at least, that I could understand) telling why.

I am not sure whether your pages 9-15 portray the standard view of the Giffen effect or your revised view of it. It would help to make that clear.

As to "4" and "5" above, I suppose it was your intent on pages 17-22 to provide those logical connections. But I could not trace the logic at all.

I suppose my lack of knowledge of economics made it especially hard for me, but I did expect to find sentences in the form I set forth in "4" and "5" above, and I didn't find any. And I expected to find sentences beginning something like, "Control theory, on the other hand, since it postulates. . . ." I didn't find a sentence like that, either.

Overall, it seems to me, you want to convey this kind of meaning: Using traditional theory, we are driven from A to B to C, and that's not good. Using control theory, we can go from A to D to G. Ain't that a lot better? But I couldn't find a clear statement of that sort.

In my sketch, Exhibit 1, I have tried to show how you might capture the reader's interest more quickly. The core idea of the paper (as I understand it) begins the second paragraph. The actual substance begins at the bottom of page 1. In your version, in contrast, I didn't find anything to tell me what the Giffen effect was about until page 19! Exhibit 1 proposes other rearrangements of content that might do better to keep the reader interested.

I hope you can leave out most of pages 1-5. On the other hand, I hope you can add at least a few sentences about the topics you mentioned in your letter: the motivation of the paper and the implications of the conclusions. For me, those two topics are exactly the topics that make a paper memorable. If an editor likes your paper but considers those topics superfluous, the editor will say so.

There is an awful lot of passive voice in your paper. Change most of it to active.

And watch your syntax. The very first sentence says that "control theory [is a] construct. . . ." A theory is a construct? Maybe economists use construct that way, but it seems to me an odd usage. In the last two lines on page 2, you say, ". . . control theory . . . is [an] . . . analysis . . . ." A theory is an analysis? In the last four lines on page 3, you say, "Recent work . . . is [an] exception to . . . assumptions. . . ." Work is an exception to assumptions? And so on.

I encountered also a great many infelicities of word usage and punctuation. I enclose Exhibit 5 to show the kind of thing I would mark up on the manuscript if I were doing detailed editing. I hope those sheets will help you.

Your spelling is shocking. If I were an editor reading a paper from an economist who didn't know how to spell economically, I would go on reading out of a sense of duty, but I would do so with prejudice. I know you said in your letter that you intended to "correct obvious mechanical and stylistic deficiencies," but it startles me that you would send the manuscript even to friends without running it through your computer's spelling-checking program. (I am supposing that you produced the MS on a computer and that you own a spelling-checking program. If my assumption is wrong, buy yourself a spelling checker today. And a dictionary.)

I have been severe in my criticisms. I understood enough of your paper (or thought I did) so that I can see how it can be important and might even shake up a few readers--if they can be enticed to read it all the way through. So I'd like to see it published. But I wouldn't be helping you to get it past an editor if I said that it needs merely a little sprucing up.

I hope I have helped.

Sincerely,

Phil R.

Philip J. Runkel

P.S. The "0931" is not part of my address. I put in the envelope so that the mail clerk will know the account to which to charge the postage.

Exhibit 1

Control Theory

and the Giffen Effect

To derive economically useful conclusions from the principle of maximization, current economic theory is embarrassed by efforts to introduce supplementary restrictive stipulations. Maintaining the assumption of the consistency of individual and market equilibrium and adding the/assumption of diminishing marginal utility, for example, turns out to yield results in some important instances that are inconsistent. One of the important instances is that of the Giffen effect, first set forth by Marshall. To explain the Giffen effect within orthodox economic theory, several economists have offered still further restrictive assumptions.

restrictive

I believe that Control Theory offers an exceptionally satisfactory way to explain the Giffen effect. Doing that, indeed, is the heart of this paper. The explanation Control Theory offers is especially satisfactory, I think, because Control Theory can also, without change or added restrictive assumptions, explain what orthodox economic theory of consumer behavior explains. I will say more about Control Theory later on, and in a still later section I will contrast some arguments used by economists to explain away the Giffen effect with the way one can use Control Theory to deal with it.

I

The Giffen Effect

The Giffen effect occurs when a consumer can maintain access to

sketch of rewrite

an economic good by purchasing two or more kinds of tangible goods at different prices. Consider, foreexample, a couple with a requirement for transportation--a higher-level economic good that can be obtained by purchasing two or more automobiles. At a particular time, let's say, their budget permits them to purchase a Porsche and a Volkswagen. But at another time when the price of a Volkswagen is higher, the couple would not be able, given the same budget, to afford the Porsche. Instead of buying a Porsche and a Volkswagen, the couple would be reduced to buying two Volkswagens. This outcome is embarrassing, because __________[I could find no place the words to fill in here).

The Giffen effect is usually explained by using bread and meat as examples of tangible goods that could be purchased to satisfy a higher-level good--in this case, a proper number of calories.

[Put here your pages 9-15, more or less. And in the middle of page 13, explain again why the outcome is contrary to orthodox theory.]

II

Control Theory

[Explain here what you consider to be the key ideas in control theory, or at least the ideas you think to be cricial to the advantages of control theory over orthodox economic theory. Tell how control theory explains figure 1.]

III

Restrictive Assumptions

[Put here your pages 6-8, more or less.]

sketch

IV

Comparison of Two Choice-Theoretic Treatments

[Put here your pages 16-23, more or less. Make it much more clear, in each item, how the two theories lead to different outcomes. I looked for an explanation like this: "Using orthodox theory, with its postulates of ____ and ____, we must conclude that _________, a clearly unsatisfactory result. Using control theory, however, with its postulates of _____ and ______, we conclude that _________, an outcome that accepts the Giffen effect and at the same time does to violence to _______." I found no explanation like that.]

V

Discussion

[Put everything else here, omitting as much of it as you can. But add the two topics you mentioned at the top of page 3 of your letter: the motivation of the paper and the implications of the conclusions.]

11 Dec 86

Dear Bill:

Today I have on my desk no class paper, no dissertation, no proposal for a dissertation, no paper from a colleague waiting for me to read it and mark it up. Looking at my desk almost gives me agoraphobia.

After this, when I come upon a paper that takes your name in vain, I'll save you some time and bile by just nominating it for the "wrong, wrong!" file.

I always like to read your comments on philosophy of science and experimental method.

Thanks to my friend Carol Slater, I have discovered that the philosophers of science have come up with an idea that I think is very useful: "natural kinds." It has some technicalities attached to it that philosophers like to worry about, but in my simple mind it simply means that if the thing we study doesn't have the quality of thingness, system, boundary, we are not going to succeed in finding lawfulness. I think you are saying the same thing when you insist on modelling. To model, you must have a thing to model. Social scientists, when they use the word, usually mean no more than a postulated set of relations among variables. They often quite ignore the thing to which the variables are attached. In social psychology, for example, researchers frequently predict a relation between an individual characteristic of members of a group--maybe even a characteristic of only one member such as the leader--and a "group product" such as whether one team will win over another. I certainly don't think it is impossible to make predictions about the entangling of the feedback functions of members of a group. But without a model of the individual and without the idea that individuals don't care much what parts of the environment they use to maintain input quantities, the usual social-psychological prediction is like trying to predict John's pulse rate from Mary's respiration rate.

I don't think the philosophers of science are making the mistake of thinking there is some objective thing corresponding to the word "water." I think most of them would say something like: H_2O (and its attendant theory) serves as a guide to reliable observation, and presumably there is some regularity in the real world that makes those observations reliable. I think you say that, too.

In thinking about natural kinds, I think the philosophers of science are not ignoring the various levels of perception (or wouldn't if they knew about them). They are distinguishing categories that humans adopt for their own comfort in thinking from categories (natural kinds) that will reward observation reliably. For example, we might investigate whether blue goes with green, is regularly found in close proximity to. They would say that a color is not a natural kind. It is only a category that humans like to pay attention to.

What's schizophrenia? I could spill out a lot of words about that, but you'd be bored.

Thanks very much for your comments on my "coat example." They will help me to get started to construct a better example. I am hoping to start revising that paper in a few weeks.

So far (above) I have been replying to your letter of 28 September. Now I turn to yours of 18 October (how time doth fly).

Yours of 18 October is a nice essay on research method. I'd like to see it as the first chapter in the first book of a new generation of methods books. We could follow it with my "Traginology," enclosed. I think we are both saying the same thing, though of course you are more sober and specific.

At one point, you said that with the input-output method, "We're just looking for more and more general ways of describing phenomena." I think I was saying the same thing on page 76 of "Inside and Outside."

Since you already searched the Social Science Citation Index, I don't feel much of an urge to do it myself. Anyway, I want to get busy putting more words on paper.

I've added a hard disk to my computer. It didn't work right at first, but the company sent me a new controller board, and I think--I hope--it is now working properly. Certainly saves a lot of juggling disks and files.

Thanks very much for sending me "A Cognitive Control System." I like it very much. How deceptively simple it is! I can't help wondering what it will look like when embedded in the other papers that will be in that book.

I am, however, happy that Levin and Fitzgerald are turning their thoughts to dynamic systems. Most of psychology, even social psychology, consists of the study of statics. At $time_1$, they say, we got this, and at $time_2$, we got that. Well, who cares about either $time_1$ or $time_2$? Why should I care what some bunch of sophomores were doing at 2 p.m. on Wednesday, April 13, 1982? What I want to know is how those people and others can manage their lives from moment to moment and week to week then or today or tomorrow. But the researchers do not tell me what those people were doing at 2:30 or 4:00 or the 14th or 15th or in 1983. Worse, they do not tell me how those people or I or anybody else could use what the researchers discovered to manage our lives. Am I supposed to give myself a questionnaire to fill out to find out what I am likely to do during the next five minutes? By the time I've filled out the questionnaire, the five minutes are no longer here. And even if I could do it instantaneously, the resarchers would still have told me only what some fraction of people the researchers have put into some category (including me) will do. They haven't told me what _I_ will do. Yet any number of them will nevertheless advise me to fill out their questionnaire to find out what I am like. They do, They have. Social-science research is very strange.

In addition to content, one of the reasons I like to read your papers is that I don't have to use up a red pencil in replying. You write with grace and clarity. Nevertheless, I do come across a small irritation now and then. On page 7, line 2, remove the hyphen after "well." On page 16, at the end of the fourth line from the bottom, put a semicolon.

On page 4, I was interested to learn about Hershberger. On page 6, middle, you mention "Hershberger's picture of control behavior." I looked back at page 4 to see the picture. Yes, it is just as you say it is. Rationally, taking the words as they flow, I should not complain. Yet somehow the "picture" doesn't have the bold colors I expect from the fact that you went to the trouble to borrow the picture from Hershberger. And maybe the description on page 4 will indeed have bold colors for other readers. Maybe I'm getting blase' about the contrasts you draw--as my dictionary says, "from habitual and excessive indulgence." Be all that as it may, I couldn't think of any way to liven up the paragraph on page 4. Sorry. Or maybe I don't need to be sorry; maybe it's fine for most readers.

Hershberger sounds interesting. Would you please send me a copy? Or if you are willing to trust the original to the mails, I'll copy it and return it.

* Various tidbits enclosed.

Your devoted correspondent and admirer,

Phil

* Traginology, following page

TRAGINOLOGY

P. J. Runkel
November 1986

Once upon a time there was a scientific field known as traginology. * The traginologists studied the behavior of trains, especially their motivation. What made trains go? That was their passionate question.

Trains ran, stopped, stood still a while, and started off again. What caused action on the part of trains? The traginologists examined multitudes of events in the environments of trains, hunting for the answer. In the early period of the science, many absurd hypotheses were seriously considered such as time of day, weather, and point of the compass toward which the train was facing. Gradually, however, more and more traginologists became persuaded that people must be important causal events in the lives of trains.

It was J.J. Whistleblurt, now known as the "father of traginology," who first demonstrated the very high correlation between the entrance of people into trains and the trains' subsequent surge into motion. Following a carefully random time-sampling, Whistleblurt spent years studying the behavior of trains in the marshaling yards at Hackensack, New Jersey. Poring over his data, Whistleblurt discovered a pattern that was far too regular to have occurred by chance. In almost every instance that people entered into a train, it would start moving within a very few minutes, sometimes even within seconds. There were a few exceptions, as one would expect in a new science, but the overall pattern was undeniable. Furthermore, he found not a single instance in which a train started off without at least one person having entered it shortly before.

Whistleblurt also thought he discerned a pattern in his data concerning the number of people entering a train and its subsequent behavior. His thorough notes revealed that in most instances when only a single person entered a train, the train would move off, but then stop again within the yards at Hackensack. As is clear from his journals, Whistleblurt was excited by the possibility that the number of people entering a train determined the distance it would travel. Unfortunately, his data were incomplete; often a train with only one person in it would go out of his sight behind a cluster of other trains or a rise in the terrain, and Whistleblurt was unable to ascertain whether it stopped before leaving the yards. He ran valiantly after the first few, but soon discovered that he was then missing observations of the start-ups of other trains. As all researchers must,

---

* From the Vulgar Latin traginare, a variant of the Latin trahere, +ology.

Whistleblurt had to make a choice between following one hypothesis or another. Wisely, he left the question of the effect of number of people to other researchers.

So was inaugurated the golden age of traginology.

There were, of course, some who resisted the new theory, despite the persuasiveness of the data. Immediately upon publication of Whistleblurt's ground-breaking book, several traginologists were quick to point out that although his time-sampling at Hackensack was carried out with admirable rigor, it was badly biased geographically. "How do we know," they asked, "whether trains behave that way in Peoria or Laramie?" Whistleblurt's students, as everyone knows, quickly disposed of that criticism by replicating his study, with impressively similar results, in a variety of cities throughout the country.

There were some, too, who went on insisting that it was something about the trains themselves that made them go, not something outside them. One of the earliest aberrant theories of this sort was advanced by A.M. Coupling, who claimed that traginologists should study the <u>structure</u> of trains. In a soon-forgotten paper, he pointed out that trains were made up of units, and that the units were connected in a way that enabled one to pull another. One unit moved, he claimed, because the unit ahead of it was pulling it, and if that logic held for the parts, then it must surely hold for the whole. Coupling was never able, however, to answer the criticism that his logic broke down at the forward end of the train, where there was no unit ahead of the forward-most unit to pull <u>it</u>.

One of Coupling's students, P.S. Towager, went so far as to claim that the outward appearance of parts of a train might be associated with its motion. He asserted that most trains contained units that could be distinguished from one another by the naked eye. He claimed that many trains included a specially distinctive unit that he dubbed the "puller." Towager conducted a study from a front window of his house, which faced upon a busy railway line. He reported that not a single train passed by without a puller, and that the puller was always the forward-most unit.

Most traginologists, devoted to a basic science that sought underlying causes and not a catalog of appearances, ignored Towager's article. A couple of brief commentaries were published, however, pointing out that observations in other places showed that specially distinctive units were not always found at the forward ends of trains, that some trains with such units spent long periods completely motionless, and--the clinching point--that Towager had mentioned nothing about such units that could be correlated with starting, stopping, running, or standing inert.

For a while, too, the offshoot "inside school" of traginologists flourished. They claimed that it was something inside trains that made them go, not something outside such as people. These traginologists spent a great deal of time clambering about in trains, opening doors, lifting lids and caps, turning knobs, and so on, as if they could reconstruct the behavior of the whole train by amassing data on the motions of all its myriad parts, large and small.

While following up Whistleblurt's attempt to explain the distances trains travelled once they got going, one member of the inside school noted that sometimes people would pour oil into the forward units of trains. He carried out a study of trains in St. Louis and found a moderate correlation between the level of oil in the tank and the distance the train was scheduled to go. He also reported that trains were never scheduled to leave the marshaling yards if the amount of oil in the tank was less than a certain number of gallons. Critics were quick, however, to point out the flaws in his study. First, almost all the trains he studied were those on or near the main lines in and out of town, not those farther off in the marshaling yard. Second, he did not actually observe how far the trains went, but merely took their scheduled trips as a surrogate measure. Third, his argument about a minimum level of oil was faulty. Since he had never observed a train to leave the station on the main line with a level of oil below the minimum he had postulated, he had no data on how far a train might go if it did go off with a level of oil below his postulated minimum. Fourth, a correlation does not prove causality. Finally and crucially, he failed to report whether people entered the trains before they started off, and therefore his research had no bearing on Whistleblurt's theory.

Though many competing theories fell by the wayside, it was nevertheless clear, perhaps especially to Whistleblurt's followers, that his theory needed further work. The question of numbers of people continued to attract researchers. Was the entry of one person sufficient to set off a train to any distance whatever? That question was never settled to everyone's satisfaction, and now, unfortunately, it never will be.

Perhaps still somewhat influenced by the inside school, some traginologists asked whether the entry, itself, of people into the train was sufficient to cause the train to move, or whether the cause was something the people did after they got on the train. Some early studies reported high correlations between sitting behavior and the start-up of the train, but the hypothesis was discarded when several later studies showed little or no correlation on commuter trips, where the trains started up while many people were standing and continued running even though some people remained standing during the whole trip.

Perhaps the most promising line of work was that pursued by the "stoppers." They pointed out that Whistleblurt had investigated what made trains start, but had not gone on to investigate what made them stop. If people entering trains made them start, the stoppers reasoned, then people leaving trains ought to make them stop. That hypothesis attracted many followers when the first studies showed that large proportions of people, sometimes all of them, left trains within a very few minutes after the trains stopped. The hypothesis was greatly strengthened when it was discovered that very few people, often none at all, left trains while they were in motion.

The enthusiasm of the stoppers was temporarily dampened by R.B. Firstling, who argued that causes should come before effects. We could claim that people leaving trains caused them to stop, Firstling said, only if we were to observe people leaving trains just before they stopped. For a time, the entire structure of Whistleblurt's theory came under a cloud. So great was the dismay in some places, indeed, that a few universities precipitously abolished their departments of traginology.

In the tradition of true scientists, however, some traginologists doggedly continued their research. Within only a couple of years, two lines of investigation brought renewed vigor to the field. First, very careful and detailed observation of the behavior of trains revealed that in almost every instance of trains stopping, one or a few persons swung to the ground, indeed, just before the trains finally came to a halt. The first of those studies brought the criticism that the trains were already slowing when those persons swung to the ground, but the undeniable and replicated data showing that the people-leaving did occur before the actual stopping behavior brought most, if not all traginologists, back into active research.

Second, the philosophical point was made that the logic of trains need not mirror the logic of humans. That point of view weakened considerably Firstling's methodological objection. Several traginologists turned to the philosophers of science for further help in their perplexities. There is no telling where that line of work would have led had not the final tragedy befallen our science only a few years later.

The breakthrough came when W.W. Slackening exposed the false assumption being harbored by the critics of the stopping studies. Slackening followed fifty randomly-chosen trains and measured their velocity every five minutes. He reported that 93 percent of the trains exhibited one or more periods of slowing that were not followed by stopping, but by increased velocity. It was clear from Slackening's study that slowing was not necessarily a part of the process of stopping; it could be serving some other function.

With that development, the science of traginology looked forward to a bright future. Then, as we all know to our sorrow, the availability of subjects for research dropped, over a very few years, to a tiny fraction of the earlier plenitude. Automobiles, busses, and airplanes diverted people from trains in vast numbers. Passenger trains became so few that traginologists could not fill the cells of their analyses with enough cases to justify statistical inference. Some traginologists advocated an organized series of case studies of the remaining trains, but most bent reluctantly to the realization that there was no hope of building a true science on case studies, no matter how thoroughly articulated they might be, and left traginology for more fruitful fields of work. Today, a few lonely but devoted traginologists can be found poring over old data, but in a few years they, too, will be gone.

So ends our chronicle of a bittersweet chapter in the history of science--the rise of traginology to vigor and promise and its sudden sad demise--a demise, we cannot refrain from pointing out, at the heedless hands of automobile manufacturers, bus companies, and airlines.

15 Dec 86

Dear Bill:

* If you don't already know about chaos, or if you haven't already read this article (you've never told me what magazines or journals you subscribe to--or have you?), I think it will give you a good time watching your own thoughts.

I was especially taken by the left-hand column on page 49 (If we were like billiard balls, social life would be impossible; indeed, it could never have arisen; it is possible because we are control systems and keep making corrections from impact to impact) and page 57 (maybe the right brain maintains some chaos). I think there is also something here to tell me what sort of thing is inherently unpredictable. I already know something about that (more than most social scientists, I claim), but I think something here will help me to know better.

Sorry I counldn't copy it in color. You'll have to go elsewhere for the color.

Phil

Scientific American, 1986, 255(6), December.

Chaos

There is order in chaos: randomness has an underlying geometric form. Chaos imposes fundamental limits on prediction, but it also suggests causal relationships where none were previously suspected

by James P. Crutchfield, J. Doyne Farmer, Norman H. Packard and Robert S. Shaw

The great power of science lies in the ability to relate cause and effect. On the basis of the laws of

predictions. On the other hand, the determinism inherent in chaos implies that many random phenomena are more predictable than had been

or unseen influences. The existence of random behavior in very simple systems motivates a reexamination of the sources of randomness even in large

* 12 pages; snippet above.

UNIVERSITY OF OREGON

Dr. Leonard D. Goodstein, Editor
<u>American Psychologist</u>
1200 Seventeenth Street N.W.
Washington DC 20036

December 18, 1986

Dear Dr. Goodstein,

Enclosed are three copies of a piece for the "Comment" department.

Reading your instructions on page 1182 of the November 1986 issue, I wasn't sure of what details of procedure and format for treating articles you meant to apply also to comments. If I haven't done it right, let me know.

I hope you will have happy holidays.

Sincerely yours,

Philip J. Runkel

Philip J. Runkel
Professor of Education
and of Psychology

I'm With You, Howard and Conway

Philip J. Runkel

University of Oregon

I certainly agree with Howard and Conway (1986, p. 1249) that "most humans feel that volitional elements are involved in their behavior. . . ." Almost all the psychologists I know, and almost all non-psychologists too, act as if they believe that they can choose to act on events in their environments in ways to further their own goals, purposes, and "interests". Most psychologists, not only non-psychologists, act as if they believe that the input-output, straight-line causation, independent-dependent-variable kind of theory espoused by almost all psychologists applies to _other_ people, not to themselves. To make an aphorism: academic psychology is the study of _other_ people.

When academic psychologists interview an applicant for a position in their department, they typically ask the applicant, "What are you interested in?" With the answer to that question, they hope to predict the kind of research the applicant will pursue if hired. (That, at least, is the way I have heard my colleagues talk at three universities.) The prediction, of course, assumes that the applicant has the power to act in the direction of his or her interests. But if the applicant is an adherent of any of the most widely held psychological theories, it seems to me that the applicant should answer something like this: "If you want to know what work I am likely to do here, don't ask me what I am interested in. Examine instead the stimuli you will present me here or the reinforcements you will give me. Those independent variables are what will control my behavior. It is true that my interests might act in a small way as intervening or moderating variables, but the main effects will come from the independent variables _you_ provide." No applicant has ever said anything like that in my hearing.

In the tradition of social psychology, researchers often want to know whether an experimental manipulation "took." For example, an experimenter might instruct the subjects in one group to compete and those in another to cooperate, and might give a questionnaire afterward asking them the extent to which they followed those instructions. We can think of the three experiments recounted by Howard and Conway simply as assessments of the degree to which the subjects accepted the experimenter's instructions. In saying that, I do not belittle their experiments. Quite the contrary. Taking that point of view raises important questions: Why _do_ subjects so often follow instructions? How _can_ they? Why should putting some sound waves into the air around them "cause" so many of them to behave in much the same way toward the peanuts? Why do those sound waves bring an effect size as high as .74 for at least one subject and as low as zero for at least one other (p. 1244)?

Why did Howard and Conway get effect sizes so much larger and p-values so much smaller than most experiments of this sort? Why did the rats of Dember and Earl (1957) choose--every one of them in one experiment and all but one in another--the environment predicted? Why did _all_ the boys in the Robbers Cave experiment of Sherif, Harvey, White, Hood, and Sherif (1961) pull on the rope to get the truck started? Why did _all_ of them hunt for the trouble in the water supply? Why does the drugstore open on time _every_ working day (barring acts of God) week after week, year after year? Why do almost all automobile drivers, during almost every minute and second of every trip, stay on the right-hand side of the road? Why are so many kinds of behavior predictable almost a hundred percent of the time?

Psychologists can easily use their favorite theories to explain the kinds of behavior I have listed. But if those theories are so suitable, why do psychologists' predictions under controlled conditions in the laboratory fare so much more poorly than the prediction of the non-psychologist who confidently goes to the drugstore early in the morning without even calling beforehand to be sure the doors will be open?

We all act to pursue goals and purposes, to maintain perceptions of inputs from the environment that will match standards for those perceptions that we maintain inside ourselves. We do that continously, unremittingly, not in spurts punctuated by periods of passivity, not in episodes with beginnings and endings so convenient for the standard experimental designs. We act to maintain standards all the way from the muscle tensions that keep us walking without falling down to conceptions of the physical world by which we judge what actions are possible and impossible. We act to maintain preferred perceptions of intensities, sensations, configurations, transitions, relationships, categories, sequences, programs, principles, and systems. I take that list from Powers (1973).

What internal standards were guiding the behavior of the subjects of Howard and Conway? No doubt some wanted to test their own discipline, some to learn more about psychology, some to assure progress toward a diploma, some to maintain cordial relations with the teacher, some to get some free peanuts, and so on. And since internal standards differ from person to person, some subjects could perceive themselves acting according to their internal standards only by conforming strictly to the experimenter's instructions, but others would perceive a match merely by signing up as subjects, with their subsequent behavior being irrelevant. What kinds of internal standards (guides for volition, one might say) might interact in the different ways with the sound waves the experimenter put into the air? How might different subjects make use of the experimenter's instructions to maintain different perceptual inputs from the environment? That is the kind of question the results of Howard and Conway bring to my mind. I hope it is the kind of question other experimenters will pursue.

I join hands eagerly with Howard and Conway when they give up the picture of "a largely passive responder to causal influences" and adopt "the picture of an active agent who utilizes his or her unique causal powers. . . ." I join them in viewing behavior as conscious or unconscious effort "in the service of imaged future goals, purposes, and intentions" (p. 1250). That's the right start. I am grateful, too, to learn about the behavior of individual subjects (e.g., p. 1244). That's another step in the right direction. The next step is to hunt for the ways we inherit or build from experience the internal standards higher in the hierarchy that can alter standards lower in the hierarchy. For a guide to ways that kind of research can be pursued, I hope Howard and Conway will read Powers (1973). If others are disturbed by the contrast between the very high predictability of many kinds of behavior in ordinary life and the usually low predictability in psychological experiments, I hope they will read Powers also.

References

Dember, W. N. & Earl, R. W. (1957). Analysis of exploratory, manipulatory, and curiosity behaviors. Psychological Review, 64, 91-96.

Howard, G. S. & Conway, C. G. (1986). Can there be an empirical science of volitional action? American Psychologist, 41(11), 1241-1251.

Powers, W. T. (1973). Behavior: The control of perception. Chicago: Aldine; now New York: Walter de Gruyter.

Sherif, M., Harvey, O. J., White, B. J., Hood, W. R., & Sherif, C. W. (1961). Intergroup conflict and cooperation: The Robbers Cave experiment. Norman, OK: University Book Exchange.

Dec. 30, 1986

Dear Phil,

You've sent me a lot of interesting stuff this month. Among other effects, you've caused a debate between Mary and me on one side and you on the other, centering on "centering around." I know exactly what centering around means. It means centering relative to something but not on it. One centers crosshairs ON a target. One centers the outer ring of a target AROUND the next-inner ring (centering ON the next inner ring is impossible). Having managed to justify my usage, I won't even try to see how I actually used the expression -- probably irrelevant to this excellent example.

Natural kinds (Dec. 11). It strikes me, too, that philosophers, having nothing much else to do, have probably gone through all this "it's all perception" stuff long ago, and are actually trying to see if experience might have something deducible to do with Reality, on the far side of perception. It's an interesting idea that lawfulness might seem to hold particularly among the elements we experience as configurations. I do hope they're not forgetting that configurations -- things -- are perceptions. Isn't a thing a model already? No, I don't think we model things that we experience. We model what we can't experience -- imaginary things which, if they really existed, would create the lawful relationships we experience, as well as the gross things we experience. But it does seem to be true that models are relationships among things. Or at least among attributes of things (the position, weight, shape, color, price of an apple). All in all I think that philosophers would get farther if they considered action as well as passive analysis of experience.

"Traginology" is a masterpiece. Send it to the Magazine of Fantasy and Science Fiction, the most literate of the lot (manuscript found among the effects of ...). Send it to a journal. It's great. It could get you assassinated.

Levin (editor of the book that will contain "Cognitive Control") is a neophyte in control theory but very interested. He called me and we had a long talk. He'll probably join the Group. His main reason for calling was to ask about a couple of the other contributions. The main one in question was by a physicist; it was really garbage. Levin will probably have to publish it, since he invited it, but he sighed a lot. I don't know what the other chapters will be. Mine will certainly be different, I guess.

Social science research is strange. Glad you feel as I do about applying mass measures to individuals. It's simply not legit, is it?

The piece by Hershberger has a funny history. He actually gave that "paper" as a little talk at the last Control Theory Group meeting, trying to get us to be less combative. I thought his idea was wonderful. Then I wrote the paper quoting what he said, as I remembered it, and immediately called him and said he had to submit it for publication somewhere, quick, because I needed the reference for my paper. He did, and now I have to send the details to Levin to finish out the references. Sort of a time-machine I-am-my-own-grandfather affair. Therefore, there isn't any "original." My only copy enclosed.

Perhaps Susan Gulik's personality came through for you, although you really have to meet her. She's a small, collected, sort of pretty young woman, who stands flatfooted and still before a group and speaks wisdom. She was a professional guitarist of great skill (I am told), but doesn't perform now. Her sense of humor is deadpan and excruciating. For example, she was explaining how control theory helped her understand stage fright. She said "It wasn't a big AHA! -- just sort of, oh." A little high mildly surprised "oh." It broke us up. Everyone went around saying "oh" that way for the rest of the week.

Your answer to her was just as nice as she deserves. And it had some awfully funny stuff in it about being a professor.

I <u>thought</u> you and Larry Richards should have something to talk about. Did you send him the Q&A about OD? He'd be interested, for content and for the method. My only problem with OD is wondering how you get it started in a company like the one I work for, which seems to be based on the management techniques of intimidation, secrecy, and firing as many people as possible (but they don't pull that on me). It seems to me that OD is for nice people.

The Chaos article (Deç. 15) may have something to say about reorganization, still the weakest part of my theory. I'm not sure what. Did you happen to notice how the figures on p. 53, the closeups, resemble the rings of Uranus? Uncanny. My experience with these mathematical discoveries has been that they float too far above the surface of direct experience to fire my imagination. Catastrophe theory, for example, seems to describe instabilities of a kind that electronic designers frequently encounter and use -- in that context there doesn't seem to be anything special about catastrophes. But you never know what will get a mathematician excited, or afterward why it did.

The latest time-machine goodie was the Howard and Conway comment. Day before yesterday, Wayne Hershberger (to close another loop) called me and asked if I had seen that article, which was in fact lying next to the telephone with your letter.

He said he was working on a comment to send to the journal. HA! I told him, the more the merrier. Looks as if our Group has its own localized Zeitgeist. Probably David Goldstein will call next, then Tom Bourbon. Your comment is excellent; I particularly liked the part about interviewing applicants for academic positions. Just reward me, boss, and I'll be interested in anything you decide should interest me. Squeak, squeak.

You might be interested in a little chronology (have an M&M):

1974: Attended meeting of American Society for Cybernetics in Philadelphia, alone, gave paper on social systems. ASC didn't meet again until 1982.

1982: Attended ASC annual meeting in Columbus, OH, with Tom Bourbon. I gave a plenary-session talk, Tom gave a paper.

1983: Attended ASC annual meeting in Palo Alto. Eight of us gave an afternoon session plus a panel: Bourbon, Marken, Ford, Jeffry, Benzon, Robertson, Mary, and I. Talked mostly to each other.

1984: Attended ASC annual meeting in Philadelphia. 15 of us gave afternoon session, demonstration session, and a continuing demo setup. Talked mostly to each other again. Got idea of having our own meeting the next year, fooey on cybernetics.

1985. Skipped ASC annual meeting (fooey executed). Meeting then cancelled, so fooey ineffective. First meeting of Control Theory Group (September), attended by Barry Clemson of ASC, and about 20 members (peak) of the Group.

1986: Skipped postponed ASC Annual meeting (fooey successful, although invited to give talk). Asked to chair session at Gordon Research Conference on Cybernetics and did; Tom Bourbon also attended by invitation (June). Second annual meeting of Control Theory Group, attended by Larry Richards (President of ASC) and about 25 (peak) CTG members. Richards asked Greg Williams (new member) if he might want to publish the ASC newsletter (Williams already ran a Bateson-ideas newsletter from his home in Gravel Switch, KY). Richards asked me to organize a morning session at next Gordon Conference on Cybernetics, pick my own speakers, for Feb. 1988 in Santa Barbara. Richards asked me to give a talk at European ASC meeting, St. Gallen, Switzerland, Mar. 15-19, 1987. Third meeting of Control Theory Group announced for Sept. 23-27, 1987. First issue of new ASC newsletter assembled by Williams, to reach all ASC members, all subscribers to Bateson newsletter, all members of CTG. Big issue with 12 papers -- you'll get one mid-January. Going to Phoenix in February; making four half-hour videos on control theory, at suggestion of Ed Ford. Switzerland in March, participate in symposium at Midwest Psychological Association meeting in May, organized by head of department of Ray Pavloski, new member. Turned down invitation for another book chapter and a talk in Toronto in June at some Systems thing. Getting picky.

If you could find some way to plot the activities in these years, do you think the curve might look a bit exponential?

I think I have to publish another book. Unfortunately, every time I try to start it, it comes out all academic, which isn't very interesting. I don't want it to be trivial; it should have some of my harder math stuff in it so I can talk about interesting models, but I really can't be scholarly -- don't know how, haven't the study time or a good library. And I want it to serve as an introduction, at least the early parts. I really don't feel up to a serious book. Oh, blah, blah, blah, I'm just complaining. Reorganization isn't comfortable. Why don't you write it for me and give me all the royalities? Why doesn't somebody just give me a lot of money? Why don't I ever win the Lottery?

I guess this is the end of the letter, Brother Phil.

Best,

Bill.

610 Kingswood Avenue
Eugene OR 97405
5 January 1987

Dear Brother Bill:

All right, you've found the exception, maybe the only one. When you center one circle "around" another, they are concentric; they have the same center. So centering one circle around another is a short-hand way of saying that the point one circle centers upon is the same point the other one centers upon. I don't think that particular use of the phrase would grate too badly upon my ear. But few writers who use the phrase mean what you say: centering in relation to something but not on it. Most do mean on it.

Your letter of 30 December has a fine bright playful tone to it. I am very glad to see that after a few lugubrious earlier letters. Good good. ¶ I recently found an author saying something like this: Recent articles "focus around" the topic of. . . . How do you like that?

I think natural kinds is an idea I badly need, and you too. I do not quarrel with your comments. But I think it helps a great deal in distinguishing between what I call the method of frequencies and the method of specimens. Helps me, anyway, and I think will help some people who will read what I will write.

I'm very glad you liked "Traginology." It has drawn an unusually high percentage of volunteered comments--about 50 percent by now. Varied comments, of course.

You did not send the Hershberger.

Glad to have the personal notes about Gulik. Your description of her "oh" makes her sound like my friend Carol Slater.

About my correspondence with Larry Richards: To what do you refer as "the Q&A about OD"? That doesn't recall a connection for me.

Well, you don't have to be nice already to participate in OD and profit from it, but it's true that you must at least have moments when you want to be nice. You'd be surprised how many people do. If, however, people start out believing that the only way to deal with the social order is to get the draw on others, hoard hole cards, push annoying people out of the way, and so on, then it takes a long time to get the people ready for the part of OD where the most profit comes. But there are gentle ways of doing that. The ungentle part comes during the profitable part where people face themselves with

Powers Page 2

the pains they have been bringing on themselves and others. But it is like what you said about conflict. The people who have not caused themselves or others much pain catch on to OD quickly, are ready for the good part from the start, and their sorrows during the profitable part are easy for them to get into and out of.

One of the troubles with nice ideas from the mathematical physicists (chaos, catastrophes, fractals, dissipative structures, and years ago set theory) and the philosophers is that some social scientists seize upon them for prestige purposes--to be in the vanguard, to hope to be one of the architects of the new paradigm, and so on. I am sure that some of the members of the CSG have such hopes. At first flirtation, I think the best use of the new idea is as metaphor. But if that doesn't lead pretty soon to what you mean by modelling, then it is better to put the idea aside for a few years than to add more redundant vocabulary to the discipline by continuing to use it as metaphor.

When you said the article was lying by your telephone when Hershberger called, did you mean the chaos article? What did he like about it?

Thanks for the account of the exponential interest in you and control theory. Very interesting. Glad to have the detail.

I am very glad you are stewing with thoughts of another book. It is time. I don't know whether it is time for other people, but I get the impression that it is time for you.

You do write interesting, meaty, and entertaining letters.

Add before set theory, information theory. Though the later did put some useful techniques into statistics.

Yours,

Phil

Phil Runkel

6 January 87

Dear Bill:

In your recent letter, you made some comments about natural kinds and configurations--"knowing" a thing is there because of the configurations constructed from the sensory inputs.

I think it is only partly a matter of configurations. (I can't believe I am saying anything youhaven't already thought of; I'm writing this mostly because I think I can use the words later sometime.)

You see something you call "tree." You walk toward it and just about at the moment your eyes tell you it ought to happen, you get bumping sensations from your tactile organs. You decide, on the basis of your interpretation of visual configuration, that you will walk <u>past</u> it. You do. As predicted, you get no bumping sensation. You feel wind brushing your cheek. You predict that if you get close enough to the tree, you will hear a rustling sound. It happens. And so on.

You see something you call a copper wire. It tastes like copper. You hit it with a hammer and it dents like copper. You connect it to a battery and a voltmeter and it "behaves" like copper. You do all those things with various pieces of stuff that look like copper wires. The sensory experiences happen in the same way over and over.

With some things, the spatial and temporal patterns sensed happen very reliably. As you get to know <u>what</u> to predict (no wind, no rustle; no battery, no movement of the voltmeter) you can even get to the point that the pattern of sensory experience happens <u>every</u> time.

With such highly reliable experience, I am going to act as if there is something out there that enables me to construct patterns out of my sensory experiences, from the energies that impinge on my sense organs. Once I form expected patterns in my neural net, they serve in the same way time after time. I can walk past the tree every time without bumping. Whatever is out there does not fool me by sending out energy that fools the pattern I have formed into thinking the "tree" is one place when it is actually another. In brief, there is something out there I will call a "natural kind."

Other "things" are not as predictable. Take temperature. As I walk along on a summer day, I get intensities from sweat glands, temperature sensors on the skin, and I don't know what all, that give me a sensation of overall temperature (heat loss). But suddenly I get a sensation of cold when I touch some ice. Suddenly I get a sensation of hot when I touch a piece of metal in the sun. One experience of temperature fails to help me predict another experience of temperature--unless I pay attention to the natural kinds with which the sensation of temperature is associated. Then I can do much better. Temperature is not a natural kind. It is a feature of the behavior of natural kinds. More exactly, it is one kind of source of intensity, sensation, configuration, and so on that I can construct when I get energies from natural kinds. No "variable" is a natural kind. (At least, that is what I claim.)

The natural kinds that are notoriously hard to predict are living creatures. You claim that we have been looking at the wrong variables. We have been trying to understand the properties of copper by measuring the brightness of the light bulb. Not a strict analogy. Hard to construct an anlogy, because copper wires don't have sensory organs. Newton told us that there is an equal and opposite reaction to every action (as in the feedback function working against the disturbance function), but that doesn't help me either. If I hit the copper with a hammer, the reaction of the copper eventually stops the motion of the hammer, but the copper doesn't bring itself back to a desired shape; it stays dented. I guess I'll have to give up trying to construct a suitable analogy.

Anyway, you claim that we should be looking for what always happens. We should look for a variable or its first, second, third derivative, that is controlled.

But it doesn't do to measure the intensity of light in this room and then measure what John's iris does if John is sitting in another room. The variable must be associated with a natural kind. And it doesn't do to measure the intensity of the light in the room and then take the average diameter of the pupils of twenty people in the room, some of who may be looking the other way, some of whom may have their eyes closed, and so on. And that is a strict analogy with what happens when a psychologist gives a "treatment" to a bunch of subjects.

It should be instructive that psychology's best successes have occurred in "psychophysics"--in studies of behavior governed by the lowest levels of control systems, where you would naturally expect the greatest uniformity from person to person. The uniformity of low-level function is what permits, for example, the construction of the Munsell color solid. But even there, the successes have come from studying individuals first, and then comparing patterns between individuals.

As you point out, control typically has more lag in the higher control systems. To detect what people are holding constant at the higher levels, therefore, you must watch the behavior over longer periods of time. In social life, I don't think there is any hope of detecting the equilibrating processes unless you watch what social psychologists call the dynamics--the moment-to-moment changes in behavior as people try to construct feedback functions to maintain desired inputs. And of course since the disturbances keep changing, the actions keep changing. When a person is holding an umbrella over his head, you can conclude that the person is using it to control some input function if you can see some invariance as environmental events change. If gusts of wind come and go, and the person's muscles act to keep the unbrella over his head, you can conclude that the person is acting to control the sensory input from rain or sun. Similarly if the angle of the umbrella changes as the angle of the sun or rain changes. And that is what you must look for also in social interaction in a group. The usual before-and-after studies are hopeless. That is the reason that consultants know so many things about interaction in groups that experimenters do not. The consultants have watched the dynamics, moment to moment.

Sometimes the utterances of consultants sound very much like control theory. "He did that because he is trying to maintain his self-esteem." Or a consultant may say to a participant, "I notice that you have not been looking at Joe for the last five minutes. I am wondering whether you want to avoid seeing the expression on his face." That means the consultant thinks the participant said something that hurt Joe's feelings, that the participant doesn't like to hurt people's feelings, and that the participant is controlling his level of causing hurt by acting to prevent evidence from reaching him.

In that example, by the way, the consultant is deliberately prodding the participant into an experience that will produce conflict within the participant. When consultants speak of "dealing with" conflict, they mean finding a new ordering of actions controlled at lower levels that will permit the person to receive more kinds of information from and about other people <u>without</u> running into internal conflict. The participant can learn to say something that will help maintain one internal standard but will at the same time hurt Joe, and then trun quickly to Joe to help him repair the effects of the hurt and reestablish trust between them, permitting continued cooperative behavior. That sort of thing, as a simple example, is what I mean when I say that the best organizational consulting enables reorganization.

It impresses me that good consultants can hold very different theories. I hear them explaining their behavior in terms of theories for which I have very little respect. I think what is happening is that the theories (mine, too, no doubt) serve more as mnemonic devices than as guides to action. That is, the consultant acts mostly from intuition, and then keeps track of the course of events by hanging memory on the rack of the theory, rarely asking about the logical fit. It is enough that the consultant can say to other members of the team or to himself or herself, "So what happened then was ..., so now I think we are ready to ...." And from knowing the lingo, the other members of the team can get a pretty good notion of the kind of bare action that took place, regardless of the kind of theoretical frame the person is using to call up the picture. The handbooks for organizational consultants are full of the most disparate theoretical viewpoints you can imagine. "This exercise illustrates how ...." and then the author will spill out a theory I think is nonsense. But I use the exercise anyway, because I can see how it will pull participants into awareness of some dynamics I want them to be aware of, and I can hang the events on <u>my</u> theory as I guide the participants through the exercise.

A consultant can be a nincompoop according to the standards of the academic experimenter, can espouse and proclaim a theory that the academic experimenters have logg ago shown not to hold water, and yet be a very competent consultant. I suppose the people who painted those wonderful pictures on the walls of caves in Spain and France had some pretty wild theories about pigments. They must have had some pretty wild theories about light, too, to paint so many of the pictures in places where they could work only by torchlight. In the middle ages, people had some very wrong notions about ballistics. But they managed to batter down a lot of walls with their cannons.

Well, where am I?

I guess I am back to my usual complaint--that psychologists study variables, not people.

And I guess I have run out of steam. I ought to round off this wool-gathering with a nice emphatic or dramatic ending, but I'd probably have to go back and rewrite everything to make it lead up to the ending. Well, you'll have to wait until another time for a decent ending.

Phil

Jan. 9, 1987

Dear Phil,

This is the first letter I will have sent out this year -- with the right year on it.

I wonder if the "natural kinds" problem can really be solved within our current knowledge about perception. Here's something that may explain my doubt, from a smarter man than I (hope I haven't sent you this before):

From: Niven, W. D.; The scientific papers of James Clerk Maxwell Volume II; New York: Dover (1965), pp. 776-785. [Out of print, dammit].

original: Maxwell, J. C.; Thompson and Tait's natural philosophy. Nature, XX (1867).

Maxwell comments on the ideas of "the two northern wizards," then delivers the following:

The Ignoration of Coordinates

> In an ordinary belfry, each bell has one rope which comes down through a hole in the floor to the bell-ringers' room. But suppose that each rope, instead of acting on one bell, contributes to the motion of many pieces of machinery, and that the motion of each piece is determined not by the motion of one rope alone, but by that of several, and suppose, further, that all the machinery is silent and utterly unknown to the men at the ropes, who can only see as far as the holes in the floor above them.
>
> Supposing all this, what is the scientific duty of the men below? They have full command of the ropes, but nothing else. They can give each rope any position and any velocity, and they can estimate its momentum by stopping all the ropes at once, and feeling what sort of tug each rope gives. If they take the trouble to ascertain how much work they have to do to drag the ropes down to a given set of positions, and to express this in terms of these positions, they have found the potential energy of the system in terms of the known coordinates.... they can express the kinetic energy in terms of the coordinates and velocities.
>
> These data are sufficient to determine the motion of every one of the ropes when it and all the others are acted on by any given forces. This is all that the men

at the ropes can ever know. If the machinery above has more degrees of freedom than there are ropes, the coordinates which express these degrees of freedom must be ignored. There is no help for it.

It seems to me that our knowledge of the world consists of empirically-discovered relationships among perceptions, and nothing else. We are the bellringers, tugging at the ropes, feeling and seeing how they behave under our efforts, but limited forever to that bellringers' room that belongs to human beings. We act and we sense; what we act upon may have immensely more degrees of freedom than what our senses report. We experience a *version* of the universe, the version created when all the degrees of freedom that actually exist are projected into the space defined by the degrees of freedom of our human senses.

If there are any "natural kinds", I believe they must be natural kinds of perceptual interpretations. To put that differently, I think that separating natural kinds into those aspects imposed by our human perceptual apparatus and those imposed by external order is, at present, impossible. "There is no help for it."

One puzzle to which I have repeatedly returned is the age-old one of why perceptions appear the way they do. In my model, they're all alike -- everything consists of trains of frequency-modulated impulses. The world sure as hell doesn't look like trains of frequency-modulated impulses. This is a problem.

Pursuing this matter into the slough of introspection, I've boiled it down to a simple sort of question; an example is, "Exactly how is *blue* different from *touch*, in my own perceptions?" The stock answer is that these are different qualities of sensation, but I wanted to know more than that. Just *how* do these qualities differ in direct experience? I'm not asking for discourses on what makes the qualities different -- I'm looking to see if there *is* a difference that I can see, however we might talk about it. So far I have failed to see any difference at all, other than the fact that *this* sensation is not *that* sensation. Other than the fact that they appear in different "places" in consciousness, there isn't any perceptible difference as far as I can see.

This is actually an encouraging result, if this sort of thing can be called a result. All perceptions could, in fact, be identical signals, without contradicting direct experience. The meaning or quality of any signal is determined by its place in relationship to all the other signals, and by the way it changes as the others change. In short, the way the world appears could indeed be the way an organized set of identical but interdependent signals, varying only in magnitude, would appear, to some observer who could receive them all. But I am probably overlooking what this implies about the Observer. "There is no ..."

I corresponded a bit with Paul Chuchland, the philosopher, about this, though not revealing my research methods. He espouses what he calls a "network theory of knowledge," in which no one datum (read: perception) has meaning in isolation. That's pretty much the solution I dimly see. Pounding is defined in terms of all things pounded upon; copper wire is defined in terms of pounding and all other actions applied to it. And the perceptual results, of course. There is no such thing as the sound of one hammer pounding without anything to pound on.

My usual route through these ideas ends up where I am now: with communication. How come I can write you letters and you can understand, or appear to understand, them? It's just a good thing that people who run power stations and fly airplanes don't worry about things like these. I'm all boggled out.

Best,

Bill

Bill

22 Jan 87

Dear Bill:

* Somebody sent me issue No. 7 of Continuing the Conversation--a title that leads one to think the inventors must have combined a desperate casting about for something not run of the mill with a belief in word magic.

I am delighted to see the whole issue devoted to you.

Before I comment on the articles, I have some requests.

1. On page 5, middle of left column, you say that "LRV is a generalization that follows trivially from control theory." I looked in Design for a Brain and couldn't find the LRV in the index. I don't have any other writings of Ashby handy. I scanned your "Purposive Behavior" again and couldn't find anything relevant. Whoops. Just now I thought of looking in Modern Systems Research for the Behavioral Scientist, ed. W. Buckley. There I found Ashby's "Variety, Constraint, and the Law of Requisite Variety." I found no difficulty reading it, but if you were to ask me how the LRV "follows trivially from control theory," I wouldn't be able to do it. Request: Can you tell me in about five sentences how it does?

2. Please send me a copy of your "Pylyshyn and Perception."

3. Please send me a copy of your "Systems Approach to Consciousness." If you don't have copies, I'll go to the library and get my own. Let me know. But if you don't have copies and will go to the trouble of making one, I'll compensate you by sending back to you ten free copies.

Comments on contents:

I enjoyed very much reading your correspondence with others.

Of the others, I thought Pavloski (p. 8) offered the most meat. I thought he got down to what you call modelling. (I'm glad I didn't have to settle for Ross's description (p. 6) of Pavloski's work; it certainly gave me no clue to what I found out from Pavloski's own article what he was up to.) But I wish Pavloski had been more specific than he was (an inch and a half from the bottom of the left column on page 9) in the sentence: " High reactors also show . . . ." Researchers often use the unqualified plural when they actually mean "a statistically significant proportion of high reactors." I do not know whether Pavloski was doing that or whether he actually meant all the high reactors.

What Pavloski says about the "intake-rejection hypothesis" is very similar to what I am trying to tell you about natural kinds (top of right-hand column on p. 9).

* Continuing the Conversation, Issue 7. All issues (#1, Spring 1985 through #24, Spring 1991) have been recreated and are posted at the website.

Valach's piece (p. 18) may be welcomed by somebody, but not me. I think it's pure guff.

I think Ross (p. 6), Robertson (p. 11), and Ford (p.11) are struggling to get hold of control theory but haven't yet done so.

Ross's first vignette on page 7 is full of superfluous ideas. He says his vignette is "an extension of the ideas of control theory. . . ." It sounds to me more like an assimilation of the ideas of control theory to current ideas from psychology that he wants to keep. I do that too, but of course I'll claim that the ideas I want to keep fit better than his.

On page 6, the second paragraph beginning in the right-hand column, Ross speaks of "ecologically valid phenomena." Well, validity doesn't apply to phenomena; it applies to conclusions or inferences or theories. But maybe that's just careless writing. Beyond that, however, "ecological validity" is a fancy term used by social scientists to mean that you will see the same phenomenon in a variety of situations where you have no notion of the range of disturbances that might be affecting behavior. It's an idea that belongs with the method of frequencies, not with the method of specimens.

On page 7, first paragraph at top of left-hand column, Ross speaks of a "powerful explanation." I wonder what he means by that. Does he mean P less than .001? As in my complaint about Pavloski, what does Ross mean when he writes "the subjects . . . can reduce error to near zero. . . ."? Does he mean that <u>all</u> the subjects did? And how near to zero? Judging from the content of Rich's experiments (as described by Ross) and from what I know about how social scientists typically measure those variables, I doubt very much that Ross meant <u>all</u> the subjects.

Goldfarb (p. 13) sounded good to me. I mean that his methods fit with my own experience, and I think they are also a sound extension into clinical practice of The Test. And his article is free from the excess baggage that Ross put into his first vignette. It would be nice if Goldfarb were to undertake a carefully measured experiment.

Goldstein (p. 10) seems to be to be using control theory more aa a personal mnemonic device rather than as a theory. But I grant his point that he can use the control theory ideas to remind himself about what he wants to do (or did do, after the fact) in all his therapeutic techniques instead of having to hang his actions on half a dozen disparate bodies of lore. I think, nevertheless, that Goldstein would have a hard time designing an experiment. For example, he locates "resistance" outside the subject and instead, in the head of the therapist--something "the therapist perceives as counter to progress." That kind of thinking is going to lead to a wrong design. Generally, I found the article dull.

I liked the arguments of W.D. Williams (p. 14). Of course, I like anything that exposes the holes in economics. Williams's piece is after the fact, looking backwards, but it is a start and has clear specifications for how one might design an experiment--I think. It would be nice to try to get some good estimates of the kinds of higher-order reference signals that people use that bring about the Giffen Effect--including higher-order reference signals that result in what we call criminal behavior. His Giffen Effect depends on people"having a budget." What if you don't "have a budget," but just go out and rob a bank? And I was happy to find that somebody had cleaned up his prose. That draft of a similar paper he sent me was horrible.

I am happy to have the bibliography at the end of the newsletter.

I will send in my $4 to subscribe to the newsletter. I will not join the Amer Soc for Cybernetics. But I am certainly willing to risk $4.

By the way, in Richards's letters to you on pages 4 and 5, he seems to be taking the tack academicians often do about "open exchanges of views" and so on; namely, that if researchers get together and yak at one another, something good will happen. They like phrases, too, like people "being exposed" to one another's ideas. I take more the view that you do in your letter to him of 28 March 86 (p. 4).

It's a problem. If you don't listen to other people, sooner or later you are going to miss a good idea. Sooner or later you will miss the benefit you would get from a person willing to criticise your own ideas. But how many bores do you have to sit through to find the very few people who will give you those benefits? Maybe there is no help for it. Whom can I trust to screen out the bores? Maybe they will screen out as a nincompoop the very person who would bring me those benefits.

The best solution I know is to keep conventions small. The CSG is now at about optimum size. If it gets bigger, you will have to sit through more bores. And the opportunities for intimate conversations where people can strive to understand one another by probing questions and by efforts to paraphrase what the other person is saying will diminish. The growing mass of strangers will encourage formality. And so on.

But if you try to cope with size by dividing the membership into specialties, then you narrow the vision of everybody and lower creativity. There are better ways of keeping size small.

But I don;t have time to think about that right now. I must go to class.

Yours,

Phil

23 Jan 87

Bill:

I wonder whether this has anything to do with the disappointing amount of attention your 1973 book got. The decline in numbers of academic and research psychologists began about 1976, it looks like. Were they then looking so hard for jobs that they didn't want to seem radical? Or was it so easy to find jobs that they didn't have to read much? I don't know what the ratio of job openings to applicants was during those years. Or maybe only a certain percentage would have read your book anyway, and when the total number goes down, that number goes down, too.

Anyway, the total academic market for your ideas seems to be declining. When you compare that with the recent rapid increase in numbers of people showing eagerness about your ideas, the increase should make you all the happier.

--Phil R

(I would expect only a tiny percentage of clinicians to get interested.)

Figure 2
Number of PhDs Granted in Health-Service-Provider and Academic/Research Subfields in Psychology: 1960–1984

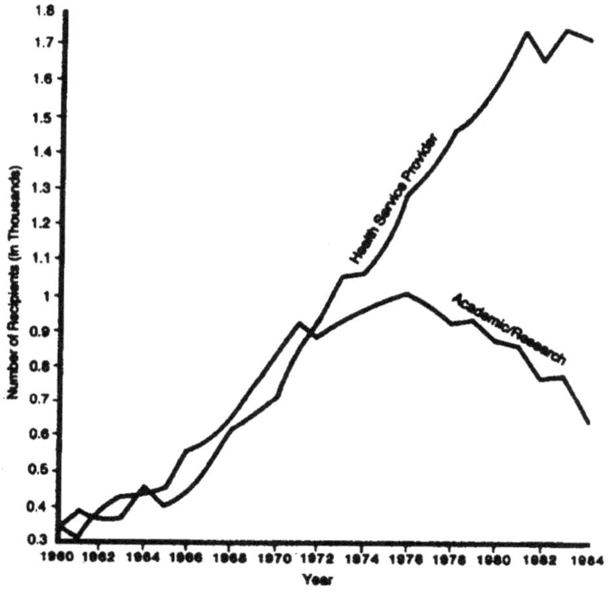

Note. Data for 1960–1982 PhDs are from *Science and Engineering Doctorates: 1960–82* by the National Science Foundation, 1983, Washington, DC: U.S. Government Printing Office. Data for 1983 PhDs are from *Summary Report 1983: Survey of Earned Doctorates* by the National Research Council, Office of Scientific and Engineering Personnel, 1983, Washington, DC: National Academy Press. Data for 1984 PhDs are from *Summary Report 1984: Survey of Earned Doctorates* by the National Research Council, Office of Scientific and Engineering Personnel, 1986, Washington, DC: National Academy Press.

Jan. 31, 1987

Dear Phil,

You're getting ahead of me again.

Natural kinds: The bundles that hang together -- aren't they exactly what we retain after going through a whole childhood of perceptual reorganizations? We keep those ways of perceiving that provide something to perceive, not just once but over and over. The most important of these perceptions are the ones we can affect, and learn to control.

But are those "natural" or "artificial" kinds? That depends on your epistemology. Ernst von Gläsersfeld would call them constructions. The implication is that there is more than one construction possible -- equally consistent, equally controllable. Are you looking for a way to find THE construction?

When I see someone else keeping something constant, it's constant in MY perceptions. Am I picking the right constancy, the one the OTHER is perceiving? Can't tell. Maybe when his is constant, mine happens to be constant, too. You hold your end of the see-saw a constant distance from the ground -- I say you're holding MY end a constant distance from the ground. If I disturb my end, you resist -- see? I don't think the answer to this question is going to turn out to be simple.

Had a thought the other day, about predicting behavior. Under the usual approach, the point of modeling behavior would be to predict it in the future. In that case, all the model has to do is predict, buy any means fair or foul. My approach also uses prediction -- but not to predict future behavior. The point is to refine the model. In my model, all the constants have to be meaningful -- relate to something physical or at least measurable. I match the model to behavior in order to find out the values of the constants, not in order to predict behavior. I would then look for some other way to measure the same constants. That's what I call learning something about the organism.

Your comments (22 Jan) about Continuing the Conversation were unerring, right down the line. Mary and I laughed about how perfectly you characterized each person. Some of them I consider to be aspirants to knowledge. I'm happy to see that they understand a little more about control theory each time they try, and I don't criticize what they do. Next time will be better. Others are hopeless. I ignore them, pretty much. And some are obviously accomplished thinkers who are trying hard to make the transition, and generally doing well, considering. I expect to learn from them. Everyone has some area in which there is a good grasp of the principles. The rest will just have to await the necessary unlearning. I'm satisfied. Think how hard it would be to understand control theory if you didn't even want to!

The curves you sent me were a bit discouraging, but there are two rays of sunshine: 1) Every new PhD in control theory has that much more influence in academic psychology, and 2) It's probably time for that particular curve to decline all the way to zero, while ours rises.

Enclosed is a POT full of papers. You may have seen one or two. I don't need multiple copies of any. If you really want to warm up the Xerox machine, make me a copy of Vallacher and Wegner, "What do people think they are doing? Action identification and human behavior." Psychological Review, "latest issue" according to Tom Bourbon, who wants me to read it. Also tell me what you think of it.

I once encouraged a guy who wrote to me when I was at Dearborn Observatory at Northwestern; he had a new space drive. That was 25 years ago. We exchanged a few letters, and I finally told him he didn't know what he was doing and please leave me the hell alone (after trying a more diplomatic approach). The last time I got a letter from him was last December. Motto: if you don't give everyone the benefit of the doubt, you might miss something that is well worth missing.

Best,

Bill

17 February 1987

Dear Bill:

I have at last read your "Systems Approach to Consciousness." *

I don't have the words in my head of the poem about carrying the news from Aix to Ghent, but they do not go something like this:

For want of a shoe, the nail was lost.
For want of a horse, the shoe was lost.
For want of a rider, the horse was lost.
For want of a message, the rider was lost.
And so on.

That came from musing on the criterion that you can tell the reference signal higher in the hierarchy by looking for the one that must control the other. You turn left at the corner so that you can get to the library. You don't get to the library so that you can turn left at the corner. But my exercise presented me with some odd results.

For want of a message, the rider was lost.

Poor rider, he has lost his purpose. What will he do with himself now?

I got even stranger results applying the exercise to going to the library:

I wanted to get to the library because I turned left at the corner.
I wanted to read a book because I went to the library.

If you phrase those a little differently, they make obvious nonsense:

I wanted to get to the library so that I could have turned left at the corner. (If it makes you happy to make left turns, go ahead. What does the library have to do with it?)
I wanted to read a book so that I could get to the library. (If you like sitting in the library, that's fine. Do you need an excuse?)

In the earlier phrasing, I seem to be saying:

Having turned left at the corner, I found myself following my customary path to the library and was seized with a sudden desire to go there.
Finding myself in the library, I thought I might as well read a book. I couldn't think of anything else to do.

* Powers, William T., (1980) *A Systems Approach to Consciousness*, in Julian M. Davidson and Richard J. Davidson, editors, *The Psychobiology of Consciousness*, Plenum Press, New York and London, 217-242.

Well, there is an example for you of the way we poke around in our minds hoping something useful will turn up. And of how we turn up a lot of uselessness. Bleagh, as they say in the comic strips.

Some writers hit on a pretty good idea and then keep repeating it in paper after paper. But I never get bored reading something of yours I haven't read before. Sometimes you are indeed repeating a topic I have heard from you before, but you always go at it from a new point of view, so I always get increased clarity. But most of the time you are elaborating, expanding, revising, and I get more than clarity; I get larger understanding. So it was with this paper.

I was disappointed that you did not say, in the end, "Consciousness appears when...." You almost did, but not quite. I myself keep feeling that I can complete that sentence, but I keep finding that I cannot quite do it. Maybe someday someone will figure out how to do it experimentally. That would be nice.

Anyway, I thought your expositions of purposes and hierarchy were rich, and your explanation of modelling expanded my understanding of what you mean.

On page 25 (the mimeographed MS you sent me), you write:

> It is not as if a single spasmodic action had to produce a predestined future consequence. The control system is always right there.... (one level of sensed consequence could be a steady approach toward some final relationship).

That's easy to say. You may remember that I tried to deal with that matter in my "Inside and Outside." It gets harder to say at the level of principles, where the "approach" is usually not steady and where consciously, at least, one often wonders whether he is indeed "approaching" or even what he is approaching. And then there is the matter of trying something that doesn't work to maintain the specified input. And then there is the matter of postponing the approach to this principle to pursue that--perhaps only because the environment seems to offer a negotiable path to this one at this moment but not to that one. It gets horribly complicated.

I agree that one must postulate that "the control system is always right there." But why doesn't it scream when we go haring off after another reference signal (haring off to reduce <u>that</u> error), postponing attention to this one? At lower levels, I agree, we get some kind of weighting so that everything gets approached to some degree at the same time, or we get some degree of conflict. I have no trouble with walking to the library, or even seeing a friend on the way and dithering for a few moments over whether I should continue to the library or stop to talk to my friend.

But at higher levels such as principles, where control is much slower, where delays and interruptions are exactly the kind of thing the level of principles is built for, I can't write a scenario, much less draw a diagram.

With a hundred principles yammering for attention, if you will allow me to anthropomorphize, how do we persuade some to be patient and quiet while we attend to others? Maybe I am thinking of error signals as rubber bands, always exerting their pull until they become slack (when error is at zero). When we see no likely feedback loop through the environment (when we judge that the stuff out there does not offer us one) to reduce the error in this perception, does some weighting from a higher level tell this level not to worry, to wait until an opportunity arises? I know I'll never build a model out of that kind of thinking, but it's the best I can do at the moment.

I have no trouble imagining the maps of the world (system concepts) that go into memory. I have no trouble imagining a kind of searching through combinations of weightings of inputs from those memories to contstruct weightings for output signals. I have no trouble imagining a weighting that means that the best action to try in regard to this error is none at all. But I can't get out of my head that picture of all those reference signals yammering for attention.

Well, my principles tell me that I don't have to understand everything, to clear up all my puzzlements RIGHT NOW. They tell me it's all right for me to have unresolved questions in my mind. See? I feel tugged at, but I don't feel yammered at. I can do other things with a clear conscience.

Yet I do often have the experience of being nagged at by an intellectual puzzle, of feeling resentment at feeling pressed to do other things (like reading a proposal for a dissertation) when I'd rather be sorting out that puzzle. Those periods of feeling nagged come and go, of course. No puzzle holds my attention year after year--every puzzle either changes its shape or gets pushed out by a better one.

Well, that's enough of that.

I always enjoy your writing style. What a pleasure it is to come upon a nice, clear, simple invitation like "Bear with me" instead of something like:

> However, it is expected that consideration of certain conditions and factors contributive to the complexity of the present problem will explicate context for enhanced understanding of the present point of view.

I am aware that in your recent writing I almost never come across any locution that rubs me the wrong way. I came across a couple in this 1980 paper. Maybe you don't commit them any more. Anyway, I enclose a few sheets on usage to convey my complaints. You did not in this paper violate my preferences repeatedly and consistently; you did it only once in a while.

I also enclose a book review that seems to be talking about reorganization.

Phil

21 Feb 87

Dear Bill:

"I say that in the study of the melodying of speech and of music we have before us not the perception of sounds but of organized movements through bodily space."

That ought to make you prick up your ears. Though I rather think you would say "both" rather than "not ... but."

Carol Slater sent me the enclosed. *

I know. You say that lots of people have said a lot of wise things, and you can't do much with that until the ideas can get modelled. Well, sometimes a wise saying can give you an idea for what you'd like to model. Sometimes dreams come before the hammer and saw.

But I also like Sudnow's skill in painting the floundering we do among our memories and among our perceptions of environmental opportunities in the search for a feedback function. He ranges in the middle levels. Most of the examples I come upon (in my own mind, too) are either down at the bottom (as in the modelling that you and Marken do) or up at the top at principles and system-concepts.

Anyway, I thought this might give you a pleasant few minutes.

--Phil R

* Undetermined excerpt from *Ways of the Hand; The Organization of Improvised Conduct* by David Sudnow. 1978 / 1981 edition.

Feb. 28, 1987

Dear Phil,

Try gerunds. My wanting to go to the library made me want to turn left when I came to the corner. Wanting to go to lunch would make me want to go straight ahead.

If you're carrying the news, you have to carry it to Ghent from Aix:

A message needs a rider to carry it.
A rider needs a horse to carry him.
A horse needs a shoe to protect its feet.
A shoe needs a nail to hold it on.

When you put it that way:

If the nail comes out pound another one in.
If the horse loses the shoe wrap its foot in your cloak.
If the horse is lame find another horse.
If the rider's missing try smoke signals.

If any corrective step fails, the step above it fails.

I think the message is: for want of a backup the battle was lost. Or is it that the bureaucratic mind, once it has settled on an Accepted Method, is incapable of varying its lower-order goals? The message probably said, "If you do not receive this directive, please fill out a Lost Message Form and return it immediately."

The problem with consciousness is that anything I can specify for the hierarchy to do can be modeled without knowing what consciousness is or what it does. What is consciousness FOR?

The control system doesn't scream when we decide to reduce a different error because, for the moment, that system's reference level is set to produce no behavior. "We" means different things at different levels. Turning off a control system means, maybe, selecting a reference image that matches the current perception, which puts the control system on hold. It's a higher order system that decides to go haring off. Howzat? Or for the moment we set the error sensitivity to zero, which allows error to exist with no action. Howzat?

Control systems don't "make decisions." I don't think that decision-making plays much of a part in behavior, except when we have learned some program for decision-making, and use it because we learned in school that we have to make decisions. A lot of what we call decision-making is probably really reorganization -- that's why it's such a muddle. We struggle to make sense, we try things out, and like E. coli, if the result is worse we try again. When we finally find a course of action that makes sense and actually works, we say to ourselves, "Now I know what I have to do." Afterward, of course, we present this as a choice we made, a decision. A shameless lie.

Reorganization is our built-in mechanism for conflict resolution. You don't have to "decide" if there's no conflict.

Thanks for tips on usage. I do try to follow them, so you're a good influence. I think my writing continues to improve, or hope it does. The comma before <u>and</u> and <u>or</u> and so on sometimes seems to put in a little pause that I like, whether or not it's correct. I'm glad to know that it's become OK to put one in after the penultimate item in a series. For while I left out that comma because it was a no-no, and didn't like it. Then I decided to do it my way, and finally everyone else caught up. What about <u>that</u> comma?

I used to think that I write the way I talk, and people have even told me I do. It's not true. Trying to read aloud what I've written is painful. The cadences are all wrong. Strange.

The supercomputer approach to perception (Science News) still suffers the defect of naive realism. There are things etc. out there, and we have to build a network that can recognize them. I would much rather see artificial brains being built that can solve the brain's problem: here is a collection of identical intensity signals related to each other in unknown ways. How can I construct an internally-consistent model that makes sense of them and lets me control them? I think that a real model is going to need actuators and sensors so it can interact with the same world we experience, and it''s going to construct its own hierarchy. I can't wait to see what it produces.

Susan Carey's idea looks good: it's a way to learn about the hierarchy of perceptions, as it gets reorganized into existence. My question has always been "Is reorganization [restructuring] random, or are there strategies built into it to make it more efficient?" I'd like to see work like Carey's done in the context of control theory -- not because I think my "levels" are right, but because, as Piaget pointed out, perception can't be separated from action. A lot of perceptual research puts the subject in the passive observation mode, and relies mainly on words as a way of finding out what a subject is perceiving.

I'm sure you anticipated that I would be a bit critical of the Sudnow stuff. I don't like to admit how things like that impress me -- have I turned into a clanking robot? Maybe I've been spoiled by my friend Sam Randlett, who is also a Master in the world of piano. When Sam speaks about the processes of playing the piano he says precisely what he means, and knows precisely what he means. As a result, he turns out competent pupils with astonishing speed -- a quarter of the time normally taken. Part of his success comes from not trying to put into words what words can't describe. As far as Sam is concerned, there is plenty of mystery and nonverbal perception involved in the higher levels of piano playing. But the only way to communicate them is to play the piano. If you want to TALK about playing the piano or any other artistic endeavor, he says, then you talk about the mechanics, from relationships on down. That's why he can teach so fast. After he has taught people how to play the piano, it's up to them what they want to express with it. He has no more patience than I have with the talkers, to whom he refers with relish as "the dolts."

Just to get even with you I'm sending you one of five papers sent to me by a woman who has appointed herself the "ombudsperson" of the American Society for Cybernetics. She says she has combined the "essence" of Gordon Pask's ideas, Humberto Maturana's, and mine into her own concept. I consider this sort of stuff to be utter GARBAGE. So how do I reply to her fan letter? I've spent my whole life trying to make my understanding clearer, simpler, more precise, more communicable. Aren't people like this (who infest the ASC) my enemies? "Dear Ms. Enemy..."

Best,

Bill

Bill

Dr. Phil Runkel
College of Education
University of Oregon
Eugene, OR 97403-1215

3/3/87

Dear Phil:

I think your understanding of the independence of H1 and H2 in tennis of degrees of freedom is correct. H1 and H2 represent two degrees of freedom with respect to the variable being controlled and this is true even when (because of the way H1 and H2 are connected to the controlled variable) there is a high correlation between H1 and H2 (which implies statistical dependence). Despite the correlation, H1 and H2 are two "independent" degrees of freedom. This could be demonstrated by reinstating (in real tine) the disturbance to the controlled variable that would require uncorrelated movement of H1 and H2 (if the variable is 10 remain controlled).

As to your other point: Yes, I have set up experiments to study intrapersonal conflict. For some reason I was never satisfied with them -- not because they didn't work, but because I didn't know how best to present the results to illustrate conflict. One of the main problems is that is difficult to induce conflict that lasts very long (unless I, myself, was the subject and was willing to "experience" the error). In fact, people quickly resolve the conflict by abandoning attempts to control one variable, leaving the experiment or (in my own case) re-perceiving the situation so that the conflict state itself becomes a goal.

Based on your interest in my description of having two people cooperate to control the same variable I am now motivated to start work on a new set of experiments and a paper to be entitled "Coordination, Cooperation and Conflict". I was planning to work again on conflict after finishing the other projects I have no time to finish because of work. But you have suggested a great idea -- extend the experiment described in the JEP paper to include demonstrations of cooperation, Coordination (cooperation when both systems are in the same body) and conflict (inter- and intra-personal). All could be nicely illustrated with the <u>same</u> control system model (not necessarily hierarchical).

Given my difficult time constraints, perhaps we could write the paper together -- I'll do the experiment and you write it up. What do you say? I can't even imagine starting work on it until mid-april. But you might sketch out the basic ideas of the paper. It would use the same task as that in the JEP article. I would argue that, in the cooperation situation both subjects have the same goal (described verbally) -- the catch is that each has control over only part of the goal (because one subject has H1 and the other has H2). The goal can only be achieved if each moves the handle appropriately and tries to achieve the same goal. This is the way I did it and it works just fine (using all three versions of H1 and H2 hookup described in the JEP paper). Conflict is easy to produce by appropriate handle connections. I'm sure you could think of a way to connect the handle so that efforts to correct a disturbance produce an uncorrectable disturbance to another aspect of the controlled variable.

I have enclosed two papers that I'm trying to get published. The one on hierarchical control (with Powers) was just rejected by JEP. Though the reviewers typically missed the mark (as usual) I agree that we could have presented this data in a far more compelling way. I plan to rewrite it with an emphasis on explaining how a hierarchical control model differs from conventional hierarchical models of performance. The reviewers felt (rightly, I think) that it was not clear why these experiments were important. (I didn't make it clear to them -- mea culpa). I do think the experiments are a beautiful demonstration of hierarchical control (which turns out to be not that easy to demonstrate), they are also completely inconsistent with "output generation" models. With the right discussion this paper should eventually get published.

The other paper has had a checkered history. I first submitted it to Psych Bulletin. It was returned, unreviewed, as inappropriate. The Bulletin recommended Psych Review or American Psychologist. I sent it to American psychologist (which I still believe is the right place for it). The reviews were quite favorable (they liked the writing and some of the points) -- but they did not recommend publication because it was nothing new or something like that. So I sent it to Psych Review; again it was returned unreviewed as inappropriate. Next it vent to Behavioral and Brain Sciences. It was rejected - three reviews were positive and four were neutral to negative. I was going to rewrite and make it more "scholarly" as the editor suggested if it were resubmitted. But I know who the editor is (Steven Harnad) -- I read his postings on my electronic mailbox at work The fellow is the worst kind of fool -- a pompous one. Rather than tie my guts in knots (Powers had done so earlier with an invited paper to B&B S) I rewrote it slightly and submitted it to Behavioral Science where it currently sits.

I would love to hear what you have to say about the paper. I will keep massaging it and resubmitting it but I think it merits publication in a broad psychological forum. I may resubmit to American Psychologist if Behavioral Science fails.

Let me know if you're interested in collaborating on the "Conflict" paper. I look forward to hearing from you soon.

Sincerely

Rich

Rick

9 March 87

Dear Bill:

Thanks. Gerunds work fine. My wanting to go to the library gives me the wanting to turn left. Good.

"... anything I can specify for the hierarchy to <u>do</u> can be modeled without knowing what consciousness is or what it does." At least in principle. Well, I want to agree but don't. Or else I agree but don't want to. I don't know which of those sentences describes me.

I can imagine a cherlescent finstophrene from Arcturus watching me as I type. "My," says Finsto (they have nicknames there, too), "what devious feedback loops those humans have, making all those little black marks on paper (what sharp vision they have!), running around putting them in front of the eyes of other people, making sound waves in the air--I wonder if they do all that consciously?" Old Finsto probably makes a similar remark about the minuscule actions of ants.

I feel as if I can't manipulate these words without knowing what I am doing. I believe, doggedly, that when I have not only a symbol (as when a chimpanzee tears some branches off a tree and runs across the greensward waving them about and making loud noises), but also a symbol for the symbol (as when the human onlooker says, "Look, he's doing that I-am-somebody-around-here thing."), then consciousness appears. We can look at ourselves looking at things. We can examine our words. We can look backward and forward. We can index this image with "past" and that image with "future."

I think those sentences make eminent sense. But nobody has yet figured out a way to test them. The experiments in teaching English to chimpanzees and gorillas are still inconclusive. And even if the language teaching succeeds beyond hope, how will we know those animals did <u>not</u> have consciousness <u>before</u> they learned English? If consciousness does turn out to be a quantum jump in the way the control systems work, my guess is that it will turn out to show an accumulation of little quanta in various parts of the circuitry (including memory) that add up suddenly to the astonishing capacity for complex language. That is, observations of gross behavior will make consciousness look more like a matter of degree than a discontinuity, though with a sudden acceleration that preceeds the appearance of language. Well, I don't spend much time in speculation like this. I am content to go on typing without wondering what Finsto thinks of it.

Your explanation of why the control system doesn't scream when it must set aside action to restore one input while it attend to another is of course "right." In fact, before I got through your first sentence, I remembered that I had written just about the same thing in "Inside and Outside." But something remains unexplained. If the "urge"--the action of the comparator--is to send a signal that sets corrective action into motinn (or keeps corrective action going), and also sends that signal upward to higher systems, then the higher system has to make a choice of the corrective action to postpone while continuing others. Where does the criterion get set? Do we relegate it to the reorganizing system? If we do, are we using the reorganizing system as a wastebasket for unexplained phenomena? A category of "other"?

I agree that "decision" covers up a lot of questions about what is going on. A lot of people would say the control system is making a decision when a program picks this sequence and not that. And no doubt, as you say, some "decisions" are reorganization. The word is probably as misleading as the current usage of "learning." For the moment, I don't know what to do with "decision" except to use it where I don't care much about being precise.

Maybe you have heard of the "garbage can" theory of decision making that is currently delighting organizational theorists. An organization is a garbage can (meaning merely a large container) into which people dump problems, people, choice sitations, and solutions. The originators of the label (Cohen, March, and Olsen) wrote in 1972:

> An organization is a collection of choices looking for problems, issues and feelings looking for decision situations in which they might be aired, solutions looking for issues to which they might be the answer, and decision makers looking for work.

"Is <u>that</u>," you may ask, "still tickling your colleagues after 15 years?" Well, remember that there isn't much that's risible in social science; one tickle has to last a long time.

You say similarly, that after some trial and error, we say "Now I know what I have to do," and pretend, with hindsight, that we have made a "decision." Karl Weick, a social psychologist, writes similarly. A plan, he says, is not so much an intention, a track laid into the future, as it is a review of where we have come and a hope for what we <u>will have done</u> after a while. Plans and decisions, he says, can be used as a message to someone (including ourselves) that something serious is happening, as advertisements to attract investors to the firm, as games to test how serious people are about what they are advocating, and as an excuse for interaction with people or about topics that would not ordinarily come together in the daily routine. Weick writes (1979);

> Plans are a pretext under which several valuable activities take place in organizations, but not one of them is forecasting. As Ambrose Bierce said, to plan is to "bother about the best method of accomplishing an accidental result."

I've forgotten what I sent you by Susan Carey. If you remember, tell me. Something about the development of some capacities in children?

If you have anything Randlett has written, please send a few pages.

Kathleen Forsythe reminds me of my mother. In many ways my mother had a keen intellect, despite dropping out of school after the fourth grade. She was very discerning, and very clever about pricking foibles and putting the magnifying glass on contradictions. She had a delightfully impish sense of humor. But she was entranced by what she called "theosophy" and "metaphysics." She liked to come across grand generalizations over which she could exclaim, "Oh, it's so true!"

I sympathize with Ms. Forsythe, because she is struggling ~~with~~ in exactly the same way I struggle. There are a lot of inspiring ideas floating around out there. ~~Somehow, they just~~ Somehow, they just <u>must</u> fit together into a more glorious whole. How can I relinquish the ideas that are merely pretty and keep those with which I can hammer and saw? What is the test for which is which? What is the test for whether one idea "fits" with another?

But I don't think precise decimal numbering of paragraphs is going to be much help.

Suppose you want to walk down to the grocery store, and you are in a hurry. Your small child wants to come along and help. If you take the child, you are going to have to slow your pace to accomodate the child's short legs. Sometimes you take the child along because you want to do something for the child. At other times you harden your heart and leave the child at home. You can't do everything for everybody every time. Ms. Forsythe, I think, is not your enemy; she wants to help. But what she wants to <u>do</u> to help will certainly get in your way and slow you down.

I get dissertation proposals every now and then that read as if Ms. Forsythe had written them. I have an easy way out. I just say that the problem is one I have no expertness in and therefore I couldn't be of help. So the person has to find someone else for the committee. And if the person cannot fill out a committee, then that seems to me the best way to screen dissertation proposals. Of course, if the person wants my comments on the proposal as it stands, then I get out my club and hit him or her with it (though I try to wrap it in flannel first).

In brief, you have my sympathy.

On the other hand, a student once came to me wanting to do an interdisciplinary master's thesis. He wanted to show how one could explain everything in the universe (so to speak) with the concept of <u>interaction</u>. He seemed to talk rationally, and what he had written had good clear sentences in it, and the paragraphs hung together properly. Although what he had written did not yet tie together physics and psychology and cosmology and so forth, who knows---? I had no evidence, either way, to decide (excuse the word) whether the fellow was a nut or a genius. So I joined his committee. I don't even remember any more how the thesis came out. But I'm glad I helped him go through the exercise.

Phil

I'm not nominating Ms. Forsythe for genius.

10 March 87

Dear Rick:

Holy Toledo! (Batman's side-kick Robin). Glorioski! (Little Orphan Annie). Think of that! (Kurt Vonnegut).

I am honored (who, me?) and delighted (wouldn't that be fun?) and aghast (can I do it?) at your invitation to join you in an experiment.

Some slight modification of Experiment 2 in your hierarchy article might do very well. (I have decided, by the way, that at least one of the proposals for altering your experiment that I gave in my letter of 26 Feb is no good.) But I got to wondering whether we could design a task that is a fairly obvious simulation of some everyday event, so that we could write, "This is like the familiar experience of . . . ."

So my brain immediately started moving symbols this way and that, hunting for way to simulate cooperation. What I will set forth here is a first try, so tell me where it is wrong or inelegant or unnecessarily frilly.

I am thinking of two people, S_1 and S_2, carrying a heavy burden (such as a large piece of furniture) along a hallway. They encounter obstacles as they go along, such as people or other pieces of furniture, and have to change their positions. Maybe they go through a doorway, and have to change from abreast to single file. So maybe one or the other encounters a slowing obstacle and both must adjust positions.

Figure 1 shows some positions you might imagine the two people taking as they carry the burden along the hallway.

They don't stop to think, "The important thing here is cooperation." They just start moving the burden, with the goal of getting it where it should go. If the burden keeps moving in the right direction, the error signal decreases. One person, if stupid enough, might pull on ahead of the other, hoping to get the job over with quickly, but then the burden would pull out of the hands of one person or the other, the burden would fall on the floor, and neither person's actions would then move it (since it is too heavy or awkward to be moved by the person who still has hold of it). So there is a maximum forward distance between Ss for the cooperation to work, so both Ss would move to maintain that maximum, regardless of impatience. That, I think, is the essence of cooperation: putting other personal goals (such as getting this over with fast) lower in the hierarchy than the perception that the task is progressing. (Almost always in human life as it is, people adopt cooperation in regard to one task at a time, though one can find groups in which cooperation overrides almost everything else other than the reorganizing system. So I think.) So I don't think we must instruct Ss to cooperate. I think we can simply let it occur in the same manner it would if two people wanted to get the sofa into the basement.

We needn't draw a picture on the screen of a hallway and obstacles and two people carrying a sofa. I propose that the screen show only what appears in Figure 2. S_1 controls the upper bar and the cross (the burden). S_2 controls the lower bar and the cross. (Use whatever symbols are convenient.) All motion on the screen occurs only horizontally.

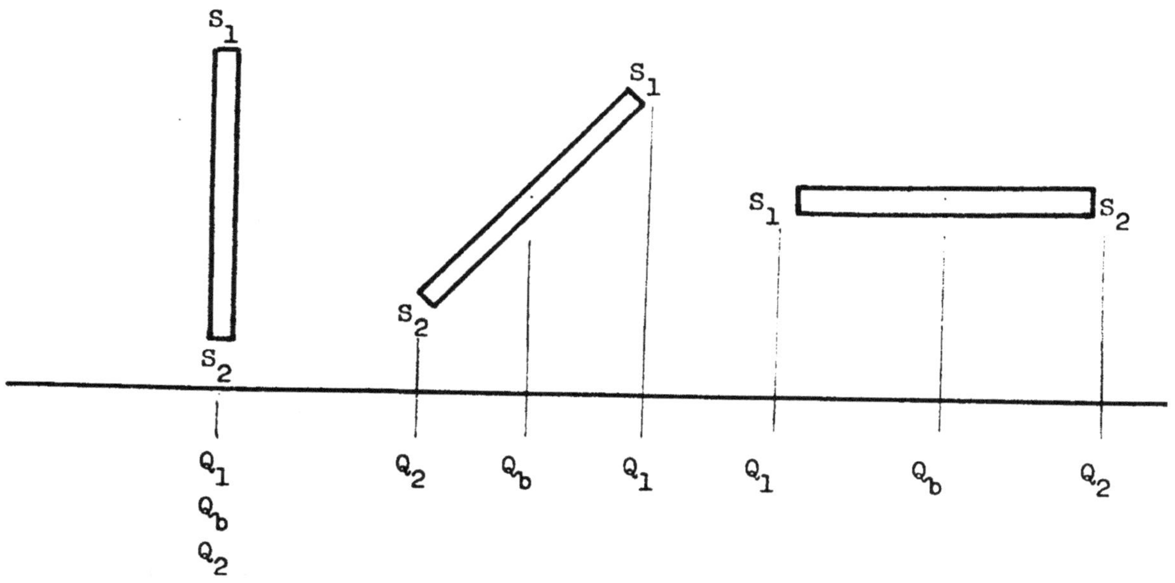

Figure 1

What is being modelled: Some positions the burden might take as S_1 and S_2 carry it along the hallway.

Figure 2

What the subjects actually see.

Actually, we could show only the cross. But I think it is a little more realistic, a little more a simulation, to show Q_1 and Q_2, since two people carrying a burden do actually see each other.

Can I write equations to simulate that task? I'll try. But I have to say that my ability to read mathematics is not very good, and my ability to write it is worse. Anyway, I'll try, and if what I write here gives you a good idea, then I'll reap a profit.

How do we get the Ss to move along the hallway? Let's let the Ss stay put, and move the hallway instead. That is, let Q_b move steadily leftward, and instruct the Ss to Keep Q_b stationary. That is, $D_{b(t)}$ is a constant negative rate; $D_{b(t+1)} - D_{b(t)} = -c$. You may wish to add some randomicity to it.
Having the Ss work against the negatively moving Q_b simulates their wanting to move Q_b along the hallway.

Then we need to admit that the Ss can meet obstacles, S_1 and S_2 separately. We can do that with D_1 and D_2. Those disturbances could be either sporadic or continuously random. I prefer continuously random, even though that is not as faithful a simulation.

I'll offer here two sets of equations. The first set does not allow the Ss to drop the burden. It cannot slip out of their hands. I'm afraid that ruins the requirement of voluntary cooperation. It is like having slaves chained to the ends of the sofa. Anyway, this is the first set of equations I worked out, and I'll put it here for what it's worth.

$$Q_{1(t)} = D_{1(t)} + H_{1(t)} \qquad \text{(Equ. 1a)}$$

$$Q_{2(t)} = D_{2(t)} + H_{2(t)} \qquad \text{(Equ. 1b)}$$

$$Q_{b(t)} = D_{b(t)} + [Q_{1(t)} + Q_{2(t)}]/2 \qquad \text{(Equ. 1c)}$$

where $D_{b(t)}$ is a constant negative rate; $D_{b(t+1)} - D_{b(t)} = -c$

$$|Q_{2(t)} - Q_{1(t)}| \leqslant m \qquad \text{(Equ. 1d)}$$

where <u>m</u> corresponds to the length of the burden.

In the next set of equations, I think I have fixed things so that if the Ss get too far apart, somebody's hands slip off the burden and Q_b stops (actually, slides leftward according to D_b).

$$Q_{1(t)} = D_{1(t)} + H_{1(t)} \qquad \text{(Equ. 2a)}$$

$$Q_{2(t)} = D_{2(t)} + H_{2(t)} \qquad \text{(Equ. 2b)}$$

$$Q_{b(t)} = D_{b(t)} + [Q_{1(t)} + Q_{2(t)}]/2 \quad \text{for} \quad |Q_{2(t)} - Q_{1(t)}| \leq m \qquad \text{(Equ. 2c)}$$

Note that the above equation specifies that

$$|Q_{1(t)} - Q_{b(t)}| \leq m/2 \quad \text{and}$$

$$|Q_{2(t)} - Q_{b(t)}| \leq m/2$$

So the next equation provides for the range of $|Q_{2(t)} - Q_{1(t)}| > m$; namely,

$$Q_{b(t)} = D_{b(t)} \quad \text{for either} \quad |Q_{1(t)} - Q_{b(t)}| > m/2 \qquad \text{(Equ. 2d)}$$

$$\text{or} \quad |Q_{2(t)} - Q_{b(t)}| > m/2$$

The last equation makes sure, I think, that Q_b stays put (actually, moves left with D_b) until both $Q_{1(t)}$ and $Q_{2(t)}$ come back to it from wherever they ran off to and take hold of the burden again before Equation 2c operates again. That is, $Q_{1(t)}$ and $Q_{2(t)}$ must not only get within m of each other, but they must do so on either side of $Q_{b(t)}$.

Can D_1 and D_2 possibly vary opposite to each other and in an amount to cause Ss to drop the burden? Yes. That is, I have written in nothing to prevent that. Maybe that's realistic, too, as long as it doesn't happen very often.

Does the second set of equations simulate cooperation--that is, allow the Ss to adopt or not adopt, as each chooses, cooperation has a higher-order reference signal?

Am I being too frilly? Is it better to do something simpler, at least for a first try?

If we wanted to get really fancy, we could imagine the Ss tugging against each other and not wanting to do that. We could load the handles so that they get harder to turn in proportion to, say,

$$\log_{10}[\, e^{\left|Q_{2(t)} - Q_{1(t)}\right|}\,]$$ for the exponent equal to or greater than m.

And if we wanted to get really really fancy, we could build a machine like one of those in the grocery store (or the video game arcade, which I have never visited). We could have the screen show a picture or diagram of the hallway moving to the left, with obstacles showing up on either side, and a picture of the two people carrying the sofa as in Figure 1. That would add a second dimension of motion on the screen. Oh, well.

It strikes me that we could send this to a journal other than the "performance" journals. Seems to me we could send it to a journal with a title something like Journal of Experimental Social Psychology.

We could claim:

1. Cooperative behavior shows in continuous graphs of data, not merely at posttest or even at widely-spaced points. As far as we know, this is the first time it's been done.

2. Ss are not instructed to "try to cooperate," but are merely asked to do a task much as one might ask a couple of friends to carry a sofa down the hall.

3. The Prisoner's Dilemma can also be run as in Point 2 (Try to maximize your winning over the long haul), but that game acts as if interpersonal interaction goes in discrete cycles. In our experiment, simultaneous and contunous behavior is plotted.

4. We show how very precise cooperative behavior can be. Not that this is news to anyone who has watched people carry a sofa or dance together, but it is rare to show data for such synchrony.

5. The data show directly how a superordinate goal (that's the social psychologist's term for higher-order reference signal) must be operating for the task to be carried out. And that shows in the graph for <u>every</u> individual, not just on the average or for a group with individual behavior buried. If other experiments continue to show that for every individual, the hypothesis of hierarchy of purpose will be pretty well demonsrated not just for most people or most occasions, but always. At least if precision of action of this sort is to be achieved. And conversely.

Maybe social psychologists are not much interested in carrying sofas down hallways, but we thought it better to start with something easily conceivable and a task that yields quick feedback. We hope social psychologists will welcome a method that can be used to test the dynamics of a model from moment to moment.

YES, I'll be glad to collaborate with you. I'll be glad to leave all the experimentation and modelling to you and take the writing for my part. I thought I'd use this occasion to try my hand at writing some equatinns, but you can ignore all that if you want. I'll collaborate on your terms. I'm in no hurry, either. You set the pace. I'll tell you if I can keep up.

Thanks! Phil R

P.S. Let's say a critic complains, "Why didn't you just watch 20 pairs of people carrying sofas down hallways? But it would be no surprise if they did it. Why make a study out of such stuff?"

Agreed, it wouldn't be very surprising to discover that the 20 pairs would do it. But how would they manage to do so? Some people believe that behavior (the behavior of other people, anyway) is controlled by external events or variables. But the environment is always changing. How, then, can human behavior be as reliable and predictable as the success of 20 pairs of people carrying sofas down unpredictable hallways?

We might post observers along the hallway to note every smallest motion. Doing that could not bring us the continuous, quantitative record the computer could bring. We might line the hallway with cameras on the floor, ceiling, and walls and analyze the tapes afterward. That might come close to what the computer could do if we knew the right ones of all those motions to analyze.

The graphs and correlations will show that cooperation need not come about by one person waiting to see whether the other does his or her part. It can occur from simultaneous action so perfectly matched that there can be no time for one person to note the other's direction of action and then initiate muscular action in return.

The data will show that cooperation need not come about from one person being the leader and the other person matching action to the leader's. Both can match action to the cooperative goal while at the same time adjusting to the actions of the other.

The data will show that close cooperation does not come about from any policy or preference for one kind of muscular action by either individual, but from continuously changing muscular actions that keep constant the progress of the desired quantity--that is, movement of the burden to the right (stationary on the screen).

The data will show that there is not some geometrical plan or vision or template the two Ss use to guide their cooperation. Either Q can remain motionless while the other is free to swing over a wide arc (corresponding to motion within m on the screen). Both Ss simply act to keep the burden moving despite disturbances.

You might catch one or two of those features--dispose of one or two hypotheses--from a video tape, but I don't think you'd catch them all, and certainly not if you didn't have control theory in mind.

P.P.S. I'm not surprised that you have trouble keeping Ss in conflict. That difficulty, after all, is what control theory predicts. Your list of three ways out is classic. I do not have a specific idea to prolong the state of conflict. The only vague notion I have is that we might try to use a higher-level reference signal--one high enough so that action takes a longer time. Maybe that's no good either, because we would be wanting to record the actions, not the gaps of inaction in between.

But one caution leaps to my mind. I have led a good many human-relations groups (sometimes called "sensitivity" or "encounter"). The role of leader is sometimes called "trainer" or even "therapist." Leaders of such groups repeatedly let members encounter conflict and then help them (usually let the group help them) to find ways out of the conflict (find paths through the environment through which to oppose the disturbance) that are different from the ways they are accustomed to using. Sometimes that just adds some useful ideas or skills to the individual's repertoire, but often it brings about reorganization. The person goes about shining with a new glory for the next week.

But diving into conflict to learn something from it has dangers. Not everyone can <u>resolve</u> (find a way <u>out</u> of) any conflict that comes along. All of us carry some conflicts with us for which we have found no solution. Some people carry severe ones. A dilemma that looks like a small one to the rest of us can trigger those people into rapid oscillation between reference signals, bringing strong emotion. And those people are almost always in terror of the emotion itself, and you get a positive feedback loop. In extreme cases, the person faints or even goes into tremors and you have to call the ambulance. That extreme case has never happened to me, thank God, but I know a few people to whom it has.

You wouldn't think that sort of thing would happen to someone sitting at a computer keyboard. And indeed, it rarely would. Most of us would do one of those things you listed and experience no ill effects. Indeed, we are all fiendishly clever at finding our way around or out of conflict. But those of us who carry severe conflict with us are not as fiendishly clever, almost by definition. (Am I being theoretically or technically correct if I say that the severe conflict reduces the sensitivity to other danger signals? Or should I say increases the sensitivity? I don't have the cybernetic meaning of "sensitivity" memorized yet.)

For most kinds of experiments we would think up, I'd say a simple test for the suitability of a subject (the subject's "ego strength," if you will allow the jargon) would be sufficient--a test such as whether the person is given to frequent defensive explanations of the reasons for his or her behavior.

So I am not worried about what you <u>have</u> done. I am, however, worried about constructing an experimental task that would somehow coerce, persuade, squeeze the subject into staying with it longer than the subject's "natural" urge to get out of it would otherwise permit.

I don't know, at least just now, how to say, "Do this, but not that." I just feel obligated to offer the warning.

I remember a colleague who did an experiment on the effects of failure. He used a puzzle I'm sure you have encountered. A piece of paper has a cloud of dots printed on it, every dot numbered. You move your pencil from dot 1 to dot 2, and so on. When you get to the last, you have drawn a picture. His puzzle had one or two places where there were *two* next numbers. That is, when you got to 5, you might find that there were *two* 6s leading off into different paths. My colleague told the subjects that if you were especially discerning, you could see which path would eventually draw a picture and which would not. Actually, however, *none* of the paths would draw a picture. He got a lot of ROTC people as subjects. He told them that his puzzle was a new test that discriminated people who were "officer material" from people who were not. After the experiment, he explained to every subject individually that the experiment had been a deception, that it was *not* a test of officer material, that it was impossible for anyone to draw a picture. All subjects but one seemed to go away satisfied. The one subject, however, came shivering in horror at the loss of his lifetime ambition. So terrible was his grief, indeed, that he could not "hear" what my colleague was saying. My colleague spent *two hours* with that subject, and the subject still went away crestfallen and worried about whether he could ever become an officer.

Reading your paragraph again, I get the impression that you are thinking of combining in one experiment the production of both cooperation and conflict. My own wish just now is to stick with cooperation and do the conflict later. But you can make a counter-proposal if you wish.

P.P.P.S. Thanks for sending the two papers. I am glad to have them both. I've read one. I'll send comments later on, after I get some other matters off my desk.

P.P.P.P.S. You said that one editor wanted your paper to be more "scholarly." I hope he or she didn't mean less readable. I should warn you that clear writing is high up in my hierarchy. The *Harvard Educational Review*, sometime back, invited me to write a review of a book I admired. They then rejected the review I wrote, in good part because it was written engagingly. They asked me to revise the review to make it more scholarly. I told them hell with it. I sent what I had written to the authors of the book. They were the audience I had really written for, anyway.

14 March 87

Dear Carol:

Well, Sudnow reminded me that we all interpret what seems likely, possible, dubious, unlikely, impossible, according to whether we can imagine ourselves participating in it. And participating isn't just producing or taking in words. It includes also all those perceptions of sensory inputs both from the outside world and from the inside world--the sensations of our muscles and glands doing this, that, and the other. No matter how unconscious we may be of imagining all those accompaniments. (I don't suppose we actually have sensations of many glands secreting, but you know what I mean. Though I think I am conscious of salivation.) Those imagined muscular participations become part of the "meaning" of the assertion or design being contemplated. The possibility of accidental nuclear war has different meanings for carpenters, viola players, computer designers, typists, jockeys, and horticulturalists.

I am convinced that a chief reason I differ from academic colleagues in how I conceive psychology--especially from colleagues who have been only academic all their lives--lies in the variety of jobs I have had and the various muscular skills I have picked up along the way. For your amusement, I enclose a list of my jobs that I dug out of my memory.

I am glad you enjoyed the Chicago conference. Many of the titles are opaque to me. I was happy to see that they scheduled times for group discussions without competing papers. I am also glad to see that the Blackstone Hotel is still doing business. I've never stayed there, but I used to eat some very good roast beef in their basement.

Love,

Phil

DEPM, Coll of Educ
U of Oregon, Eugene 97403
20 March 87

Dear Rick:

Thanks again for sending me the two papers. I have now read one: Marken & Powers: "Hierarchical Control in Human Performance."

You said in your letter that you were planning to rewrite that paper. So I thought I would kill two birds with one stone. (1) I would offer you some help in rewriting it. (2) I would give myself some practice in writing about this sort of thing in preparation for doing my part in our forthcoming experiment on cooperation.

I enclose an editing of the Marken & Powers. I have actually rewritten several pages.

I don't claim that my rewritings and alterations are right or best. What is best depends not only on the writer, but also on the reader. You can estimate the modal ways of thinking in some audiences better than I. Also, some of my alterations may suffer from defects in my understanding of mathematics, control theory, computers, or your experimental method. In brief, I offer my alterations as proposals for your consideration, not as now-you-know-the-right-way-to-do-it.

As to journals, it occurs to me that journals such as the Journal of Mathematical Psychology might be interested. I have never looked at that journal; maybe you have. Also Psychometrika and Psychonomics. I'm not sure I have that second name right; maybe it is Psychonometrics. I'm not even sure those two journals still exist; I haven't looked at them in years. I have asked our reference librarian to send me a list of journals like that. If the list gets here today or tomorrow, I'll enclose it. Oh, also Behavioral Science; for that one, you must write an abstract that fits the article into J.G. Miller's scheme of living systems.

I urge you to send copies of the MS to Michael Posner and Steven Keele (whom you cite) here at Dept. of Psychology, U of O, 97403. I like to send my stuff to people who write on similar topics; sometimes I fall into delightful exchanges of correspondence. Posner, as you may know, is a renowned expert on human performance, one of the most cited psychologists in the USA, and a member of the Amer. Acad. of Sciences. If he were to begin citing your work, a few more people might look up your articles.

I think I've told you that my wife suffers from Alzheimer's disease. I think she is now in her seventh year of it. I think she often exhibits the neural hierarchy at work very clearly. I don't remember a lot of the phantasmagoria of her behavior, but here is an example. She will say something like, "When they get here ah he goes over here you know now I want to be sure all the cupboards cups c c into the backs of packs of it belongs in the festing firsting fir firn. . . ." By now her voice is trembling and so soft I have to put my ear to her mouth to hear what she is saying. Her chest is heaving with the effort of trying to find the words. Then she will burst forth loud and clear, with no hesitations, "Oh, I just can't fit the words together right!" Some system that deals with words is somehow "looking down" at another system that also deals with words, and somehow the higher one works much better than the lower one. But obviously the higher one is calling an end to the struggles of the lower one.

In my editing of Marken & Powers, I have kept closely to your organization of the paper; I think it is fine. I have made only minor shifts of content from one place to another.

Overall, I daresay I have altered and written with my typical colleague more in mind than yours. I justify that liberty by my hope that your papers will reach a wider audience than experimenters on human performance. Maybe even Discovery or Sci. Amer. some day--who knows? So I have tried to make the prose more easily comprehensible to readers not familiar with computers or experiments in human performance. Even aside from whether you will send the paper to JEP-HPP, B&BS, or Educational and Psychological Measurement (actually I'd put the latter very low on the list), I should think you would want to send copies to members of the CSG, and surely a lot of those people would be glad of some added explanatory phrases.

I have tried to point up, to make stand out more clearly, the features of the work that I think you and Powers should brag about. And in the first pages, I have tried to suck in the reader faster than I thought your paragraphs did.

I think you and Powers are getting blase about your work. It probably looks so obvious to you, and you may be ~~shy~~ about rubbing readers' noses in the assumptions underlying the work, ~~and that~~ but which, to me, are the seminal and triumphant posts on which you should hoist your banners. Excuse mixed metaphor I have written in a few sentences here and there to flaunt the banners, and at the end I have written several paragraphs of that sort.

If I were writing your paper to be sent to the kind of journal in which I have published in the past, I would screech things like "Here is a quantitative (not merely statistical) demonstration of the age-old postulation of internal hierarchies!" and "Look what you can do with the right theory! You can build a working model of the human animal by using only two simple equations that fits more closely the behavior of every subject than all your multivariate regression equations can fit only the average of a hundred subjects!"

Be that as it may, I have also edited your paper for clarity, frequently by changing passive voice to active (see ACTIVE AND PASSIVE VOICE in the enclosed sheaf of pieces on usage). I also offer changes in phrasing, single words, and punctuation.

As I said, you are the authors, not I, so accept or reject my alterations as you wish.

I am thoroughly delighted with your paper, and I enjoyed testing my understanding by trying to improve clarity and emphasis. I hope I have helped.

Your apprentice,

Phil

Philip J. Runkel

Mar. 28, 1987

Dear Phil,

Back from Switzerland, head full of pictures. We attended the cybernetics meeting in St. Gallen for five days, then Mary and I went on a four-day train-trip around Switzerland (13 different but wonderful trains!). Our first time in Europe. During the meeting and on our first day of travel there was snow every day, but the day of our trip up to Wengen at the foot of the Jungfrau the sky was absolutely clear. We changed our plans on getting to Wengen, which is an ancestral home for me, but is now a tourist trap with 2000 hotel rooms. Nothing old left. We went on up to Kleine Scheidig right at the base of the range and marveled at the scenery, but again were driven off by the tourists -- all skiers. On to Grindewald, then back to Interlochen and onward to Lucern, where we stayed two days. A long day. On the last night Mary contacted an old school friend who lives in Mannedorf and we went to spend the night with them on the shore of Lake Zurich. Turns out that the husband is a full-time collector of ancient oriental (Indian) art, a very rich, very nice, totally amoral fellow who showed us perfect hospitality while explaining that a lot of pieces have to be smuggled out these days. He has some 600 sculptures inside and outside the huge house. Odd experience. We finally left Zurich Flughof at 10:20 AM for Amsterdam, then left Amsterdam at 1:00 P.M., arriving in Chicago at 2:30 PM. Our biological clocks are still trying to reset. Amazing how much you can get done when you wake up at 3 in the morning.

There were six of the CSG at the meeting: Frans Plooij from The Netherlands (new), Toto Grandes, Dick Robertson, Ray Pavloski, Mary, and I. Plooij, Robertson, Pavloski, and I gave talks. Mine was a 2-hour interview (by our inside man, Larry Richards, currently president of the ASC), before a plenary session -- the first time that von Foerster, Maturana, and Pask (big wheels) had actually been present while I spoke about control theory. The more standard cybernetic sessions were sometimes interesting, but mostly Scholasticism. I think we made a little more progress; at least the cyberneticists now understand that control theory isn't about controlling people.

At a business meeting on the last day, von Foerster, the grand old man, tried to kill support for Continuing the Conversation in its new format. He spoke about "diluting" the cybernetic content, and I spoke about "purity," reminding him that the CSG contains many members of the ASC who consider their work part of cybernetics, too. It was all very delicate. Finally the question was put to a vote, and von Foerster lost, 25 (approx.) to 1. Signs of change. Greg Williams will be pleased.

After the meeting we all went to a library in St. Gallen and stood around looking at 1100-year-old books. I wondered if anything I am doing will be read or known about 1100 years from now. We were told that one monk wrote, at the end of the oldest book, "People who don't know writing think it's all in three fingers, but I can tell you that at the end of the day, the whole body hurts." Another good one: "I have put the last period in the book and now I will dance." The monks had it in for the Protestants in St. Gallen: when Protestant services were about to start, the monks would open up their tavern.

Frans Plooij says that many Dutch scientists are using my work -- it seems to be accepted in many places. He is a perfectly splendid young man who will advance our work by a huge amount. He wants to set up a book project with a German outfit that collects perhaps 50 people together to work intensively for a week with half a dozen authors, a think-tank format, with the product being a scholarly work of high prestige. The "scholarly" part cows me, but the rest sounds great. This would happen about two years from now.

On to the mass of literature I found when I returned.

Criminetlies, Zero! (Little Annie Roonie). Blow me down! (Popeye). Nov Schmoz Kapop! (The Little Hitchhiker).

Somebody is going to have to sit on you and give you a lecture about programming graphics. Your ideas are great, but they'd keep Rick busy for six months working out the displays and animations. I like the idea of attaching the experiments to a real situation, but we're going to have to rely on the reader's imagination a little more if this is going to be done any time soon. Remember that we can't SHOW these events to the reader without going to video tapes, which wouldn't be a bad idea. But that's full-time work, more than either Rick or I have available right now. I expect that you and Rick will find a compromise.

Your suggestions and rewrites of the Hierarchy paper are really helpful. We will take full advantage of you, given our natural inclinations to independence.

Consciousness. I agree with you that consciousness seems extremely important, but I mean it when I say I can't model it. My models run without it. All content of consciousness seems accountable in terms of learned mechanistic functions -- complex ones, but still mechanistic. What I can't account for is the fact that somebody KNOWS about that content. There is an Observer. The Observer can also act, arbitrarily, inserting goals into the system at any level without regard to the organization of the system, and probably doing a lot of other things too. But the Observer is not rational: rationality is a mechanistic process. Thinking is done by the brain, not by the Observer. As far as I can tell, there is no process of thought or action that can't go on either with or without consciousness of it by the Observer.

I have found one phenomenon of consciousness that has proven to be interesting and useful: point of view. It seems that we, the Observerws, can experience from the point of view of any level in the brain's learned hierarchy. If you look at configurations in the environment, after a while you will realize that everything is made of configurations -- nothing else exists. Then you can switch to seeing as a relationship-perceiver, a sequence-perceiver, a sensation-perceiver, and so on. What's interesting about this is that according to my model, these various perceptions are hierarchically related: when you're perceiving in terms of relationships, for example, the relationship-perceptions are being constructed from the behavior of sequence, configuration, transition, and so on -perceptions (isn't that an easier place to put the common hyphen? It probably wouldn't always work.).

In other words, all the perceptual signals not being attended to are still present, still representing the constructed world as usual, even though they aren't in awareness. They have to be, in order for relationship-signals to be present. And we can then extrapolate, and suppose that all the higher perceptions are also present, even if not in awareness, and further that all the associated control systems are acting, even if unconsciously. The higher contrtol systems account for what you inexplicably want to do about the relationships, if anything. When your point of view is working at a given level, the reference signals reaching that level appear in consciousness as the "right" states of the perceptions of that level.

Most of us (academic types) spend most of our time looking at the world from the program level. We talk a lot. We think a lot. It's easy to get stuck at this level. When your consciousness works at this level most of the time, moral principles (for example) dictate what you program about, but your only experience of this is a feeling that you're doing the right programs. Your system concepts are unconscious, but they still determine what principles apply and hence what programs you execute. The Superego.

A friend of mine (Kirk Sattley) and I got curious about this phenomenon about 30 years ago, and set out to see if there was any limit to the movement of a point of view from one level to the next higher one. The method was very simple. One person picked something to talk about and talked, while the other listened for the point of view from which the talking was happening. Don't ask me how you can talk when you're attending to configurations -- you can. But we didn't use my levels, even in their truncated form that existed then. We just started with any topic and looked for the attitude, viewpoint, opinion, feeling, that the talker was showing ABOUT the subject matter.

For instance, the talker says "I don't know what we're trying to do; this isn't getting anywhere." The listener encourages the talker to say more about that. You mean you're sort of lost, trying to figure something out? "Yeah, it seems we're wasting our time, I feel foolish."

Eventually the listener thinks he has a handle on the background viewpoint, and says something like "Describe how it feels to be wondering what you're doing. Good feeling? Bad? Is it OK to be doing this?" and so on. If the idea is right, this leads the talker to begin describing attitudes toward the former subject matter, at which point, of course, the talker can't be IN those attitudes. "Yeah, I was thinking all the time, 'This isn't going to work.'" You can't see the viewpoint you're in (I forgot to mention that principle). There is an abrupt, and I mean ABRUPT change in feeling-state, attitude, and content of speech when this shift occurs. If it doesn't occur, you just keep trying until it does. Twenty minutes us usually long enough for jumping two or three levels. Sometimes it's long enough to go farther.

Of course the question was, "Does this process have any end-point or does it just go in circles?" It turns out to have an end-point, although circles are often encountered. Also, even if you don't get to the end-point, the experience of going up one or more levels is very clear and easily remembered -- and highly therapeutic. Part of the brain's map has become clarified, and mostly it stays clarified. I think this approach is the essence of psychotherapy, the short-cut to Nirvana. What happens afterward supports this idea: there's a long period of rebalancing when you "come back." Things keep going "pop!" for days.

The highest level reachable is not describable. Kirk got there and I got there, and in the intervening years I have met several other people who, by one means or another, got there. I met one at the cybernetics conference -- made a remark that drew a delighted grin from him, had a suspicioun, asked, and verified that he accidentally transcended his mathematics once and hasn't been the same since. We agreed that there wasn't much to say about it. At that level you can see the entire path you followed, but there is nothing either to think or say about it. It's just there. The thoughts about it are there, too. You're not identified with them any more. The job then is to modify your brain's model of itself to include this fact about awareness. Because there are many paths to the top, you have to do this over and over to fill in other details. But here I am trying to tell you about it.

I don't talk about this very much. The reason I'm doing it now is the description you gave Rick of the way Margaret talks. It rang a bell. When she was struggling to make something other than hash of her words, her awareness was locked into the system that was trying to make the words. She was being a trying-to-make-words system. Behind or above that focus of attention was an attitude of trying that originated in a higher level. Finally she gave up, and her awareness popped up one level, whereupon she described to you, fluently, the attitude from which she had been operating: I can't get the words to come out right. So there is nothing wrong with her ability to make words or to use them to describe what is going on inside her. What's wrong is the system that lies just above making words, the one that has forgotten how to try in the usual organized way.

Of course I don't KNOW any of this. Obviously she has lost some kind of brain function. But the Observer is not a brain function, I can tell you that from experience. Or if it is, it's one that I have never been able to formalize. At any rate, Margaret has a mechanistic problem, and I wonder whether she couldn't find a way to function anyway. Maybe not, I don't mean to be cruel. But maybe these ideas will give you a third ear to listen with. I hope I'm not just tantalizing you.

Well, you can see that I've been engaged with the problem of consciousness for some time, and that what I think I know about it would be extremely hard to put into a computer model. I think there's more to it than signals whizzing along axons. The brain is a sort of interace, I think, between consciousness and the external world: the brain detects the external world as a collection of intensities, and for seventy or a hundred years proceeds to try to construct a world that makes sense of those intensities. The brain's constructions are the objects of consciousness: they are not themselves consciousness. Or that is how I would put it now.

I wish I knew what the Observer is. That means, "I wish I had a way of fitting the experience of the Observer into some kind of model my brain could hold." I don't know why I wish that. Maybe that's one of the Observer's reference levels. I'm down here in the hierarchy trying to figure out something in the hierarchy's terms, and that is probably futile. But you can see that my brain has some interesting data to work on.

I think that many people have had experiences like the one I described. They have dressed it in all varieties of theoretical clothes. This is probably what religion is basically about. Also the Eastern philosophies. The brain has to try to make sense of these experiences in terms of other things it knows. Now we know control theory. Maybe that will help, eventually.

Last topic. If a decision isn't arbitrary, it isn't a decision but a deduction or a simple intention. Only the Observer can really make a true decision. Decisions are not rational. The only time we have to make a decision is when we're in conflict -- when we have reasons for going two different ways at once. If nothing changes in the system, the conflict will simply persist. To break the conflict you have to go up a level and change a goal -- and if your mind resists doing that, you have to go up another level to find out why. At some level, all you have to do is Observe, and the conflict will be resolved, quietly and immediately. The Observer says, "Do <u>this</u>." And the decision is made. I think that decision theory is funny.

Sam hasn't written anything. You might send for Greg Williams' tape of his talk.

I saw Kathleen Forsythe at the conference, and managed to avoid getting her stuck to me. Told her that my emphasis was different from hers, much as you suggested. Your comments had Mary and me laughing -- so appropriate.

Best,

Bill

2 April 87

Dear Bill:

You are very generous of you to tell me about your experience of satori, the noumenon, the Cloud of Unknowing. And to offer it because of your sympathy for Margaret.

I have not been there. I have clambered up a few steps, maybe several a couple of times, but I have not been at the top. I was glad to hear that you found the path in hand with another person. All the tracts on mysticism, as far as I know, say you must do it alone. But the easiest times for me have come in small groups when we were all doing what you and your friend were doing: Is this what you mean? Could it include this? Does it feel this way? While you are doing that, are you also doing this? Are you now seeing yourself doing this thing, that you could not see yourself doing half an hour ago, on top of that other thing? Those were times that left the "point of view" hanging on for days afterward.

It is a wonderful feeling. The whole world is more open to view. The sunshine is both brighter and softer. I break into running as a child will do. I feel as if I can talk to birds.

It doesn't go away. You can get immersed in a routine and forget it, but you can sit quietly for twenty minutes and find your way up the steps of viewpoints again. But I can't explore new paths by myself. During one period of my life, I got up at six every morning and prayed. It was a nice peaceful way to start the day, but that's all it ever amounted to for me.

I do not, however, know how to connect any of that with Margaret. There are times, other sorts than the example I gave, when she can say that she is looking down at herself, but I don't know how to make use of those moments. Her memory is too short. The experience is gone the next minute. I've tried now and then to induce that looking down. But I suppose my urge to do so comes at the times that are hardest for her to do it. And again, the experience is gone in a moment. Many times, to check whether she is hearing me, I will ask, "What did I just say?" and get no iota of what I had said. Often, her memory is just long enough to repeat my last five words, namely, "What did I just say." Maddeningly irrefutable logic!

But I thank you for your love and your urge to help.

Aside from that, your paragraphs about points of view are helpful to me theoretically. I don't think I ever brought the matter up with you, but it seemed to me that the stuff, the capacity, the function of one level can be applied to the output of other levels. For example, I can perceive categories both of sensations and of principles. It doesn't have to be in words or conscious, in either direction.

Maybe that's wrong. Maybe I can categorize only words about principles, not the principles. Maybe if I take an action that will rectify error for both principles A and B, maybe I am not categorizing them as being similar or interconnected; maybe I am just acting and noting the effects and lo and behold, this act kills two birds with one stone, much as I might move a cursor on a screen and discover that it alters two variables at the same time.

Maybe I can apply the stuff of one level to others as long as they are below the level from which I am acting. Maybe from programs or principles, I can perceive transitions of relationships--matrix algebra.

Be all that as it may, I caught on some time ago that we cannot be aware of the level at which we are acting. Maybe it's something like logging in to one drive of a computer and operating in another. (That's meant as whimsy, not serious talk.)

Your remark about academics being in love with the program level reminds me of the paper I sent you in which I dug out assumptions about method from an article I read. Methodologists, in particular, are always looking for the right procedure, routine. In the book McGrath and I wrote, we said that all research methods had strengths and weaknesses; no one was best. One of the readers the publishers got to assess the MS was a famous author of books on research methods. He was outraged that no place had we told the reader the right way to do research. I enclose a few words by McGrath (from another book of his) setting forth our point of view. I think McGrath's point of view on the matter is close to what it was in 1972, but mine has changed. I don't mean that I now know the right way to do research; I mean that I have more subtle ideas about the purposes various strategies can and cannot serve.

Standing back and looking at our behavior from a higher level is a very useful skill to have. "Oh, would some power the giftie gie us." Schools should teach people to do that starting at about the third grade. They don't. They just have the students memorize Robert Burns.

There is a fellow named Chris Argyris, a consultant and professor of organizational psychology, who has made a big thing of writing books about the difference between "espoused theory" and "theory in action," a fancy way of pointing to the difference between talking and acting, between what you think you do and what others see you doing. He has long claimed that it is extremely difficult to teach people what he calls "double-loop learning", a fancy way of saying standing off and looking at your behavior from another level up. It is not easy, but I don't think it is nearly as hard as Argyris thinks. I think he has not gone about it the right way. You get people together with others they work with and care about, and then you do that paraphrasing and questioning. After a while the person catches on, and it is a break-through. It's never as hard, thereafter, to do it again. But it often takes a lot longer than 20 minutes. Sometimes it takes days (not concentrated), or weeks or with some people even months. And tears.

I hope I have caught on to what you were saying in your letter of 28 March. But you will let me know if I have not. And of course I have not caught on to all that is in your head and never will. We can only toss words back and forth and hope that they make us feel good. Too bad we are not close enough so that I can pound you on the shoulder now and then.

I am sure that some of the images I have of the functioning at the various levels of the hierarchy are pretty far from the images you have, but I'm not going to ask you to write out more words. I'll get clues gradually as we exchange letters and as I think about the matter myself.

Well, I'm glad you are willing to use the word "decision" when there is a conflict. But I agree with you that the "decision theorists" put "subjects" through a lot of puzzle-solving and build a lot of airy castles on the results. I don't pay any attention to that literature. In our book on organizational development, Dick Schmuck and I have separate chapters on decision making and problem-solving because he likes it that way. But the chapter on decision making is not on deciding by individuals, but on whata group can do to come to agreement about what to do. So you wouldn't call that decision making, either, nor do I.

One of the nice things about computers . . .

Well, those words remind me of a story my brother told me many years ago. It is apropos of nothing, but it's a nice story. Kenneth at that time had been patronizing regularly a restaurant run by a GReek family. One day, when he went to the cashier's counter to pay his bill, he saw a letter lying there and saw that it had a GReek postmark on it. He remarked to the proprietor about it.

"I see that letter has come all the way from Greece."
"Yes."
"I suppose it is written in Greek, too."
"Yes."
"It must be nice to know Greek."
"Sure is. Then you can write to your friends."

So one of the nice things about computers is that you can write out your account of your trip to Switzerland and send the whole account to all your friends. I had a good time reading about your experiences and thoughts. I have never been out of the Western hemisphere nor farther south than Peru. Nor farther north than Calgary.

Oh, dear, what can you say to a thief who is your host?

My AHD says: "scholasticism. The dominant theological and philosophical school of the High Middle Ages, based on the authority of the Latin Fathers and of Aristotle and his commentators." Maybe you meant "scholastic. 1. ... academic. 3. Pedantic; dogmatic." But maybe you did mean antediluvian. (This in response to your sentence "The more standard cybernetic sessions were . . . mostly Scholasticism.")

Good heavens! Do *Cyberneticists* get "control" mixed up with controlling *people?* Naturally behaviorists and organizational theorists and managers do, but *cyberneticists?* Maybe they think of it as controlling the output of a machine and therefore controlling the machine and therefore controlling people?

25 to 1! Good.

You have a copy of a piece I sent to *Continuing the Conversation*. In other news of publicatinns, you see from the enclosed that *Amer Psychol* turned down my comment. I hope they accepted Hershberger's. And I'll be very surprised if some other published commentator makes "substantially the same point" as mine about academic psychologists not acting according to their own theories when they might apply to themselves.

I envy you examining those old, old, books. Thanks for telling me about the postscripts. Often, when I read translations of old books (very occasionally) or hear about them, I yearn for those good old medieval days when scholars were less stuffy. But they no doubt had their own brand of stuffiness, of which the Galileo episode and the Spanish Inquisition are among the worst examples. Come to think of it, I think I'll stay in the USA in 1987. (Not in some other countries.)

We look at those old books in museums (few of us *read* them) because 1100 years ago there were few books, and few of those have survived. I'm sure a lot of those old books you saw wouldn't sell three copies today at $9.95. They are more marvelous for the ink than for the words. And with all the acid in the paper nowadays, no book can last more than two or three decades. Only those that are pickled or put on film can last longer than that. But the main point is your thousands of competitors. You have heard, no doubt, that more "scientists" are living today than lived in all history up to (fill in some recent year). Or something like that. Sorry, friend, you'd better aspire to 100 years, not 1100.

Glad to hear you are becoming known in the Netherlands. And the project by the German publishers sounds wonderful. And you can be plenty scholarly enough.

Thanks again for the travelog.

I'm very glad you found my editing of the Marken-Powers useful. I'd hate to go to all that work to no effect.

Now to the Marken-Runkel experiment. For the first time, I think I have caught you speed-reading. In view of the pile of mail you must have found waiting, I'll excuse you this time. But it is possible that I somehow omitted a couple of pages from the copy of my letter to Rick that I sent you. The pages are numbered. There should be seven pages and a sheet with two figures on it. Anyway, at the bottom of page 1, you will find: "I propose that the screen show only what appears in Figure 2." Though I modified that by one bar in my letter of 14 March. I am *not* proposing animations! That later kind of talk was just castles in the clouds.

What I want to know from you is whether my four equations on page 3 (modified by my recent letter about angular motion on the part of the subject) come anyplace close to proper specification.

So tell me that.

Again, I am grateful and honored that you care so much about Margaret and me. It's a feeling like being hugged. How fine it is that such a friendbhip can grow from letters and one phone call. Thank you.

Phil

610 Kingswood Avenue
Eugene OR 97405
4 July 1987

Dear Bill:

Hey, hey. Good, good.

Your two letters on simultaneous equations are very helpful. The last one, replying to my MS pages, more immediately so. I have already revised those pages.

I am progressing steadily with the book. I recycle every now and then, of course, but I keep going two steps forward for one back. I keep having the feeling that it will be either wonderful or absurd. It can't possibly be merely pedestrian. Actually, I know better than to have such high hopes (high, because surely it can't, it just can't, turn out to be absurd). I often have the experience--don't we all--of writing something I think is unusually perceptive only to find readers picking out the sentences they can fit into their old ways of thinking and somehow not seeing the other sentences, apparently, at all. Despite such intellectual realism, however, my spine keep tingling with delight as I move from section to section. Surely readers will find something somewhere in it to be a disturbance they will have to cope with.

I enclose the table of contents (so far) and chapter 1. You can postpone reading them as long as you like. The little chunk I sent you was from chapter 4. I am now drafting chapter 6.

Curve fitting in social science is very different from curve fitting in engineering. In engineering or physics, even though you may generate the data and then wonder what curve will fit best, you do postulate a curve eventually and then see how well the data fit the curve. In social science, you also generate the data first, but then you don't bother to wonder what the curve might be like. You just assume (usually, almost always) that it will be a straight line, and you make the line fit the data by declaring that the right line is the one from which the mean squared deviation is minimum. You don't bother to look for meaning in constants or coefficients at all. It is very common in research reports in social science to see authors bragging about "beta weights." To get a beta weight, you convert the original raw scores to differences from the mean in units of standard deviation. That is, you move the origin to the means, which does away with any constants, and then you convert the observable units to standard deviations, which disposes of any practical information, and you have the form $y = bx$, in which **b** is the correlation coefficient, and in that simple form **b** is simply the slope, but in units of standard deviations, and **b** is called the beta weight. Researchers crow when they get a high beta weight. But it is impossible to know what it would mean if one were to try to make a prediction that included a constant or a coefficient in observable units. It is fantastic in the original meaning of the word.

I am very glad to have your paragraphs on trying to solve equations recursively instead of simultaneously. I know I'll use that someplace.

Yes, there are often interaction terms in the linear "model." They are explicit in analysis of variance and the like. I quoted Brown's remark about their ubiquity and their multiplicative increase as the number of variables increases. I'll be dealing with that in my chapter on "fine slicing."

Yes, I am going to say what all this means. I have already said it in chapters 2-5, and I will say it again in most of the other chapters. I hope it will become clear as I say it in two or three different ways in every chapter. In brief, the method of relative frequencies tells you where to find contingencies, covariances, percentages of this that you get with that. The method of specimens tells you about human nature. Both are very useful. The method of relative frequencies tells where you are likely to find the greatest incidence of ughitis this year and maybe next. The method of specimens tells how you can help this particular person get rid of it or avoid getting it.

Well, well. Now I know why Rick Marken never replied to those letter I wrote him after he invited me to collaborate with him. I'll be eager to see your dissertation on reinforcing bacteria.

But I do worry about all the energy you are putting into battling the behaviorists. I know they are still thriving, and shouldn't, but they are not the largest sect among psychologists. Of course, I must admit that their underlying rationale permeates the ordinary talk of other psychologists <u>and of the public</u>. It turns my stomach. Educated people in other fields talk as if reinforcers, rewards, and punishments were facts to be accepted the way we accept water running down hill.

But that's not the only way people talk about things as true that are not because of what they learned in psychology class. Most social scientists and just about all the popular writers on the subject write as if research using the method of relative frequencies tells you what <u>all</u> the subjects did and what <u>all</u> people will do, even though the plain fact in the research reports is that only <u>some</u> of the subjects conformed to the hypothesis. There is a lot about that in my new book, also.

Anyway, if you want to plunge once more into the breach, I hope you have got hold of a current text or have looked at a few issues of a journal devoted to behaviorism so that you can be aware of all the ways those people use to patch up the holes in their garments. Current behaviorism, especially in the form advocated for practical use, is a far cry from its pure Skinnerian form. I have heard a couple of those people claim that reinforcing always works in practical life, that when it does not, the reason is that the practitioner just hasn't done it right, hasn't recognized the proper reinforcer, or something. I am reminded of the early arguments about Freudian theory. When it didn't predict right, you looked back with hindsight and discovered that you had been looking for the wrong neurosis; it was the other one all along. Indeed, when you saw a prediction failing, it was often not the prediction that failed, it was you, because your own neuroses prevented you from seeing that the prediction had actually succeeded! Somehow the reinforcement people often sound like that.

I'd like to know two or three particular readings that have outraged you and Rick. I'm willing to spend two or three hours catching up with the current scene.

Thanks for the example of the whirling ice-skaters.

And I thought all this time that surely you must have the fanciest word processor on the market or must have rewired whatever you had bought. My Perfect Writer does subscripts and superscripts very easily. To get x_1, I just type x@-(1).

So now back to THE BOOK.

Your friend,

Phil

Philip J. Runkel

8 July 87

Dear Bill:

Someplace you wrote about having to have as many equations, if you are going to solve them simultaneously, as there are variables on the right, leaving the same variable always on the left, as we were taught in elementary algebra. For example:

$$y = a + b$$

$$y = ab$$

I have looked and looked, and I seem to have lost the paper on which you wrote about it. I can't find it in my letters from you. It may have been in that first CT issue of Continuing the Conversation. And I seem to have lost my copy of that. If that's where it was, and you tell me so, I'll write to Ford for another copy.

I suppose your argument is that if you can't solve some equations to get points that will plot in a curve on graph paper, you can't test the prediction as a <u>model</u>. The widely used "linear model" in social science is a single equation:

$$y = ax_1 + bx_2 + cx_3 + \ldots + e$$

where "e" is "error," by which social scientists mean anything not explainable. The x's are estimated by least-squares regression. And "linear" means that a straight-line plot (regression line) is assumed. In other words, another equation such as

$$y = x_1x_2$$

is ruled out. It is simply assumed that such an embarrassing and difficult thing is not going to happen.

But with a single equation, you just toss in all your measurements (regardless of outliers, too), and if they cluster about a straight line more than you'd expect by chance, YOU HAVE SOMETHING! And another assumption you have to make to believe you HAVE SOMETHING is that all those points not actually on the straight line are somehow "tending" to be there. Isn't that strange?

Is that the way you were thinking in what you wrote?

Phil

July 19, 1987

Dear Phil,

Long time no write. No particular reason -- trying to work on the book, taking a mental vacation, writing a few short things (one is included, a commentary for Behavior and Brain Sciences). I appreciate watching over your shoulder while you communicate with members of the CSG. I guess it's "system" now, instead of "theory," by cultural drift.

Simultaneous equations. I don't recall either where that particular argument was put forth. Byte articles? First version of book chapters? That rings a bell. The physicist says you haven't got a model, just one equation. Anyway ...

Here's another stab at it. Consider an organism with an input variable x and an output variable y. If we look at the environment we can see that x depends on y according to the path that links the output back to the input. That dependence might be rendered as

$$x = C + ay + by^2 + cy^3 \ldots$$

A general polynomial. This tells us that the environment obeys some empirical law. What it doesn't tell us is what y will be. Of course we can treat y as an independent variable, and for any way of varying y we can predict how x will vary. But that gets the experimenter into the act -- why vary y one way rather than another? We can't predict x unless we can predict y. We can't predict the rate of reinforcement merely from knowing how it depends on rate of bar-pressing, because we can't predict the rate of bar-pressing without knowing the rate of reinforcement, and how it affects the organism.

Now suppose we treat y as a dependent variable and x as independent. We would do this in proposing a model of the organism. We say that the organism's output, y, depends on its input, x, according to

$$y = ux.$$

Aha. We now have a second relationship between x and y independent of the first. We have found a second constraint on the relationship between x and y. With two constraints, both x and y are determined. Now we know that

$$y = u[C + ay + by^2 + cy^3 \ldots].$$

We can now express y as a function of the coefficients of y on the right, a,b,c ..., and the constants C and u. And when we know y, we can divide it by u and get x as a function of the coefficients and constants.

Finally, if we now imagine that C is an independent variable, we can predict how both x and y will vary as C varies. In other words, we are predicting both the input and the output in this relationship between organism and environment.

What's wrong with curve-fitting isn't the goodness of fit, although one prefers good fits to bad ones. It's the assumption that the independent variable is independent. In the formulation R = f(S,O) -- response is a function of stimulus and organism properties -- the assumption is that the stimulus is independent of the response. Us feedback guys know this isn't true. There is a second constraint on the relationship between S and R, namely S = g(R,D) -- stimulus depends on response and disturbances. In principle we can solve these two equations simultaneously, and see how both S and R depend on disturbances. That's what we do in control theory. There's another independent variable, x* -- the reference signal inside the organism. So we find that S and R really depend jointly on D and x*.

So this is what I was trying to get at.

One added thought. There are cases in which it's hard to see how an independent variable (say, score on an intelligence test) can be affected by a dependent variable (performance on academic tests). The difficulty may be that the independent variable really is independent. But it may be that you're overlooking the real input variable, and seeing only a disturbance. What if we ask about the effect of knowing one's score on an intelligence test? Now taking the intelligence test is a disturbance affecting one's assessment of one's own abilities. Obviously, that assessment follows from taking the test, to some degree. At the same time, it affects how one approaches academic tests. If I'm told I'm very intelligent, I may take that to mean I don't have to study as hard as other people. Lack of study will lower my grades. This evidence will join the evidence from the intelligence test, to alter my assessment of ability. So both the assessment of ability (the actually effective "input") and the performance on academic tests (the "output") are affected by the intelligence test.

Sometimes there really are independent variables, but sometimes a controlled input variable is overlooked.

In the middle of August, Bill Williams, Greg Williams, and Tom Bourbon are coming to my house for 10 days of work. Tom and I will work on programs and demonstrations for teaching control theory; Bill Williams and I will work on his economic models; Greg Williams will bring his recording apparatus for some more oral history of control theory. It will be exhausting but fun.

Is Hugh Petrie really going to visit you this summer? I haven't heard from him about attending our meeting in October. You'll be getting literature about that pretty soon, FYI. If you want to send a bunch of copies of something you're writing, for distribution to the group, UPS them to me and I'll distribute them.

I'm actually writing an outline for the book, by the way. You're a good example. I hardly ever do this on paper, instead letting a mental outline take shape over many rewrites. Making a real outline is going to make the book much better organized, and has the added advantage of putting off actually starting the writing. For some reason, thinking about writing the book makes me very tired. I need to simplify my life.

Hope some of this made my intentions clearer.

Best,

Bill

Bill

24 July 87

Dear Bill:

I hope things are all right with you.

I don't know whether I have already mentioned to you that I am writing a book, a little book, I hope, on method in social science. I have taken the document I called GENERALIZING and the methods parts from INSIDE AND OUTSIDE, rearranged them, discarded some, written more, and so on, and I am well into it. I am working on the draft of chapter 5. Maybe there will be about 12 chapters.

* Please help me with the enclosed chunk, the part between the red marks. Have I said it right?

Gratefully,

Phil

* 870724_gen_ch4.pdf —enclosure at this volume's web page.
The start of *Casting Nets and Testing Specimens*?

July 30, 1987

Dear Phil,

OK, I see what you're getting at. Maybe I have a few more things to say on the subject of multiple equations in view of the exerpt from your new book (yay!).

The additive model proposes that an output, y, depends simultaneously on some set of independent input variables, x1 .. xn (an explanation of _why_ we expect that relationship, by the way, is what I would call a theory. The theory leads us to expect to observe a relationship y = ax1 + ... mxn). (Don't some statisticians speak of interactive terms? The rotations of factor analysis, I had thought, were meant to eliminate the cross-products of orthogonal terms, blah, blah, blah.).

In physics and engineering, there is seldom a need to "postulate" that two variables are related by a curve. It really works the other way. First, the relationship between the variables is carefully studied through experiment, so the relationship exists as tables of numbers. If you can't find a clear stable relationship you just keep trying until you can -- you don't publish until you can. Then a mathematical form is found that meets two criteria: (1) it comes as nearly as possible to passing through all the data points, and (2) it has theoretical significance. Obviously, if there are n data points, a polynomial of degree n - 1 can be made to pass exactly through all the points. But that polynomial would have no theoretical meaning.

A physical theory gives meaning to both the variables and the coefficients of an equation. For example, Newton said that for any piece of matter, the acceleration _a_ of the piece is proportional by a factor _m_ to the applied force _f_. The coefficient _m_ is called the "mass" of the object, in the familiar equation _f_ = _ma_. A second relationship proposed by Newton is that gravitational attraction is proportional to mG/(r^2). That coefficient m is the same mass, and now we have a second constant, G, the universal gravitational constant. Knowing _m_ from other experiments with various pieces of matter, we can evaluate G. And so it goes. Every element of an equation in physics or engineering is named and takes its place as a theoretical entity in the whole picture. The same coefficients show up over and over in many different contexts. There are no coefficients that are left nameless (except numerical constants like pi or 2).

In the approach you're laying out, the coefficients are simply what they need to be to give the best fit to the data. They have no meaning that could be transferred to any other context.

Suppose we have a theory that proposes a perceptual signal p that represents the taste of chocolate. We propose that this variable arises from some combination of four intensity variables, x1 (sweet), x2 (sour), x3 (salt) and x4 (bitter). That is the theory part -- proposing that just four independent intensity inputs give rise through some kind of combination to a dependent perception of a given taste.

Through biophysical experiments, I suppose it is possible in principle to determine how a given sensory intensity-signal would depend on the amount of the input variable that gives rise to it. Experiment would lead to four independent relationships stated as pairs of numbers. These would no doubt be nonlinear relationships, but for convenience we would probably choose to find the straight line that fits them best over some range of normal intensity inputs. This would give us four equations involving four coefficients. The coefficients would be called the "sensitivity" of each kind of input receptor: p1 = (s1)(x1) and so on.

Now our theory says that p results from some weighted combination of p1 .. p4. We have four more coefficients to determine, w1 .. w4. If our data were very nearly noise-free, we could find these weighting coefficients by measuring p for four different combinations of values of the x's, and using the s-coefficients already determined to calculate the corresponding signals p1 .. p4. We would then have four equations in four unknowns, and could find the values of the weighting coefficients that satisfy all four equations at once. Of course since this is experimental data it would have some small uncertainty, so we would have to settle for finding the coefficient values that lead to the least squared deviation of the predicted values from the observed values.

Now we would expect these same four coefficients to predict the degree of chocolate-taste perceived for any combination of the four kinds of inputs. One particular kind of input pattern would lead to a maximum of the chocolate taste, given constant total stimulus intensity. That's "real chocolate."

Suppose now that we used a different substance, say lemon juice, that stimulated the four kinds of receptors in a different pattern of intensities. The chocolate-perception would fall off. But in the same way as before, we could find four new coefficients relating p1..p4 to p; these would be the ones yielding the maximum response to lemon juice. We wouldn't expect the first-order coefficients to change, since they describe presumably built-in sensory receptor characteristics. Only the second-order coefficients would be different.

Now if we imagine _two_ sets of second-order coefficients, representing independent second-order perceptual systems, we can see that "chocolate" would yield the most signal pc in the second-order system having chocolate-weighted coefficients, and that lemon juice would yield the most signal pl from the other second-order system that uses a different combination of coefficients. Perceptually identifying the substance now becomes a matter of seeing which second-order signal p is the biggest.

That's how engineers and people who work with system models use equations. You'll notice that determining that a relationship exists is a rather trivial part of it.

(Since the brain has no other way to identify tastes, of course, it can assign any label to either signal -- for example, the visual color that goes with it. "Ah, there's the yellow taste again!". Let's make yellowade, by adding some transparent-gurgly and enough white-gritty. It's perfectly natural to think in terms of correlations, when you come right down to it. Ever notice how odd white chocolate tastes? Wrong correlation.)

This example is probably closer to what you're talking about than my previous discussion of simultaneous equations. But what I said before still is important.

On p. 15 you point out correctly that in the simple additive equation, nothing is pinned down -- anything can happen. For instance, you might find that all the coefficients are zero, or so close to zero that you realize you have it wrong: y doesn't depend on those x's. Of course if you're not willing to admit being wrong, then you simply treat the coefficients in a relative manner, disregarding their magnitudes, and point with pride to the correlations, which don't reveal the size of the regression constants unless you choose to finish the calculation. Also, with no model in the background, how are you supposed to know what constitutes a "small" coefficient? Is 0.02 small? Only if y is significant when it is larger than, say, 10. If a value of y of 0.001 is significant, a coefficient of 0.02 is quite large -- unless the range of its corresponding x is from 0.00005 to 0.00007.

On pp. 16-17 you show that if the three equations are treated as simultaneous, you discover necessary relationships among the xs. That's just a way of saying that you find that a particular set of coefficients is necessary to satisfy all the equations at once. Once all the coefficients are known, you can calculate y for any combination of xs. Given three values of y, you can solve for the necessary values of the xs: given the coefficients, only one set of xs will produce all three of those values of y. Actually it's quite arbitrary to call one set of symbols "coefficients" and the other "variables."

On p. 15 you really should make those three y variables into different variables, y1, y2, y3. To get the additive equation, you add them together to get y = i(y1) + j(y2) + k(y3). Clearly, to evaluate i,j, and k, you need at least three different sets of evaluations of the variables y1 .. y3. The only time you are required to solve simultaneous equations is when you claim, for some extraneous theoretical or observational reason, that y1 = y2 = y3, which is what the first set of equations on p. 15 subtly says because of using the same symbol y in all three equations.

In my last letter I talked about simultaneous equations in a different context -- maybe you don't even want to bring it in at this point in your book, if at all. I was talking about multiple relationships that hold simultaneously between variables. Here are two variables, x and y. We can see that there is a link from x to y, making y depend at least in part on x. By studying this link we can arrive at a description of its form: y = f(x). The functional form f, whatever it turns out to be, describes the link.

But we also notice that a second link exists, one that involves a quite different physical path. Studying this second link, we find that x should depend on y according to a different function, g: y = g(x).

Now we have y depending on x in one way, while x depends on y in a different way. The question is, what will be the actual relationship between x and y when both links are in effect at the same time?

If x = 5y + 7 and y = 7x + 5, we can solve to find that x = -16/17 and y = -27/17, I think. In other words, when these two particular constraints are in effect, there is only one pair of values that can exist. If these are input and output variables, we can only conclude that input and output must be (or become) constant.

Suppose the first equation is changed to x = 5y + d. Now the solutions are x = -[(25 + d)/34] and y = (5 - 7d)/34. I guess. We find that even though d enters only into one equation, both x and y depend on d. We have discovered a relationship that wasn't obvious in the original description of the two links.

This sort of thing happens all the time in systems analysis. You analyze a system by trying to describe each link between its variables in isolation from all the other links. Then you solve the whole set of equations, if you can, to see what states of the variables will satisfy <u>all</u> the conditions that relate them at the same time. The result, when this method works, is a description of the <u>actual</u> relationships that will hold among the variables, as opposed to the relationships seen when only one link is considered at a time. If time-variables are involved, you get the simultaneous behavior of all the variables. Usually you're surprised -- not that the result matches the real behavior, but that it possibly could, considering what you started with.

Going back to the first two equations, x = 5y + 7 and y = 7x + 5, let's see what would happen if someone tried to predict the values of x and y by analyzing these two relationships sequentially. Suppose y starts out at 0. Then x = 7. Going to the second equation and setting x = 7, we find y = 54. Going back to the first equation and setting y to 54, we get x = 277. The next value of y is 1944, the next value of x is 9727 -- no need to go on. The problem is that we're violating our own rule, which says that these two relationships hold simultaneously. We think that one variable can change in one equation without changing at the same time in the other one. After every step, the values of x and y in the two equations are not the same. If we don't wake up to what's wrong here, we might decide that the coefficients 5 and 7 are too big to describe the real system, and select values small enough to let the process converge. Thus we would miss not only the point, but the line, plane, cube, and so on.

Notice that we would have the same problem if we put d back into the equations.

I claim that a real model is one that contains enough independent relationships among its variables to force a solution to exist among the system equations. That solution can then be compared against the actual behavior of the variables. By forcing a solution to exist, you stick your neck out as you ought to, risking disproof.

The additive "model" is not a model, because it forces no prediction. I don't count the prediction that the past will repeat itself as a scientifically useful prediction. Or better, we always predict that, so it's a trivial prediction.

Brown's analysis of the literature comes out worse than I had suspected it would. 0.27! Egad. 0.24! Gadzooks.

Are you going to say somewhere what this all means? I mean, what it says about 33 to 67 percent of the published knowledge about human nature, which is presumably the base on which future scientists will build and on which current scientists are building?

* * * * *

Rick and I have gone around three more times on the bacterium-reinforcement paper, and I believe that it has finally reached an effective form. I'll see that you get a copy. We are now planning a larger paper on the general idea of operant conditioning, reward/punishment, reinforcement, etc. These concepts are completely wrong, irrelevant, misleading, and disgusting. Behaviorists have cheated and lied. They have argued like lawyers, not scientists. I find the whole behavioristic movement utterly repulsive. It will probably take six tries to get the new paper into shape, just because the more Rick and I learn about how behaviorism is conducted, the more outraged we become. I'm a man of peace in here somewhere, but sometimes I really feel like clubbing someone with this olive branch.

If you want a nice example of a non-causal simultaneous interaction, think of two ice-skaters revolving around a common center, leaning back and holding hands in the middle. While one is keeping the other from falling, what is the other doing?

On to other matters -- trying to get a database program in shape for Tom Bourbon's visit on Aug. 13, so we can set up a coherent set of demo-experiments.

Best,

Bill

Bill.

P.S. I can do subscripts, but it's such a hassle to set them up that I just didn't bother. Too much to do.

610 Kingswood Avenue
Eugene OR 97405
23 August 1987

Dear Bill:

HELP! I know it is dastardly of me to ask you to divert your attention from your book to mine. But I have no one else to whom to turn. If you don't want to puzzle over my puzzle now, I can wait a few weeks. If you don't want to take time for it at all, I can simply omit the section from the book. Or maybe if I let it sit and puzzle over it a month from now, it will all come clear. Anyway, if you can help, I'll be grateful (as usual). If not, I remain your friend.

The last thing I sent you on simultaneous equations was for chapter 4 on linear causation. What I am sending you with this letter is a section for a piece of a chapter on simultaneous causation. I think that will be chapter 12. I * enclose also a revised table of contents.

You will find the details of my request for help in a set-off paragraph in the middle of the attached section.

Where are you now with the two articles you told me about in your last letter? Give my regards to Mary.

Yours,

Phil

Phil Runkel

* 870823_SimultaneousCausation.pdf —enclosure at this volume's web page.

Aug. 26, 1987

Dear Phil,

Just a slight problem with the equations -- all is well.

The first equation, $Y = aX + b$, is OK if we consider only one-way deviations away from $X = 0$. As X increases, the amount of rain in the face increases, with some constant amount of rain b leaking past no matter what.

The second equation is the problem. The amount of rain in the face depends on X; now what does X depend on? It depends on the action of the person, D.

$$X = kD.$$

The more the person Does, the farther to one side (X) the umbrella is held.

Now what does the person's action D depend on? The amount of rain in the face, Y, relative to the amount desired, Y*. So

$$D = j(Y - Y^*).$$

Ideally, we would like all constants to be inherently positive so we can see the directions of effects from the signs in the equations. If Y exceeds Y*, then D will be positive, as we have written this equation. But the more D, the more rain in the face, since X increases as D increases, and Y, rain in the face, increases as X increases. So an excess of rain in the face leads to even more. To make an excess of Y lead to a negative D and thus less rain in the face, we would have to put in a negative value for j, or else leave j positive and swap Y and Y* in the parentheses. The latter allows all constants to be positive, so we now have

$$D = j(Y^* - Y).$$

Combining the second and fourth equations, and using K to represent the product of the two old coefficients j and k, we now have

$$X = K(Y^* - Y). \qquad (1)$$

The first equation was

$$Y = aX + b \qquad (2)$$

This is the simultaneous interaction boiled down to the minimum of two equations.

We can now solve for the amount of rain in the face, Y, by substituting (1) into (2):

$$Y = aK(Y* - Y) + b, \text{ or}$$

$$Y = (aKY* + b)/(1 + aK). \qquad (3)$$

Also, we can solve for the umbrella angle X:

$$X = K[Y* - (aX + b)], \text{ or}$$

$$X = KY* - KaX - Kb, \text{ or}$$

$$X = K(Y* - b)/(1 + aK). \qquad (4)$$

Suppose that the two constants a and K are VERY large. In that case 1 + aK is the same as aK, so we have the approximation for equation 4

$$Y = (Y* - b)/a \qquad (5)$$

This, however, is a lie, because if we say that the person wants zero rain in the face, so that Y* = 0, we get

$$X = -b/a.$$

That's a lie because in this case negative values of X, the umbrella angle, can't cancel the minimum amount of rain b -- they increase it again. The problem is that equation 2 really applies only to one-sided deviations, whereas it seems to imply that negative values of X would <u>subtract</u> rain.

But this is OK too, because it shows that we have to think, not just turn the mathematical crank. What we find, given that the minimum useful value of X is zero, is that it doesn't do the person any good to set Y* below b, the amount of rain in the face that occurs with the umbrella facing the wind.

If b is an insufficient amount of rain in the face, the person can increase Y* and experience a Y in excess of b.

When you set up equations like this, always make the reference level (Y*) explicit. You can always make its value zero, later, if that's what's called for. If a person responds to pain, P, you say he responds to P - P*, the difference between the amount of pain experienced and the amount desired. Etc. Usually you end up wanting to swap the variable and its referencve velue in order to keep the constants positive.

Howzat?

Best Bill

610 Kingswood Avenue
Eugene OR 97405
1 September 1987

Dear Bill:

I am giddy with gratitude to you for sorting out for me this second pair of simultaneous equations, and so promptly. Thanks very much.

Of course one of the equations should have Y* - Y in it. I wonder how long it will take me to learn the first principles of control theory.

I have eight chapters in good enough shape to go to preliminary readers (friends) and the other eight pretty well blocked out with some good-sized batches of text for most of them.

Sincerely yours,

Philip J. Runkel

610 Kingswood Avenue
Eugene OR 97405
14 September 1987

Dear Bill:

I've asked you this question before without getting an answer. Now I need an answer. Question: Don't several existing machines have feedback loops with sensors and reference signals such that they maintain (the builder's) desired perception? Notably the rockets that focus their sensors on a planet, or the artificial pterosaur? Maybe you know of a few other notable examples.

Thanks.

Yours,

Phil

Philip J. Runkel

16 September 87

Dear Bill:

I need some help in making sure my references are properly detailed in the book I am writing. Please tell me:

An entry in your résumé says that Powers, Clark, and McFarland was reprinted in "General Systems V." Was that the Yearbook of the Society for Gen Sys Something-or-Other, or what? Please give full reference, with pages if possible.

The same entry says "Part reprinted in Smith"(etc.) with a gap between "Part" and "reprinted" as if a number were typed too lightly on the typewriter. Please clarify. (Smith, by the way, was here at U of O for several years. Well, I expect you knew that. I used to have his book. I threw it away before I knew I should read what you write.)

Can you give metthe page numbers in Krippendorf?

Is Levin and Fitzgerald out yet?

Thanks.

Phil

Sept. 17, 1987

Dear Phil,

Sneaking in a quick reply while working on -- ta-dah -- chapter two. I am finally ON THE RIGHT TRACK.

In answer to your outstanding (i.e., debited) question: no, there are no control systems that maintain the builder's desired perception. The control system doesn't know (a) what perception the builder desires, or (b) what the builder's perception is at any given time. The control system can control only its own perception relative to its own reference signal.

Of course the builders can do their best to give a control system a sensor that creates a perceptual signal they think depends on the environment the way their own does, and they can set a reference signal in the device that represents a state of the device's perception that is presumed to be like the builder's reference signal. But then they have to turn the control system loose, and what it does is from then on no longer connected to the builder's perceptions or reference signals. The control system works as it does because of the way it is organized, not because of the way the builder is organized.

Often, of course, we happen to turn out a good design, and we find that our own perceptions are nicely stabilized right where we want them by the control system. But we and artificial control systems age at different rates and in different regards. No matter how good the control system, pretty soon we look at what it's doing and say "No, no, not that way, this way." And we have to reach into it and tweak it.

A home thermostat is supposed to keep the room at a temperature comfortable to its user. But it doesn't: it controls only the temperature of the air right around its sensor, over there on the wall where a draft hits it when the window is open. When the draft leaks in, the thermostat shivers and turns on the furnace, raising its sensor back to 72.00000 degrees as it was told to do. Of course for it to achieve that, the rest of the room has to be at 101.8834 degrees, because of the draft. The thermostat controls its own perception, thank you, as Rick Marken would say.

Furthermore, the temperature-setting knob can go out of calibration as springs weaken and knobs slip. So the user confidently sets the reading to 72, when in fact this puts the stationary contact where it can be closed only by a temperature of 51 degrees or lower. So the owner sends for the furnace repairman (the owner doesn't know any control theory) who cheerfully informs him that he needs a new furnace, driving the owner into bankruptcy and depression, all because he assumed the control system knew what he wanted, and still wanted what it used to want.

Now if you want to speak metaphorically, yes, we do manage to set up systems that act as if they were controlling our own perceptions relative to our own reference signals. But that is a metaphor, and we control theorists no longer have to resort to that sort of thing, do we?

Best,

Bill

Bill.

610 Kingswood Avenue
Eugene OR 97405
22 September 1987

Dear Bill:

I am glad to hear the you are ON THE RIGHT TRACK. I know that is a good feeling.

But I didn't get the kind of answer I wanted to my question about space ships that keep their cameras pointed toward Saturn and the pterosaur that adjusted its wings to counter every air current. I fear I phrased my question awkwardly.

Granted that those devices do not have all the necessary sensors and repair systems to keep their bodies in good repair. The pterosaur flew beautifully the first time, but in Wash DC it hit a severe draft and the neck broke off. It didn't have the sensors to tell it that the stress was exceeding the strength of the "bones" and "muscles" in its neck. Well, what do you expect? I don't think your computer models have in them sensors and repair systems to cope with parts of the computer that wear out, either. And it seems to me that the pterosaur came very close to performing the way it "wanted" to, not the way the designers "wanted" it to. All the designers wanted was for the thing to stay up there until called down. That seems to me parallel to building a two-legged robot that would actually succeed in maintaining its balance while walking on very rough ground. I think that would be a high achievement in modeling.

The point of my question was not whether I could put my reference signal into a robot built by the present state of the art and expect the robot to go on doing what I intended it to do for the next 25 years. I was not asking about impressing my desires on the robot. I think my question was much simpler than that. I'll try to ask it clearly: "Does the space ship and does (did) the pterosaur contain feedback loops?" And a corollary: "Did the pterosaur, at least, contain a hierarchy of at least two levels of feedback loops?" I guess I mean my question to find out who else knows about the kind of feedback loop you talk about in addition to you and Marken. I know that very simple feedback loops have been used in industry (e.g., thermostats) for a long time, but I also understand that almost all of them have been intended by their makers to produce what you would call irrelevant side effects, and that an essential part of them, from the designer's point of view, is a reference signal that is adjustable from outside, so that the designer can compensate for troubles that are bound to develop in such simple devices.

So I am asking whether devices such as I have mentioned

have acted, at least for a brief while, as if they were pursuing their own purposes. Is that a pseudo-question? Is the answer yes, depending on how short a time you are willing to accept as a demonstration of performance?

I am not in a hurry for a reply to this. What you wrote tells me at least that the question is not as simple as I had thought when I wrote. That in itself is a help. I would like to have an answer to this, but Christmas or Easter will be OK.

I wish you good flow with the book.

Yours,

Phil

Philip J. Runkel

Dear Phil, Sept. 26, 1987

Enclosed is my reply to Dan White's little nasty in CC10, * which I presume you receive. I wrote two others first, to get rid of the indignation -- most of it.

I guess I really didn't grasp your question -- or else, maybe, I wasn't sure you COULD be asking that question. Practically everything I know about control systems, or at least about basic control theory, came from the branches of engineering that actually build such systems. A process-control system that runs an automated plant will have sensors all over the place, and many levels of control loops. The computers that run the top levels of the control systems (lower levels are often autonomous except for their reference inputs) are designed to control very abstract variables, for example, a "figure of merit" that represents how near to optimum the performance of the plant is, a perception drawn from complex functions of dozens of variables. Hierarchies of control abound, in other words.

As to the question of whether such systems ever behave for any period of time as if they were pursuing their own purposes, the answer is clear: sure they do. Just think of the spacecraft that went to Jupiter, and the way it operated at a time when the round-trip signal transmission time to Earth was one hour and 25 minutes. All the human operator could tell the spacecraft was something like "Now lock your sensors onto Icarus and commence rotating your body through an angle relative to that direction at so-and-so many seconds of arc per second." There was no possibility of telling the spacecraft how to do that -- only the goal to be achieved could be transmitted in time. So the human controller could change the goals of the spacecraft only every hour and a half*, with the spacecraft selecting and executing all the necessary subgoals by itself, with no specific directions.

You've heard about star-trackers "locking onto Canopus", one of the favorite reference points. A spacecraft doing that carries in its memory the approximate coordinates of Canopus, approximate because the spacecraft's own attitude is not known very accurately. So a little control system goes into operation: it starts a small telescope hunting for an object of a specified brightness within some search area. When its photocells record the right brightess, it starts comparing the brightnesses detected by four photocells clustered together so each sees one quadrant of the field of view. The imbalance in brightness signal between opposed pairs of photocells is used to run an electric motor that moves the telescope relative to the body of the spacecraft, one control system for each axis. This centers the image of Canopus in the field of view, and from then on the

*or 1½ hr behind real time, anyway.

* CC10 = Continuing the Conversation, Issue 10. All issues (#1, Spring 1985 through #24, Spring 1991) have been recreated and are posted at the website.

electric motors hold the image precisely centered.

Here's the cute trick. Suppose now the command is to turn the spacecraft's body to establish a specific angle between its axis and the line of sight to Canopus. There is a sensor in the mounting of the small telescope that reports the angle between the space-craft's axis and the axis of the telescope. If that angle is too large, thrusters are activated that start the body of the spacecraft turning in the direction that makes the angle smaller. When the angle gets to the required (reference) amount, the thrusters stop the change of orientation.

But wait, you say, the telescope is mounted on that spacecraft, so when you turn the spacecraft, doesn't that turn the telescope away from Canopus? Of course not. That's what the optical control system is for. It uses the body of the spacecraft to push against in order to slew the line of sight of the much less massive telescope, so the telescope stays rigidly locked onto Canopus while the spacecraft turns under it. The movement of the spacecraft is just a mild disturbance that the star-tracker easily counteracts. The spacecraft's control system uses rocket thrusters to control the angle sensed at the mounting of the telescope; the tracking control system uses an electric motor to change its angle relative to the star, sensed as the position of the star's image in the field of view. Both systems happen to affect the same physical angle between the mounting of the tracker and the hull, but the spacecraft cares about that angle while the star tracker doesn't.

With three star trackers, each locked onto a different star, the spacecfrat's body can be turned relative to all three lines of sight to point in a known direction in space. It's just as if there were three long rods extending from the spacecraft and rigidly anchored on each star, forming a stable frame of reference. So now a second-level control system can point the spacecraft in any direction by giving it three reference numbers, and the two lower-order control systems will automatically do what's necessary to achieve the specified direction while keeping the framework locked in place.

Spacecraft are jam packed with feedback loops, sensors, comparators, actuators, and multiple levels of control starting with the servos that turn the main rockets in their gymbals and extending to just about everything movable.

There's another level to the question. All these artificial devices get their highest-level reference signals from Earth by radio. The operators on earth also get telemetry telling them about all the controlled variables at all the levels, and can interfere as they please. Lots of action on the spacecraft is carried out open-loop, simply because control engineers haven't seen the basic organizations that I have seen, and still do a lot of things the hard way. I think that some day my way of seeing complex control as control of perceptions will find its way into control engineering, but it isn't there yet, that I know of. The control systems that are now designed are often a lot messier than they have to be; they sort of just grow, without much conscious grasp of the sorts of relationships I talk about. Of course all the principles are there, embodied in the machines, or else the machines wouldn't work. But there's a lot of patchwork as well.

You can tell that control engineers aren't thinking the way I do yet, when you realize that the Three Mile Island accident was caused by an engineer's installing a valve position sensor in the command line, instead of making it sense valve position, or better yet, water flow. Even engineers still have one foot in the old world.

But there's another level still. No artificial control system is yet truly autonomous, because all such systems are designed to carry out an adjustable human purpose: the top reference signals are manipulated by human controllers. Even while operating on their own between adjustments, the control hierarchies are bound to maintain the highest level of perception at the last setting of those highest-level reference signals. The lower-level reference signals can vary as disturbances come and go, so at the lower levels the system governs itself. But at the top level it doesn't.

For those Mars-exploring robots people are tentatively designing, we need a really autonomous hierarchy of control. We need a noticing-machine with curiosity that we can plunk down on Mars and set loose. I once suggested to Jacques Vallee that this is what UFOs are. Exploring an unknown planet can't be done by telling the robot what to look for: we don't know what to look for. What we need, beside the usual lower-level control systems for maintaining temperature, moving around, and looking here and there, is a reorganizing system that will cause the robot to try out different perceptions, different reference-levels for those perceptions, and different modes of actions to correct the errors. We specifically don't want to tell the robot what to perceive at those levels, because it is supposed to be discovering the new, not recognizing the familiar. When it comes back we'll read out its memory and try to figure out what it learned to perceive and control on Mars.

So my idea of an exploration-robot is something that hasn't been built yet. It's something pretty close to what I think an organism is, aside from reproduction and materials of construction. We still have to supply some top-level reference signals, telling the robot the sorts of things it is to find painful and pleasurable; we will have to substitute our own guesses for the processes of evolution, unless we have a lot of time to kill and want to simulate that, too.

I would really like to try building a robot that works under the control of a curious noticing-machine. The fact that I don't know how is only a minor obstacle -- the biggest one is that I have no money and won't live long enough. Other than that...

So anyway, somewhere in here I may have come closer to answering your questions. Did I?

Best,

Bill.

p.s. References.

Powers, W. T., Clark, R. L., and McFarland, R. L.; A general feedback theory of human behavior, Part I. In Bertalanffy, L. and Rapoport, A. (eds), General Systems V, 3-73, 1960. Reprinted from Perceptual and Motor Skills, Monograph Supplement 3-VII 11,71-88, 1960.

Powers, W. T., Clark, R. L., and McFarland, R. L.; A general feedback theory of human behavior, Part II. In Bertalanffy, L. and Rapoport, A. (eds), General Systems V, 5-83, 1960. Reprinted from Perceptual and Motor Skills, Monograph Supplement 3-VII 11,309-323, 1960.

Powers, W. T., Clark, R. K., and McFarland, R. L.; A general feedback theory of human behavior, Part I. Ch. 32 in Smith, A. G. (ed), Communication and Culture; New York: Holt, Rinehart, and Winston (1960), pp. 333 - 343. Reprinted from Perceptual and Motor Skills, Monograph Supplement 3-VII 11,71-88, 1960.

Greg Williams borrowed my Krippendorf and I won't have the page numbers until I get it back.

Levin and Fitzgerald is not out yet. I left a message today for Levin to call me Monday and let me know what's going on.

I didn't know Smith was at U of O.

Let's see - guess that's all I owe, or care to owe, for now.....

From the constructed viewpoint of a control theorist.

Dan White, you're basically right. If all I knew about control theory were what has been published in psychological and mainstream cybernetic writings, and if all I knew of science was the sort of travesty practiced in the branches of the humanities that have called themselves scientific, I think I would opt for the Buddha, too. Not that I reject the Prince -- he was a swell guy. But he was just as one-sided as, say, B. F. Skinner is. Or as your letter in CC10 was. I'll get back to that.

It's not really fair to argue against today's control theorists by imputing to them the beliefs, aspirations, and philosophical stances of the very sciences they are trying to revolutionize. The control theorist does not believe that "scientific method" as now used with respect to organisms is worth much. The control theorist is, true enough, concerned with quantitative analysis, but is also vitally concerned with the human capacities for perceiving the qualities of experience, from simple intensity to system concepts. Imagination, insight, creativity, and feeling are all part of human nature and we control theorists try (with varying success) to integrate them into our understanding of human nature. Control theory -- real control theory, not that "programmable functions of stochastic machines" junk -- probably gives us the best medium for understanding constructivism, for making it real, illustrating its premises, and saving it from solipsism. At least one control theorist, me, has known that the word is not the object from the age of 16, after reading first A. E. van Vogt and then, God save him, the Good Count's entire impossible book. I don't know how old you are, Dan, but I could guess that might have been before you were born. It's getting to where I did practically everything before anyone else was born, sigh.

Now, this one-sidedness. Eastern philosophers and holy men have told us a great deal about perception and states of inner being. They have shown us, crudely, how to reach certain states of consciousness that seem better than the state of mindless desire (and inner conflict) we normally go around in. Western philosophers and holy men, on the other hand, have focused on action: learning how to do things that have predictable and thus useful consequences. That's basically what the experimental method is about: producing results.

Western science, however has been under the impression that the "results" they learn to produce have objective existence; thus Western science has failed to understand that what we know of reality is subjective, internal, perceptual, and interpretive, and that therefore what we learn to control through acting is of the same -- human -- nature. In fact Western science has pretty much abandoned the study of the subjective, although that abandonment is gradually, it seems, being abandoned. Control theory shows, I think beyond doubt, that action is organized not around its effects on an objective world, but around its effects on perception. Some sort of external reality gets into the loop, imposing its rules by telling us through demonstration that some acts work and some don't; we can experience those rules as they project into the world of perception. But those rules don't mean what Western objective science has traditionally thought they mean. They may have a basis in what is Out There, but they aren't Out There. Modern physics, oddly, is gingerly approaching the same conclusion.

Eastern philosophies have pretty much ignored action -- in fact some of them urge us to abandon it altogether. I don't think that is a very wise suggestion. When you just sit, observe, and describe, your mind is cut free of all constraints but those that limit the kinds of perceptions human beings can construct. Perceptions then become fluid, a million-dimensional Rubik's Cube that the mind can twist and turn into an endless variety of shapes and appearances.

This capacity, of course, is the essence of being human, but when it operates without constraint it loses its value. A good idea doesn't look any better than a damn fool idea -- there's no way to tell the difference. You can only tell the difference by acting, doing something with your muscles that alters your perceptions. That's how you find out that some patterns persist and some make no sense; that's how you discover the regularities that allow you to control what happens in your experience. That's how you separate useful ways of perceiving from useless ways. That's how you discover that there are independent agents in the world other than yourself. So while the Eastern preoccupation with perception has taught us a lot in areas that Western science has shunned, it has also led to a lot of blather. Without action, you can never know the extent to which you and the other guy are talking about the same thing. It's really amusing sometimes to see the way certain philosophers go on about aesthetics and other abstractions as if their words have one and only one objective meaning, exempting communication from the principles they apply to everything else.

Bucky Fuller's image of the cyborg is also amusing, because it wraps up so vividly all the mistakes of the so-called mechanistic approach to organisms. It's just the sort of comic-book image that would frighten a person who knows nothing of mechanism. And it's painfully faithful to the conventional scientific mechanistic view of behavior, which was constructed largely by people who don't understand machines, especially modern machines.

To anyone who does understand machines, and I can claim to be one of those, Fuller's image is ludicrous because it so relentlessly directs us to the superficial appearances of artificial contrivances, and so completely misses the point. A machine is only an embodiment of a law of form, a law of organization. A student of behavior, at least in my field, who uses machine models as illustrations doesn't think that people are like those simple machines. Once the machine has shown that a principle does in fact have some force in nature, it's served the theorist's purpose: the principle can be cut loose from the machine and applied, one hopes, to any situation in which the same relationships can be found.

To get more specific, there is no "analogy" between control mechanisms and organisms. Control theory is a set of principles that applies whenever an active system affects and is affected by its environment at the same time. There are a few other provisos -- the gain around the loop must be at least ten, most of the gain must occur in the organism rather than in its environment, and the whole system has to be dynamically stable (you'd have to learn something about control theory to understand those ideas). That's all you need to know to have a reason for trying to apply control theory. You don't build a machine and then show that it has a few superficial resemblances to an organism, and then announce triumphantly that organisms are "just machines." That would be dumb.

There's one thing that machines do for us theoreticians that is of great value, when we can figure out how to get them to do it. They can tell us whether an explanation has been laid out completely enough to work. It's very easy to propose a big tangle of relationships and say "There, that's my explanation." It's a lot harder to verify that the tangle of relationships would actually behave as you claim it would. That's quite apart from the question of whether the relationships have anything to do with the phenomenon you're trying to explain. The first question should really be, is my explanation an explanation, or is it just a lot of words? If you draw a diagram of a model, you ought to be able to deduce the properties of that model before you even think of applying it to anything else. This is hardly ever done outside the hard sciences (and doing it is what has made the hard sciences work so well). This is why we use mathematics in constructing our models. Words are just too slippery; a statement made in words has almost whatever meaning the listener would like them to have, so there's no way to test verbal statements to see if they "really" would imply what you hope they imply. Mathematics is different. When you make a statement in mathematics, you can work out what it implies even if the implications are too complicated to follow intuitively, or are counterintuitive. That's why mathematics was invented by human minds. Of course mathematics is subject to the GIGO rule; lots of elegant mathematics has been applied to stupid premises. But nobody said that mathematics generated wisdom. It just keeps a theoretician honest.

I don't see any particular difficulty in merging the worlds of quality and quantity. Every quantity is a quantity of something identified by its qualities: every quality has to be present to a quantitative degree, as is beginning to be suspected even in the worlds of AI and linguistics. When you compare a World and a Grain of Sand, you're seeing similar qualities, but if you couldn't distinguish quanitities you wouldn't understand the meaning of the comparison, which derives its force from a difference in quantity. The world is made of significant similarities, from the human point of view, but it is also made of significant differences and amounts, from the same point of view.

I suppose that short of Nirvana, human beings will always be a bit egotistic. It seems to me that the height of egotism is to assume that one's private understandings suffice to explain what everyone else in the world has in mind. Seek out your own salvation with diligence, Dan, and leave mine to me and those who have bothered to learn what I mean.

Oct. 1, 1987

Dear Phil,

My book is still moving along.

Your book seems to be moving along faster than mine. Grr.

I will, of course, be happy to read your entire book and comment on whatever I am competent to comment on. What I say will be of no help to you in terms of testimonials -- who ever heard of WTP? Somehow I think YOUR endorsement would do ME more good than the other way around. But I'll read your book if my doing so will help you.

I don't know if this will ever figure into your present book, but your concept of predicting and anticipating fits well into what I call "model-based" control. It's like knowing that there are no more steps to go down in the dark. You walk down the model steps in your head, which you can "see," manipulating your real legs at the same time as your model legs. In the extreme form, model-based control can completely substitute for real control, if the model is detailed and accurate enough. The same signals that make the model behave correctly then make the real world behave correctly. Even the best mental models, naturally, have to be recalibrated from time to time. You can walk around the house with your eyes shut for five minutes without bumping into anything, but not for an hour. You can't even make five minutes if someone is moving the furniture.

One advantage of the concept of model-based control over that of predicting or anticipating is that it's not so obviously intellectual. Predicting sounds like something you have to be conscious to do. It suggests rational thought, logical deduction. I wonder just how much of that kind of predicting or decision-making or choosing people actually do. I'm not aware of making many choices, of weighing alternatives and picking the best one. Not much of the time -- it happens that way mainly when I'm designing something and there is a conflict between doing it right and doing it cheap. The rest of the time I neither predict nor choose: I just try to minimize the distance between me and my goals by finding as many multi-purpose behaviors as I can (right now I'm practicing touch-typing by looking only at the screen). Simon got a Nobel Prize for discovering that managers seek goals instead of maximizing -- satisficing, he called it. Maybe that's a matter of style -- I'm a satisficer, while other people are more into being optimizers. Optimizing takes a lot of prediction and decision.

I'm not saying that people don't engage in prediction -- anyone who looks at the sky and thinks it might rain is doing that. Prediction definitely belongs in there somewhere. But I think that formal prediction is really learned -- it's an exaggeration of the kind of prediction we naturally do, a sort of hypertrophy of a minor natural function. Prediction doesn't help us a lot with control except under unusual circumstances (why do I have the feeling I'm overlooking something obvious?).

The basis of prediction, as you say a lot of times, is the idea that ceteris will remain paribus, excuse my pig-Latin. One reason that decisions are risky is that we know disturbances are bound to arise and screw everything up. People making risky predictions are almost always making a mistake, aren't they? After all, a single person can't take advantage of the numbers: he generally gets just one chance at each important prediction. There's no such thing as the chance that selling my stock short a week ago would have netted me a profit this week. I win or I lose, and if I lose I don't get to try again. Probability doesn't apply to a single trial. Same for getting into college, getting employment, and squeezing between two trucks in heavy traffic. Generally speaking, individuals really shouldn't make predictions. They should learn to control what they can control. All the stuff about group influences on risk-taking really applies to cases in which there really isn't any risk. You should try those studies again, making each person put up a $1000 bet that the choice will be right -- with a $100 payoff for picking correctly. I'll bet the result would shift a little. Most subjects wouldn't play, being sensible folk. The real risk-takers are a little funny in the head, aren't they?

Well, you say most of this very well, although as your editor complains, you're being a little too gentle. You say, to keep from scaring people off. I wonder.

When you extend anticipation and generalization to sequences and programs of automatic action, I begin to drag my feet a little. One has to watch out for applying a high-level perception to what is really a low-level behavior. Miller, Galanter, and Pribram discovered the if-then programming loop, and proceeded to apply it all the way down to spinal reflexes. It can be applied to anything at all, once you get started from the program-level point of view. Here comes a falling raindrop. Test: Has it hit the ground yet? No. Operation: fall some more. Test: has it hit the ground yet? No..... Yes. Splat, exit. So even gravity operates like a program. Only it doesn't.

A sensation is a weighted sum of intensities. It is therefore invariant with respect to which set of intensities happens to be present. Warmth is warmth, anywhere on the skin. AHA. Even second-order perceptions generalize. But do they? No, they calculate weighted sums. You can't recognize that as a generalization unless you're capable of perceiving generalization as a phenomenon. If you had only two levels of perception and control, you'd know nothing about generalizing. I think I'd restrict the control of generalizations to the principle level. Let programs be programs, categories be categories, etc. .

I think that the main thing we anticipate when we act is low error. When I grab the doorknob, turn it, and pull, I am not surprised that the door swings open. But neither do I say to myself, "If I act in such and such a way, the door will open." I just set the reference level for an open door and my (by now) automatic control systems make it be open. I would be surprised if it failed to open, but usually before that surprise can have any effect, my cerebellum or whatever has simply increased the pull and yanked the door open anyway. Prediction implies the possibility that something isn't going to work. When control systems are acting, most actions adjust to conditions immediately, and things work, although after an unpredictable amount of effort. We don't really care how much effort we use to open a door, if it isn't too much. So we don't have to predict how much effort will be needed. Whatever is needed will be produced. Is that a "prediction?" Well, if you look at it that way ...

The TV repairer isn't just "systematic." It's using a theory. The theory of what makes a color television set work is more complex and detailed than any medical theory of what makes the body work. There are voltage regulators and frequency regulators and phase regulators and extraction of information according to when as well as where it occurs: a good TV repairman uses this theory to deduce that a capacitor is leaking or a resistor has changed its resistance. The theory is complete enough to show how the system will behave with bad components as well as with good ones. The only real generalization the TV repairman makes is that the theory of television will apply to this TV set, too. From there on it's a process of deducing which component, if it went bad in a certain way, could produce the symptoms being observed.

Of course some TV repairmen, too many nowadays, just replace everything until the symptom goes away. Expensive.

Generalizing, I agree, is a universal human characteristic. Let's make sure, though, that we restrict this concept to levels of organization at which generalizing actually occurs.

Particular to general, general to particular. I see it this way. We observe phenomena A, B, and C. For some reason, we classify them as examples of "the same thing," ignoring the differences that enable us to distinguish among A, B, and C. We also observe X, Y, and Z, classifying them together for the same reasons. We now have the ABC category of phenomena and the XYZ category. Now we think we see some regularity holding between a member of the ABC class and a member of the XYZ class. If we try to explain this regularity without investigating just why it holds true, we end up making statements like "ABCs influence XYZs." In other words, we invent a law of nature relating the names of the classes. I claim that this mode of explanation is pre-Galilean -- it's the way people explained why heavy things fall faster than light things, why affinities exist between certain substances, why people with small heads have choleric dispositions, and so on. This is the mode of explanation that existed when people thought they could discover laws of nature by finding the right names for things and looking for relationships among the names. In other words, I agree that people do reason from the particular to the general and back again -- but that doing so is mostly a reflection of ignorance. That is not the route to making true discoveries about nature.

Acting as if, as I've already been trying to warn you, is an intepretation of an observer, not necessarily a way of acting of the observee. In some cases, this interpretation is in agreement with how the system does work. The heart acts as if it is a pump because it IS a pump. But when we breathe, we may act as if we will discover that we are surrounded by air containing 21 percent oxygen, while in fact we just breathe. If there isn't enough oxygen we discover that soon enough. The observer too easily puts his own knowledge and point of view into the system being observed, with no proof that the system itself thinks or experiences as the observer does. I am dead set against this kind of metaphorical treatment of behavior. Why use metaphors when we could be asking how the damned thing actually works? Maybe at bottom we can't get away from metaphors, but some metaphors are hardly distinguishable from literal truth, and I think we should be as literal-minded as possible. I don't claim that organisms are like control systems. They are control systems, literally.

The latter parts of Chapter 1 do get more literal, and I suppose that you are on the way to saying everything I'm being impatient to hear. Around page 22 you get back to general and particular, and start making noises more in line with my objections. You're more willing than I am to admit the value of certain kinds of generalizations -- "exploratory studies" -- or at least you're sounding that way. I wish you would be a lot more specific about the "quo bono" question (more culture, there): who benefits from the use of generalizations? I claim that the state and the corporation benefit a lot more than the individual does.

On page 32 you hit my wavelength: social sciences have been confusing purposes with actions and actions with statistics. More, more! What is prediction of social trends, anyway, but an attempt to divine other people's intentions? The future does not simply follow from the present. It is caused to happen exactly as it does, barring meteor strikes and hurricanes, by what people intend to experience. If you know what people intend, you know pretty much what is going to happen. Scientific prediction is usally based on the rule that once you turn the apparatus on you have to stand back and let events unfold by themselves. The human race isn't bound by that rule. If human beings see events trending away from the direction they want, they don't wait until the experiment is over and the future arrives: they get in there and push like hell to correct the course. What do you call it when people make predictions of who will lose the race, and then run alongside the chariot pulling and pushing and throwing rocks at the driver to make sure the prediction works? You probably don't call it prediction. I sure don't.

Research, you claim, is one of the ways we get ready for

future experience. Zatso? I think research is the way we learn how things work, so we can explain events and cause the ones we want to happen whenever we want them to happen. We aren't passive observers, most of us. Research that only enables us to guess, sometimes, what is going to happen if nothing unusual happens really doesn't get us very far. Something unusual usually happens. If we do the kind of research that teaches us how nature works, even human nature, we don't have to worry about predicting what is going to happen. Whatever happens, we can deal with it when it comes, because we understand how things work. We don't have to predict that the Mississippi will silt up in the year 2050. Once we understand the silting-up process we can see it going on right now and stop it, or figure out how to make a profit selling silt and speed it up. Maybe there is a place for passive prediction -- I'm sure there is. But it can't play any major part in deciding what the future is to be, and creating it (again, I'm having a sort of lopsided feeling -- too heavy on one side of the argument, I guess).

I trust that the summary will change. I'd like to see you conclude that generalizing is one way of knowing something about cases we haven't examined yet, but that making good working models is an even better way.

I think it's going to be a really terrific book.

Best,

Bill

Bill.

610 Kingswood Avenue
Eugene OR 97405
15 October 1987

Dear Bill:

I have finished the manuscript for the book. You have read some of it before, but even those parts are rearranged and clarified. And you helped with the two places I ventured into mathematics. Thanks. You can see what it looks like now from the enclosure, which contains a table of contents and the first and last chapters. When I say it is finished, I mean that it is ready for criticism. When it goes to the publisher, I don't want it to have the gaffes in it that I can't see but you can.

I hope you will want to read the entire manuscript and tell what what you think needs fixing. I'll be very grateful. If you do want to read it, let me know and I'll send it. I know you are busy with your own MS and no doubt a dozen other things. Well, I'll just hope.

Now, however, I must be a little impolite. It costs a surprising amount to have copies made of 500 and some pages. So I have had only four copies made. If you are willing to read a copy, please tell me whether you can read it within the next month. If you want to read it but cannot make time right away, I'll be glad to send you a copy that comes back from an early reader. (Actually, I am sending this advertisement to only a dozen colleagues. Maybe only four will want to read it at all!) I'll be able to use the late-arriving comments as well the early ones, because it will take me a while to make revisions and find a publisher.

I repeat. I'll be grateful.

Sincerely,

Phil

Phil Runkel

610 Kingswood Avenue
Eugene OR 97405
6 November 1987

Dear Bill:

I am very happy indeed that you will read the whole of my TWO METHODS. I would feel very shaky going into print without your having read it. Thanks very much, also, for your long letter about chapter 1. I will be interested to see to which chapters bring forth fewer words than that and which bring forth more. I expect the number of your words will tend downward as you go along, because you will be writing notes like "See my comments on chapter 1."

You and Carol Slater give me more words per page of mine than anybody else. One reason she does it, I know, is that she is using the occasion to examine her own thoughts. I hope some of what you write to me will help you think about your own book. I am very glad to hear that your work on your book continues steadily. When you are ready for me to read a piece of it, I'll be eager to do so.

Your letter has a great deal in it. I'll answer the topics in the order you put them down.

You said my book seems to be moving faster than yours. Some of the words that appear in this MS, you will find, are words I first set down in January 86.

You asked whose words quoted in an advertisement, yours or mine, would help the other person's book more. You forget how small are our professional circles. The number of people in any discipline that are known outside that discipline is typically smaller than the number of your fingers. Usually those names are the same names that are known by the public at large--Carl Sagan in astronomy, Joyce Brothers or Skinner in psychology, and so on. Within a discipline, the same principle applies to the sub-specialties. I am known by people "all over the world," as we like to say, for my work in organizational development. That probably means more than 500 people, but certainly fewer than a couple of thousand, in USA, Canada, Australia, South Africa, Israel, a few countries in Europe, Japan, and maybe one person in India. Every now and then someone calls or writes who has read something I have written (it happened yesterday--someone who had read something I wrote in 1970!), and those occasions make it feel as if gee, there must be people ALL OVER who know about me. But it is like encountering your friends in the Chicago airport. You encounter them because they also lead the kind of lives that take them through the Chicago airport. Well, there may be ten thousand people (judging from the sales records I know about) who have seen one or more of my publications over the years, but I'd guess maybe a tenth of those people would recall my name if you said it to them.

The American Psychological Association once made a survey of the extend to which their journals were read. They sent out a batch of reprints to some thousands of people and asked "Do you remember seeing this article?" It turned out that the average article in an APA journal was read by one person. And of course the distribution was Poisson; most articles were read by almost no one.

The circle of people who know me as a psychological methodologist is smaller than the circle who know me as an OD person. I am well respected by those few, but it is a small circle. The circle of methodologists is itself small. And the circle of people who know me as a social psychologist is still smaller than that.

You are widely known in the small circle of cyberneticists with a particular leaning. I suppose there are hundreds, maybe thousands, who know your name and don't know mine. Probably each of us can help the other not as much with our names as merely by saying something quotable and being identified as "author of ____."

Thanks for your explanation of model-based control. It's not a new idea to me, but it is a good idea to use a separate name for a sequence or program of sufficient complexity. I guess I have usually used the word "map." And I agree that it is difficult to keep readers from thinking you are talking about conscious maps in language or other symbols when all you mean is a sequence that will run off or a program containing only a few choice-points and where the choices can be made unconsciously. That's the reason I made a special point of talking about both conscious and unconscious "anticipating" in chapter 1 and the way one shades into the other.

I agree wholly with you that conscious "predicting" in a reasoned way with language and other symbols must surely be a small part of our lives. It probably happens mostly in chess, and even there the player converts detailed predictions of moves into promising "positions." As you know, computer people and people who design "information systems" are now complaining that humans are incapable of using the amounts of information that computers can now deliver to them.

But even without computers, people usually reject a lot of information. I have lost count of the meetings I have attended where the members seem to agree that a "decision" must be reached (a vote taken) before the 5 o'clock going-home time, even though most of the members seem to agree that some important information is not at hand and even though nothing would be lost by taking the vote tomorrow or next week.

I think most of our "predicting"--weighing choices and picking the best one--is done after the fact, after the choice in action, not before. We are wonderful and compulsive about explaining things to ourselves (I don't think we could be the thinkers we are if we did not have this compulsion) and especially compulsive about explaining ourselves to ourselves. When you ask someone, "Why did you do that?" you almost never get "I don't know" in response. The person can always think up some explanation within a second or two. Indeed, an odd feature of our culture is the almost hundred percent frequency with which people think they are obligated to have a reason for things.

You said that probability does not apply to a single trial. Of course not. I am repeatedly astonished that so few social scientists know that.

Yes, most of the experiments on the risky shift have been done with groups in which there was very little to lose. I was careful to point that out. On the other hand, money isn't everything. Some people in some groups would rather pay $1000 than admit to being persuaded by that bastard across the table.

You drag your feet at extending "anticipation and generalization to sequences and programs of automatic action." I think you and Carol Slater are making the same criticism. She says that a fish anticipating water doesn't have the same meaning as a driver predicting that people will drive on the right-hand side of the street. I have stumbled, she says, over what philosopher-logicians call "opaque" and "transparent" contexts. (I won't bother you with the details.) I think you are both wrong. I am saying that evolution deals with the neural net and with the rest of the body all as one piece. Flippers versus feet, the opposable thumb, the rotatable neck, eyes of one kind or another, and neural nets of one capacity or another are all ways of being ready to deal with various arrays and combinations of what can be sensed. I like the words "being ready" or "anticipating" better than "generalizing," because they have a little less of the intellectual flavor. Unfortunately, I don't know a word that is equally balanced between the lowest and highest levels of control. I suppose the reason we don't have such a word is our heritage of the mind-body split.

I agree that metaphors are always dangerous traps--that you can make a falling raindrop sound like program-following. And I don't claim to be able always to avoid the trap. So I'm glad to have the warnings. But in this case I am not ready to admit that I have fallen into one.

I agree that I cannot talk about generalizing unless my neural net contains the level of principles. But if a rock in the sun feels hot to my finger, I am going to expect it to feel hot to my toe. Once I can think at the level of principle, I can construct a sentence like that. I am certainly not saying that the expectation of what my toe will experience is what social scientists mean by "generalization." I am, however, saying that if evolution had not provided me with the capacity for that expectation, it could not have provided me later (so to speak) with the capacity to talk like a social scientist. I claim that our higher capabilities are made up by multiplying and recombining the physical units that provide the lower capabilities. It seems a principle of evolution that new structures are built by reshaping old ones, multiplying or subtracting them, too, as suitable. It seems to me your multiple-level "wiring diagrams" are testimonials to what I am saying.

You say, "I think that the main thing we anticipate when we act is low error." Huzzah! I couldn't agree more. I don't remember just now how much I said about that under "Diagnosis" in chapter 1, but I say that several times in later chapters.

"We don't really care how much effort we use to open a door if it isn't too much." I think I say the same thing in a later chapter when I say that often, in ordinary life, we don't care as much about predicting what people are going to do as we care about predicting what they are <u>not</u> going to do--that they are not going to mug us on the street, not going to vote against us in committee, not going to fart at the tea party. In brief, <u>not</u> going to put <u>large</u> errors upon us.

Certainly, a good deal of "generalizing" is pre-Galilean. And as you have often said, so is a lot of social "science." I read an article yesterday in which the author (1) defined OD, thereby, as Korzybski pointed out, substituting words for things, (2) quoted some phrases from another author who described the values he claimed to characterize OD--not saying how many OD practitioners gave what kind of evidence of hewing to those values under what circumstances, but just claiming that those values characterized OD "in general,"--and thereby substituting more words for more things, (3) quoted still another author who had discerned different values in different nations among people employed by IBM by looking at their answers to a questionnaire, thereby substituting still more words for still more things, (4) said that the values listed by the one author sounded like certain of the values listed by the other author (though using fancier language than "sounded like") and (5) said, therefore, that OD consultants working in one country would have more difficulty than when working in another, thereby predicting behavior from the similarities he saw among words.

I think that is a good example of your complaint about research that finds a relation between names of classes. It reminds me of my brother's recipe for getting rid of a cold. Get your self a duck and identify your cold with the duck. Hit the duck over the head with a mallet, and your cold will be gone.

Now, going back to metaphors, don't sell them short. The history of physics is full of advances inspired by analogies and metaphors. Sometimes you seem to be saying, "Yes, but we ought to be past that by now!" But we'll never be past that. We will always face stages in which we must explore and grope and do a lot of wrong things.

Yes, predicting social trends is divining people's intentions. Indeed, the Institute of Social Research at the University of Michigan makes an annual survey of the intentions of consumers and does a pretty good job of predicting economic trends with it.

You said that research is not one of the ways we get ready for future experience; instead, you said, it is a way we can cause the events to happen that we want to happen. That is what I meant; we can be ready to use more effective oppositions to disturbances than we could without the research. I don't know whether I can say that better in chapter 1. I don't remember whether I said it better later.

You said that you hoped I would say that making working models is a better way of knowing something about cases that haven't happened yet than "generalizing" is. I do say that in a couple of the chapters on specimens.

So those are my thoughts on what you wrote in your letter. While writing this, I suffered from conflict. If I don't reply in detail to Bill's long letter, I will be dishonoring the effort he put into it. If I do reply in detail, I will be adding a lot more words to the 500 I am already asking him to read.

Your verbose friend,

Phil

Nov. 20, 1987
Dec 4, 1987

Dear Phil,

Herewith my comments on and inspired by The Book. I did not adopt the principle that you suggested; namely, that the length of the comments be scaled according to importance in relation to my comments on Chapter 1. I think the snows would come and go ...

Chapter 2: Relative frequencies: what the method will do.

There's a theme I wish you would start in this chapter and carry through (please don't think you need to do so -- you can take all comments of this nature as requests to handle certain subjects in a paper or another book). The theme is reliability of statements that depend on more than one "fact." I think there's a hidden uncertainty in the method of frequencies that is (may be) only partially accounted for by the normal statistical methods.

Suppose you find that 30 percent of male US citizens in a given sample are fathers (p. 2-2). The statement "X% of male US citizens are fathers" is uncertain because of sampling errors, but there's also some uncertainty in determining whether the subjects are male, US citizens, or fathers even if you measure the entire population. The probability of truth of the final statement is the product of all the probabilities that the component statements are true: all apparent males are males AND all apparent US citizens are US citizens AND all apparent fathers are fathers, AND the calculation of 30 percent had no errors in it.

This problem gets worse if you then try to apply this knowledge in a slightly more complex way. Let's find out if male US citizens who are fathers and are registered to vote actually vote more than those who are not fathers. Are all voters listed in the registration book really registered voters (not in Chicago!). Are all registered voters listed in the books? (not in Georgia!). Are all votes recorded? Etc. Now we have the uncertainties in determining registration and actual voting multiplied by the uncertainties in determining maleness, citizenship, fatherhood, and arithmetic correctness, multiplied by two universes of sampling errors -- can there be any truth in what remains?

In polling, the membership in a population on the left side of the hypothesis is in question, but so is the right side of the hypotheses: whether people who apparently are or do X really are or really do X. In causal experiments, people present when stimulus X was exhibited may or may not have seen and understood it, so there's uncertainty in the exposure, as well as in the right side of the hypotheses. In correlations the uncertainties apply to both sides of the hypotheses. I think I'm talking about uncertainties that aren't taken into account, but maybe they are.

I guess I'm asking you to discuss two subjects: (1) taking ALL sources of uncertainty into account, and (2) drawing conclusions from a series of statements ALL of which have to be true to make the conclusion true. I think it's not only possible but likely that statistics-based statements of the kind normally applied to individuals have a calculable and high probability of being false when applied to every single individual in a group, even though the statement is statistically true of the whole group. You mention this in relation to average height and average sex (later). I think you could make a lot more of this.

I think that when you examine the "richer lodes" that the method of relative frequencies is supposed to uncover, what you come up with are isolated facts. Maybe sometimes this is all you want -- how can I avoid the expense of mailings to credit-card holders who are unlikely to buy insurance? (Compare that interpretation, by the way, with what you say on 2-23, bottom). That's a one-shot, one-purpose deal. But such isolated facts always have a low probability of being true, and when you try to put such facts together into a systematic body of knowledge, what you get is a collection of deductions that most probably are false concerning either individuals or populations. Are scientists doing science, or helping institutions take advantage of mass phenomena?

The method of relative frequencies is what people have always used when they don't understand why things work. I've called it the witch-doctor approach: you don't know what the critical ingredient or process was, so you don't dare leave anything out, even whirling three times in a circle widdershins.

I think that the method of specimens is always preferable to the method of relative frequencies. Even knowing just one causal factor, and why it works, you can predict an effect on a population. If you know that most people in cities live too far from their work-places to walk the distance in the available time, you can predict that they will ride something to work. Isn't this really how most statistical studies begin? You have an idea about some real causative factor, which if it existed ought to show up as an observable effect, for reasons that you can explain in the manner of the method of specimens. The statistical approach, it seems to me, is an attempt to find verification of this suspicion in one stroke, while simultaneously excusing failures of the suspicion. I'm right even when I'm wrong. But the method of specimens requires that you investigate the apparent cause more closely, to make sure it really works in all cases. Then if you turn to statistics, you can expect high correlations, and the failures tell you where to look for other causal factors.

In short, I think the holes in the net are too big.

I do think that the way you introduce internal standards and disturbances is clever and effective. You make it sound as if everyone knows about these things.

Chapter 3: Relative frequencies: sampling

Isn't the rationale behind random sampling the idea that by being unsystematic, you stand a chance of having all the influences of which you know and understand nothing cancel each other? Or is that just the general idea behind statistics?

I think you have done a masterful job in this chapter of showing how difficult it is to take a random sample! Also in showing what is wrong with accounting for more and more variance by adding conditions. Cronbach's complaint is like the one I offered above: assembling constructs into a network in this way is impossible.

You can't make a dependable numerical estimate of probability of replication without random sampling, it sez. Hmm. Is this a conclusion from experience or does it come about because a non-random sample may be non-random in the respect of interest, too? But if you don't know why the effect appears, couldn't a random sample reduce the effect seen? Something trying to come through here, but I can't pull it all the way out. Let's see. You postulate that people choose positions in a queue according to the sex of the person in front of them. The people actually line up boy-girl-boy-girl etc.. Now you pick a random sample from this infinite line of alternating people. What you find is boy girl girl girl boy boy girl boy .. The alternation is gone, because of the random sampling. Another example: on odd days, people with odd street addresses can water their lawns. You see where that goes. My intuition is trying to tell me that you can destroy systematicity by random sampling. Anything to this?

Can even a random sample save you from a population that is changing systematically (p. 3-12)?

Unlistable groups. People with $300 to $500 in their checking accounts. People wearing plaid shirts. People driving on the right side of the road. People arguing about a bill. People reading your book. Baseball, etc., teams. Congress. Prisoners. In general all the people engaging in some fairly normal activity, such that the individuals doing it change from time to time, and sometimes are not doing it. If by a population you don't mean a particular collection of individuals, then you can only be talking about the <u>conditions to which anyone in the same situation is subject.</u> For example, what percentage of people with less than $300 in a checking account pay a service charge the next month? Answer: essentially all of them whose banks have a minimum balance of $300. Second answer: the same proportion as there are banks that require a minimum balance of $300. So this study would sound as if it is revealing something about the population of persons, but actually reveals something about the population of banks. What percentage of prisoners stay in their cells from 9:00 PM to 5:00 AM? And so on. I'm not sure you have dealt with this question of whether the person or the situation is being measured.

Of course answering this question takes us inevitably back to the method of specimens. To ask what it is about the situation that leads anyone in it to behave in some particular way is the same as asking about common reference levels and disturbances. Everyone driving on the right side of the road steers to the right when the car approaches the left lane marker, except those making a left turn and those about to be eliminated from the population. As soon as you start trying to understand why someone does something you're into the method of specimens.

The junk box is replete with wonderful jokes. It's a perfect way of showing that we can't get knowledge from random samples that don't have <u>some</u> sort of formal or informal organizing principle selecting the data before it's analyzed. As you say, one must always make some decisions. If we're going to develop a science, those decisions have to be made according to an explicit principle, don't they? The problem in the psychological sciences is that the most important decisions are made informally and covertly, according to subjective criteria that the researcher brings into the studies from a whole life history, and very often on the basis of word-association.

The problem with guessing that objects of higher densities will be harder to compress than objects of lower densities is that this guess is a stab in the dark based on a vague idea. It doesn't arise from a network of already-established relationships, but only from a private hunch based on informal experience, outside the boundaries of the study and even the discipline. The chances of discovering a solid relationship by guessing and getting lucky are not great: the informal test will not distinguish aluminum from uranium, because heavy machinery is needed to compress either one measurably. There is no substitute for learning the details of a subject, although it seems to me that psychologists are always trying to avoid doing this.

Mary's already noted that denser things squash ~~more~~ LESS, not ~~less~~ more (p. 3-23, end of 2nd graf).

You might mention that these stabs in the dark are often called "theories." I have a theory, says someone, that dark-colored items from the junk-box will be warmer than light-colored objects. This theory is merely a proposition to the effect that if we measure temperature we will find this difference. I don't call this a theory. The theory would be an explanation of the underlying relationships that lead us to expect this observation: that dark-colored objects absorb more radiation than light-colored ones. A conclusion one could draw from the theory is that light-colored objects must be cooler; then we check out the junk-box to see if the real items behave like the theoretical ones. Method of specimens again.

Chapter 4: Linear causation

This is a great chapter. I would like to see in it the distinction between flow charts and system diagrams. A flow chart goes

```
Press the lever ----> eat the food   or

Hit the nail ---> All the way in? --yes-->exit
    ^                            |
    |                            |
     ---------no----------
```

The latter is the Miller-Galanter-Pribram TOTE unit (test-operate-test-exit). Notice that in the TOTE unit there is an exit -- what is going on in the left part of the diagram when we exit? Wrong question -- this is not a diagram of variables in relationship to each other, but of actions taken by some unnamed agency of unspecified organization. A flow chart shows what is done, but not what does it or how. A flow chart looks on paper like a system diagram, but it is not a system diagram.

On page 4-20 I'm not sure if you're using the simultaneous equations right. First, if this space is three-dimensional we might as well use x,y, and z. The first two equations below are equations for different planes: if there is a solution, it is the line of intersection, which can be expressed as a projection onto the z-x plane and another on the z-y plane.

(a) $z = 1 + 2x + 3y$ and

(b) $z = 2 + 3x + 2y$.

To solve for the z-x line we eliminate y:

1: multiply (a) through by 2 and (b) by 3. Multiplying all terms by the same number doesn't change the equations:

$$2z = 2 + 4x + 6y, \text{ and}$$
$$3z = 6 + 9x + 6y.$$

2. subtract resulting upper equation from lower:

$$z = 4 + 5x.$$

To solve for z-y line, multiply (a) by 3 and (b) by 2:

$$3z = 3 + 6x + 9y \text{ and}$$
$$2z = 4 + 6x + 4y.$$

Subtract lower from upper to get

$$z = -1 + 9y.$$

What this tells us is that if the behavior z is legitimately representable under various circumstances by (a) and by (b), then the only conditions under which both relationships can hold true are those that are described by the two derived relationships, which together define a specific line in space. If the real behavior is constrained by both three-dimensional equations, it can only actually fit the line of intersection.

The coefficients reported by Brown are even worse than what I had imagined. Can you translate them into probability that a statement about a relationship would be true of an individual?

Where do the names of personality traits come from? Aren't they another example of what I spoke of as informal criteria? Suppose you try to be specific about the attributes of behavior you expect to see if a person is "generous." If he has more money than another person, he will give the other person some money. If another person says something that is apparently in error, he will give the other the benefit of the doubt and interpret the saying to make sense. And so on and so on. When the list is complete you have an explicit definition of all the symptoms that reveal generosity.

Now it's extremely easy to determine whether any given person is generous. You simply observe to see if all the symptoms are displayed. If half of them are displayed the person's generosity is 50% of the maximum possible. The only problem is that you have no idea what is causing this generosity to be seen. Making the list totally explicit and exhaustive shows the circularity of which you speak for what it is.

Chapter 5: Relative frequencies: substituting people

I already mentioned the question of whether one is studying people or properties of their environments (the banks and the minimum balances). I think that question is relevant in this chapter, too. There are really two reasons why we might find people to be interchangeable: (1) they have common characteristics because they are all of the same species (this is what is hoped for) or (2) because they have common characteristics that remain totally unsuspected, their behavior reflects only causal connections in their environments.

So I'm raising the question of what it means if people are found to be interchangeable. My suspicion is that the more nearly interchangeable they prove to be, the less we are learning about them and the more we are studying environmental constraints without knowing it.

I absolutely love your analysis of the eight possibilities: the Ignoble Eightfold Way. How many studies would survive this kind of analysis?

Needless to say, I approve of all your comments in this chapter. "Does it predict for all people?!"

Chapter 6: Relative Frequencies: correlations and change

Profound observations here. I particularly like your criticisms of the idea that correlational regularities "betray some natural necessity at work." This idea could be expanded, or tied to later comments, by talking about models that do purport to show "necessity" in behavior. But it's probably a different kind of necessity.

If you put together a control-system model with all the parameters specified as numbers, then of necessity the model will behave in a certain way: relationships will occur among controlled variable, disturbance, output, perception, reference signal, and error signal. Given the model and the numbers, there is only one behavior it can produce as disturbances change, and it must produce it. That's the kind of necessity I mean here.

This kind of necessity concerns the consequences of making assumptions. If you make assumptions explicit and quantitative, you're committing yourself to a necessary outcome. There's no way left to cheat. All that remains is to see if the behavior of the model matches the behavior of the system being modeled.

Does the necessary behavior of the model reflect some necessary behavior of nature? If you can show that precisely the same component relationships can individually be found in the natural system, and if you're willing to guess that nothing else of importance is entering the picture, then yes, nature necessarily acts the same way. Now the primary assumption is that nature is self-consistent -- laws that apply to the parts keep applying when the parts are hooked together. Of course there is no enforcer of this necessity. All we mean then by necessity is consistency. Science depends on a self-consistent world.

Re: Least Mean Squared Deviation: Enclosed please find poem: author known, affiliation and place of publication not. Courtesy of a friend.

As far as I can see there can be no rebuttal of your arguments but to say, "It's a complex subject and this is the best we can do." Implication: nobody else can do it better. You will certainly hear this if you haven't heard it already.

I've written, I think, about an application of the ideas in this chapter. Behavior is to a large extent the process of acting on controlled variables that are affected by independent disturbances. Because the effect of the action opposes the effect of the disturbance, there is a very high negative correlation between action and disturbance, right? Wrong. The high correlation is between the effect of the action and the effect of the disturbance. If we measure the action and the disturbance at the source, in the most obvious and available way, we will not find such high correlations. Consider the driver and the wind. If we measure the force generated where the hands meet the steering wheel, and the velocity of the wind, we will find some correlation but not a very high one. To get the high correlations of which control theory boasts, we have to translate both measures into a sideward force on the car. The hand-force is amplified and run through a non-linear power steering booster and a nonlinear steering linkage, where it cocks the wheels, distorts the treads, and applies, nonlinearly, a sideward force to the car that is strongly dependent on the car's speed. The wind velocity has a direction that adds to the car's velocity vector, producing a sideward component that varies about as the square of the wind-car relative velocity. If you apply all those corrections you will get essentially a perfect correlation, assuming the car actually goes straight. If you just measure hand-force and wind speed, you will find about what is expected in normal correlation studies.

Chapter 7: Using words

Oh, dear. What can we do about it? Taking words away from students of human behavior is like taking away their eyes and ears. I think it might help if we could be clearer on the difference between words and their meanings. My levels of perception are all silent, as I think I've told you. Excluding the words themselves, the meanings of words are to be found in the perceptions to which they point. If they don't point to any perceptions but other words, they are probably pretty useless, giving the impression of meaning without having any. I exclude linguistic terms like "adjective," of course. How do we find out what meanings people are controlling? By letting them control, and applying disturbances. If they oppose the disturbances it doesn't matter what they say they are doing.

You do talk around this idea, but you could be more explicit. The meanings of "strawberry" and "pistachio" are not words: they are tastes, textures, temperatures, and colors. I advocate taste-tests, which can reveal things like "This is terrible strawberry and great pistachio -- too bad, since I like strawberry better."

I think we need to investigate how people use words to designate experiences, rather than just diving in and using the linguistic habits we started learning at the age of one or two. Everybody assumes far too much.

Good chapter. My reaction is to recommend that everybody shut up for a few years.

Chapter 8: Fine slicing

This chapter sums up many of the points already made. You might make it more consciously a summary, but I don't think you need a special section.

PART TWO

Chapter 9: Specimens: Natural kinds

I think we have to be careful about specifying the domain we're investigating. If we are investigating "nature," then we take our perceptions for granted, become naive realists, and enquire into the properties of matter and energy in large or small units. The question of natural kinds doesn't arise: everything is a natural kind.

On the other hand, if we're investigating human beings, then we have to take into account the fact that human beings perceive with human equipment. Now there is no question of nature just being there: we are now dealing with a universe of perceptions, and what we want to understand is how the brain assembles these perceptions layer by layer into an experienced world, starting with an assemblage of individual signals each indicating only intensity. From this viewpoint, natural kinds can be only the modes of perception that human beings share -- they have nothing knowable to do with the presumed outside natural world.

I think that the philosophers you've been citing are confused about these two disparate viewpoints. They're trying to find something Out There that makes some things natural kinds In Here, and others not. I think they're really trying, without knowing it, to discover categories of perception common to human beings, the same thing I've been trying to do in arriving at my ten levels. But they're trying to find the naturalness in specific things that are perceived, which is wrong. The naturalness of natural kinds is in the way we represent reality to ourselves in (I think) specific identifiable ways that are independent of <u>what</u> we are perceiving. If we are looking at configurations, then all configurations are of the same natural kind: configurations. All systematic smooth change is of another natural kind: transitions. And so on through the levels I've developed.

The method of specimens concentrates on the real natural world, and so takes the observer's perceptual organization for granted. This is the proper point of view from which to analyze behavior from the outside in terms of control theory. Even when we talk about the neurology of the brain and its role in perception, we're still being naive realists. The only point where we have to back off and consider the nature of perception is when we are looking for things to investigate. Then we take the world apart and see it as a collection of perceptions. Every time we discover that something is a perception, we acquire a new dimension of the world to investigate, and can go back to the naive-realist mode with something new to do.

You seem to be saying that at the lower orders we all perceive alike, but at the higher orders are different from each other. Again, you have to be specific about the domain. If we're talking about perceptual categories, then we are all alike at all the levels with respect to the kinds of perceptions we can experience. We expect all people to experience intensities, and we expect all people to experience system concepts.

Within any one level, however, high or low, there is no reason to think that any two of us perceive alike. That doesn't matter, though. What counts is that we find ways of translating from one person's world to another's, in a way that is so transparent that we don't even realize we're translating. I say, "I see a green apple -- do you see a green apple, too?" You say, "Yep, that's a green apple, all right." What we've done is establish a translation scheme. Whenever I have the experiences I call "green" and "apple", I can expect you to say words I recognize as "green" and "apple," and furthermore to show me configurations, sensations, and actions I consider appropriate meanings of those words. What we have NOT done is to establish that the signals in my head are like the signals in your head, or whether your words and actions sound and look to me as mine sound and look to you. Neither have we managed to discover whether or not there is some common input transformation that we are BOTH applying to the incoming information, so that greeness and appleness are related to but not at all like whatever is out there causing these perceptions.

Obviously if we're going to do science we can't spend a lot of time worrying about matters like these. But if the science has to do with human nature we have to spend some time at this.

Anyway, my point is that at the lower levels we seem to be more alike in what we perceive than at the high levels, but that's probably only because the translations are easier at the low levels: fewer steps are involved.

The discussion of invariance may need expansion. The Dember and Warm subjects all gave different plots -- doesn't that show variance, instead of invariance? I think this is a good opportunity to talk about levels of perception. At a low level of perception, the plots all all different because they fall in different places in the diagram. But at a higher level, they are all alike in that they all lie on straight lines. The data points for any subject have a relationship to each other that is invariant with respect to which subject they represent: straightness. This is one of the features of perception that human beings find significant. At a still higher level, the lines that show differing slopes are invariant with respect to the form of the symbolic representation we can use for all of them: the power law. If one of the lines had been curved it would have stood out because of our perceiving in terms of this higher-level invariance.

The Gestalt examples are good. I think it would be good to emphasize that in all the different ways of perceiving the dots, the faces/vases, and so on, the lowest level perceptions remain precisely the same. This is a demonstration of how the brain creates perceptions by altering its interpretations of the same inputs.

It seems to me that the rats in the figure-eight runways showed a preference for visual complexity, but not for novelty. If they had wanted novelty they would have been switching to the plain corridor occasionally, just for a change.

The concept of the "pacer" stimulus should probably be treated in the same breath with Skinner's concept (and demonstration) of "shaping." There is something in here of importance to learning about reorganization. In order for reorganization to work, the environment must be such that a small change has a small effect -- this is a sort of continuity requirement, that applies to Koshland's bacteria as well. It seems that there must be a way of reaching a new level of skill through a series of small changes in behavior. This implies something about the organization of behavior and of the environment as well. You have to be able to tell whether a change took you closer to the goal or farther from it. If the route is a random series of wild ups and downs, there's no way to tell if you should retain the change or make another change right away.

I'm saying that I don't think "pacer stimulus" is the right description of what's going on, although it's related to the right idea.

The main point, that the rats are being treated as individual specimens, is made well.

The question the personologists should be asking is not whether people are high or low in ethnocentrism, but whether that scale is a dimension along which they exert control. To show that it's a relevant dimension all you have to do is find a reference level -- all the reference levels can be different, high or low. Unfortunately, personologists tend to treat categories like this as objective, so they mix people who don't care one way or the other with those who actively maintain positions in the middle of that scale. No wonder their correlations are the lowest in the business.

Natural kinds: see my remarks at the beginning of this section. The concept of a "boundary" belongs in the world of configurations, and so is an example of a natural kind, configuration level. If you're perceiving configurations, they retain their character in every place and time as long as you don't start using a different configuration-recognizer. Four dots makes an invariant square -- until you decide to see them as two triangles fastened hypotenuse to hypotenuse. "Thingness" is conferred by thing-perception functions, not by Reality.

More ideas. To say that the property B is shared by all members of some natural kind A is not to say that every A is identified ONLY by that property B, nor that property B is typical ONLY of natural kind A. We identify natural kinds as a function of many attributes at the same time, each determined within some degree of accuracy. No individual object has precisely the same mix of attributes as another individual object we treat as being the "same" natural kind. Ergo, natural kinds are reference signals, not perceptions. Another way to say this is that natural kinds exist only in our mental models of the world, as ideals.

"Gold" refers to a particular mix of attributes, which mix is closely approximated by objects from various purified samples. We go beyond experience when we claim that there is a substance called gold that has one and only one combination of attributes, each of which has a mathematically precise value on a measurement scale (meaning that if any attribute deviates quantitatively from this reference level by however small an amount, even a trillionth of a billionth of a percent, the substance is not gold). I think that the search for natural kinds is a search for Platonic Essences, because as soon as you introduce a quantitative scale, you realize that the idea of natural kinds admits of <u>no deviations of any amount whatsoever</u> from the ideal amounts of each attribute. Is an object made of "gold" if there is one atom of silver somewhere in it? The more I think about this, the more it looks like trying to objectify a subjective mode of classification. Why not just admit that we know about nature only in the form of idealized models? I'm happy to think of the control-system model that way.

Now I'm not sure what you are getting at by this whole discussion of natural kinds. I guess I would claim that all observations come down to evaluating variables, and that "natural kinds" are just category-level perceptions, or reference levels. The "kind"ness is an artifact of perception.

On p 9-31 I don't think you want to say that "every person must exhibit the same rules for organizing behavior." I know what you mean but the reader won't. I think you're trying to say that every person will prove to operate in ways that are described by a single model, as all curves were described by a power law -- but not that all people will obey the same rules, or behave alike -- do the same behaviors.

Re control groups: in the method of specimens, you <u>can't</u> do a "control experiment." Every experiment is a real experiment. I have done one in which a person is asked to repeat a tracking run, and when the second run is done, the behavior of the cursor from the run just finished is played back on the screen, so the handle is actually having no effect on the cursor. So this is a "control" experiment, isn't it? We're seeing the same experiment with one factor left out, the feedback. In fact it's a new experiment, that shows the lack of control even though the subject feels in control. What is new is making the cursor move without being affected by the handle. (The way the cursor would have moved if connected to the handle, which can be reconstructed from the data, is very different from the way it moved with intact feedback -- the behavior is in fact very different, although it feels the same). If you used two people, one with and one without feedback, the one without feedback wouldn't do anything even vaguely resembling what the person did, without feedback, when the previous run was done with feedback.

Dependent and independent variables. In modeling a behaving system, we first analyze the system into isolatable functions each of which can be described in terms of one dependent variable that is a function of a set of independent variables. In other words, we analyze to the point where we are reasonable sure that ordinary lineal causality holds true. In fact we have to analyze the system to that degree. This allows us to represent the parts of the system as a set of functional relationships. Then we reconnect the parts expressing each connection as an equation relating several cause-effect fragments. Finally, we solve the equations as a simultaneous set, or failing the ability to do that, we simulate all the parts and connect them by making one part of the simulation depend on the appropriate others. The behavior of the solution shows us whether we have the connections right, and whether the functional representations of the parts are adequately accurate. The solutions almost always reveal modes of behavior, relationships among observable variables, that we could never have imagined to exist. If the model is a good one, we find those same relationships in the real system.

The moment you see a system as a collection of simultaneous dependencies, the system as a whole ceases to be representable as a simple independent-dependent variable relationship. However, it is still possible to ask about the effect of an independent disturbance applied to the system. If the model works correctly, it will predict the effect on some other system variable -- but it will predict effects on all system variables, all changing at once in proper relationship to each other. The only real difficulty with the conventional formulation is that it encourages us to see the effects of disturbances as the result of some simple chain of events with one event at the origin and one at the destination. It's just too elementary to tell us how the system is really working -- why, under particular circumstances, we observe an apparent dependency, and why under other circumstances we don't. The effect of stabilizing a variable is only one of the possible relationships that can be predicted, although an important and unexpected one.

Precision, 9-36. The control model predicts tracking behavior very accurately. This means that it predicts the errors as well as the general shape of the behavior. The model's cursor does not track the target precisely; if it did, we would observe zero correlation of the model's error with the subject's error, because the subject's error does fluctuate. In fact, we can adjust the model so that its errors show correlations of 0.9 and more with the subject's errors. The model doesn't show us merely that the subject "is a control system." Its predictions are not merely qualitative. It shows us that all the details of the subject's behavior are quantitatively what they ought to be if a control system with particular parameters is operating. A proper control model of the person with the umbrella would predict how much rain in the face would occur. If the control model predicted that there would be no rain in the face, it would be wrong.

There's a missing subject, I think: prediction. The method of relative frequencies is used to predict when a person will do something again -- the circumstances under which the behavior might repeat. The models growing out of the method of specimens are used primarily to explain what a person is doing right now. The model is refined by continually testing it while surrounding circumstances are changed, until it finally arrives at a form that always fits the behavior actually observed.

This looks very much like a "prediction," but that's not the main point. We can see this when the person's organization changes. As long as the model's behavior continues to match that of the subject, we are satisfied that the model continues to be correct. When the subject's behavior changes form, however, we don't necessarily take this as a failure of prediction. What we do is look into the model and ask, "Is there any simple way in which the model could change that would generate the same change in observed behavior?" If there is, we can then make the claim that we are using the model as a way of measuring changes in parameters. If we can continue to fit the model to the behavior by continually altering the parameter, we can see those alterations as a perception of something corresponding that is changing inside the subject. The model remains valid, and in fact becomes more general.

The simplest example I can think of is a reference signal. Tracking behavior is ordinary studied under conditions that encourage subjects to maintain a stable reference condition. Sometimes, however, subjects will change the reference condition to a new value. We can make the model continue to represent the tracking behavior by altering its reference signal as necessary to reestablish the fit. We can then claim that the model's reference signal is telling us about the subject's reference signal. Another example. Subjects, even highly-practiced subjects, will occasionally make the cursor deviate by a large amount from the target for a period of less than one second. Analyzing the shapes of these deviations (and knowing subjectively what is going on), we can show that the control model will behave the same way if its output connection is briefly reversed, causing a moment of positive feedback. After about half a second, another reversal restores control. So we discover that control systems in the subject and the model can be adjusted in a new way: the sign of their error-responses can be flipped back and forth between positive and negative. This allows adaptations to environments with different kinds of responses to action. It also occasionally causes control to fail. So the "failure of prediction" is actually a source of valuable information that strengthens and extends the model instead of proving it wrong (if the failure isn't too drastic).

Chapter 10: Specimens: Using the environment

To the excellent opening discussion I would like you to add a comment on another flaw in the concept of conditioning. The assumption is apparently that more behavior is always better. At the very beginning of a conditioning experiment this may be true: a little response in the right direction is better than no response. But as the process continues, there had better come a point where further reinforcement leads to less "improvement," or the target will be overshot. All these "theories" to the effect that repetition "strengthens" neural connections or responses overlook this factor. They all take it for granted that the strengthening effect will stop when the response reaches the magnitude that just attains the appropriate final state. If neural responses just kept getting bigger and bigger, after a while you'd have an organism throwing itself madly around inside the cage upon the slightest stimulation. I tell you, these theorists just don't think their own ideas all the way through! Nor are they averse to magic: how come "satiation" or "inhibition" occurs just in the nick of time? The only reason for introducing such magically helpful phenomena is to save the idea that learning involves the "strengthening" of responses. Note (A) pasted here

This, of course, in addition to all your other brilliantly-stated objections.

Acting on the environment

I think you're missing an opportunity here to continue the critique of the first section (maybe this comes up later). If we act on the environment to counter disturbances, the result is an illusion of cause and effect: the disturbances seem to cause the behavior, and in a way they do. I think it's useful to show how it can be that many generations of scientists have thought they saw stimulus-response phenomena: they did see them. But because they were using the wrong model, they misinterpreted them as simple input-output causation. The observations were basically OK: these were not crazy people. They were smart people in the grip of a convincing illusion.

I wonder if you aren't making a mistake in this section by bringing in "patterns" of action. In fact we maintain patterns of consequences of action, but there is seldom any resemblance between the variations in our pushes, pulls, twists, and squeezes (my own version of the "fundamental forces") and the patterns that result. Think for a moment, in detail, of the forces one applies to the crank that brings a bucket up from a well. the pattern that results is a repetitive circular motion of the handle, but the forces we apply to the handle vary all over the place in direction and magnitude. The handle prevents the radial forces from having any effect; only the tangential components count. So we can be very sloppy about applying the forces, and if you try this you'll see just HOW sloppy.

When you count coins you see a pattern of coins sliding from one stack into dollar groupings in other stacks.But think in detail about those movements and the forces that create them: do they bear any resemblance to the outcome? When you raise an arm and "point at" something, in what direction are the forces you apply to the arm? Not toward the something, for sure.

I think that by emphasizing the patternedness of action you tend to negate the point finally to be made, which is that these are patterns of perception and not of action. All you have to do is keep disturbances in mind -- even when action patterns do seem to resemble their consequences in an undisturbed environment, that correspondence disappears when random disturbances enter. The action patterns become just as random as the disturbances -- but the pattern of consequences remains the same.

This also applies to tool-using and language. Am I trying to be too pure? Maybe, but I think we have to be very persistent in avoiding slipping back into the old concepts -- and particularly in keeping a reader from losing an uncertain hold on the new idea.

All the "acts" in this section are really consequences of acts. They all amount to bringing one perceptual situation into being as a means of affecting another perceptual situation -- the means of doing this remain anonymous, and highly variable. It's manipulation of the perceptual world we're talking about. Isn't that the fundamental message of control theory?

It would be nice to unify the concepts of doing and not-doing. The unifying idea is that all variables are quantitative, and can exist anywhere on a scale running from zero to maximum. You can say that there are things that we want people not to do, but that's a rather confusing idea. The control-theoretic way of saying the same thing is that we have reference levels for certain behaviors of other people. We try to influence them to make more if it is there is too little of it, and less of it if there is too much of it. If the reference level is set high, we tend to encourage the behavior; it it is set at zero, any of that kind of behavior is an error that we try to reduce, by reducing the amount of the behavior: we "want them not to do it."

Operational and conceptual definition

This is superb -- is this distinction your own? This is exactly what I have been searching for a way to say. Everybody knows that some people are smarter than others -- that's the conceptual definition, which is nothing but a subjective impression brought into psychology from one's previous life. Intelligence, however, is a score on an intelligence test. The unspoken assumption is that this score has something to do with relative smartness. The informal subjective notion is all that gives meaning to the number called IQ. There is absolutely no reason to think that this number has any relationship to the impression of smartness. The findings of psychology are hopelessly contaminated in this way: if these findings were presented with strict objectivity, we would say "The effect of factor A23 on factor z19, in the presence of condition q." Then it would be clear that the results aren't about anything at all. All the meaning is injected subjectively and informally, and one could even say unconsciously.

When an objective psychologist looks at the data without taking his own subjective additions into account, he is looking in a mirror and not recognizing himself: the vampire effect.

In the summary, it sounds as if people only take action when their internal standards are threatened. I think you should make it clear that they also, at the same time, are adjusting internal standards as a way of bringing patterns of perception into being. Many of these variations have no external causes: those are not just reactions to disturbances, but represent creative purposive acts demanded by higher levels in the system for reasons having nothing to do with fending off disturbances: writing a symphony, for example. Watch out for making the model look like a fancy stimulus-response organization that acts only when set into motion by the environment, a la Descartes.

Chapter 11 Specimens: Control theory

The front part of this chapter needs strengthening. I think it is essential to follow the course that Marken set. First we must establish control as a phenomenon. This is not a theoretical matter. We have to show that organisms actually do stabilize external variables of all degrees of complexity against disturbances, maintaining them recognizeably near reference conditions that we can identify experimentally. We have to show that the relationship among controlled variables, disturbances, and actions is a real relationship, a directly observable fact of nature. No theory is needed in order to do this. The fact is that organisms do behave in this way. This observation has nothing to how they could behave this way and still be physical systems.

This is precisely where psychology went astray. Psychologists observed this phenomenon, although they didn't observe it very competently, and chose to disbelieve what they saw because it went against principles they had decided to treat as holy and superior to the data. Essentially all the contortions of psychological theories and philosophies of science have been generated exactly to explain how it is that behavior can appear purposive yet not actually be purposive. I think the miserable record of the life sciences hinges on this fateful choice to ignore the data.

In any case we control theorists have to establish the reality of the observations first. Then we can raise the question of finding a theory that makes sense of them. Fortunately, this theory exists in mature form: it is called control theory. Control theory is the body of analytical methods that has been developed specifically to help us understand the operation of systems that behave as organisms do in relationship to their environments: closed-loop systems of causation. This theory, in turn, leads us to new interpretations of old data, and suggests new ways of exploring both behavior and the nervous system. It suggests a model of the nervous system that is consistent with the many levels of apparent organization that we see in behavior.

So first we have the phenomenon of control. Then we have the theory of control systems. Then we have the model build on that theory to account for more and more of behavior. Control theory is not simply the proposition that organisms control things. That proposition must be treated as a report on a phenomenon, different from the theory that illuminates the phenomenon. Control theory explains control behavior.

In relation to the discussion of Gestalt psychology, I have to tell you about a stunning insight of Sam Randlett's, which he offered at the cybernetics conference I just got back from. The discussion concerned the apparent bifurcations of perception found in the Necker Cube, the stairsteps "illusion", and the faces-vases figure-ground example. You can't see both possibilities at once, the speaker claimed. Sam said that in the faces-vases example you actually can. What you imagine is two people with their noses exactly pressed into the corresponding dents in the vase. WOW! DOUBLE WOW! The speaker, a mathematician named Lou Kaufman, and a truly terrific person, said "It's so obviously right that I don't even need to try it. You've just destroyed half of my talk." Imagine: this illusion has been around for half a century or more, and Sam solved it -- in fact did away with it. Could it be that by finding the right higher-level perception we could dispose of ALL phenomena of that kind? The relationship of the faces to the vase does the trick. What will let you see both cubes? Note (B) pasted here

Anyway. In the rubber-band game you can emphasize that the game demonstrates the phenomenon, and the theory explains it. The more you can show that the phenomenon really does exist, the easier it will be to convince the reader that control theory really does explain it. In the diagram, you can emphasize not just the connections that are observable on the right of the dividing line, but the relationships among action, disturbance, and controlled variable that are seen even if the area to the left of the line remains blank. The phenomenon is found on the right side of the dividing line. The model includes what is on both sides of the line. The theory lets us calculate how this arrangement will behave under various assumptions about its parameters and about variations in the disturbance and internal standard.

Circular causation

David Goldstein came up with a nice image for the operation of the closed loop. The customary view entails tracing effects from one point to another and back to the start. This inevitably leads to the idea that something is happening in the loop only in one place at a time, the place where your attention is. You forget that while you're looking at the error signal traveling toward the output, there is already an output effect traveling toward the input and an input sensory effect travelling toward the comparator. What David said, with typically nice simplicity, was "Oh, I see -- it's like a wheel." All points along the rim of the wheel move at once. Another image might be the border of lights around a theater marquee: the whole circle of lights appears to flow simultaneously all around the marquee -- you don't have just a single light zooming around and around.

I think that's better than the image of a "field" in which every part affects every other part. That image suggests lines taking shortcuts across the loop, like the spokes of the wheel.

The hierarchy

My reaction to that long exerpt from my book is that I sure used to underline a lot. I'm starting to do that again -- got to cut down on all this yelling.

There's a rather subtle point about my hierarchy that may or may not be too difficult to get across -- I know you don't want to dwell too much on the specific levels. The point is that each new level represents a new type of perception, I think exactly in the spirit of Bertrand Russell's Theory of Types (his solution for paradoxes). You can, for example, analyze configurations into smaller configurations, and so on into ultimate tininess, or see any configuration as an element of larger configurations, and so on to hugeness, but you're still perceiving at the level of configurations. When you think, however, of a series of configurations that appear rapidly one after another with only small changes in any one step, what you get is not just more configurations, but motion, or as I term it more generally, <u>transition</u>, a completely new dimension of experience that has nothing of configuration about it. All modalities of perception contain transitions, and they are all dependent on configurations in that modality.

Why do I see a hierarchy here? It's because the sense of motion is derived from the sense of configuration, but configuration is not derived from a sense of motion. Whether we see motion or not depends entirely on how the configuration perceptions are behaving, but the reverse is not true. This is simply a matter of fact that anyone can verify by looking around. It's a phenomenon that I noticed out there in the real world, and only then realized was actually a property of perception and not of that external world at all.

Exactly the same jump from one type to another occurs when we analyze configurations not into smaller configurations but into <u>sensations</u>. Now we have dark, light, color, gradient, edge, curvature, and in other modalities other sensations, which even though they make up all configurations, are not themselves configurations. Unless the sensations are present we cannot detect any configurations, but the converse is not true. Again, a hierarchical dependence of one type of perception on another, a dependence that on naive inspection appears to be a property of that world out there, but which is obviously not a physical property of the world once you notice how it relates to configuration. It's a property of perception. Gestalt psychologists noticed examples of these phenomena but I don't think they ever saw the hierarchical structure that extends across all modalities of experience.

In exactly the same way, sequences are derived from and depend on transitions, relationships are derived from and depend on sequences (and here I have to add " -- and lower-order perceptions" because of such things as static spatial relationships), and categories are derived from and depend on the lower-order perceptions that are their members. The same hierarchical dependence is supposed to exist between all the levels in the model, with each new level bringing in not just an elaboration on the previous type, but a completely new and unexpectable type. There is no generating principle from which the nature of one level can be derived from the nature of lower levels. This is why I believe that I have noticed some real phenomena. This is why my hierarchy is not like any other that has been proposed. All the others I know about are based on some organizing principle such as size or complexity. Mine isn't. Mine was derived from a very long, very close, and very skeptical examination of the apparent external world. The fact that I have found only ten levels in over thirty years of looking should tell you that this kind of skeptical inspection is not an easy thing to do. For that reason, I don't expect anyone to grasp the meanings of my proposed levels easily. There is far more to them, they are far more fundamental, than any of my friends have realized. That's why I have tended to minimize them: not because they aren't real or important, but because I know that others hear only the verbal descriptions and can't easily use them as guides to repeating that skeptical inspection for themselves. So far nobody else has done that.

I wish that others would try this, but I'm afraid that the words themselves get in the way. Ed Ford called me a few weeks ago and said he thought he had an eleventh level: categories of system concepts. Well, yeah, at the category level that's how an eleventh level would be imagined -- how else? -- but point to what you mean. Which part of the external world, what aspect of it, are you talking about? That's where you'll find an eleventh level if one exists. You won't find it inside where you push words around. It will be right out there in plain sight just like all the others. Only you won't see it at first -- it's just the way things are, it's part of what you still take for granted. No, I had to tell Ed, I won't buy it until you can tell me what those words mean -- that isn't a word.

So there is a lot of structure behind these levels of perception. On top of that, these are also levels of control. The same hierarchical dependency shows up, precisely because control is control of perception, not action. In order to bring about and maintain any given transition, is is necessary that configurations be altered. To alter a configuration it is necessary that the sensations making it up be altered. And to alter the sensations it is necessary that the intensities that are summed to generate sensations be altered. I have tried to make sure that this same necessity relates control processes at all the levels, although I become uncertain that I've succeeded at the higher levels. At the highest levels this exploration begins to seem too much like trying to see down the inside of your own nose. But I think that the principles that we can see more easily at the lower levels can help up at the higher levels. AFter all, what we're looking for is right there in plain view where anybody can see it. The problem isn't seeing it. It's noticing it.

I don't know how prepared you are for all this. I guess I just wanted someone to know what still lies in this model to be discovered by the people who are trying to understand it. In my head the model is an extraordinarily beautiful tight-knit structure of relationships that make mutual sense in almost uncountable ways, even where its form begins to get misty. This is a true picture of how we are organized, of how we experience a world, of how and why we act to affect that world, even of what that world is. I really can't conceive how it could be wrong in any major way. Of course. If I could conceive that, I would change the model. Whatever the reality behind my delusion, I feel that in my head is a huge structure that we haven't even begun to explore. It grew from the seed of control theory, but it also grew from accidental discoveries that are quite outside the mechanics of control theory. These discoveries wouldn't have made any sense without control theory -- maybe that's why they haven't been seen before. But I can't really explain how this seed crystal developed as it did. I've been a spectator to a remarkable subjective phenomenon. How I long for others to witness it! Considering the crucial role of control theory, I don't see how it could ever have occurred before, except perhaps partially and abortively. Will I ever be able to show it all to someone else? The ghastliest possibility of all is that I will not be able to, that I will never know another person who can say yea or nay to what I think is the truth.

I hope some of this will help to make Chapter 11 communicate better this all-but-incommunicable concept of hierarchy.

Reorganization: very fine, and appropriately brief. The Test, likewise. Have you tried the coin game? I think it demonstrates something about the Test, but also about reorganization in the person trying to apply the Test.

Chapter 12: Simultaneous causation

Very good examples: clear. Causation is a problem primarily because people forget about (1) simultaneous multiple causation, and (2) different causes of precisely the same outcome. It's the old error concerning implication: if A implies B, then doesn't B imply A? The physical world obeys the true principle of implication, in which only one combination of observations is false: that a sufficient cause appears but its effect does not. Turning the wheel to the left will cause the car to veer to the left, but if the car veers to the left, you could very well be turning the wheel to the right -- in a strong enough crosswind.

The maintained loop: It's easy for us to focus so much on the stabilizing effects of feedback control that we forget to mention that the very same system can cause systematic variations. The conductor waves his baton in an unending pattern of movements that hardly ever repeat, and each position of the baton is under control -- would resist disturbance, and in fact always resists the disturbance of gravity. But the baton nevertheless moves, because the conductor is waving the reference signal, the set-point around which the baton's position is stabilized. We don't want to create too much of an impression that this is a static system, a homeostatic system. That's one of the major misconceptions the cyberneticists have, and one of their reasons for not accepting control theory. The organisn doesn't just defend the status quo. It makes things happen, not according to the ordinary laws of physical dynamics but according to its inner wishes.

Rain-in-the-face. Some strong suggestions. The biggest problem is that you don't seem to have closed the loop because you don't do that until the numbers appear. Let's complete the solution before using any numbers.

Y = rain in face
X = angle of umbrella, compass direction
W = angle of wind, compass direction

$x - w$: angle of umbrella relative to wind

$Y = a(X - W)$ rain in face depends on angle of umbrella relative to wind.

Y^* = desired amount of rain in face

$Y^* - Y$: shortfall of rain in face

$D = j(Y^* - Y)$: action depends on shortfall

$X = kD$: umbrella angle depends on action

Now we have the basic set of equations:

(1). $Y = a(X - W)$ (see diagram below)

(2). $D = j(Y^* - Y)$

(3). $X = kD$.

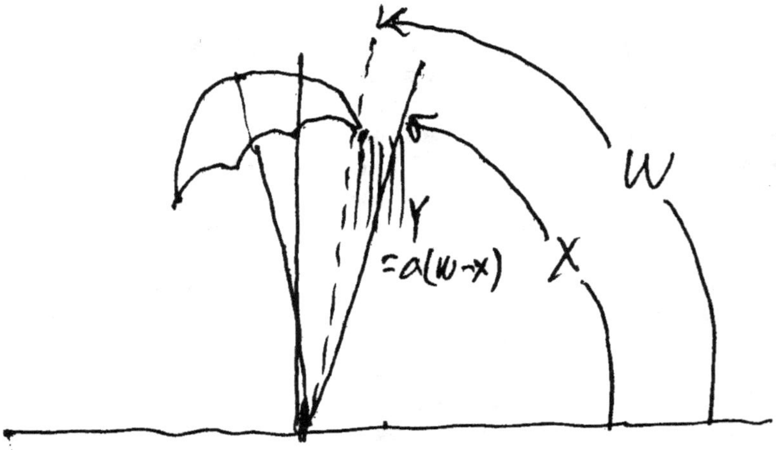

First substitute (2) into (3):

$$X = kj(Y^* - Y)$$

then substitute that into (1):

$$Y = a[kj(Y^* - Y) - W]$$

Multiply it all out:

$$Y = akjY^* - akjY - aW$$

Move the term in Y to the left side:

$$Y + akjY = akjY^* - aW$$

Extract the common factor Y on the left:

$$Y(1 + akj) = akjY^* - aW$$

Divide by the parenthesis:

$$Y = \frac{akjY^* - aW}{(1 + akj)}$$

That's the amount of RITF.

Now do it all over starting with X = kD

$$X = kj(Y^* - Y) \qquad (3) \text{ into } (2)$$

$$X = kj[Y^* - a(X - W)] \qquad (1) \text{ into } (2)$$

Multiply out:

$$X = kjY^* - akjX + akjW.$$

collect X terms on left and extract common factor of X:

$$X(1 + akj) = kjY^* + akjW$$

Divide by parenthesis:

$$X = \frac{kjY^* + akjW}{(1 + akj)} \qquad \text{Angle of umbrella}$$

NOW we can put some numbers in.

Let's say that one unit of effort generates one unit of umbrella angle, so k = 1. This is just a matter of units of measurement.

Then let's say that one unit of relative umbrella angle generates 20 units of rain in the face -- again, a matter of choosing units of measurement. So a = 20.

Now the units are fixed, and we have only the error sensitivity j to choose. This number says how many angle units the umbrella will move for one unit of rain-in-the face shortfall. Let's pick j = 50.

Finally we have to choose values for the two independent variables, but let's plug in the other numbers first.

$$Y = \frac{(1)(20)(50)Y^* - (20)W}{1 + (1)(50)(20)}$$

$$X = \frac{(1)(50)Y^* + (1)(20)(50)W}{1 + (1)(20)(50)}$$

OR

$$Y = = (1000Y^* - 20W)/1001 \text{ or}$$

$$Y = 0.999Y^* - 0{:}02W$$

$$X = = (50Y^* + 1000W)/1001, \text{ or}$$

$$X = 0{:}05Y^* + 0.999W.$$

OK, let's say the reference amount of rain in the face is 1 unit. Y* = 1. Let's say the wind angle is 10 angle units: W = 10.

Y = 0.8 rain units, and

X = 10.04 angle units.

Now let the reference amount of rain be 20 rain units and

the wind angle still be 10 units:

Y = 19.78 rain units, and

X = 10.99 angle units.

Sø by moving the umbrella angle to a slightly larger angle the person raises the rain on the face from 0.8 unit to 19.78 units, within about 0.3 unit of the desired amount in each case.

We can let the person's error sensitivity j be large enough that the 1 added to the denominator can be dropped, allowing some cancellations and the "approximate control system" equations to be found:

$$Y = \frac{akjY^* - aW}{akj} \quad \text{or}$$

$$Y = Y^* - W/kj, \text{ and}$$

$$X = \frac{kjY^* + akjW}{akj}, \text{ or}$$

$$X = Y^*/a + W.$$

You can work out the results: they're the same.

The rest of this chapter is fine, although you'll want to use the new numbers, which I chose to avoid problems with negative numbers.

Chapter 13: Experiments

To add to your collection, I enclose a drawing -- I don't know if I've describe this to you, but will do it again anyway.

Imagine a person holding a two-dimensional joystick (x and y continuous motion). On the screen are two dots. The right dot mirrors the movement of the joystick exactly.

The left dot deviates from a center in the left portion of the screen in a complex way. First, the direction of deviation is always the same as the direction of deviation of the joystick from its own center of motion. In polar coordinates, the angle is represented directly. But in the radial direction, the position of the left-hand dot is determined by the deviation of the joystick outside or inside a circle of constant radius. The farther outside this circle the joystick goes, the faster the left-hand dot moves away from its center of motion along the established angular direction; the farther inside the circle the joystick is, the faster the left-hand dot moves toward its center of motion.

The result is that the only way the left-hand dot can ever stop moving radially is for the joystick to be exactly on the reference circle.

The enclosed traces show what happens when the person uses the left-hand dot to generate geometric (or any other) figures on the screen (very slowly!). You can see two concentric squares and a larger triangle. On the right is a complete record of all the joystick positions that were used in making those figures, one after the other. No matter what the person chooses to draw, the joystick remains near the same circle, moving slightly inside and outside to move the left dot radially, and moving in angle to move the left dot in angle.

This is the best illustration I have yet devised for showing that it is perception, not action, that is controlled. There would be no possible way for an observer seeing only the record of joystick positions on the right to guess what figure was being drawn.

I presented these results at the International Meeting of the American Society for Cybernetics in St. Gallen, Switzerland, in March of 1987. The experiment has never been published or written up. Be my guest.

Chapter 14: Social Psychology

Your discussions of the Test in real life are fine. You make it sound like a reasonable procedure. Maybe the coin game could go here? Or have I started skimming and missed it?

A thought. Institutions persist because in their files they keep a list of reference signals that define the goals that anyone who works for the company is supposed to adopt. The structure of the institution is embodied in all the means available for people to carry out these goals: forms, routines, reporting systems, and so on. People come and go; those who cease to adopt the goals go, and those who newly adopt the goals come. This process has the unpleasant effect of making institutions seem to have a life of their own, and their operation is not necessarily in anyone's interest. This is one reason I think that making corporations into pseudo-individuals was a very poor idea. The goals that are adopted by the people have no morality built into them: if a corporation were in fact a person, that person would probably be under treatment as a psychopathic or sociopathic personality, and would be considered a threat to public safety.

Controlling others: you say everything that I think needs saying on this subject. If I were writing it I would foam at the mouth more.

On reward: control theory shows what "enough" means, a concept that is missing from the picture of Economic Man used by economists. Supply and demand work only when people don't have enough.

Chapter 15: Action research

This is mostly for your colleagues. Not that I don't consider us to be colleagues; we're just not that kind of colleagues.

Chapter 16: Possibilities

I know that this book is slanted toward those most likely to read a book by Runkel, but it might be worth mentioning some possibilities of a broader nature.

Over 300 years ago, the life sciences began constructing a model of living systems along the lines being followed by natural philosophers interested in physics and chemistry. This model contained no closed loops, and assumed simple Newtonian causation. All that was developed in the following three hundred years assumed the correctness of this model.

But that model is wrong. The wrongness of the model accounts for the fact that the life and social sciences are dismal failures as sciences. Intelligent individuals have managed to learn a lot about the practical arts of living together, but all that has been learned was what could be accomplished within the same conceptual framework that existed before Galileo or Copernicus -- in other words, progress has occurred at about the same rate that the physical sciences progressed before Galileo: very, very slowly.

Control theory represents the same sort of step out of the old world that occurred in physics in the early 1600s. It is the equivalent for the life sciences of the invention of Newton's laws of motion, of the principles of conservation of energy and momentum, and of the development of the mathematical tools that changed physics and chemistry into quantitative sciences. The sciences of behavior, therefore, start their modern development now.

Chapter 17: Summary

Maybe the foregoing belongs here. Maybe you don't agree that it belongs.

If this book is actually read, it will be a major event in the social sciences. I fervently hope it gets the attention it deserves. Thank you for writing it.

Best

Bill

Bill

Note (A)

But it is not my purpose to give a thorough criticism of reinforcement theory. Anyway, I can't think of a simple, easily-explained example.

Note (B)

I let it stand because I can't see them both at the same time. Also because it makes the ~~xxxxx~~ point anyway for most readers. And it is OK for Randlett and

Powers to say phooey in something they write.

Jan 23, 1988

Dear Phil,

Just got back from the Gordon conference, which I won't attempt to describe in any detail. Basically there is a group that resents and fears quantitative science, probably for good reason, and has trouble distinguishing control theory from behaviorism (we were actually referred to as behaviorists by some). Some pretty tense moments, everyone feeling intimidated and defensive on both sides. But slowly, slowly, I am persuading them that it's only me, Bill, not a monster in a white coat. Sigh.

The new presentation of the rain-in-the-face equations looks good to me, and your careful explanations are (a) all correct, and (b) communicative. There is only one problem that needs to be corrected, and it's a matter of rhetoric, not content. When you put the numbers in, it turns out that the discrepancy between wind angle and umbrella angle needed to produce the right amount of rainintheface is only 0.04 units out of 10, which is uncomfortably precise.

All right. You've just called.

So the answer to that is to decrease the magnitude of a from 20 to, say, 2 or 5, and increase the magnitude of some other variable accordingly. In other words, muck around with the constants until we get rid if the hypersensitivity to small differences. Then the results will be more convincing.

After a long and slightly tipsy discussion, the concomitants of which are seriously affecting my typing (invisible due to the magic of word-processing) Mary and I have had a brilliant, if temporary, idea. The next meeting of the control systems group will be the last. The control-system group, having reached the verge of institutionalization, will now dissolve after the last week of September, 1988. A cyberneticist, Doreen Steg, has offered us the use of an island off New Jersey for our next meeting (post-1988), and I think we should take her up on it. But when we do, it should be under not only a new venue but a new name. I don't much care what it is, but it should signify a new direction. Our interest (as I will argue) is not in control theory but in human nature. Those who understand control theory have, of course, a particularly sharp tool to use in these pursuits, and when "we" meet I fully expect control theory to play a major part. But we must now go beyond being control theorists and become students of human nature.

Suggestions are solicited. I think you may get the idea I would like to convey, which is that stagnation lurks around every corner. Let me know what you think.

Best

Bill

610 Kingswood Avenue
Eugene OR 97405
24 January 1988

Dear Bill:

I felt very good talking on the phone to Mary earlier and to you later.

Three things enclosed. First, the opening of chapter 13, where I have put my description of your "circle demonstration." Please tell me whether I have described it properly. Thanks.

Second and third, sheaves for you to send to de Gruyter if you wish. The letter I had from Treville Leger, Executive Editor, was on the letterhead of Aldine de Gruyter, to whom I had sent my query. The content was the standard prose for rejecting manuscripts.

One sheaf contains title page, contents, some pages from chapter 1, and all of chapter 17. You will note I am trying out a new title. The table of contents is revised. The first three pages of chapter 1 are revised. The last pages of chapter 1 and all of chapter 17 are about the same as what you read earlier.

The other sheaf contains comments from readers, excerpted or abridged. Eight other people have the complete MS but have not yet said anything to me about it.

And here is another item. A while back, Mary sent me a couple of pages from a book by Mary Bateson. I quoted some of it in my MS. To get permission, I wrote to Mary Bateson in care of Knopf. They said try her literary agent. I wrote to Gerard McCauley Agency. They said try Amherst College. I am about to write there. But maybe one of you knows how to reach Mary Bateson?

Your friend,

Phil

Phil Runkel

31 January 88

Dear Bill:

Here is the revision of rain in the face with new numbers for the constants. If you have other things you'd rather do than go over this again, feel free. I'm 95% confident it is OK. And if it isn't, nobody but you will ever figure it out anyway.

I tried about a dozen combinations of constants,and that was enough so that I could tell by interpolation and extrapolation what I would get if I tried other combinations. The combination I use here gives good separation, I think,among the three examples.

Thanks again very much for your help.

Pages 15 to 19 are very much as they were except for a few improvements in phrasings. The new numbers begin on page 20.

Phil

Dear Mary:

I am certainly sorry to hear that the CSG will not be coming to Oregon next September. Doggone. Phooey. Grumph.

Well, please arrange now to come in 1990.

For example, that first place I told you about that I like very much (can't remember the name just now) was filled up for Friday nights. If you were to reserve the dates now, no doubt the dates you prefer would be available that far in advance.

Phil

Feb. 6, 1988

Dr. Treville Leger
Aldine/de Gruyter
200 Saw Mill River Rd.
Hawthorne, NY 10532

Dear Dr. Leger,

I'm writing to redirect your attention to a book manuscript that has already been through your hands -- Two Methods was its title at the time, author Philip J. Runkel. I was one of those to whom Runkel sent the entire manuscript for criticism.

The reason he sent it to me is that large parts of the second half of the book were written in support of concepts I presented in my 1973 book, Behavior: the control of perception (of which you are aware, as it is still in Aldine's catalogue). Runkel is presenting these ideas to an audience I would not know how to reach; I consider his work a major contribution, from my own selfish standpoint.

It is, however, a contribution to another and wider field: behavioral and social research in general. Runkel's manner is deceptively mild and informal. In fact his analysis of standard statistical methods as they are used and misused in many fields is a bombshell, accurately aimed with full professional awareness of the target. Runkel is an expert on methodology in the social sciences, having written respected books and papers on this subject for many years. Many of the faults he has found in standard methodology concern methods he advocated and taught through his whole active professional life. How he was able to pull himself up short and undertake such a major reconsideration of his own views is still beyond me -- it is an extraordinary sign of strength of character and intellectual honesty. It is even more remarkable to me that he has been able to undertake this radical revision without becoming a fanatic, carefully sorting babies from bathwater. He presents his position firmly and with great clarity, but without taking any dogmatic stances. With regard to his uses of my work on control theory, I can only say that he is a superb teacher.

So I hope you will ask Runkel to resubmit the manuscript. He sent the enclosed materials to me after I told him I would write to you.

On another matter, I am working on a new book. That is nothing new; I've been working on it for about ten years, with little progress. Now it seems to be taking the shape I want. I will of course offer you exclusive first refusal on it, as my way of showing gratitude for your long loyalty. You may have noticed that sales of BCOP have been rising a little lately. My work is becoming more widely known; it's been the subject of half a dozen doctoral theses, and is taught in as many universities. It's also becoming fairly well-known in Europe. There is a "control-systems group" now, consisting of about 60 scientists and social workers mostly in America but some in England and Europe (and even one in Thailand), who use my ideas in their work. Our fourth annual meeting will occur in late September of this year -- about half the group will attend.

I can't predict when I will have anything to show you -- it may be one or two more years. I still work in industry to make a living, and can't devote full time to my real work until I retire, which will be in three more years. When that time comes, and assuming you find the book worth publishing, I will offer you a proposition: if you could undertake to get the manuscript of my first book onto an IBM-compatible disk (Wordstar or XYWrite), I will undertake to revise it. I'm making a conscious effort not to repeat what is in that book in the new one -- that's one condition that has slowed my progress on the new one.

That's more or less for your information, as there is no action called for right now. My main message here is to encourage you to have another look at Runkel's book, which I think with just a little encouragement and ageing will become a classic.

Respectfully yours,

William T. Powers
1138 Whitfield Rd.
Northbrook, IL 60062

Feb. 6, 1988

Dear Phil,

The new equations still look fine, and the numbers are more believable. Enclosed is the letter I have sent to Leger.

Best

Bill.

610 Kingswood Avenue
Eugene OR 97405
17 February 1988

Dear Bill:

I got the CSG Newsletter today. Your summary on the asymmetry of control is an elegant little argument. Little in the space it takes, BIG in its implications. Thanks for doing it. At the bottom, you say that E is generally less than unity. The reason for that, I think, is entropy. OK? But suppose the environment affected by K is other living creatures. Then I suppose either they get out of one another's way, or they spend a lot of energy trying to "control" one another?

I see you have $1500 toward a new computer. Good! And congratulations on drawing a nice award from your admirers. I hope what you have now, along with the next installment, will enable you to get what you want.

Yours,

Phil

Phil Runkel

* Dear Robert Lord and Paul Hanges, 16 Apr 88

I have been reading and thinking about your paper, "A Control System Model..." for some time now, and I guess I'm as ready as I ever will be to comment on it. I started out with the idea of writing a commentary for Behavioral Science, but that format seemed too confrontational, and also beyond my expertise. Far better to write you informally, since, although I have many problems with your paper, I am really delighted with your basic assessment that control theory provides a useful model for integrating concepts in the social sciences and psychology.

I'm sure that you, your editors, and reviewers feel that you have presented an accurate picture of how control systems work and how to go about applying control theory in the behavioral sciences. What, then, are you to make of what follows? I hope for the best, knowing that by now you have a heavy commitment to your own analysis, but perhaps are willing to revise your thinking, given sufficient reason to do so.

My main concern is that in offering control theory as a means of integrating various concepts in the behavioral sciences, you have fallen into the trap of altering control theory to fit the concepts rather than rethinking the concepts in control theoretic terms.

Control theory (which did not, as you seem to believe, develop from work in cybernetics) is a robust and sophisticated branch of engineering. It deals with purposive systems; systems which have intentions designed into them. You seem to believe that purpose or intention requires consciousness. How then do you conceptualize your home thermostat, which has built into it the intention of maintaining your house (or at least that part of it where the thermostat is located) at exactly 72 degrees? What is that little computer most cars have nowadays? Part of a regulatory device that is intended, and is itself designed to intend, to regulate the composition of exhaust gases by by varying the air/fuel mixture in the carburetor. These devices reflect the intentions of their designers and users, but the designers and users aren't around and the devices are carrying on all by themselves. They are control systems. The theory behind their design is explicit and quantitative.

Cybernetics began when Norbert Wiener met a physiologist named Arturo Rosenblueth, who was working in the lab of the man who invented the concept of homeostasis, Walter B. Cannon. What happened then was not a discovery but a recognition: that living systems showed the same characteristics as that particular kind of electronic device called a control system. That is not too surprising, since the aim of the engineers originally had been to design systems which could do a particular kind of thing that until the 1930's only people (and other living systems) could do: maintain a particular state of affairs (such as the pressure in a boiler) at a desired - intended - level. This version of history, among other things, again challenges your assertion that control

* A control system model of organizational motivation: Theoretical development and applied implications Robert G. Lord and Paul J. Hanges Beh Sci 32, 1987 www3.interscience.wiley.com/journal/114043846/abstract

theory does not explain behavior that is "only reactive", or that is "produced solely by automatic or unconscious processes", since it clearly applies to homeostatic functions, etc.

I won't go into what happened to cybernetics, except to say that most of your references appear to be psychologists and social scientists, who, if they learned their control theory from each other and from cyberneticists, have lost quite a bit in the translation. There is one exception, who you cite quite often, and that is William T. Powers. (And I should put this card on the table - I'm his wife). Bill (as I'm sort of used to referring to him) spent quite a bit of his professional life designing and building artificial control systems. When he talks about control theory, he has an expertise that cannot be matched by anyone in the behavioral and social sciences. You have several dozen references in you article. How can you possibly judge their relative value? My (admittedly biased) assertion is that Bill Powers' version of how a control system works is better informed than any of your other sources, and it is from his perspective that I am writing this letter.

I'm not going to do anything here except discuss your diagram 1 (and by implication, diagrams 3 and 4). Your version of control theory, as expressed in this diagram, shows several functions and interconnections which are inconsistent with any model of a real control system. You call it a "general flow diagram", which, according to my definition of such things, gives you license to draw any boxes and arrows you please. But is that freedom worth what you are giving up? In Bill Powers' somewhat similar diagram, every function and every connecting signal can be represented mathematically, and their interactions expressed as equations. From general diagrams such as his, an engineer can proceed to circuit diagrams of actual components, and eventually to construct an analysis of a real, working control system. From such diagrams computer programs can be written that simulate the performance of such systems. Above all, from the theory represented by such diagrams, experiments can be devised that test the model. As scientists, I hope you appreciate the potential value of working within the limits of such a model: it defines the boundaries of what is possible, and avoids the trap of untestable hypothetical constructs. One of Bill's colleagues (a psychologist) has written computer programs that simulate a control system performing a certain task. His intention was to simulate a reinforcement model on the same task and compare the two. After many weeks, he was forced to conclude that the concept of reinforcement is descriptive only, and cannot be modelled. Unfortunately, the same thing is true of your flow diagram.

1. System boundary. The sensors and effectors of a control system are not located inside the system. They are at the boundary. Sensors are transducers, which convert events in the physical world into neural signals. Effectors convert neural signals into output forces (muscular effort). There is no other way in or out of the system. Why does this matter? See 2.

2. Input from a disturbance. All physical events must pass through sensors at the system boundary. Your disturbance does not go through a sensor. How does it get into the system? ESP? (literally).

3. Standard. Your reference signal (unlabelled) comes from this box and goes to the comparator. Incoming events come from the sensors and also go to the comparator. You show them going to the standard box, which is incorrect. The standard box itself is confusing. How does it generate a reference signal? According to your diagram, the disturbance sets the standard, but as I just said, that is not how a control system works. Outside disturbances cannot directly affect reference states or positions or standards.

In a simple artificial control system, the reference position is set by an outside agency. This is commonly and confusingly labelled in engineering diagrams as the reference input. This is not the same as the inputs to the system through the sensor. In a thermostat, the outside agency is a human being resetting the dial to a desired temperature. The output of the person, a finger movement, acts to set the reference level of the thermostat, which then turns the furnace on and off as a result of comparing the sensed temperature with the reference temperature. It is not, by the way, comparing a finger movement to a temperature. It is comparing its own representations of these phenomena.

In a programmable thermostat another layer of control is interposed. A person sets the controller for various temperatures at various times of day. The output from the controller sets the thermostat's reference level to different temperatures at different time - by itself.

Now take this model inside a living system. The lower level system receives reference signals from higher levels in the nervous system. The source of those signals is the output from the next level up. If there must be a box labelled "standard", it must also be labelled "output" - the output of another (higher) system. Better to forget that box called standard.

4. Decision mechanism. This was the first thing that caught my eye when looking at your diagram. "What on earth is THAT!" Well, it's a box you added to handle one of those concepts control theory isn't developed enough to deal with. Cancel that - sarcasm will get me nowhere. Seriously, this is a major problem with your diagram. This is interjecting a totally arbitrary function because you felt it was absolutely essential for a social science model to have such a mechanism. And I agree. If you want to model people who make decisions, there has to be a place in the model for decisions to be made. So, what is a decision? Choosing one course of action over another, so that the output varies according to which decision is made. Yes? No.

Here we come to the really tough part of understanding the

control model. What is it that is being controlled? The answer flies in the face of everything social and behavioral scientists are taught to believe, because it is not behavior. It is perception: it is one's perception of the way things are in comparison to the way one wants them to be. To make it a little more palatable, it is the perceived outcome of behavior. It may be that the desired perception is to see certain actions being performed, but it's much more likely that what is desired is certain (perceived) results. Actions must vary in order to achieve results, because events in the real world affect actions in unexpected and constantly varying ways.

Where does decision-making come into the picture? In order for a decision to be wanted in the first place, a reference level for decisions has to be established. This is generated by the level above, and is the output of that level. I am a manager, and it is my role to make decisions. O.K. Now what decisions? Decisions that will have outcomes that enhance my role as manager. What are they? Bigger profit for the company? Good morale in the work force? A promotion for me? I must choose strategies that I think will produce whichever outcome I want. Maybe only one of these matter to me. Maybe they all do. Maybe achieving one conflicts with achieving one of the others. Designing a strategy that achieves all the wanted results may be impossible. But the point here is not what strategy is designed, it is that the output of decision-making is designing a strategy. And that is setting the reference level for the next systems down in the hierarchy. In order for a strategy to be carried out, certain sequences of events must be put into action. And in order to do that, reference levels for the next lower level of systems are set, and so forth. The manager's actual output, at the lowest level where his control hierarchy actually produces outputs that affect the world, are probably very minor - a bit of writing in a memo, a few words spoken. If these outputs specify for other people what their actions are to be, then your aim is to see people performing these actions. If your aim is to see certain results of people's actions, then you need not specify their actions.

If decision-making is established as the operation of the strategy level in the control hierarchy, then how do all the other levels decide what to do? The answer is that they don't. In a well-functioning control system the decision is already made: do whatever it is that has to be done to reduce error. This ordinarily works so well that there is no sense of error at all: result follows intention so smoothly and immediately that one characterizes the whole process simply as "doing". But what is this "doing"? Not simply producing the required outputs, but continually varying the outputs in relation to what is going on in the environment. You can't walk uphill using your muscles the same way you use them when you walk downhill; you can't open a door that's already open; you can't start to type a letter until you put paper in the machine. Actions operate in an environment that is continually varying, continually being disturbed. There is no way for a behaving system to anticipate all the environmental effects on its output. This is why a control system does not control its outputs as is so commonly believed. It varies its outputs as necessary in order to achieve inputs which match a desired perception.

Incidentally, this is the beginning of questioning your assumption that control theory can only handle one goal at a time. A multiple-level system with numerous units at each level can and does handle many goals at once. What it cannot do is handle competing goals that simultaneously demand different outputs from the same subunits. That situation is called conflict, of which much more, but not in this already much too long letter.

5. Cognitive change loop. This is included in your diagram but not explained in the text. I'm not sure what it's supposed to be. It seems to be an output from the decision mechanism. In my scheme of things, the decision mechanism is a control system unit at a quite high level in a hierarchical system. Its output is the setting of a reference level for the system below; that is, it is a signal feeding into the comparator of the system below. There it will be compared to signals from the input, and the resultant error will drive that system's output, setting the next lower reference signal, etc. There is one other path an output signal can take, and I think this is your cognitive change loop. In the Powers model, the output of a control unit can be fed back into the input of the same unit, or into the input of a unit one or more levels down (by way of reference signals and output, not directly) without any output actually emerging into the environment as actions. (The details in such a loop are provided by memory). This he calls the imagination connection. Without actually doing anything, a control system with a loop like this can imagine the consequences of one or another action. The vividness of such imagination, its "realness", depends on how low in the hierarchy such a loop occurs. A rather high level loop, mostly verbal, could probably be termed cognitive. Evaluating several such imaginary outcomes will show which one appears to satisfy the reference condition with the least error, and that is the course of action one will decide to take. As it proceeds, its effects, and the environments's effects on it, are continuously monitored and adjustments made accordingly. This, of course, is feedback.

I do need to make some comments about your use of the term feedback. While you distinguish between feedback in the control-theoretic sense and in popular usage, you seem to come down on the popular side, in which feedback is primarily seen as a property of the environment: it is "provided". It can be "infrequent", or "vague", or "too costly". You're talking about error signals here: quality of information relative to a reference condition (often, specific, cheap). Feedback in a control system, by definition, is always present and always strong. That is part of what defines a control system.

It's a touchy process, transferring control concepts from an individual to an organization. If individuals are assigned functions – sensors, comparators, effectors – this fails to take into account that each individual is himself a control system, and does not receive reference signals from higher in the organization in the same way that a lower level in an individual hierarchy does. See my diagram about this. I am enclosing several alternative diagrams which I believe better express the known and required properties of control systems. I very much hope you will consider adopting them in place of your own, which I believe perpetuates many misunderstandings about a model which is far more powerful and complex than you have recognized so far.

If you've made it to here, you might be interested in knowing that there is a small group whose purpose is to study, experiment with, use, and expand the Powers control model. Meetings of this group (which would welcome your participation) have drawn representatives from a number of fields: experimental, social, clinical, and physiological psychology, sociology, economics, social work, microbiology, management science, music--there is a unity of concept that crosses disciplinary boundaries. It is true that there is little published research; most of those from whom research might be expected are working full time at something else, are unfunded, and their work has not met the acceptance you seem to have achieved. I happen to believe that this is because of what I criticized in your work: that you have adapted (and distorted) control theory to fit prevailing concepts, whereas at least some of our group have taken a more confrontational approach, attacking the conventional wisdom where control theory contradicts it (and reaping a harvest of rejections from the journals). Obviously your approach is more successful, but at the cost of being incorrect in a number of respects, as I have attempted to explain. I can only hope that you can see the value of revising some of your thinking, so that the control model you apply in your work more accurately represents the theory on which it is based.

Sincerely,

Mary A Powers

Mary A. Powers

April 16, 1988
1138 Whitfield Rd.
Northbrook, IL 60062

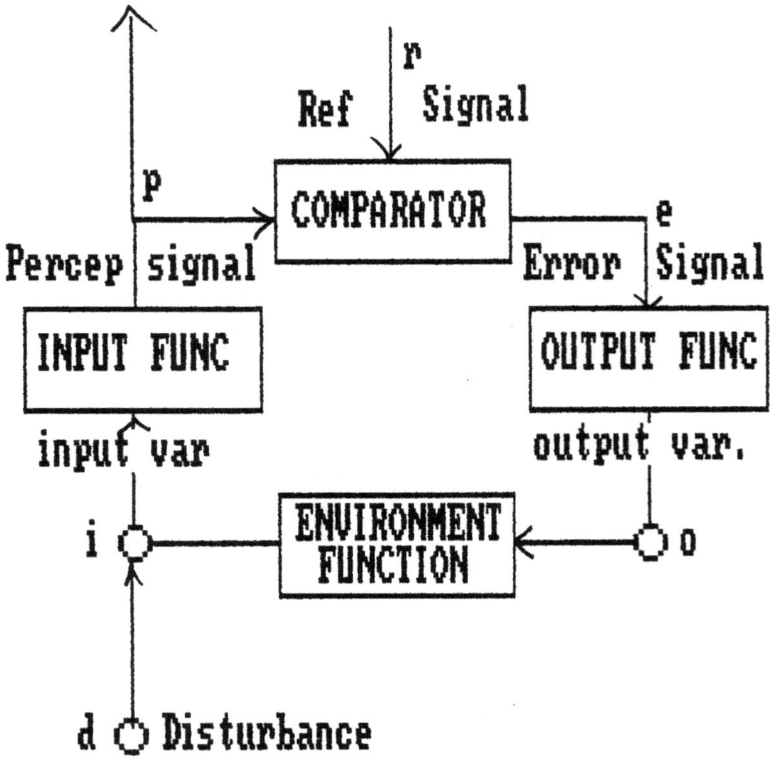

Fig. 1: Generic control-system diagram

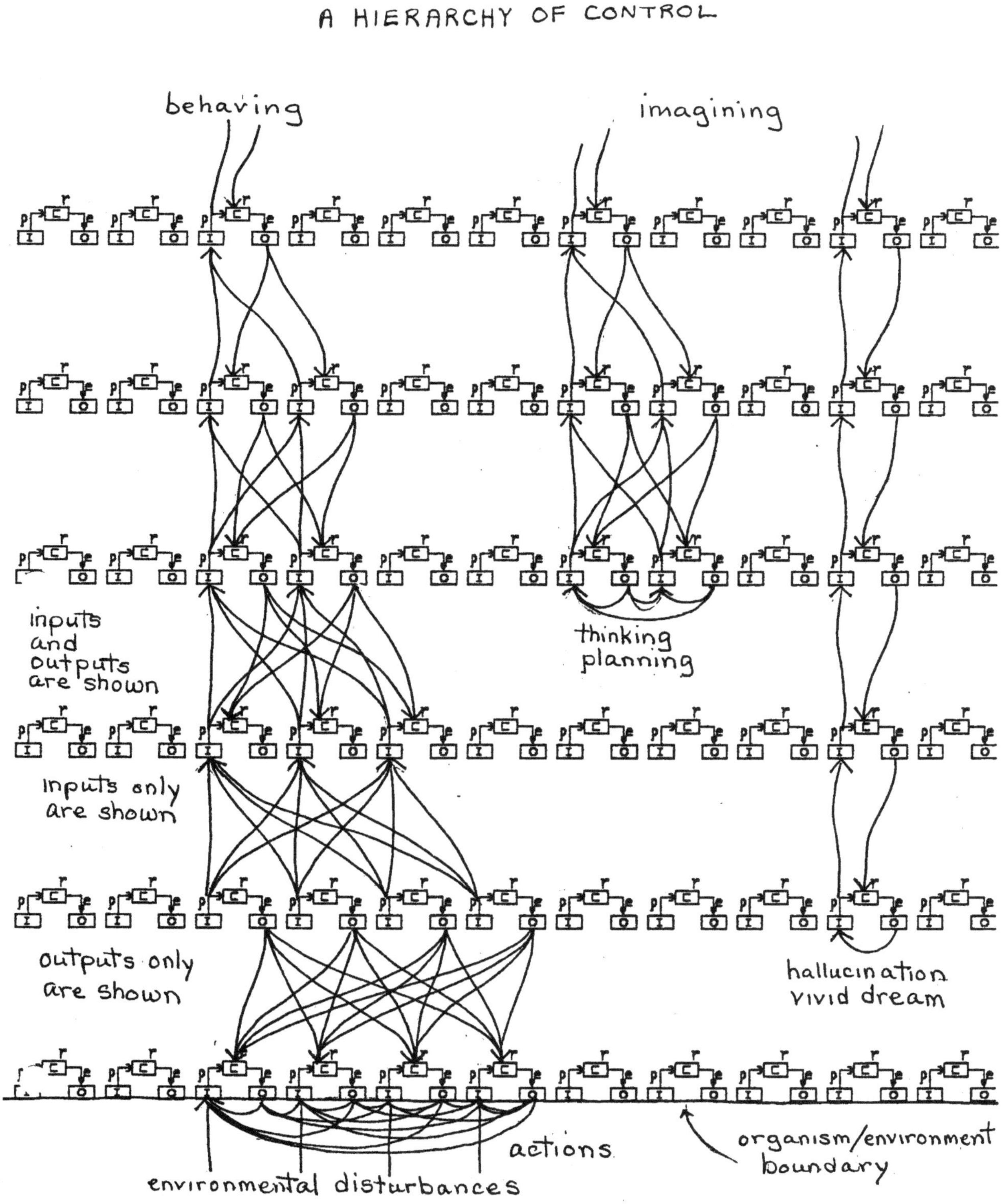
A HIERARCHY OF CONTROL
behaving
imagining
inputs and outputs are shown
thinking planning
inputs only are shown
outputs only are shown
hallucination vivid dream
actions
organism/environment boundary
environmental disturbances

This is a metaphoric use of control theory as applied to an organization or any other situation where people have roles that resemble the components of a control system (input, output, reference, comparator. A very different picture emerges when each individual is recognized as a separate system (next diagram.

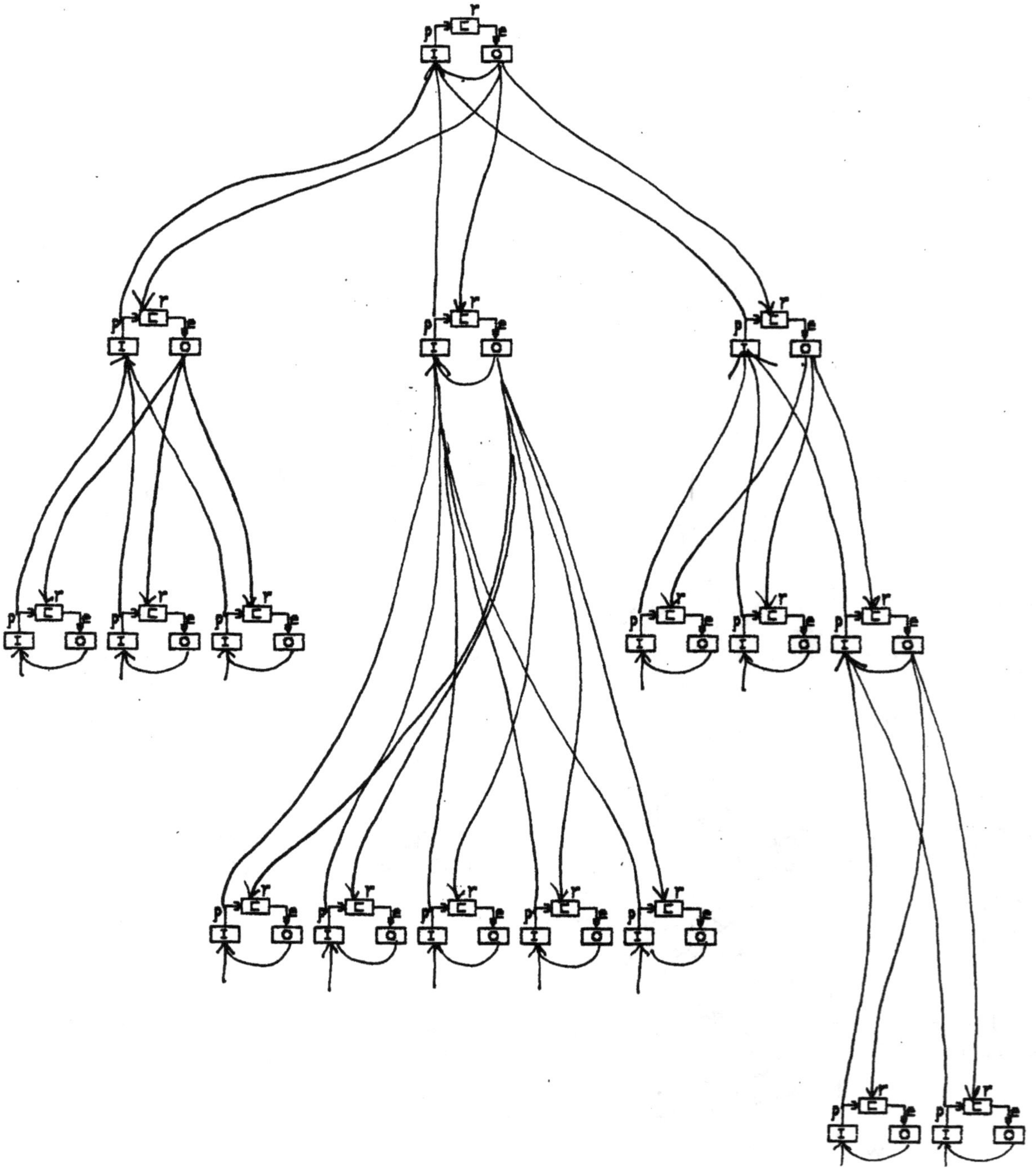

Here is a diagram of an organization taking individuals-as-control-systems into account.
How individuals handle orders from above depends on the relationship of those orders to each individual's own reference levels. Orders are inputs, not goals.
In this particular example, the top person is receiving input only from those immediately below, with no independent sources of information. Also, there is no communication sideways. Other organizations would show other inputs and outputs among their members.

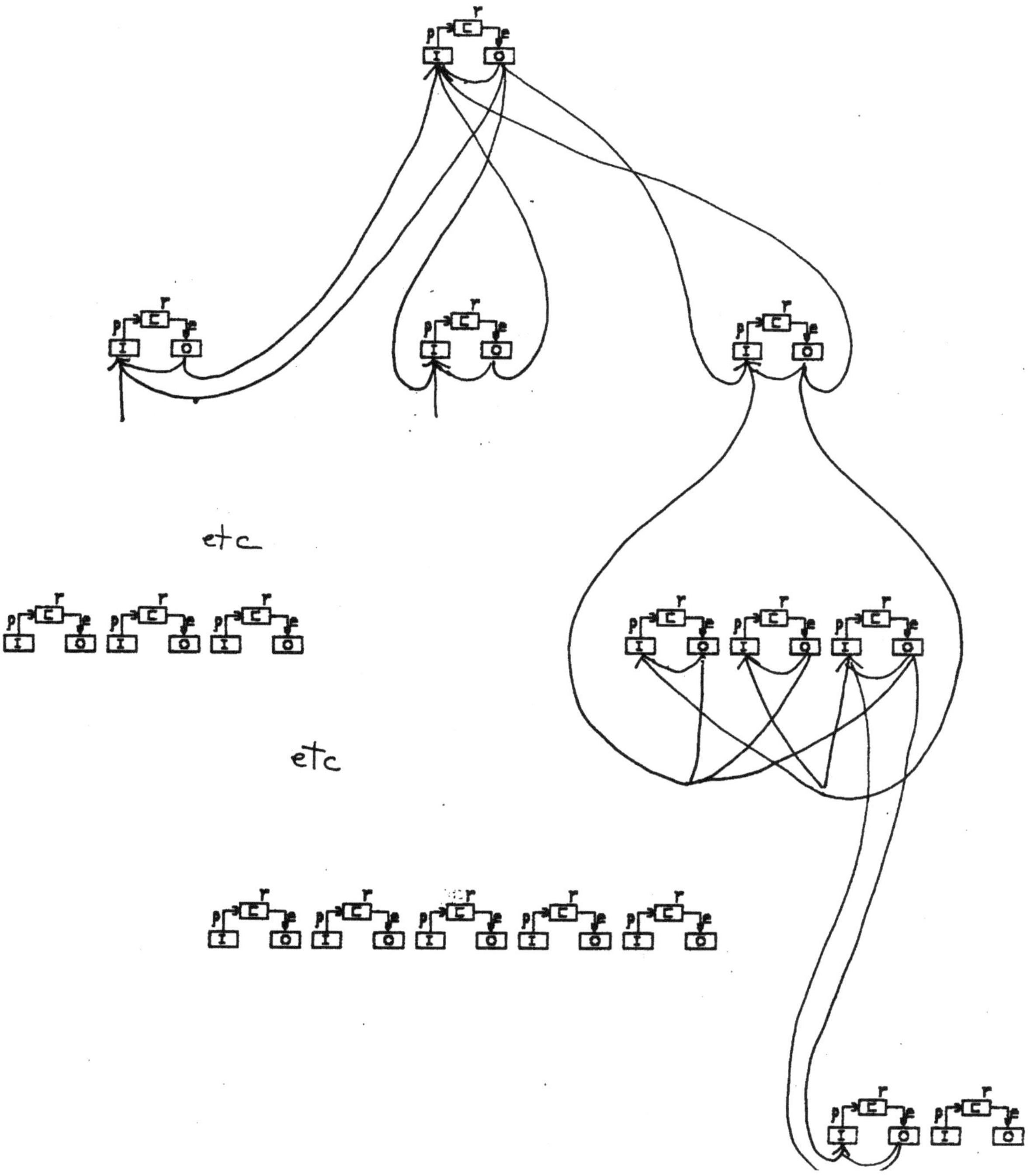

24 April 88

Dear Mary:

Margaret went into the nursing home Friday noon, and it is now Sunday noon. I'll go visit again after I type this letter. I feel very disconnected. I keep wondering what I'm doing and what I ought to be doing and whether anything is doing. Reality seems to be something I once knew something about, but I've forgotten what it is. The world seems dreamlike. Many of the objects in our house have no purpose any more. Maybe this letter will sound disconnected.

I am typing this on my 20-year-old typewriter. My computer is falling down on the job again. It turned itself off yesterday. It turned itself on again when I gave it a cold boot, but I'm sure it will turn itselfoff again before long. It doesn't even say excuse me. So I guess I'd better stop hoping for the best and buy a new one. I've been worried about money lately, but lo and behold a friend has offered some. What a relief.

I think your letter to Lord and Hanges dealt very clearly with the points they ought to think about. I think you picked them out xxxx unerringly and explained with (to me) great clarity. I fear, however, that the first point for which you chided them will make it difficult for them to understand your letter. That is, they seem to have the view, as most social scientists do, that all theories are simply up for grabs, and you take various features that sound good to you from the various theories you read about and put them together to make your own theory. If you want to make a machine that flies, you take some wheels from this vehicle, a wing from that, a couple of comfy seats from this, and ignition system from that, and so on. You send the description to a journal to find out whether other people think it looks pretty.

I'm not exactly sneering. I've done some of that myself, and I expect I'll do it again, though I hope my writing will be less xxxxx larded with that sort of thing than it has been in the past. I hope at the very least I'll be able to tell the difference between trying to describe how things might <u>work</u> and offering some analogies, metaphors, or directions in which to look for hints of evidence that might enable some people (not all) to get ready for another try at testing how things might work.

Obviously, L&H don't understand what you mean by <u>model</u>. That will also get in their way.

I think you must have struggled with yourself a good deal to decide what <u>not</u> to berate them with. If you had picked out for comment every inaccuracy, wrong assertion, and piece of nonsense, your letter would have been at least five times as long. I admire your discipline in limiting yourself to the points you chose. You chose to hit them over the head gently—to say, "Hey, wake up." I would have been tempted to beat them harder—to want to punish them. But that would help nobody.

To you, however, I can complain without doing any harm. Only one of the features of their article that betrays their view of models and diagrams is their "venn diagram." A venn diagram is supposed to be a graphic way of writing logical implications. That is, this set of things is a subset of that set of things, where the "things" are logically defined concepts, not real things. Their diagram is just silly, and does nothing that a string of words would not do.

I guess the thing that annoyed me as much as anything else is the evidence that L&H didn't bother to read Bill's book with much care. Citing the book, they say that Bill proposes *seven* levels of system, and they say that the highest level is that of *principle*. Even someone who merely looked at the headings in chapter 13 wouldn't make those errors. When readers come at a piece of writing with very different assumptions from the writer's, I don't get angry if they misinterpret some things or even skip over some of the sentences. But when they report the number of items in a list without bothering to count them, I get angry.

You mentioned to L&H their mistake in claiming that control theory doesn't explain pursuing more than one goal at a time. Bill talks about that at several places in the book. If L&H read the book a year or two before they began writing their article and didn't bother to review it, I suppose they could have forgotten about that point.

I think a lot of social scientists write about other people's writings with just about that degree of care. And we poor readers too often suppose that they have portrayed the other writers' ideas accurately.

I'm not surprised that *Beh Sci* accepted the article, because it is unlikely that the editor or any of the reviewers know much about control theory, either. But I must say that I am somewhat surprised that those gross errors got past *two* authors. Wouldn't you think one of them would have caught the other out? Woudn't you think one of them would have said, "Hey, I was just checking Powers's chapter 13, and there are *nine* levels"?

I'm glad you wrote to L&H in a persuasive tone instead of a vituperative one, and I admire your skill in doing so. I hope you get an answer. It would be wonderful if the answer were not "Here is why we are right," but "Oh, now we understand better."

Phil

Dear Bill and Mary 13 October 1988

It was certainly good to get to know both of you eye to eye and hug to hug. I am reporting to several people about my trip. Here is the report going to all.

* * * * *

My recent trip was the first time I had ventured into the wide, wide world since 1980. It was in the later part of that year that Margaret began to need close care. During those eight years, I made only two overnight trips. During the last four years, someone had to remain with Margaret every minute, so I very rarely even left the house after supper. When I ventured upon this recent trip, therefore, I felt as uncertain and anxious as if I were departing for the wilds of the Amazon.

Margaret must have felt the same way in May of 1980 on the eve of a trip to visit her sister in Montana. She said she was anxious about finding her way through the Portland airport. That from a woman who had traveled alone, by airplane and ship, over most parts of the Western Hemisphere.

I found that some routines of traveling had changed during the eight years, but most things went smoothly enough. Nobody seemed to think I was behaving strangely. A check-in agent at O'Hare airport told me I was bound to get lost if I persisted in looking at the signs hung near the ceiling instead of listening to him. Actually, I got lost only twice, and those times were when driving on the highway.

I went first to visit friends in Champaign IL. The Hastingses and the McGraths were happy to see me, and I them. Neither of the Hastingses were very well, sad to say. I was glad to contribute a little variety to their careful lives. The McGraths were both fine. I visited the house where Margaret and I lived with Margaret's parents and my mother in Champaign. The little trees we planted had all grown far up into the sky. The yard I remembered as an open greensward with a little hopeful tree here and there had become a dark jungle. Few repairs had been necessary on the house in the intervening 24 years. The McGraths took me to a hotel dining room that Margaret and I had often enjoyed.

Then I visited my brother, Kenneth, in Wausau WI. I was happy to discover that he was pounding the piano, at the age of 80, with as much verve as ever. I found that I reverted very comfortably to the role of little brother. Kenneth gave me a piano lesson, beat me in a game of chess (I was able to hang on for more than 20 moves!), and told me some interesting lore about recorded music. He knew of some excellent restaurants in that still-small town. It made me feel very good to be with my generous, conscientious, and intelligent brother after so many years.

Then I went to a lodge on the fringes of Kenosha WI to a four-day conference on "control theory." Only a couple of dozen people attended, just the right size for a good conference. (I gave up many years ago attending meetings of the American Psychological Association with their swarming thousands.) I had been corresponding with some the members of the Control Theory Group for a couple of years, but this was the first time I had met any of them. I enjoyed talking with all of them. Several gave interesting short talks. I was suddenly called upon to give a talk, too. I got good applause, so I guess I got my sentences in the right order. A couple of people had brought computers and demonstrated several fascinating psychological experiments with them. William Powers has written a program that makes it easy for anyone to design an experiment and then tell a computer how to carry it out. Flabbergasting.

I feature I especially liked about the conference was the variety of people there. Academicians would call it "interdisciplinary." People who prefer shorter words would call it "a nice mix of people."

I enjoyed the whole conference very much. Every afternoon was unscheduled, so there was lots of loose time to get into small ad lib conversations or to go for a walk without feeling pushed by someone to do something else.

Then I drove back to O'Hare, slept in a motel nearby, got on the airplane the next day, slept in a motel near the Portland airport that night, and the next morning got in my car, which I had left there, and drove back to Eugene.

I did not enjoy entering again the big city swirl. Signs slapped at my eyes everywhere: Do this. Do that. Not here. Keep moving. Turn left. Turn right. Turn both ways. Watch for this. Watch for that. Watch for things to watch for. Most of the human-made environment had the standard urban surface slickness and the standard urban disrepair and the standard urban disruptions of excavations and rebuilding. The menus offered the standard mixes of calories. The vending machines offered the standard fat-soaked crunchies. The coffee was the standard number of hours old.

The people seemed standard, too, reliably classified in their standard varieties. I was especially depressed by the variety I observed in the United Airlines' Red Carpet Rooms. There I saw a large proportion of neckties, blue or gray suits, and businesswoman dresses. I don't object to a blue or gray suit; I had one once myself. But I find it depressing to see so many in one room. A few people were dressed as if they were just back from safari. Maybe they were. Camels are not allowed in the Red Carpet Room.

Almost everyone was spreading papers on the refreshment tables and making check-marks, doing arithmetic, reading letters. The room was furnished with semi-private booths with telephones. They were all full of people busily talking into the phones, often loudly enough for me to hear--whether I wanted to or not. I thought it odd that almost all of the people talking on the phones seemed so ebullient.

I am glad to be back in Eugene, where nobody minds if I relax, even in public.

* * * * *

I looked in the yellow pages of the Portland telephone book and found three possible places for the next conference of the CSG. They will send me descriptions, and I will forward them to you. So far, I have received the description only from Alton Collins. It is enclosed. I have been there. It is a beautiful place, with excellent services. They will pick up people at the airport for an extra charge, as you will see from the leaflet. Unfortunately, Friday is their busiest day and night, and their Fridays are booked for the next two years. You can see from their note that next September is pretty full, too. Maybe they have four days in October starting on a Saturday.

I don't know anything about what our members think is a good time to meet. I must leave that to you.

I also called the Convention Bureau in Portland. They acted as if they knew of places not listed in the yellow pages. They will send information, they say.

Should I look for something near Seattle?

As I get information, I will forward it to you.

Sam Randlett send me a lot of instruction sheets on signaturs and fingerings. How nice of him. He and Greg W asked for copies of my MS. I sent them.

When I started to write this letter, my computer wouldn't come alive. I punched this and that with no success. I turned everything off and turned it on again. Everything worked fine. Machines of many kinds like to do mysterious things like that to me.

Love,

Phil

610 Kingswood Avenue
Eugene OR 97405
18 November 1988

Dear Bill and Mary:

You charged me with writing to some science writers to persuade them to write about experiments on control. I have not succeeded in doing as much as I had intended. I looked at my bookshelves, and all I could find was Richard Attenborough. I looked in my recent copies of Smithsonian and Scientific American (I don't keep back copies) and found nobody who seemed suitable. Some writers for Science News might be suitable, but I don't know whether they are independent writers. Should I write to the editor and ask? Should I go to the library and look up some agents and ask whether they have writers who might be interested?

Richard Attenborough would actually not be very suitable. I don't think he does short writing for a living; he produces science shows for the BBC. His home base is England, and he probably wouldn't want to cross the Atlantic to sit before your computer.

Anyway, I did draft a letter. I wrote as if I were writing to Attenborough. Please tell me whether it seems to be the kind of thing you had in mind. And give me any advice you wish about finding some writers.

Yours,

Phil

Phil Runkel

610 Kingswood Avenue
Eugene OR 97405
20 November 1988

Dear friend Carol :

Once again, you have picked out some writing that I can understand and that I am glad to have. (What do I mean by "understand"? I cannot know whether my thoughts while reading are like the authors' thoughts as they wrote. I can mean only that my thoughts while reading weave themselves together in a way that feels good to me.) Bogen and Woodward make a distinction (data versus phenomenon) that I could not have stated before I read their article and that I am glad to be able to state now. It fits in with a good deal of what I have thought about and makes it easier to think about.

I liked especially their arguments on page 336 where they used parapsychology as an illustration.

That is not to say that I think they have THE ANSWER.

So that you won't have to go to your files to refresh your memory, I'll quote here a few key sentences from their last few pages.

> We can think of the traditional picture--according to which science explains facts about what we observe--as motivated by two considerations. The first is the unexceptionable idea that (1) we should have good grounds for believing that those explananda which we require a theory to 4explain are (roughly) true. The second is the idea . . . that (2) perception and sense-experience . . . have an epistemologically priveledged status regarding the justification of beliefs about the natural world and that the most secure and convincing grounds for belief that something is the case is that one perceives it to be the case. . . .
>
> Our view is that while claim (1) is indeed correct, claim (2) is fundamentally misguided, at least when it is understood as the claim that we lack secure grounds for belief in the existence of entities we cannot perceive. For the most part, phenomena cannot be perceived and, in many cases, the justification of claims about the existence of phenomena does not turn, to any great extent, on facts about the operation of the human perceptual system. Nonetheless, we are justified in believing claims about phenomena as long as data are available which constitute reliable evidence [where reliability means reproducibility]. Thus, the proper strategy for philosophers . . . is . . . simply to focus on the relevant bubble chamber data and the complex

> considerations (having to do with correcting for the neutron background and so forth) which were relevant to establishing that the data were reliable.
>
> It is overly optimistic and biologically unrealistic to think that our senses and instruments are so finely attuned to nature that they must be capable of registering in a relatively transparent and noiseless way all phenomena of scientific interest. . . .

But I wonder whether, in distinguishing data from phenomena, their argument could be phrased equally well as distinguishing lower levels of the scientist's perception from higher levels. They seem to be making an implicit distinction between "perception" and "inference." In control theory, those processes are not sharply distinguished, just as "sensation" and "perception" are not either. Instead, making sense out of incoming energy patterns is done in a hierarchy of processes, every level making use of the interpretations from the levels below. Every level gets "perceptions" from levels below, the lowest level getting its perceptions from the sense organs.

Even in the simplest, most "direct" observations, the "data" are not very "direct." When you watch a ball rolling down an inclined track, I suppose a single datum would be found in the excitation pattern on the retina at one fleeting moment. But we interpret the succession of the changing patterns as motion--as the "transition," as Powers would call it--of the image of the ball at the top of the incline to its image at the bottom.

So I am wondering if the distinction made by Bogen and Woodward is more a matter of degree than of two classes of concepts. And I am wondering whether they are guilty of reification.

In a typical tracking experiment in control theory, a target cursor on the computer screen moves, and the subject operates a key or a handle to cause another cursor to chase the target cursor in some pattern--perhaps to maintain a constant distance from the target cursor no matter how the target cursor may move. Control theory says that everybody is capable of doing a thing like that and can do it with very great precision (not just better than chance). (Also at much higher levels. If you take samples over scientists and over moments, you will find scientists maintaining the "perceptions" we call theories with much greater constancy than chance.)

I suppose B&W would call data the traces on the screen of the target cursor and the subject's cursor--the changes in position of those cursors. Those sequential positions are not the "phenomenon" the existence of which the investigator wants to test. Rather, the investigator wants to test the constancy of the distance between the two cursors: their relation. The eye does not register the distance directly. That relation must be "perceived" higher in the neural net.

It doesn't matter in the slightest, in testing the theory, how the connection gets from the subject's hand to the cursor, nor how the connection goes from the experimenter's design to the cursors on the screen--as long as those connects actually connect and are not fraudulently described. I think B&W would say that, too.

So much for my thoughts about B&W's distinction.

I was interested, too, in the comments of B&W about error. They interpret error in the way physical scientists and engineers do, not in the more extended way that social scientists do.

There was an article in American Psychologist a few months back in which the authors sought to show that physicists use statistical inference to justify their findings in the same way psychologists do. He displayed some tables showing how physicists examined assessments of physical constants and used probable errors to throw out the measurements likely to be most in error. Doing that uses the same logic, he claimed, as using meta-analysis to throw out psychological experiments most likely to be doing the wrong thing with the null hypothesis. He didn't seem to be able to see that ascertaining the accuracy of a measurement has nothing to do with whether the measurement supports a theory. Do philosophers know the difference?

A thing that struck me about B&W and that has struck me about every piece of writing by philosophers that you have sent is that the authors make no note of the fundamental difference in entropy between living and non-living things. B&W use neutrons and humans as illustrations as if they were both the same natural kinds. That is, I would expect them to say that not only are gold and iron the same sort of natural kinds and not only are humans and giraffes the same sort, but all four are the same sort. I welcomed what you sent earlier about natural kinds, because that idea seemed to me at least to permit philosophers (and the rest of us) to distinguish living from non-living things.

I see from your scribbles on the copy that you, too, were annoyed by B&W's use of "data" as a collective noun--their writing "data is" instead of "data are."

Love,

November 21, 1988

Dear Phil,

I've been letting Mary handle the correspondence with you for about 6 months. Ridiculous. But so is the state I've got myself into. You've seen the Chapters, or at least those you could print (replacements without troublemaking graphics under separate cover -- your printer isn't Epson-compatible). Where am I going? What do I want to say? How can I write a book without just writing the same things I said in 1973? Do I really want to write another book? Do I want to write *anything*? All I'm doing is making one start after another, without any clear vision of what the whole book is going to be or who will have dunnit in the end.

I should have started out by saying how much pleasure I have in thinking back over actually meeting you, watching you work as you loosened up with the group, talking back and forth with you instead of in monologues. You are exactly the person I thought you were. It was doubly pleasureful for me to see you making contact with my other friends at the meeting. I never got around to asking you what you think of the organizational design of the Control System Group. I'd like to know.

Don't take my complaints too seriously. Things will even out, I'm sure. If I can ignore the protests (not *another* five chapters of a book!), maybe I will finally find the attitude I'm looking for. There will only be false starts until I find it.

The letter to "Attenboro" looks like a good way to go about it. Mary is going to send you a list of possibles.

In some piece of writing, I commented that there are actually people who think that invented realities and imagined models are *more* real than simple silent experience. In your letter to Carol you quote what seems a direct example, in Bogen and Woodward. "For the most part, phenomena cannot be perceived ..." -- what an extraordinary statement! They are redefining "phenomenon" to mean "what we imagine or deduce to be the case" as opposed to "what we observe." This usage, I think, defines what is wrong with intellectuals.

It may be that the difficulty lies, as I think you suspect, in their pejorative term "epistemologically privileged status." They talk about beliefs and explananda, justification of beliefs about the natural world, belief that something is the case, claims about existence, evidence, and "phenomena of scientific interest." All these terms speak to me of a person so busy *talking* about experiences that the experiences themselves are just a springboard from which one can reach higher levels of verbal abstraction. One bounce and we're done with *that* (scented handkerchief brushing away the traces).

The concept of levels of perception is probably, as you say, one factor that is missing. But I have always suspected that before the lower levels can even be seen *as perceptions*, it's necessary to get out from under words, language, reasoning, deduction, all that ponderous machinery of thinking. I think we have to become aware of the way we push our patterns of thought toward preselected conclusions, slipping cleverly from one meaning of a word to a different meaning, skipping blithely over holes, switching the train of thought around difficult spots as much as following it to its foreordained destination. Only then can we see that models and other kinds of explanations are no more then plausible imaginings, some more plausible than others. When plausible imaginings are carefully constructed, and when they are tested against nonverbal experience as frequently as possible, they can become powerful tools: *viz*, physics, at least prior to quantum mechanics. Well, I suppose even after, although I'm reluctant. When imaginings lose their anchors in experience, they turn into intellectual games and we lose the ability to choose the best imaginings.

My impression of meta-analysis is that it represents the next stage in abandoning all hope of understanding behavior.

Now it's Thanksgiving Day, and as usual the date on this letter is out of date. So --

Thanks for coming to the meeting
Thanks for being a friend.
Just, you know, thanks.

Bill

26 November 88

Dear Bill:

I made a mistake in one of the comments I wrote in your MS.

You reported a correlation in one experiment in the .7s, and then a correlation between the same two variables, in a later experiment that permitted more variation in one variable, in the .9s.

I wrote an explanation in my own words.

I now realize that I have been thinking about that wrongly all these years.

Statisticians and psychologists call that effect a lowered correlation because of "restriction of range." But that is not what does it. It results from restriction of significant figures.

For example, suppose you have two variables, and each can range from zero to 100. First, suppose your actual data, too, range from zero to 100. You could get a very high correlation if the data fall right, even 1.00. Second, suppose one of the variables ranges only from 57.00 to 58.00. But suppose you have measured that second variable by the hundredths of a unit, so that the data fall 57.00, 57.01, 57.02, etc. Obviously, you can get the same perfect correlation--from the points (0, 57.00), (1,57.01), (2, 57.02), (3, 57.03), etc.

In your case, your computer screen was allowed only a certain number of points. Its space was not continuous. The variables could take on only xx the values that you allowed those points on the screen to stand for. So you couldn't (or didn't) record the equivalent of 57.00, 57.01, etc. So for a single value on one variable, you had several different values on the other variable, and that brought the correlation down.

I have never encountered a psychologist or a statistician who knew about significant figures. I learned about them not in college, but when I was an engineering draftsman.

Phil

December 1, 1988

Dear Phil,

Short note. I'm enjoying the usage book by you and Margaret. It's full of nice funnies, and some (gulp) surprises. I have to disagree with you on a couple of points, just for fun:

Executive summaries. We put such summaries on our reports to management at the Sun-Times. The executives interpret the subtitles to mean "summaries for important people." We interpret them to mean "dumbed-down summaries that even an executive can understand."

Mary says, "Now, now, Bill, you mean reports on technical subjects written to be understandable by non-technical executives." Maybe so. I notice that in technical journals like InfoWorld (how do you like XYWrite's hyphenation?) there are executive summaries after reviews of a large array of computers, printers, and so on, written clearly in plain English. I interpret "executive summary" in that case to mean "summaries for the kind of people who ask you to come to their offices to read the instruction book to them."

Models. The main meaning may once have been a miniature version of the real thing. In engineering and computing there is a new meaning, which you glean from my usage. A model is a representation of a real thing, that reproduces the major internal relationships and behavior of the thing without necessarily looking anything like the original. (Howzat comma?) Hence, mathematical model.

Positive and negative feedback. You already knew about this! Smart authors.

Today's note: there is another possible explanation for the increase in correlation as the variable varies more (although yours is pretty good). When a control system is controlling behavior, it is counteracting a disturbance (in all my experiments). Suppose that the handle movements are made of one component that is very systematically related to the disturbance, and a second truly random component with a small amplitude (hand tremor, etc.). The handle movement will correlate very highly (and negatively) with the disturbance, because the small random component is negligible compared to the systematic component.

When we look at the cursor movement, however, we see the result of subtracting the large systematic component of handle movement from an equally large disturbance of essentially the same form. The cursor movement is what is left over. The handle tremors appear full-sized in the cursor movements because there is nothing in the disturbance pattern for them to cancel. Now that small random component is much larger in comparison with the total cursor movement, even though the tremor is of the same size. That is why the correlation of cursor position against anything else tends to zero for simple compensatory tracking.

In pursuit tracking, the cursor is made to move along with a target, so it now has a large systematic component of motion. The same random handle movements appear with the same size as always, but now add to an excursion which is comparatively large. Therefore the correlation between the model's cursor (which has no tremor) and the subject's goes up.

For the same reason, the correlation goes up even in compensatory tracking when the disturbance is made more difficult. The subject -- excuse me, I'm trying to substitute "participant" -- is then making larger errors because of a systematic lag in correcting the error. The model recreates this systematic lag accurately. Now the truly uncorrelated small tremors are added to cursor movements that are larger because of a larger systematic error. Again the correlation of the real cursor behavior with that of the model goes up -- there's a bigger systematic component of cursor motion to predict, while the random component is the same size as always.

But your point about significant figures remains valid.

I suppose I really have to go and get a haircut now instead of pounding happily at the keyboard for the next hour.

Best,

Bill

February 18, 1989

Dear Phil,

After reading that review by Shaw of Weir's book (early one morning before going to work) I left Mary an anagram to solve: CUT RATE P.R.. A good clue might be "Psychologist speaks execrably." Unfortunately I left a worse clue and she didn't get it. Let's see if you can.

When I read stuff like this, I always make the mistake of thinking it must make sense, and I waste a lot of time trying to see how the author filled the holes left as the argument progresses -- the undefined terms, the assumptions never backed up with examples or evidence, the reasoning alluded to but never spelled out, and so on. In fact people like Weir aren't capable of developing a serious explanation of anything (and evidently Shaw, failing to recognize that, is another). These people aren't scientists, but myth-makers. If they can come up with a story that seems to move along and lead somewhere, they don't worry if they assume impossibilities or omit crucial parts of the explanation. They're exactly the type who make up stories of resurrection, never wondering whether the dirt and germs get resurrected too, or how the body can see Heaven with nothing but holes in a skull. If a little magic, or a lot, is needed to make the story work, throw it in -- after all, it's a work of fiction.

I don't see much difference between the kind of reasoning we see in articles like this and the kind that allows the ayatollah to scream "blasphemy" and utter death threats. Weir is defending the faith in no more rational a manner, the faith being that behavior must somehow be caused by external events, even if we have to let perception work backward through time to get the job done. It's just silliness. I'm sure that Weir would accuse me of blasphemy for suggesting that organisms control what happens to themselves. Unfortunately, there are almost as many gullible people willing to follow Weir's kind of story as to swallow any of the other kinds. And they all get nasty when you doubt their faith.

Shepard's article was a little better, but the same problem is there. In this case we have a Catholic priest who has met a beautiful enticer with whom he would like to dally, but his commitment to the faith prevents it. Gibson was another myth-maker who knew how he wanted the truth to come out and didn't let mere logic or facts get in his way. I loved the part about the earth's "level solidity" affording walking for humans, but its "frangibility" affording burrowing for moles and worms. I suppose its arability affords farming it, and its vendability affords buying it, and its appreciability affords admiring the view. Gibson never seemed to be able to follow a line of reasoning more than one step. What invariant of nature it is that affords human beings the perception of levelness or solidity?

Actually the brunt of Shepard's paper is a thorough refutation of Gibson's views cast as being inspired by them and expanding their validity. The priest is pointing out that God not only affords temptation to us, but also our response to it. The next step is to hop into bed. But no, they hardly ever do, do they? I suppose they usually catch themselves just in time, and give thanks that the Great Satan didn't catch them in his wily rational trap.

Shepard's article reminded me of the many psychologists who are perfectly willing to let perception happen with no relationship to the brain or nervous system. The metaphor of "resonances" is Shepard's way of bypassing the lower levels of perception, letting the higher levels somehow tremble to the Aeolian touch of reality without ever existing as crass neural impulses.

Telling stories is OK if you plan to check up on them somehow (unless they're meant just as entertainment, in which case you wouldn't check up on them). Gibson's story is that the real reality is really there and our brains simply pick it up. Fine, good, OK. Now how are you going to find out if that is true? To see if that is a true statement, you would have to have some way of checking to see if the "optic array" gives us a picture of reality that is just like the actual reality. That means you need a way to know about reality that doesn't depend on your own or anyone else's optic array. Gibson put a lot of store in "tangibility", as if touch weren't a sense. If you can touch it, it's real. But how do you know you're touching the same thing you're looking at? You don't. Gibson doesn't. Nobody does. We just assume, and try to make our senses cohere in terms of each other. There aren't any other terms.

Once in a while I wake up and look at all these solemn people posturing and pronouncing and making up their tales, and I think, "Why, you're nothing but a pack of cards!"

Love,

Bill

February 23, 1989

Dear Dr. Shepard,

Phil Runkel has been urging your work on me in the past couple of weeks -- thank goodness for friends who insist on trying to overcome my stubborn lack of scholarliness. His first attempt consisted of your 1984 paper on Ecological Constraints (1984); my reaction was that you had refuted Gibson but were having trouble overcoming your former belief in his work. Then Phil sent me two more papers, your Psychological Relations paper (1981) and your Psychophysical Complementarity chapter (1981), and I stopped quibbling. Both papers filled me with involuntary admiration, not only for the ideas but for your clarity of expression. It is, unfortunately, refreshing to read the work of a psychologist who is a master of language instead of its slave.

An aside concerning your paper on Stevens' Power Law. I read Stevens' 1961 Science article, and shortly thereafter wrote a letter to the editor in which I pointed out that there was very little difference between Fechner's law and Stevens', give or take a logarithmic transformation. The letter came back (unpublished) with a copy of Stevens' comment on it: balderdash, the power law has nothing to do with logarithms, I should read his article more carefully and learn some elementary mathematics. I was furious. Of course I had no inkling that this criticism could be expanded as you did, to derive a fundamental statement about the relationship of perception to the external world. What a beautiful piece of work. When I read that paper I began to take you seriously.

I was not, however, prepared for the "complementarity" paper. You have hit on what I have been looking for, a communicable attitude toward modeling the brain in connection with models of the external world. For many years I've understood that perception is a phenomenon that takes place in a brain, and that the world we experience is only what a brain can extract from the assumed external world with a somewhat limited complement of types of sensors. So I've always been a constructivist. But my brand of modeling also cheerfully assumes the laws of physics and the principles of engineering even though I know they are constructions; most of my experiments look as "objective" as any behaviorist could wish (although my conclusions are not acceptable in that camp, and behaviorists completely miss the point of my experiments). In your brilliant formulation of "equal epistemological status" you have resolved the nagging inner conflict I've felt concerning my methods, and at the same time have shown me how to make clear to critics (and to myself) what I am trying to do.

You've already seen my 1973 book, Behavior, the control of perception. I wish you would go back and read it again. I'm sending you a reprint of a chapter from a book published by Ozer in 197 -- uh -- 9, which elaborates on my conjectures (as of that year) about perception in a more thorough, or at least more wordy, way. I hope you will focus on the general proposition concerning a hierarchy of perceptions -- not because my definitions of levels are right, but because I would like to have your serious criticism of this proposal, in general and in detail. The levels I define have been derived from a very long critical examination of experience but not from the kind of dogged systematic experimentation at which you excel. I'm fully prepared to give up that structure in the face of any evidence against it or (more to my liking) see it modified to fit facts found through a more orderly and public approach.

Here are some other general ideas you might think worth commenting on:

A case can be made for saying that all perceptions are one-dimensional -- that is, that they can change only in the direction of "more" and "less." The <u>nature</u> of a perception, or its <u>kind</u>, is given by the neural computations that derive it from incoming information from the environment or from lower-level processes; that relationship can be as complex as you please. But the perception itself, the signal that arises from one specific neural information processor, can vary only in amount. In other words, a neural processor, once organized, is a single-purpose device; it does not emit different signals to mean different things, but only one signal that always means the same thing. Different kinds of perceptions arise from functionally distinct (although perhaps interacting) processors. This idea, which I'm sure has Selfridge's "pandemonium" model in its ancestry, leads to seeing perceptions as being composed of attributes, each of which is representable as a single neural signal that can vary only in frequency to indicate the amount of one attribute. That's good, because that's the only way neural signals can change.

This picture is completely unsatisfactory if we think only in terms of one level of perception. That worried me for a long time, because my ambition is to construct not just a workable model, but a believable one -- one that fits direct experience, not just abstract theory. An object made of color, shade, edge, curvature, texture, and so on is not just a collection of these individual attributes; it has a quality in experience that is quite different from the qualities of the attributes (I, too, admired the Gestalt theorists for a time). It has location, shape, orientation, relationship to the background, and so forth. But -- aha! -- those things are <u>also</u> attributes; they or something like them could be expressed as one-dimensional variables, too. The trick here is not to think generally, but to consider specific cases only. There are lots of things you can <u>say</u> about "objects," using words, but when you <u>experience</u> a particular object and put the words aside, by golly you can find the attributes of objectness that vary, each one only in one dimension: <u>more</u> or <u>less</u>. As you squash a clay sphere into a clay pancake, the sphereness smoothly decreases and the pancakeness smoothly grows. One analogue attribute gradually shrinks in magnitude and another increases, until finally we are forced to reclassify the set of perceptions that now exists (leading to a new experience of class membership and a new label). That is where the phenomenon of hysteresis comes in -- not in the analogue perception but in the digital classification. The classification may be a digital process, but it works by placing arbitrary boundaries in a continuum of change.

These ideas have a relationship to the concept of "invariants." If you think of a perceptual signal as being a function of several lower-level variables, there will be ways in which those variables can change that leave the signal unchanged; that concept defines an invariant, which must always be defined in terms of an operation orthogonal to a function. The outline of an irregular object is invariant with respect to translation, but not with respect to rotation about a vertical axis. So _any_ function of multiple variables creates an invariant. This is a little like your discovery of the g-function's role in the "power law." Gibson thought and many others evidently think that perceptual invariants are things that don't vary. Not so. The value of a function of many variables will vary under any operation that is not the invariance-defining operation, determined in turn by the nature of the transforming (perceptual) function. Therefore the invariants we experience are created by the way we organize our perceptual functions; however we organize them we will experience invariants. Presumably, we end up with an organization in which the invariants can persist for more than a few seconds. Invariances define variables.

I can agree with your acceptance of Gibson's idea of affordances, but only by interpreting it in a way that I think Gibson would not have accepted. If we assume a real universe in which regularities exist, then any set of constructed perceptions that begins with raw sensory stimulation will contain regularities that reflect _but do not duplicate_ the external regularities. We can discover regularities, but only in terms of the way the world is represented in a nervous system containing a given set of perceptual functions. The "g-function" can have _any_ form, so we can't reason backward from experience to the true nature of reality. Well, maybe not _any_ form, but there's certainly no reason to think that affordances are things like "level solidity" or "frangibility," which themselves are human perceptions afforded, presumably, by some deeper structure in reality that we can't know directly. I don't think that Gibson ever got completely free of naive realism. If that's the term I want. Of course, who has?

Naturally, the concepts of sensory stimulation, a nervous system, a brain, and so forth are already constructions; we don't get any closer to objectivity by using them to explain other experiences. But if we're careful in using such ideas, if we make systematic models built on them, we can achieve at least consistency between the models of physics and models of the brain -- giving them, as you say, equal epistemological status (I'm still reverberating to that illuminating concept).

By the way, the concept of neural analogues implies something that ought to please you. Neural signal generators probably share one property with other physical phenomena: their outputs can't jump instantly from one state to another. If attributes of an object are represented by neural analogue signals, then presenting an object that really jumps instantly from one state to another will give rise to perceptual signals that must necessarily change from one magnitude to another by traversing all intervening magnitudes. Because neural signals consist of trains of discrete impulses, the "magnitude" of such a signal can be defined only over some short interval. The shorter the interval the noisier the signal will be, the worst case resulting from observing only for a couple of milliseconds. If we perceive anything as a continuous experience, the signal embodying that experience must be averaged over some period of time to be seen as continuous. The same averaging will impose smoothness on all signal transitions, even though the actual input changes "instantly" (very rapidly). So if a "position" attribute-signal changes from one magnitude to another, all intervening positions will be briefly "represented." Voila, the Phi phenomenon.

I must bring this tome to an end, but not before expressing an earnest hope that you and I can work together. There is very little I could contribute to your work with perception, but I think that if you were to become fluent in my brand of control theory you could put your work into a context that I believe you would find useful. Practical experiments with human control processes are necessarily linked to an understanding of the perceptual processes involved. I'm very good at understanding control processes and devising experiments to test my ideas, but I've always been limited in my knowledge of perception. For the most part I didn't see much going on in that field but blather, and so more or less ignored it (thus, as subsequent events proved, substantiating my prejudice). That idea has changed now; I think that your understanding is compatible with control theory where all the behavioristically-oriented stuff (while possibly of interest in revealing phenomena) is so philosophically different that no meeting of minds can happen. An understanding of control theory and the experimental methods that go with it will give you all sorts of new ideas about how to investigate perception. And what you discover (and already know) will certainly advance control theory toward being a realistic model of human nature.

Best regards,

Bill Powers
1138 Whitfield Rd.
Northbrook, IL 60062

610 Kingswood Avenue
Eugene OR 97405
17 June 1989

Dear Bill:

* Please tell me, so I can tell my brother, whether you wrote science fiction in the 50s.

As to people who take theories as "perspectives" on behavior, all with more or less equal claims on one's attention, sometimes those people actually believe that, but sometimes they may actually prefer one theory, even strongly, but remain faced with the necessity (so they believe) of fitting it into a course of lectures or into a contemplated book. Then it must be categorized and compared with other theories; that's called scholarship. Few are as bold as Robertson, willing simply to throw away all those other theories they have been promulgating all these past years. Not to speak of having to brave the criticism they would get from colleagues for "going overboard." As to the pains of change, I am reminded of a professor at the University of Illinois at a time when the University Senate was discussing whether to convert from two semesters and a summer to three equally-sized "trimesters." The new plan would have changed the time-period from 15 weeks to 14. The professor stood in outrage and cried, "Do you realize that I'd have to rewrite my lectures?"

I'm not making any guesses about Robert Lord. I was merely reminded of the hazards of the profession.

** I am eager to see the new ARMDEMO. My computer at home is a clone of the IBM-XT. I know it has no graphics card, because I bought it that way intentionally. But I told you I had run the program on a computer on campus. I even told you the brand-name, though I have now forgotten it. It is slow, but does well enough. I'm going to run the program again soon to learn more about it. I might be able to use it in teaching.

How do you explain to the uninitiated that your program is more marvelous than Pac-Man? In computer games, the operator chases after a target the program moves. In your program, the program chases after a target the operator moves. But how should the uninitiated marvel at that?

I like your new paper. Glad to be chosen as a critic. I have no complaint about length, organization or balance of topics, or general tone.

Send it first to <u>American Psychologist</u>. Some time ago, I sent you a list or two of journals I thought might be receptive. Some of them were new. You might, by the way, forward those lists to Bourbon; he is also looking for likely journals.

* The answer is yes. See page 177.

**ARMDEMO and several other programs for DOS and Windows are available free at www.livingcontrolsystems.com. *Living Control Systems III: The Fact of Control*; by William T. Powers (2008), ISBN 0964712180, features updated, more interactive versions of these simulations for Windows interwoven with an explanation of PCT.

I do have a flock of small comments on your paper. I have put red numbers in circles on the enclosed copy. Comments below are numbered correspondingly.

1. At this point, some authors would insert a few dozen citations. Maybe some other persons to whom you have sent the paper will say something about citations at this point. I do not claim that it is vital to stick in citations here, but if you choose to do so, you might do it this way: "There are too many to list here, but examples are ...." Maybe cite one each from Bourbon, Marken, Rijt-Plooij, Robertson, and the Smiths--maybe two from the Smiths. *

2. "There are even more, many more, of this sort. A few arbitrary examples are ...."

Anderson, James J. (1985). A theory for attitude and behavior applied to an election survey. Behavioral Science, 30(4), 219-229. Copy enclosed.

Fischer, Kurt W. (1980). A theory of cognitive development: The control and construction of hierarchies of skills. Psychological Review, 87(6), 477-531. Copy enclosed.

Vallacher, Robin R. and Daniel M. Wegner (1987). What do people think they are doing? Action identification and human behavior. Psychological Review, 94(1), 3-15. Copy enclosed.

"An early example of compatible theory in a systematic text was"

Krech, David and Richard S. Crutchfield (1948). Theory and problems of social psychology. New York: McGraw-Hill. Some words of mine about the book, with quotations from it, enclosed.

You may notice that I have picked out very respectable journals and book. You probably can't see it from Northbrook, but I'm blushing.

I enclose also a paper by Heise, which seems to belong in neither category. He names control theory, but doesn't understand the need for at least two equations. He falls back on inferential statistics. I see that I inadvertently copied the first page with my sticker of notes still stuck to it. Excuse it, please. I think I already sent you the article by Klein on work motivation--an article you would want to cite only as an example of what not to do.

3. Some readers will take the unqualified "organization" to mean General Motors. How about "internal organization"?

4. How nice to see here the proper use of the singular "system" after "kinds of."

* The Smiths, i.e. Thomas J. Smith and Karl U. Smith contributed all of Continuing the Conversation, Issue 15. All issues (#1, Spring 1985 through #24, Spring 1991) have been recreated and are posted at the website.

5. "... observation of a behaving system and from interaction with it." I hate the dangling preposition. And I would write "Each of those features ..." since the demonstrative adjective points backward at what you have finished doing (listing). And "When those conditions..." and "In that list...." But suit yourself.

6. I would italicize, but suit yourself.

7. "... to particular aspects, features, or parts of the organization of a living creature."

8. "... psychology, where, although the same causal chain is assumed, ..."

9. Model? This does not seem to me the meaning of the term you prefer to reserve for it.

10. Here it is customary to cite: "B. F. Skinner's (1492, p. 93, for example)...." I do not have any reference at hand for Skinner. Dewey's article on the reflex arc was, I think, in the Psychological Review in 1893. I seem to remember a paper in which you quoted from it.

11. "I joined the field ..." and "... of those groups."

12. I know I am here and there being more picayune than necessary. Be that as it may, I would write, "Control theory went beyond Skinner, first, by ... loop and, second, by ... system; namely, the reference signal."

13. Citation for Minski.

14. "... and I have pictured the progression here as more orderly than scientific progress ever is."

15. Well, those who have espoused such a view often say that they deserve the appellation "scientific" just as much as those other guys. The articles and rebuttals often sound like: "We are the real scientists!" "You are not! We are!"

16. The reader who is not scrutinizing every word may not note the distinction between "effect" and "action." How about: "the muscular actions that produce the effect must also...."? And citation with page number for James. I seem to remember you did that, too, in another paper.

17. Some people like to call this "equifinality."

18. "... about all the physical forces in the environment that will...."

19. "Control systems unravel...."

20. "... control proceeds cyclically, in a progression of events in sequence round and round...."

21. Cite the piece in which he said he didn't mean it that way, with page number.

22. I verified your algebra, in case you care. Should you give, in addition, the solution that begins with r = b/k + d as one of the specifications? I don't know. Maybe most readers who get this far will actually want to do it themselves, and most who wouldn't want to will have given up a couple of pages back.

23. Maybe quotation marks around "reinforcer"? But maybe it would confuse some readers. And maybe it would add unnecessary complexity to take space to translate reinforcement into perceptual input.

24. Whoops. Every proposed theory gives a new (to some extent) picture and new terms. Putting it the way you have it here encourages the reader to yawn. "Ho hum, another author with a bee in his bonnet." And (next paragraph) many proposed theories claim to predict effects others don't.

But I don't know what to offer instead in this context. The virtues of control theory that most impress me don't fit onto your pages 10 and 11 very well. What I admire most about control theory (as of 8:17 p.m. on 17 June 1989) is (1) its precision: correlations within a few points of 1.0 and 0.0, as appropriate (in the method of relative frequencies, that's called validity. I never did understand the concept of validity--though I pretended to myself and others that I did--and now I know why), (2) its generality: it works with that precision with every living creature properly tested--so far, anyway, (3) its ability to be modeled, (4) its underlying simplicity, nicely expressed by your statement in the documentation to ARMDEMO that the modeling required only 40 lines of the program, (5) the way it clears up previously fuzzy concepts such as learning, action, perception, memory, "relevant" variables, etc., (6) its fit with common-sense observation: no action when things are the way we want them, higher standards overriding lower ones, the reorganization attending insight and revaluing, the ease and smoothness of most action, etc., (7) its implication to give up, in general, trying to predict particular acts--another thing observable by common sense--as well as its specification of the stringent conditions under which particular acts can be predicted, and (8) the way it clears up--for me, anyway--the anomalies and undesirable assumptions in traditional research method.

But only the first two of those virtues seem appropriate to your pages 10 and 11, and to explain them would interrupt too much the flow of your exposition. (You do touch on (1) on page 13.) Sorry I can't propose some substitute sentences.

25. I haven't followed the current literature on reinforcement--though I sent you an article from the Amer Psychol by a reinforcement theorist who was complaining about people who had not read the recent literature. So I don't know how "usual" the description is. Ed Walker noted the rise and fall of activity under continued "reinforcement" in an article in 1964. Actually, I first heard about it when I sat in his class about 1952.

Walker, Edward L. (1964). Psychological complexity as a basis for a theory of motivation and choice. Pages 47-94 in David Levine (Ed.), Nebraska symposium on motivation, 1964. Lincoln: University of Nebraska Press. Here is what he said on page 85 after describing some experiments yielding learning curves:

> For the sake of the argument I am certain to get, let me take the position that the appropriate "learning" curve shape in runway studies, conditioning studies, and selective learning studies is one that rises and falls to zero or to a steady level below the maximum performance. The curve that rises to a steady maximum and remains there indefinitely is likely to be rare. The reason that we see few "learning" curves of the postulated type is that most experimenters know in advance what a learning curve is supposed to look like. As a result of this knowledge, they stop training when the "asymptote" is reached, or, if they obtain a curve which does not fit their conception of what one should look like, they find a great many other ways to respond to the illegitimate child other than to publish their sin against respectability. They throw away the data. They restructure the apparatus. They change the parameters of the study. They change the design. This process is known as the establishment of experimental(er) control. Sooner or later they manage a situation in which they obtain the "right" answer. I can attest that this process is carried out in good faith and under the assumption that in so doing, one is behaving like a sound, rigorous, and careful experimentalist. I can attest to this because I am one of the sinners.

26. Don't leave this hanging! Give at least one sentence to an example. Or omit the paragraph. Or say it in a way that doesn't tempt me to call it "snide."

27. This paragraph may be impenetrable for readers accustomed to thinking of environmental events as stimuli without asking themselves what it is about the event that the subject may care about. If a child at a railroad station runs away when the engine thunders in, what variable is that event disturbing? Too much noise? Bad smell? A threat of being run over? Not wanting to meet a hated uncle who will get off the train? Maybe you could sharpen the idea that an environmental event can disturb

numerous variables, and the investigator must find out whether the subject is controlling any of them, and if so, which.

For example, I would phrase the line I have marked "x" something like "... correlated beyond chance with some environmental variable to which the experimenter would like to give attention is usually sufficient for the experimenter to call that variable a stimulus." And "... as if the causes of the disturbances (such as the arrival of the train)...." Or something like that. But it would be nice to use an example that would allow the last sentence of the paragraph to fit in neatly.

28. I don't know what kind of "terms" you are talking about. Foot-pounds?

29. Cite three or four examples: Powers, Marken, Bourbon, etc.

30. "Even in the 'higher' realms of behavior, however, the ideal is not...." (To make the connection with the earlier mention of tracking experiments and the distinction from them.)

31. Uninterpretable.

32. "... investigation in which little balls rolled down...." I hope most readers will know about Galileo; I am often surprised at the varieties of ignorance I encounter.

You will remember that in my "Casting Nets and Testing Specimens" I compared Robertson's experiment with that of Frey and Stahlberg. (Frey and Stahlberg may not be mentioned in the version you have. After I sent them a copy for criticism, they asked me to identify them.) Frey and Stahlberg's paper was published. They used many more assumptions, many more subjects, many more days, much more work on the part of the subjects, and so on, and got much poorer results than Robertson. Apparently the editors and reviewers of Frey and Stahlberg did not think the hypothesis was trivial. I think your imputation is correct. Frey and Stahlberg must have been working at something important, because look at how hard they worked and the mess they left behind! Robertson and colleagues couldn't have been doing something important, because look how easy it was. Same reasoning, I guess, by which you claim to be doing something wonderful for students when you make life hard for them.

33. Surely you don't mean "putative" here? Maybe "intended" or "aspiring"?

34. Some reviewers and editors will interpret this as hubris. How about something like: "I hope future generations can learn control theory first. Then they can learn about the old way...."

35. To many psychologists, "apply" means to use theory to make choices in practical affairs such as how to teach school or how to write a law. How about "When using control theory in research, it is a mistake to turn first to the higher levels of the individual's neural organization."?

36. "... the use of control theory in studying higher levels...."

37. "... and elaborated by Powers, 1973)"

38. Is it too sweeping to say that cognitive psychology studies mostly the shapes of output functions? Or hunts for them?

39. "... with no instruction except its reference signal"?

40. Which Williams? I want a copy.

41. Unlike a blueprint, a goal often changes as the person learns, during the pursuit of it, about goals that will satisfy even better a higher-order standard. That's something I have repeatedly to teach people who work with groups to solve problems in organizations.

42. "These are merely convenient labels...."

43. Different from what?

44. If control theorists have found themselves with ideas about consciousness while thinking in the terms of control theory, then there must have been some "relating" there somewhere. How about something like "Though control theory has so far made no specifications about consciousness...."?

45. "... most control systems within an individual...."

46. Psychotherapists and others--novelists and actors on the stage, no doubt. I enclose an advertisement for a book by E. J. Langer that no doubt illustrates your point.

Consultants in "group process" must develop the ability to "stand back" and watch their own actions. I think maybe this "standing back" or leaping to a higher level can take one to only one level at a time. I'm not sure I can actually act and watch at the very same time. It is like trying to watch the muscles steer the bicycle. My fingers hit the keys on the piano more surely when I am listening to the sounds than when I am actually watching the fingers. When I am working with a group, I think I actually watch what I have just done or said, not what I am doing in the present moment. If my mind, while I am talking, suddenly steps back to hear the "higher" implications (the more usual term is "deeper") of what I am saying, my mouth stops, and I must start again. At least that's how it has stuck in my memory.

Maybe one can flip-flop back and forth among lower levels very rapidly, even in split seconds, but in leaping among higher levels, maybe one must stay at one level longer before taking the next leap. Something like finding one's bearings or realizing where one is, perhaps.

I do not mean any of those comments as a revision of your paragraph.

47. Could you here refer to a couple of your own writings for elaboration?

48. Sometimes you seem to get mixed up between colons and semicolons. A century or two ago, the colon was called a "full stop." Nowadays, however, the semicolon does that. It replaces a period and capital letter, but ties the two sentences more closely together than would a period and capital letter. The colon serves as a pointer or a replacement for "namely."

49. So why does providing a worker with prompt feedback about the effectiveness of the worker's work often fail to produce the desired result? Because the "desired result" is the <u>boss's</u> internal standard, not necessarily the worker's--a point many researchers seem to miss (not all do). This sort of thing is implicit in the opening sentence of your next paragraph.

50. Here, too, change the flavor somehow, or omit. Reminds me of the claim of the Freudians that people who don't believe Freudian theory are prevented from doing so by their neuroses. Sounds too much like: if you're against control theory, you are a hopeless diplodocus and only a fossil at that. That may be true, but nobody likes to be called a diplodocus. Instead of dire warnings about becoming obsolete, how about an inspiring invitation to future glories? Here is an invitation, though not very rabble-rousing, that I wrote in the latest version of my methods book:

> To researchers in social science who seek zealously the human nature, I propose that it is time to turn to methods--and to the corresponding theories--that can enable us to predict correctly at least 98 percent of the time. It is time, however, to give up trying to predict particular outwardly observable acts and to study instead the perceptual consequences of acts, unpredictable though the acts themselves may be. It is time to cease substituting what one subject did in an experiment for evidence of what another subject did. It is time to stop relying as heavily as we do on what subjects tell us and what we tell the subjects. It is time to give up the assumption of linear input-output causation and adopt instead the assumption of circular causation in feedback

loops. It is time to investigate the control of perception.

But I'm sure you can do better than that. Anyway, that was oriented toward method, not toward content theory.

Thanks again for sending me the paper. I always look forward to your writings as bringing me both bread and cake.

Your brother,

Phil

24 June 89

Bill:

* In my last letter (criticism of your paper), I wrote out some things I admire about your theory. I should have kept a copy so that I don't have to start from scratch next time I want to do that. But I can find a copy neither in my files nor in my computer. If you have not yet thrown that letter away, please send me those pages; I'll return them to you if you want. Thanks.

Phil

* It seems Phil found his copy after all. June 17, 1989 from Phil features the criticism of Bill's paper.

July 4, 1989

Dear Phil,

Your little note got here too late. We had our monthly Polish cleaning ladies in, and just before that I did a massive putting-away project. In the course of all that, not only your notes but the marked-up copy of my paper disappeared. Fortunately, I had already been through both and made all the changes in the original in my computer. But unless there's a miracle, your list of things you like about control theory will have to be thought up all over. (Speaking of ending sentences with prepositions, have you heard "Mommy, what did you bring that book I didn't want to be read to out of up for?").

I do remember your reiteration of a question about Lord, which I seem to forget every time I write. Yes, it is That Lord. I have sent him a copy of the mss. for comment. Haven't heard back yet. He may be a possible, with a little education.

My article for System Dynamics Review (invited through Bill Williams' contacts) was accepted and will appear in the next issue. Not only was it accepted, but George Richardson, the editor I dealt with, sent it back all edited and retyped (significantly improved) "to save you the trouble." Richardson is a <u>definite</u> possible. He treats me like a revered ancestor. He's being considered for a tenured professorship in Public Administration, which he plans to approach from the control-theoretic standpoint. I've forgotton which University -- SUNY Albany, I think. When I get back from vacation (July 6 - 22) I will receive a package from the search committee and will do something nice for George back.

I'm working on a Forword for the "Selections" book. I have to tell Greg that I don't like the heading on the Bibliography: Publications, William T. Powers, 1957-1988. What a shame, only 31 years old, too.

See you at the meeting.

Best,

Brudder Bill

610 Kingswood Avenue
Eugene OR 97405
20 July 1989

Dear Bill:

Thanks for looking for my letter. I am glad you made the alterations to your MS before you lost the marked-up copy. After receiving your letter, I looked harder for the one I had written to you about. I found it hiding in my computer in a little-used directory. In case you care, I enclose that letter's page 4, bearing my list of the virtues of CT. I also enclose that letter's page 7; please answer items 38 and 40.

Yes, I had heard about the book "read to out of up for." But I had forgotten how it goes, so I am glad to have you spell it out for me again.

No, you had not forgotten to tell me about Lord.

I'm glad you found a new suitable journal: System Dynamics Review. I've never heard of it. I sent Marken the name of a new journal some time ago; I don't know whether he passed it on to you:

Methodika, published by
C. J. Hogrefe
12-14 Bruce Park Avenue
Toronto Ontario M4P 2S3
Canada

I know you are not much interested in method, but to me (as you know) theory and method are inseparable. With theory (or metatheory) I include the low-down assumptions lots of researchers never think about, such as linear versus circular causation.

The editing and retyping the editor of SDR did for you is certainly a very rare courtesy. I have never heard of such a thing.

When you send your "Foreword" to Greg, I hope you will spell the word that way, not the way you spelled it in your letter. But don't blush. A man with whom I have often been co-author once spelled it "forward."

Greg's heading for the bibliography is hilarious. Reminds of me of George Washington. When he had a new suit tailored for him at the age of 62, his waist was only 36.

I've had another nibble from a publisher on my methods book. On 19 May 1988, I sent a query to University Press of America. On 10 June 88, they asked for the whole manuscript. I sent it to them on 20 June. On 3 August, they sent me a reviewer's comments and asked if I would revise the MS accordingly. On 9 August I wrote them to say that I did not wish to write the kind of book their reviewer thought I ought to write. On 30 January 1989, I wrote to them to say that I had shortened the book from 600 pages to 338 and asked if they wanted to see the new version. They wrote on 3 February to say to send it. I sent it on 13 February 1989.

Time went by and time went by. I had also sent the shortened MS to two other publishers who had told me the book should be shorter, one on 13 February and the other on 1 March. Didn't hear from any of those or from two further publishers whom I had asked whether they wanted to see the new MS. I began to think they had all colluded to act as if Runkel did not exist.

On 22 June, an editor from Rowman & Littlefield called on the telephone. He said that Univ Press of Amer was their "sister company," and that UPA had sent my MS to him, saying that they liked the MS, but it wasn't the kind of thing they could market, but maybe Rowman & Littlefield could do it. The editor at R & L said that he had concluded the same thing: he liked the MS, but he didn't think R & L could market it. He said, however, that he knew the psychology editor at Praeger, and would it be all right if he sent the MS to his friend at Praeger. I said please do. (Incidentally, I had sent a query to Praeger on 31 March 1988 and had got no answer.)

The psychology editor at Praeger wrote to me on 11 July. He said he thought the book "unique and interesting," and would I sent him information about etc. etc. Well, it was refreshing to hear somebody recognize the book as unique. Too many reviewers have tried to fit my MS into their traditional frames and concluded that everything they couldn't fit was ipso facto wrong. So I sent off the requested information yesterday.

Well, I've got this far with a couple of other publishers who wrote at last to say that the editorial committee just couldn't figure out who would want to buy my book, however marvelous it might be. But maybe this time it will get through. Let's hope.

I am steadily gathering momentum on my next book. I have collected and indexed in my computer a small mountain of notes and references. The mountain is about two-thirds the bulk it will eventually be. I have also written a draft of the first chapter. I enclose a copy. As you no doubt can guess, I wrote this first draft of the first chapter primarily to test whether my thoughts were sufficiently organized so that I could tell myself what I thought I was going to write about. I may eventually even discard the entirety of the draft and write a new one. Depends on how well I have predicted my own behavior, always a dubious undertaking.

I am not asking you for comment on the draft chapter. I send it merely as news about what I think I am working on.

It strikes me that you have your revolution pretty well organized. For active researchers, you have yourself and Marken and Bourbon and the Rijt-Plooijs. Maybe for semi-active you have Robertson. For interpreters, you have yourself and Robertson and William Williams and me (and no doubt several others, such as Whosit & Scheier, whom you would just as soon not have). For publishers, you have Greg W and your friend at System Dynamics Review. And no doubt I have overlooked a few. And I am sure those lists can only grow, not shrink.

Love to Mary also.

Phil R

Phil

Phil attached two marked-up pages from his letter of June 17, 1989, focusing on the long paragraph on page 4 and items 38 and 40 on page 7. The selections are shown as fragments here for identification only.

~~theories claim to pre~~

But I don't kr
The virtues of control
your pages 10 and 11 v
theory (as of 8:17 p.m
correlations within a
(in the method of rela
I never did understanc
pretended to myself ar
(2) its generality: i
living creature proper
to be modeled, (4) its
your statement in the
required only 40 lines
previously fuzzy conce
memory, "relevant" var
observation: no actic
higher standards overr
attending insight and
action, etc., (7) its
to predict particular
sense--as well as its
under which particular
clears up--for me, any
assumptions in traditi

~~But only the f~~

37. "...

38. Is it
studies mostly the
them?

39. "...
signal"?

40. Which

41. Unlik
person learns, dur
satisfv even bette

July 25, 1989

Dear Phil,

Just got back from a vacation. Saw the Very Large Array near Socorro, our grandchild (Derek) near Durango, and my son (Denison) getting married near Boulder. All firsts. This computer seems very strange after resting my eyes on distant terrain for so long.

Note (A) pasted here

The problem of spelling "Foreword" (blush) has been sidestepped by Greg Williams, who informs me that an author's preliminary remarks in a collection of works are called the "Preface," which even I can't misspell. Rick Marken has written the Foreword. The book is essentially finished -- there will be copies at the meeting. It looks pretty good to me. Greg and Pat are doing a truly meticulous job of proofing, including one person reading the originals aloud while the other checks the text. The cover illustration will be a blowup of part of that multilevel control-system diagram that Mary made. A very carefully-produced book.

The introductory chapter of your new book fills me with joy. It is a masterful application of control-theory ideas that avoids all preposessing technicalities while making the basic ideas seem fresh and familiar. I hope that in the chapters that follow you will maintain the same tone of assuming the tenets of control theory without proselytizing. In the other books you've written lately, there's been a lot of enthusiasm for a new idea; now it's time to treat the idea as if you take it for granted. Maybe you feel that some teaching of the principles is still necessary. If you do, I hope you will be matter-of-fact about it. Maybe you can explain -- perhaps in the Forward -- that the theory can be learned in more detail elsewhere. I love to see the theory just being <u>used</u>. And I can't wait to see how you use it. This is going to be a publication of prime value to the Control Systems movement. It will do more to validate control theory than any amount of abstract pedagogy could do.

Note (B) pasted here

It would be pleasing if the first book were finally accepted. I suspect, however, that it was (they were) only a warm-up for the one you are working on now. That's right.

Loose ends:

38. Cognitive psychologists <u>think</u> they are studying output functions; they are really studying systems that compute reference signals that demand certain outcomes to be perceived. "Buy 100 shares of IBM" means "perceive that 100 shares of IBM have been bought." Never mind dialing the broker or filling out a form and licking the stamp -- they don't worry about <u>how</u> the decision is supposed to be realized. So I'd say that they study the formulation of high-level reference signals, not outputs. Or perhaps that they study how people formulate statements

O|C

describing reference signals. They seem to think that the highest mental process is verbalization.

40. I had proposed listing Greg Williams as the author of the collection of my works: Williams, G. (ed). I was overruled.

During the vacation I felt some new points of view starting to form, somewhere in here. I can't put a finger on them, but the Foreword, or Preface, hints at what's going on. One result is that I looked at the list of editors that you sent with a total lack of interest. I really don't feel like knocking on doors any more, not just to get published <u>somewhere</u>. Opportunities will present themselves. I will probably send the Higher Realms paper to the American Psychologist, as several people have suggested. But I think I'm about ready to let others do the publishing of that kind. I should start writing straight to the converted, trying to teach whatever I haven't taught yet. The most important task now is to get everyone's level of expertise tuned up as high as it can go. I want to become nonessential, so I can just muck about with ideas that strike my fancy. The gap between what I know and what the members of the CSG know is modest, but the gap between me and the normal life scientist is enormous and daunting. I think I'd rather deal with the establishment through intermediaries. I think I'd like to pretend that the establishment doesn't exist, just ignore the standard literature and hire bridge-builders instead of trying to do it myself. I find these thoughts liberating.

Note (C) pasted here

So back to work. I really need to retire so I can get something done.

Best always,

Bill

Bill

thought I had sent this a week ago --

Note (A)

Glad you had a good vacation. I envy you the sight of the Very Large Array. I was going to go to Pasadena this month to see the direct transmissions from Voyager II from Uranus (?), but then I had to have a new roof put on the house and ran out of money.

Reading aloud is really the only way to do a good job of proofreading. Another thing I have to keep teaching people.

Note (B)

I'm very glad you like the prospectus (as publishers like to say) for my next book.

I'm not surprised you are confused about how many books I am writing—you have seen so many versions of various chunks. But at the moment, there are only two images in my mind of what might appear between covers: (1) the methods book that is now with the editorial committee at Praeger, and (2) the book on social psych of which you have just read the tentative first chapter. This will be the BIG one—though of course publishers will tell me to cut it back.

As to toning down my "enthusiasm for a new idea," I'll keep it in mind, but I make no promises.

Note (C)

I agree heartily with everything you say here. To put it another way, do what you feel good doing. Some people feel good doing battle with the forces of evil. Some don't. Some feel good doing it in spurts, doing other things between spurts. And so on. I think your assessment is correct that there are now others willing to do battle with the establishment, and if you feel better thinking your own thoughts, instead of pitting your thoughts against the establishment's, then do it. I think I was feeling some of your feeling when I was urging you to give up battling with the Skinnerites.

Why Should Anybody Pay Attention to Control Theory?

P. J. Runkel
August 1989

Q. Why bother with still another theory about human behavior? We have lots of them. Isn't one about as good as another?

A. No, some theories are better than others. Do you know a psychological theory that can predict the behavior of one particular person? I don't mean on the average, or the behavior of a statistically significant number of people or to a statistically significant degree. I mean, can your "just as good" theory actually predict almost exactly what one randomly or arbitrarily selected, particular, single, actual person will do, every time?

Q. Of course not. No theory can do that. Well, psychophysics can. But not theory about actually moving around and doing things.

A. Control theory can. But here I must not mislead you. I don't mean that control theory can predict anything and everything. But control theorists are not capricious, vague, or mysterious about what they can predict. Control theory precisely specifies the sort of thing it can and cannot predict. Particular acts are in general impossible to predict. You cannot predict with much reliability whether a particular rat will crawl into a garbage can, bite your finger, or press a lever. It is true that under severely restricted conditions, it is possible to predict particular events pretty well. If you make the rat very hungry, imprison it in a small box, and give it no way to get food except by pressing a lever, then you can predict pretty reliably that it will press the lever. But in ordinary conditions, no. You cannot predict well the particular acts of particular individuals.

But you can predict well, once you know a person's purpose, what the perceptual consequences of the person's acts will be, and you can predict that very accurately.

Q. That sounds pretty vague to me. Can you give an example?

A. Suppose George wants to catch Scruffy, the cat. In the first place, you cannot usually predict where Scruffy is going to run, so how in the world can you predict where George is going to run? Suppose you had a motion picture of Scruffy's running and were asked to "predict" George's running from that, how would you be able to predict just where George would change direction, speed up, slow down, and so on? Not to speak of predicting the necessary muscle tensions in George's body that would keep him upright, compensate for centrifugal force, turn his head so that he could keep an eye on Scruffy, and all that? And if you could not predict any of those components of chasing the cat, how could you predict how George will fare in chasing Scruffy?

Well, if George wants to catch Scruffy, that means that what George wants to see is his hands firmly around Scruffy's body, and to reach that perception, George has to perceive the distance between him and Scruffy getting smaller and smaller. Isn't that simple? It turns out that predicting that reduction of "error" between where George does see Scruffy and where he wants to see Scruffy is what does the job.

Q. What job?

A. The job of predicting what George does.

A. Seems to me you are actually getting farther away from predicting what George will do to catch the cat. Just how can control theory predict how George will catch Scruffy?

A. Well, to test the ability of control theory to predict George's chase, we would have to record very accurately just where Scruffy puts his feet and just where George puts his. So if someone would kindly build a large floor covered with pressure-sensitive spots and wire them to a computer, then we would be able to record exactly where the feet of George and Scruffy fall and see how closely the theory comes to predicting George's footprints. We would, by the way, expect the theory to come within a very few percentage points of the distance between George's feet and Scruffy's. Unfortunately, building a floor like that is expensive.

Q. So?

A. So we have done the next best thing. We have reduced that floor to the size of a computer screen and substituted a target-mark for Scruffy and a cursor for George's feet. Then we let George, instead of using his feet, use his fingers on the keyboard or on a joystick.

To make the chase realistic, we have the program not only move Scruffy (the target) around the screen, but we also have the program put unpredictable variations into Scruffy's behavior, just as would happen in chasing Scruffy around the back yard. And we also put random variability into the way George's fingers affect the cursor on the screen. That happens in the back yard, too. Bumps on the ground, gusts of wind, the laundry on the line getting in the way of George's vision--that sort of thing would put unpredictable variations into George's chase, and he would have to compensate for them as he runs hither and thither.

Q. You mean the movement of Scruffy on the screen and also the effect of George's fingers on the pursuing cursor are both going to move unpredictably?

A. That's right.

Q. And yet George is going to catch Scruffy?

A. That's right. It does happen in the natural world, you know.

Q. Well, how will you show you have actually predicted it? And how accurately you have predicted it?

A. You let George chase Scruffy around the screen, and you continuously record where the two of them are--well, say every thirtieth of a second. You also use control theory to build a "model" of the chase. In other words, you use control theory to write a program for the computer that will enable the pursuing cursor to act in relation to the target just the way George causes his cursor to act--that is, a program that will try to reduce the distance between target and cursor, no matter where the target goes and no matter how the pursuing cursor's motion is disturbed by those unpredictable variations. The program must do that, of course, without being given any information about what the real George actually does. Then you compare the "behavior" the computer produces with what George actually does. Do you know a theory that will permit a program like that to be written?

Q. Well, I suppose a computer expert could write one that would chase a target. I don't know how the expert could make it chase the way George would, though, if you won't tell the expert where George put his feet.

A. Yes, I agree that a computer expert could write a program that would chase a target. The expert's program, without a psychological theory, would be many times longer than a control theorist's program, and the chase would probably not remind us much of George. Do you know a psychological theory that would help the computer expert write such a program?

Q. Well, if there is one, I haven't heard of it.

A. Neither have I, except for control theory. And now here is the crucial test. We can take exactly the same program written for Scruffy and George, change the unpredictable variations we gave to Scruffy and to George's cursor, then put Charlotte at the keyboard instead of George, and get the same high degree of match between model and human that we got with George! That is, control theory enables us to write a program that not only will chase Scruffy in the same way George does (without being given any of George's footprints), but will produce the same accurate match when we give Scruffy new zigs and zags and put in a new human, Charlotte, for the program to match. Granted that I have omitted a few technical details, what do you think of that?

August 9, 1989

Dear Phil,

I wrote you an undisciplined 8-page reply and just had the pleasure (the next morning) of deleting it. All. Now I can address your request for suggestions about your piece. I have a theory that when one doesn't write for a while, little bits of ideas tend to accumulate anyway and start drifting into corners and under sofas along with the general lint. Before any real writing can start again it's necessary to sweep all that junk together and dump it onto paper, which can then be disposed of neatly.

That was a lead-in. Yes, some theories are better than others. But what I was writing about yesterday (from a paper-in-progress on human nature from the standpoint of control theory) was that some theories are of a different kind. When you compare the predictive power of control theory with that of traditional theories, the effect is to imply that the only way to choose between theories is in terms of their ability to predict behavior. But when you do that, you soon have to back off and explain that we can't really predict the details of action, but only of consequences. Others claim they can do that without control theory (Skinner claims he can do it without any theory). All that's hard to discuss in a non-confusing way.

The comparison I would like to make, and am trying to make in this in-progress paper, is between prescientific and postscientific theories. A prescientific theory proposes that some consequent regularly follows some antecedent under some conditions. The theory is accepted as right if statistical analysis shows that this proposed sequence can in fact be observed. In other words, a prescientific theory is an attempt to guess a natural procedural rule: do this, and that will happen. People used to make theories like this informally, basing them on personal experience and common sense. Now they use statistics and formal experiments, but the method is the same.

A postscientific theory proposes a model of reality at a level that underlies observed sequences of events or procedural rules. The entities of the model are unobservable. So are the forces, influences, fields, resonances, and what-have-you that relate one entity to another. In other words, such a model is imagined. But it always has connections with the observable world. One set of connections consists of objects that we can see and manipulate experimentally. These connections give us the ability to affect the supposed underlying system. The other set of connections gives the underlying system the ability to affect variables that we can observe -- sometimes just meter readings, but usually some natural phenomenon. The model of what lies between these sets of connections must transform what we do to the world into what the world does to our observations. Such a model is not a miniature replica of the real thing; it is all we know or can guess about the real thing.

A postscientific theory therefore tries to explain why it is that the relationships proposed by prescientific theories are or are not observable.

If "prescientific" and "postscientific" are too aggressive, we can just say "empirical" versus "model-based" theory. I'll do that.

I think you're on the right track in pointing out that empirical theories aren't very good at predicting, except in the long run. They're useless for dealing with individuals. But if you want to use the model-based approach in contrast, you have to be delicate about it, because "strictly scientific" psychologists actually sneer at the model-based approach. They don't realize that it is precisely this approach that has made physics so successful, and precisely the lack of it that has kept psychology in the prescientific era.

The reason for the sneering is that attempts by psychologists to build models have been uniformly incompetent. The use of models is reduced to proposing "intervening variables" like fear, ambition, cognitive dissonance, and so on. To the empiricist, such intervening variables look totally unnecessary -- and worse, there is no way to test them directly to see if they really exist. And the empiricists can rightly point out that the use of intervening variables hasn't improved predictions; if anything, they're worse.

People have tried to make models just as long as they have tried to make empirical theories. Aristotle proposed that objects are maintained in flight by air rushing in behind them in accord with the principle that nature abhors a vacuum. When the original impetus is exhausted, the objects just fall straight down. That's a model. We don't actually observe the proposed underlying processes, but if they existed they would -- Aristotle said they would -- account for what we do observe. The ancient Egyptians explained that the sun is really a god in a chariot making a daily journey across the sky. If there were a god doing that, and if gods doing that looked like the sun, then that model would explain what we do see (the glare hides the chariot). People have been making models as long as they have been doing anything that we know about.

The problem is that they haven't been very good at making models, so the empirical approach has worked far better. In the behavioral sciences that has remained the case until the advent of control theory and related "systems" approaches, which introduce the model-making methods of physics to the life sciences. You will note that where the life sciences have been most successful, they are simply doing physics and chemistry inside the organism.

Why is the model-making method so much more powerful when it is used right? Not because the models refer to something "real," but because of the high standards that are maintained in the successful sciences. The electron was not accepted into the physics-model until it had been given properties that explained everything that could be attributed to electrons. No exceptions. No "under these circumstances." No "80 per cent of the time." The model had to work all the time and it had to predict observable phenomena quantitatively, to the limits of our ability to measure. The model had to be, as near as we could tell, perfect.

Most physical scientists, of course (with recent exceptions), think they are "discovering reality." Let them. Whatever they think they're doing, they're making models in their imaginations. But they're doing it in a disciplined way that demands perfection. They would say that nature is governed by exact and immutable laws, so they're simply trying to improve their analyses to make them into closer and closer approximations to the actual exact laws. It's nature that is perfect, as it can be only and exactly what it is. The effect is the same: the models are worked over and reworked and tested with ever-increasing finesse, any discrepancy between the model's behavior and what is observed being reason enough to work on the model some more.

In the behavioral sciences, the long centuries of failure have resulted in a different view of natural laws: organisms are inherently variable and inexact. What's the point in demanding perfection of models (or any method) when nature itself is largely random? The result of this view has been a drastic lowering of scientific standards.

****

Well, I'm not going to try to finish that paper here. I think you see the main points developed so far. Maybe you will want to talk to your imaginary questioner about the proposition that we can actually come up with exact understandings of how people really work (translation: a model that we can afford to be very fussy about). If your questioner is like many real psychologists I have met, the reaction will be primitive: to most people the idea of really understanding human behavior sounds like a challenge to the gods, an invitation to be struck down for presumption, and a threat to expose all the nasty secrets that people believe they alone have to hide. The state of psychological theory is the direct result of not trying to make theories that work all the time. But you will know what to do with that idea.

More later, maybe. Got to mow the lawn.

Best

Bill

610 Kingswood Avenue
Eugene OR 97405
17 August 1989

Mr. Wm. T. Powers
1138 Whitfield Road
Northbrook IL 60062

Dear Bill:

Now I need formal permission from you to reprint. I follow here the form letter provided me by Praeger.

I will be grateful for your permission to use the material specified below in a book I am writing entitled Casting Nets and Testing Specimens, and in future editions thereof, to be published in a limited scholarly edition in hardback by Praeger Publishers, a Division of Greenwood Press, Inc.

Description of Material:

432 words from your chapter in Davidson and Davidson (Eds.), The psychobiology of consciousness, New York: Plenum, 1980. The excerpts are attached.

The diagram you displayed at the meeting of the American Society for Cybernetics in St. Gallen, Switzerland, in March of 1987. A copy is attached.

I already have permission from Plenum for that piece.

I request permission for all language rights throughout the world. If you are unable to grant full world rights, please tell me to whom to write.

I will give full credit to your publication in a citation and in the list of references, as well as in an acknowledgment of permission at the usual places.

For your convenience, I am including a form for release on the next page and an extra copy of this letter for your records.

Thanks very much.

Sincerely,

Philip J. Runkel

SAGE Publications, Inc.
2111 West Hillcrest Drive
Newbury Park, California 91320

The Publishers of Professional Social Science

(805) 499-0721

September 14, 1989

Philip J. Runkel
610 Kingswood Avenue
Eugene, Oregon 97405

Dear Dr. Runkel:

C. Deborah Laughton requested that I send you the enclosed reviews, coded A, B, and C, of your proposal for CASTING NETS AND TESTING SPECIMENS. We hope they are helpful to you.

Good luck with your book.

Cordially,

Frances Borghi

Frances Borghi
Editorial Assistant

Enclosures

Bill and Mary:

These comments are typical of some others I have received.

Thank goodness they are not like some others.

I'm not surprised at any of the unfavorable reviews. Nor will you be.

I think many critics are right when they say that the book is not sharply focused on any one group (stereotyped) of readers. I can't help that. --Phil R

TO: C. Deboreh Laughton
RE: Review of <u>Casting Nets and Testing Specimens</u> REVIEW B

This is a very unusual book. It raises legitimate criticisms of mainstream approaches to research, but doesn't answer a number of important questions that seemingly should be part of the critique. As examples, (1) the discussion in Chapter 3 ignores variability around central tendencies, which speaks to homogeneity of the sample with respect to the characteristic measured; (2) the author leaves "aggregation confusion," being unclear about the level at which inferences should and are being drawn(The author seems to want to infer at the individual level; (3) strength of effect is also ignored. It provides information about the likelihood that people will behave as predicted; and (4) the arguments about simultaneous causation ignore a major literature on the issue of simultaneous v. finite lag causation. The examples confound occurrence with measurement; even if events occur more rapidly than we can measure, they do not necessarily become simultaneous(My view is that even in cases of reciprocal causation, there is still a finite causal lag and events cause later events).

Perhaps my philosophical position just differs from the author, so I may have been particularly critical. I suspect, however, that most readers will hold an orientation generally similar to mine(how's that for assumed similarity!) and finish the book unconvinced about the value of the alternative. Nonetheless, I thought the first part of the book was the strongest, for it should force readers to think critically about their methods.

As I moved onto the "specimen" parts of the book, I was less favorably impressed, for I felt the <u>methods</u> orientation was given up for a philosophical one that was hard to follow. I in fact found myself asking questions the author thought readers might ask like "Why would scientists want to focus primarily on events in which outcomes are highly predictable? They are simple in structure and likely well understood." I also felt that the author belittled researchers trying to understand complex phenomenon by classing them as trying to impress colleagues. As is obvious, I came away unconvinced, perhaps not understanding but clearly not buying. What is missing for me here is a more extensive discussion of why research hasn't taken a specimen focus and how a specimen focus will explain complex human behavior.

With respect to potential audience, I am uncertain. It seems to be pitched at upper level undergraduate students or beginning graduate students; I found it straightforward but somewhat repetitive and in some instances simplistic in its points(which suggests to me that the goal was to "bring home" the central argument even to a somewhat confused reader). (Example: the number of times the author repeated the "casting a net, compiling a catalog or history..."). Since the book doesn't attempt to cover topics of a traditional methods course, its niche would have to be as a supplemental book to get students to reflect on the "big picture" (your Qu4) and consider an alternative as they learn traditional methods and statistics. Such a use would take advantage of its critical position with respect to "traditional" approaches and inferences processes. Thus, unless it is priced reasonably, I think it stands little chance of being used. On balance, I might cite it as a possible reference for a graduate level methods class but would not require it, because I didn't feel that it did a good enough job of balancing its presentation or selling its arguments. I think my students would be confused by it. More importantly, since I, no traditional laboratory researcher, found it difficult to accept, I'd expect others to have even stronger reactions.

Finally, the writing style is a strength of the book. It reads easily(although, I personally would like to see it trimmed considerably to lessen repitiveness, shorten some of the long examples, and make the points more succinctly). If it in fact is pitched to persons just entering psychology, upper level undergraduates or new graduate students, then perhaps the pace, detail, repetition, and examples are fine(but I'd try to get someone at that level to read it and comment on these points).

Review A inserted here out of sequence to save a page.

Review B continues on the following page with Specific Comments.

Review A

1. The entire methodological argument should be deleted and the section on control theory should be expanded. As an introduction to control theory, without the evangelical tone, the book might be useful. You should get the opinion of someone who is familiar with control theory on this.

2. The writing style is too chatty and imprecise. It would be easier to follow if the author made it clearer from the beginning that the book was leading to control theory. Most readers will quit long before the main point is made.

3. The book does not provide enough background for undergraduates. They won't understand what the author is arguing about. Graduate students can find better books.

<u>SPECIFIC COMMENTS</u>

Chpt 1	p11	19:1 not 20:1
	p16	It's not clear to me why regression doesn't fit the "how humans as a species function"
Chpt 3		I thought the discussion here spent too much time on issues of "which is the appropriate level for discussing inferences", particularily since it didn't draw from the broad literature on aggregation, level of analysis, etc.
Chpt 4	p3	if the manipulation works, Ss should worry more than control Ss
Chpt 5	p6	Bus examples violates homoscedasticity assumptions for drawing inferences (i.e., descriptive coorelation is ok, but prediction isn't)
Chpt 8	p3	Why assumption 3? Do most scientist look for "universals"?
	p7	The example supports lagged causation; my behavior builds trust for my partner's <u>future</u> behavior
	p8	Multiplicative relations can be included by product terms. Others that the author says can't be done <u>can</u> by defining thresholds and recoding variables.
Chpt 9	p3	It seems that the author's beliefs are not scrutinized in the same way the work he criticizes is scrutinized (and dissected)
	p10	If one can't predict in real world settings the level of complexity that would be a "pacer stimulus", the whole phenomenon becomes just a post hoc explanation of behavior??
	p25	The correct regression model would be a reciprocal feedback model, which would generate different solutions than the one presented.
	p26	<u>paragraph 4</u>: The conclusion is much too simplistic
Chpt 10		I've lost the "methods" thread of the book
Chpt 11		Does the author believe there is a bias against the work he describes? If not, why does much of it appear in "low quality" journals or is unpublished?
Chpt 12		<u>beginning</u>: Reads like a consistency theory discussion
	p11	The example of operational definition is misleading.
	p32	The cooperation example is not a good one, for the goals are "mixed motive"
Chpt 13	p3-4	The "action research" examples may or may not be action research, since it's not clear that they attempt to examine theory.
	p5-6	typos- note redundancy

I have reviewed Casting Nets and Testing Specimens as you have asked. I found it to be an interesting but very idiosyncratic manuscript. The author has decided that traditional research techniques have only limited validity and applicability and has developed some alternatives. This is all fine and good but I think the author has overreacted. It is clear the there are problems with traditional methods. What is not as clear is how much of a problem these difficulties pose for social and behavioral scientistd. The author does not present sufficient evidence, in my opinion, to warrant such a radical procedures as proposed. This is the basic and fatal flaw of the manuscript. The author could have made the case of alternative approaches without having to throw out everything that came earlier. The polemical and argumentative nature of manuscript is bound to turn off many readers as it did me.

The writing style is an odd mixture of very informal prose mixed with technical presentations. It was never clear to me who was the target of the book; was it professionals, graduate students or undergraduates? I think the author had not made up his mind about this and it shows in the manuscript.

I would describe this book as an interesting but flawed approach to research methodology. I think the author shows some brilliant insights but these are lost in the rest of the text. The writing is very uneven with some sections being very technical and at a professional level and other sections appropriate for an undergraduate.

This appears to be a very personal book. The author expresses a lot of his own feelings and opinions about the current state of research. This approach is not satisfactory for a textbook but is very good for a book at the professional level. Unfortunately I can not tell which audience tl author is targeting. I would not adopt this book for a class nor encourage Sage to offer a contract.

Review C

FYI — I really will write to you some day! Bill

January 14, 1990

Dear Thomas, Gillian

Might as well drop the formalities right away; after all, we both know Phil Runkel.

We are in very similar boats. I am working, but wish I weren't, but my wife would be carrying the load if I weren't, so I am. We have a son playing belated catchup in Mechanical Engineering at Boulder and two other grown kids who will need us as a backstop for a few years still (although they don't like to admit it). What I want most out of life is to be able to devote all my energies to control theory. We're getting so close to major acceptance! But the time isn't quite yet. Certainly the money isn't there.

Your first three propositions sounded like my brand of wishful thinking. I haven't found any place where I could make a living doing control theory stuff -- I don't even have a Phud. You'd have a better chance, but the truth is that all the established places are defending themselves against control theory, not hiring people to promote it. When you apply for a grant, the people they call in to evaluate the proposal are the very ones whose life work would be invalidated by control theory. Even though they don't understand it, they can smell that threat a mile away.

The fourth proposition, however, perked me up. There are lots of people in the Control Systems Group who would love to have an Institute devoted to control theory. I've thought about it for years. The problem is that I'm not very worldly (practically a hermit), and nobody else in the group has much savvy in that regard, either (at least not combined with time to do anything about it). But if you think you have the know-how and the contacts to make some progress in that direction, you will find lots of support in the CSG.

There's only one catch: if you acquire the money and other necessary means, you'd have to run the place. I certainly wouldn't do it, and I don't know anyone else in the group who would want to. You can't administer an institute and get anything done at the same time. Presumably, we would eventually hire some lackeys to serve as president, vice-president, and so on (as you might guess, I don't see organizational functions as a social hierarchy. Damn commie pinko.).

Of course as we stand now we don't have much to sell to a layman with money. We're still pretty much building the foundations of a new science -- the interesting parts come later. It's odd that you should write now, however, because lately I've been thinking more and more about how to understand human affairs in terms of the higher levels of control, the ones we really haven't studied much yet. That isn't <u>just</u> because social scientists find those the most interesting levels of the model; it's mainly because I want to see if we can't find a new way to approach human nature at those levels, more or less before it's too late. I've exhorted members of the CSG for several years to try using control theory in its own right instead of tacking it onto some other approach with which they're familiar. All the best researchers in the group have actually started doing that. Now I want to try the same thing myself at levels where I normally get by on common sense and cultural wisdom.

I presume you'll be coming to the next CSG meeting with Phil. At that meeting, I think I'll deliver another homily on this subject, this time with the theme of "burning bridges." If control theory is a revolutionary way of understanding human nature, then let's each have a revolution. Everybody came into this business (you don't mind if I practice on you, do you?) from some set of well-organized beliefs, some of them acquired in school but most of them acquired through contact with a family, a community, a church, or a culture (or all those). This is where we got most of our higher-level organization: our customary strategies for doing things, our principles, and our concepts of who we are, what a society is, and so on. We get hardly any of these ideas through personal understanding or conscious choice. Most of them came dissolved in our Pablum.

For most people those higher-level concepts are extremely vague, and where they come from is even vaguer. We learn that we're supposed to be honest, not murder, and not swear. The reason we're given is that it's wrong to do these things, and furthermore if we do them we'll get hit (or at least made to feel equally demeaned). So ethics and morals seem to be handed down through a combination of mysticism and coercion. I don't think that's a good enough way of understanding these subjects or of making them important in human affairs in the way they ought to be important.

One level higher our teachings are in even worse shape. The world is full of entities, some of which we can see and some of which we can't. The most important ones are invisible: God, Science, Society, Law, Democracy, and the like. I call these "system concepts." The ability to perceive and control things in this category is the most important ability of conscious life. Whatever we choose as principles is ultimately governed by how they add up to a coherent system concept. Principles in turn determine how we will choose and organize our strategies, our logic, our language, and in general our practical decisions and actions. System concepts therefore sit at the top of everything we think and do.

I've just been re-reading John Dewey (the "Early works"), with considerable disappointment but growing understanding of what's wrong. When I first read Dewey I was pretty much like him: I tried to reason everything out in words, and thought I was pretty good at it (I thought he was, too). Now I can see that Dewey was just playing with words, trying with an inadequate tool to understand and explain issues that can't be handled at the level of words. I realize now that his words appeal in a silent and subtle way to the meanings of words that one is supposed, as a member of a language culture, to "just know." At the bottom of his explanations and discussions is nothing by a sea of words whose meanings are assumed to be not only common knowledge, but objectively meaningful. It's easier to see this when reading old Dewey than when reading someone modern, because Dewey relied on images and unspoken meanings that have gone out of style. He speaks, for example, of the "quale" of sensation -- does that bring a sparkling image to your mind? Not to mine. But he thought it meant something.

Seeing Dewey struggling to deal with terms that we no longer believe in or use makes it plain to me that the problems he was trying to solve were mostly not problems at all, except in language. I've always (well, not _always_) tried to look behind words to see the phenomena to which they supposedly point. There are great gobs of words that point to nothing but other words -- it was a great comedown for me as a verbally precious youngster in my late twenties to realize that much of what I said so glibly didn't mean _anything_. I'm much quieter these days.

To realize the full potential of the control-theoretic model (goes the homily) we must each examine very closely and even mercilessly the higher-level concepts we have always taken for granted. This will be difficult because it implies a critical look at our own most important levels of organization. The biggest problem in developing control theory at the higher levels will not be that of convincing others that what we say is true. It will be that of convincing ourselves that we would be better off to have a coherent and organized understanding of the things we have each accepted as right living and right thinking. Our religions, our concepts of science, our most precious philosophical or scientific or common-sense understandings, our expertise -- those are the things we must try to understand in a new way. If we're not willing to put everything out for skeptical inspection, we simply won't get anywhere at the higher levels. We'll just go on playing the same games as everyone else.

Control theory doesn't tell us what strategies to use, what principles to adopt, or what system concepts to use for organizing our existences. It tells us _that_ we do these things. It makes sense of those things by showing how they are related to each other in human perception and action.

I am particularly stirred to get this message across by the events of the last few months in Eastern Europe and the USSR. All these people are discovering that they can't live without freedom. But they don't know what they mean by that. They know that they are experiencing something good: they know _what_ is good, but not _why_ it's good. They don't have any organized understanding of these matters. Things aren't a lot different on this side of the ocean.

Control theory tells us that freedom is the only natural state of an organism. Organisms are not built to be controlled from outside. That is simply a biological fact. Attempts to control another organism result in conflict; if the organisms are evenly matched in intelligence and strength, the conflict can only escalate. Or one wins and one loses. There are lots more losers than winners. Conflict involves positive feedback; the result is that winners get strong and win more. This results in a few people have a lot of power and most people having none.

Human beings try to adopt similar system concepts in order to avoid the penalties of conflict. It doesn't much matter what the system concepts are or what principles they give rise to and are supported by. When you think carefully about the conditions under which conflict arises, you can see that it is a natural outcome of one organism trying to control another. Conflict destroys everyone's ability to control what matters to himself. Adoption of similar system concepts and principles will lead to practical actions that can coexist. It keeps goals from becoming mutually exclusive at the most important levels of organization.

So control theory's message won't come in the form of a design for some new Utiopa -- do everything our way and everyone will be happy. It will come in the form of an explanation of how we work. Its criticisms of existing societal orders won't be cast in terms of which ideas are bad ones and which are good ones. They will be explained by showing that some ideas require human beings to work in a way that they don't work. The brunt of the message is that we must make all our levels of organization consistent with each other, both inside individuals and among them. That can be done in many ways; one has to be willing to alter anything at any level that creates inconsistency, which is only to say conflict.

Of course this means that the control theorist must avoid arguing in terms of which moral system is more justified than which others, or which political system, or which religion or anti-religion. Those are empty issues from the standpoint of control theory. The control theorist will be trying to show how the design for any social system one wants to try has to be put together in order to have a chance of lasting. To do this the control theorist has to shed the idea that one system (one's own, of course) is more rational, more valuable, or more sacred than another.

What societies need to understand is that any system must be organized to allow for the actual properties of human nature. The control theorist's contribution is to lay out what those properties are -- the properties that remain the same no matter what people do, think, or hope for.

End of homily, first draft.

I have this vague idea of how we can find a new way to talk about human affairs at the higher levels. I know that some CSG members are already interested in this kind of project. My homily is just a guess at how it might turn out. We need something to get us off the ground here. Any ideas?

I haven't written to Phil for a long time -- I guess I figure that he's in the bag. Hope you don't mind if I send a copy of this to him.

Anyway, I think you can see that I, too, have some thoughts about influencing the course of our evolution. Control theory ought to be good for something.

Best regards,

Bill Powers

1138 Whitfield Rd.

Northbrook, IL 60062

Thomas Gillian was a student of Phil Runkel's.

16 March 90

Here is another salvo in my battle to get you to cease battling with the behaviorists.

Maybe you will want to read only the parts I have marked in red. Or maybe not even those.

I hope you are well and happy.

—Phil R

*

Behaviorism, Neobehaviorism, and Cognitivism in Learning Theory: Historical and Contemporary Perspectives, by Abram Amsel. Hillsdale, NJ: Erlbaum, 1989. 105 pp. + xiv. $19.95.

Systems and Theories in Psychology (4th Edition), by Melvin H. Marx and William A. Cronan-Hillix. New York: McGraw-Hill, 1987. 509 pp. + xvi. $39.95.

–1–

Looking Backward to See Ahead

A review by Howard H. Kendler, *University of California, Santa Barbara*

<snip...>

Amsel's Complaint

Amsel is an angry young senior citizen. He objects to the intellectual lynching of behaviorism and neobehaviorism at the hands of cognitive psychology, not because he is one of the designated victims but as a matter of principle. This

<snip...>

lynching has taken the form of the negative advertising—the flagrant distortion of an opponent's point of view—in order to enhance cognitivism at the expense of behaviorism. Unlike Dukakis, Amsel is not willing to take the abuse lying down. He vigorously attempts to set the record straight by demonstrating that cognitivists have misrepresented the meaning of behaviorism, ignored and distorted history, and committed a variety of methodological sins.

<snip...>

Amsel has written his book as a participant in the theoretical warfare between cognitivists and S-R neobehaviorists. As a consequence, he sees the issues through the colored lenses of a S-R

Dawkins, in the Blind Watchmaker, says that an explanation, to be easily useful, must use terms at the next lower (reduced) level.

* Psychological Science Vol 1, No. 2, March 1990 pp. 107-113
This article runs 7 pages. I show only the snippets Runkel marked.

–2–

Where Did Everybody Go?

A review by Robert C. Bolles,
University of Washington

<snip...>

psychologists.

Without arguing about the last point, I am still left wondering what happened to psychoanalysis. Thirty-something years ago when I was a student, everyone was talking about Freud and his ideas. Even experimentalists read Freud in those days—everybody did. Clearly something dramatic has happened, and the interesting thing is that no one wants to talk about it. What happened to all the Freudians? There were so many of them that they could not have all died off. We can see that there are a lot of behavior modifiers around, but I am left wondering whether they replaced the Freudians, or are themselves converted Freudians, or perhaps Freudians in disguise. History and systems authors do not seem to care about this sort of thing, but is it not a vital part of our history, maybe half of

(BOLLES, continued on p. 112)

<snip...>

the neo-Hullians. Everything is left, frozen, just as it was about 20 years ago. There is no hint that anything important has happened. The truth in this case is that all hell has broken loose. Learning theory is unrecognizable. Hull would have been unable to read a present-day research paper; none of it would have made any sense! Learning in Hull's time, and for 20 years after his death, had to do with the reinforcement of S-R associations. Today relatively few theorists think about S-R associations, or believe in a reinforcement process (procedure maybe, but not process). Further, in those times learning theory was center stage, the heart of all psychology. Today learning theorists are off in a little corner by themselves, fussing with each other about methodological details that the majority of psychologists do not care

<snip...>

And upon consideration we rediscovered Pavlovian-type S-S associations.

The problem that Amsel complains about, that no one respects S-R reinforcement theory anymore, is not because it cannot deal with cognitive matters, but because it endorses the S-R unit that many of us now find unattractive. The reinforcement idea has a very similar history. It was much more mechanistic (scientific) than the pleasure-pain principle, which is why Thorndike proposed it, and why after a time learning theorists bought into it. We bought not so much Thorndike's idea as the ideas of Hull and Skinner. Reinforcement became very big. But suddenly there were awesome problems with autoshaping, with avoidance learning, and with the superstition experiment. These problems were so horrendous that a lot of us started abandoning the reinforcement idea. We became rather comfortable without it. So the conflict is not between mechanists and cognitivists, it centers on the substance of the S-R reinforcement position. The real problem with S-R reinforcement theory is that it clings stubbornly to the S-R formula and the reinforcement idea. If Amsel could abandon these two minor planks of his platform, then he and I could join hands and both be behaviorists again.

<snip...>

as equipotentiality, then I would be happy to join hands with them and be a fellow general process theorist.

There is another problem that weighs on learning theorists of all persuasions, and that is our loss of centrality. Thirty years ago learning theory was the heart of psychology; today it is not. What happened? Should those of us who are still in there think about getting out? Interestingly, 30 years ago, Marx and Amsel and I were all doing essentially the same experiment. We manipulated the rat's drive level and conditions of reinforcement, we carefully counted the number of reinforcements, and we observed how bar pressing or runway running was affected. The outcome not only mattered to learning theorists, it mattered to everyone in psychology. It was not only a central problem for the rat runner, it was important to S-R reinforcement re-

March 20, 1990

Dear Phil,

I'm not battling with behaviorists -- only with what they believe. Behaviorists probably wouldn't say "only," although they're perfectly capable of slipping into the use of terms like "believe" (as in "superstitious beliefs"). I'm in correspondence with two of them now -- Gary Lucas and William Timberlake, of Indiana University (Bloomington, IN). They're trying to make working models, and have come up with some feedback ideas, so I've sent them my latest "operant behavior" model (with a mouse this time, drawn by Mary, instead of a chicken). I'll talk with anyone capable of supporting one end of a conversation while still allowing the other to proceed.

The argument between cognitivists and behaviorists (Bolles/Amsel review) is not the same as mine between control theory and behaviorists. Maybe cognitivists have "polluted" the meaning of behaviorism, but I haven't. I think I understand behaviorism very well. The cognitivists denigrate the "dependent variables" of behaviorism as "colorless movement" and "glandular squirts," which is to say that they object to them on aesthetic grounds. I object to them because they don't exist. They do not "depend" in the assumed way. You don't even need control theory to prove that.

I have read Watson, as Amsel recommends; Watson's works are based on the assumption that behavior is a dependent variable, and that environmental events are the independent variables. Says Bolles/Amsel, " ... the major message of behaviorism, conveniently ignored by cognitivist critics because of the questions it raises and problems it poses for rejecting behaviorism, is that knowledge claims of psychology cannot meet the standards of natural science methodology unless behavior is employed as the dependent variable."

There's the problem laid out plain. The behaviorists have always claimed to have a lock on the only true natural science methodology (they assume, incorrectly, that it's the same methodology that physicists use). This is the methodology, of course, that says you vary the independent variable and look for a correlation with the dependent variable. What the behaviorists conveniently overlook (aside from the initial false assumption) is that this way of viewing behavior doesn't meet the methodological standards of the natural sciences <u>either</u> (by which I mean physics and chemistry). The predictions made on this basis don't predict worth a damn.

"Behaviorism is a methodological approach that demands public behavior serve as the dependent variable in psychology." This demand has created a mind-set that is almost impossible to surmount. When a behaviorist looks at a behavior, he sees an <u>effect</u> and ignores the obvious role of the same behavior as a cause. He sees an effect because of imposing an interpretation on observation -- a mental model. Well, we all do that. But what makes this a pernicious imposition is that the behaviorist denies that there is any subjective interpretation involved; the subtle mechanism of the denial is to call such agreed-upon modes of observation "public." They are not public; they are private. It makes no difference whether a pair or a multitude of behaviorists adopts the agreement. Each individual still has to agree to use the interpretation, use it, and reject any other.

The way you find out that this is a private view is to find another view that can equally well be adopted without altering the observations. Control theory gives us another view. But behaviorists reject control theory because it is another view, one that they believe to be wrong simply because it opposes the interpretation they take as the only scientific one and mistakenly claim to be an objective one. In the name of objectivity, they have made their own subjectivity unconscious and unmentionable.

Bolles says that equating behaviorism with Skinner's version "... leads to the absurd conclusion that behaviorism opposes the use of abstract theoretical constructs when interpreting behavior...". In a backhanded way, therefore, Bolles is saying that behaviorists and cognitivists make the same mistake, which is to confuse the invention of abstraction generalizations with making real theories or models. He's objecting because the cognitivists mistakenly claim that behaviorists don't also indulge in this erroneous conception of science.

It's the use of abstract constructs that stands in the way of psychologists when they try to understand the control-system model. They simply can't grasp the idea that it is not an abstract construct. Because they treat control theory as just another construct, they compare it with existing constructs using the same criterion they always use: verbal plausibility. They don't ask how well it works, because the idea of a theory working is all but unknown to them. And they certainly don't ask whether the components of control systems physically exist -- they never claim that even for their own constructs.

But control theory is a literal description of how an organism works. Never mind whether it's a correct description: that's another subject, the subject of testing models. It's a literal description because every component of a proposed control organization, including boxes and arrows, inside and outside the organism, is supposed to represent the operation of some observable thing. When I speak of reference signals, I'm not just talking about an arrow in a diagram or an algebraic variable. I'm proposing that inside the brain there are real neural signals that we could measure, that act in neural circuits to establish reference levels just as they are established in real electronic devices that we can take apart and study. When I draw a line and label it "perceptual signal," I'm proposing not only that such signals could be found in the brain, but that these signals are identically what we experience when we experience perceptions. These are strong and falsifiable propositions about a real physical system.

Amsel, like most True Scientists, is irate about anthropomorphisms. This attitude has been handed down from biology. "Instead of employing animal psychology to explain human behavior, they use human psychology to explain animal behavior," Bolles complains. But if you turn back a few pages in any biology text where you find a statement like this, you will find an exposition of the many and detailed similarities between animals and human beings -- so many that biologists regularly conclude that there are no important differences other than those of complexity.

Given these extensive similarities, what do they conclude? Not that animals are like people, but that people are like animals. There's a subtle difference. Being human, we are each in a position to observe things that can't be seen in any other organism, even another human being: the way colors look, tastes taste, pains hurt, pleasures please. The stream of thought and reason. The appreciation of forms and symmetries. The efficacy of will. We have 24 hours'-worth of evidence every day that there are phenomena occurring inside us that are invisible to others, yet perfectly real and valuable to us. This is the <u>only</u> evidence we have about these internal phenomena -- we can't even observe them in another human being. We know, however, that they occur in one sample of one kind of organism. On what basis are we to conclude that they don't occur in <u>all</u> organisms? Is there any scientific justification whatsoever for asserting that either other people or animals don't have similar private experiences, which are just as valuable to them as ours are to us? If we have to draw conclusions on this subject, shouldn't we use whatever evidence is available?

But this isn't how it's done. The argument starts not with human experience, which is directly observable inside any human observer, but with conjectures about animal experience, of which we observe nothing. It's said to be absurd, for example, to suppose that a frog can think frog thoughts and feel frog feelings. No reasonable person would suppose otherwise (this is how the argument usually goes). Therefore, animals don't think and feel. But we know that animals are like human beings in a vast array of regards. Therefore, human beings don't think or feel, either, or if they do, it's just a mechanical side-effect. Human beings are like animals, and their so-called subjective experiences are illusions with no explanatory weight.

This line of thinking is closely connected to the insistence that behavior is a dependent variable. If you can account for behavior by finding the independent variables on which it depends (manipulable by an experimenter or fixed by inheritance), then there is no need for supposing that any internal process could have a causative effect on behavior as well -- Occam's Razor. Intentions, reasoning, desires, thoughts -- these are epiphenomena, side-effects, and play no causal role in behavior. Such conclusions are the direct result of insisting that behavior is a dependent variable.

The animal activists arouse biological scientists to rage not because they make intellectual errors (such as supposing that frog thoughts are like human thoughts), but because they point to the <u>real</u> reason for biologists insisting that animals are simply soft machines. The reason, of course, is in the way biologists and other experimenters in the life sciences treat animals. Animals do not scream in pain: they vocalize as a result of noxious stimulation. Animals do not cower in fear or attempt to escape at the approach of the experimenter: they exhibit flight responses to looming-stimuli. So it goes. If animals were supposed to share some version of all the subjective experiences we each know that a human being has, a great deal of research would become morally repugnant. But we need to do this research (which is always valuable, even if we can't interpret the results), so it follows that animals don't have a subjective life anything like ours. And of course, because of the overwhelming evidence of similarity, neither do we.

I consider this standard scientific view of both animal and human consciousness to be not just confused, but pathological. I think it has crippled the life sciences by making the begging of questions a formal part of scientific reasoning.

I do agree with Bolles on one point. The cognitivists have not shown anything wrong with behaviorism; they have simply abandoned it. That leaves all the phenomena discovered by behaviorists in limbo -- neither explained nor explained away. The cognitivists have done nothing more than shrug off what they can't explain. I find this attitude irritating beyond support. People have criticized me for spending so much time thinking about operant conditioning; they say, "Nobody thinks that's important any more, why are you wasting your time on that old stuff?" My answer is that it's important until we understand the phenomena; just turning to something else is no answer. Control theory can explain all the substantive phenomena that behaviorists have explained in terms of drives, reinforcements, and so on. But it's not enough to say that we <u>can</u> explain it. We have to <u>do</u> it in such a way as to leave no room to doubt that control theory does a far more convincing job than any behaviorist explanation has done.

I don't buy this "orthogonal" garbage. It's all one system. If we're to understand it, we have to bring the whole thing and all the phenomena associated with it under a single consistent theory. Otherwise we go back to doing our own thing and not worrying about thinking six contradictory thoughts before teatime. "Microtheories" are for dilettantes.

So it isn't just behaviorist ideas that I battle, Phil. It's a whole history of hubris and dishonesty in the life sciences, and the resulting failure to develop even the rudiments of a real science of behavior. As far as I'm concerned, we're starting from zero.

So that's my answering salvo. Now you have to try to guess whether you hit a battleship or a destroyer and where to aim the next round.

Best,

Bill

5070 Fox Hollow Road
Eugene OR 97405-4008
25 September 1991

Dr. Jerry Suls
Spence Laboratories of Psychology
University of Iowa
Iowa City IA 52242

Dear Dr. Suls:

I am of course deeply gratified to receive your letter of 12 September about my book Casting Nets and Testing Specimens. I am especially pleased to have comment from a fellow social psychologist who, like me, has had "nagging doubts for years" about our methods.

Let me know how the ideas strike your students in your methods course. Is the course for graduates or undergraduates? I often think, by the way, that the ideas in my book are actually very simple and must surely sound like ordinary common sense to anyone who has not been propagandized for years with the reasoning of agricultural statistics. The reason the ideas are so difficult for you and me is that we have had to fight our way out of all that wrong-headedness after believing it devoutly for years. I know two other people who are using the book, or the ideas in it, to teach methods. I'm sure they would be glad to hear from you:

Dr. W. Thomas Bourbon
Department of Psychology
Stephen F. Austin State University
P.O. Box 13046
Nacogdoches TX 759662
Office phone: 409-564-2974
Home phone: 409-568-1426

Dr. Richard J. Robertson
Psychology Department
Northeastern Illinois University
5500 North St. Louis Avenue
Chicago IL 60625
Home phone: 312-643-8686

As to Carver and Scheier, yes, I did read their book. It is like most theorizing about the psychology of personality and social life; the authors collect the ideas from the literature that seem good to them, put them between covers, and that's that. I don't say it is immoral to do that; I'm all for free speech. But that is not a productive way to do science. You don't build testable theory, especially theory testable with quantitative precision, by throwing together some "good ideas."

So yes, you are right about one of my objections: though Carver and Scheier have used some phrases from control theory, their own research has not turned to testing a model with individual subjects. To sharpen my point, here are a few sentences Wm. Powers wrote in a little newsletter in 1987:

> Control theory is a set of principles that applies whenever an active system [living or non-living] affects and is affected by its environment at the same time. There are a few other provisos--the gain around the loop must be at least ten, most of the gain must occur in the organism rather than in its environment, and the whole system has to be dynamically stable.

I don't think Carver and Scheier would know how to begin to understand those sentences.

And yes, C&S misinterpret Powers. I'm sorry I cannot now give you specific instances. I looked through my records and through the correspondence I have saved with Powers, and all I can find now is a note I made to myself a few years ago after having read C&S:

> Tries to show how existing research connects with Powers. Not very convincing to me. And I think misinterprets some ideas of control theory.

There are also the books by William Glasser; they, too, claim to rest on Powers's work. They have become popular in the sense of attracting followers to "apply" them in their work. Many teachers of teachers have picked up Glasser's ideas; he himself has conducted numerous workshops for teachers. But again, this is a case of an author snatching the parts of Powers's writing that sound good to him and omitting or distorting the rest. Again, I want our society always to allow its citizens to do that; I want free circulation of ideas and of proposals to modify them. But again, that's not the path to science. I don't even object to a "scientific" journal publishing speculations like that, especially if the article is labeled speculation, not theory. I would like the article to begin something like, "I wonder whether if maybe...."

So I didn't mention William Glasser, either.

You may wish to correspond with Powers:

Mr. William T. Powers
73 Ridge Place, CR 510
Durango CO 81301

There is a group of enthusiasts of control theory called the Control Systems Group. Its publishing arm is operated by Greg Williams; I have asked him to put you on our mailing list:

Mr. Greg Williams
Route 1, Box 302
Gravel Switch KY 40328

The group also puts out an occasional newsletter. If you want to receive it, write to:

Mr. Edward E. Ford
10209 North 56th Street
Scottsdale AZ 85253
Phone 602-991-4860
or 1-800-869-9623

So that's all I can think of just now. Thanks again for telling me of your pleasure in the book. Write again if you feel the urge.

Sincerely yours,

Philip J. Runkel

PHILIP J. RUNKEL
5070 FOX HOLLOW ROAD
EUGENE, OR 97405

12 November 1991

Dr. Tom Bourbon
Dept of Psychol
Stephen F. Austin State Univ
P.O. Box 13046, SFA Station
Nacogdoches TX 75962—3046

Dear Tom:

Excuse the formal inside address; I do it for archival reasons.

Most of my previous letters to you have obviously been typed by a printer connected to my computer; this one obviously is not. My computer has been on the blink for several months. It works fine until printing. Then it prints a line or a few lines or a page or two and then stops. It will start up again only after being started up from scratch. We have tried a different printer, a new cable, a new interface in the machine, re-installing the word processor, etc etc etc. Nothing helps.

Be that as it may, ...

I have before me your letter of 10 March 1991. In it, you said that a person performing a task must be holding a reference level for performing (maybe completing?) the task. You said, as an example, that the person in a tracking task must (if it is to be a "task") not merely observe the relation between cursor and target, but must intend to control it. The person must not merely recognize that the distance, for example, exceeds the distance you are requesting the person to use as a standard limit, but must take action to limit it. And the person must not do that merely once or for half a second, but must go on doing it for the while that you requested.

You said that if you stop a moving target, the person will act as if the task has been interrupted—that is, as if the person is frustrated from carrying out something he or she wants to carry out. This sounds very much like the effect known in social psychology as the Zeigarnik effect. That is, if you interrupt a task a person has started, the person will want to continue it, but beyond that, the person will remember doing that task bettwe than the person will remember tasks the person carries through to completion. The discussions of the Zeigarnik put a lot more emphasis on the remembering than on the urge to complete, but I think the two have to go together.

But your point I think is a fine one: that any kind of coherent activity requires a reference for doing it. If you are swimming, you must see to it that you maintain your relation to the water. If you want to get somewhere, if getting from here to there is your task, then getting out of the water and walking will suit you very well. But if your task is to swim a while, you are going to be annoyed if you find yourself trying to swim in the sand. Thirty years ago or so, a couple of social psychologists wrote a paper about the "interaction goal," by which they meant the purpose for which two people would carry on a conversation or other kind of interaction. Two conversants must not only have standards for linking words and conveying meanings, they must also have at least one standard that keeps them communicating. The interaction goal, in my mind, must be a sub-variety of cooperation. In organizational development, we speak of work in groups as having two aspects. One is the "task"—the purpose people agree upon when they come together, the thing they want to accomplish. It is sometimes called the "convening task." The other is the "process"— he interpersonal dealings by which people organize themselves while they arepursuing the main task, the procedures or customs that are "the way we do things around here." Both aspects require agreements about what is to be done and how it is to be done. Often, special agreements must be made about how to tell whether things are getting done. I suppose those remarks are relevant to what you were thinking about in March.

Earlier, you had sent along communications from and to Andy Papanicolaou about interpreting conditioning via control theory. I am not a good person with whom to discuss conditioning. I have always felt an active distaste, almost a resentment, with the ideas of Pavlov and those who followed him. I am irrational about it. I have a very hard time making myself focus my attention on writings about conditioning. Again and again, I find that I have skipped parts of sentences. But I'll risk a comment or two.

In your letter to A.P., y ou say that the "generalization gradient" pertains to empiricism versus ideal forms. Does it also pertain to the perception of thingness? Seems to me that gradients specify boundaries and enable us to have a perception of a perception that it has given us a "thing."

You spoke of helping, when you provide some part of the corrective action to correct an error another person is trying to correct, as cooperation. I wouldn't call it that. You could provide that kind of help without the helped person knowing that you were doing so. I apply the word "cooperation" only when two or more persons know they are working jointly on a task to reach an agreed end-state. This is not a matter of what "is." It is a matter merely of how to use the word. The important point is that we have two different tasks or states of affairs here, and they derserve different words.

Sorry I have been so long in writing. The new marriage and the new house have swept me up into a new world and riveted my attention for long periods. But the house is gradually getting into order, so I hope to have more time for more usual activities before long. I hope to get back to working on the next book before too long. But I hope my computer can get fixed soon.

REGARDS!

Phil

Dr. W. Thomas Bourbon
Department of Psychology
Stephen F. Austin State University
P.O. Box 13046, SFA Station
Nacogdoches TX 75962--3046

5070 Fox Hollow Road
Eugene OR 97405-4008
3 February 1992

Dear Tom:

Congratulation on your new job. I am sure you will welcome more time to pursue your research. To what extent do you think somebody there will undertake to tell you how to go about it? Another thing: I have the impression that you often enjoy your teaching. Will you be doing any teaching?

(I myself would not like to be without a class to go to, but I have a very strong preference for just <u>one</u> class. One is enough. With more, I can't keep all the students straight in my mind. I say things to Amy when the right person to have said them to was John.)

You say the new position will be funded by a grant. Does that mean you will not be on a permanent budget, heaven forfend? Well, if you are delighted, then I am delighted. How about your family? Are they delighted? What are you going to do about hurricanes? It is a good long distance to Galveston, so I suppose you will be moving your residence. When you send out your change-of-address cards, be sure I am on your list.

Thanks for the clarification of "aid" versus "cooperation." I had never given thought to aid before. Glad to do so. Thanks for relaying the kind remarks by Andy Papanicolaou. I'm always glad to hear when I've pleased somebody.

Yes, Claire and I have every intention of showing up at the 1992 CSG conference.

* Enclosed is the MS you sent me of "Models and Their Worlds." I still think it is superb. I am proud to be mentioned in it. I suppose it is too late for me to have any further influence on it, but I cannot refrain from making some comments, among which are some proposals for small alterations, even though my proposals may be futile. (You can see that I am indeed fond of hearing myself talk--or of seeing myself write.) You will find some small marks and comments on the MS itself. Here I will make some comments that require more words.

On page 3, the reason I have written "partly" is that here I think you skirt philosophy of science. Your statement can mean that we affect the physical environment: we divert a river;

* See references on page xxxi. Bourbon... (1993)

we smoke up the atmosphere; we make a footprint in the mud. In that case, we provide part of the effects on the environment. The environment provides some of its own effects by volcanic eruptions, striking comets, sun spots, and excavating ants. You might also mean that we see the world the way we think it is because of the way we see (behave). So our understanding of the way the world is has the character is has because of the way our nervous systems go about giving us "understanding." I wasn't quite sure which way you were leaning when you wrote that sentence. On the other hand, I don't think you should go into a two-page digression at that point. Maybe it is just as well for the reader to do a little wondering, as I did.

Pages 11-12. I think your strategy here is very clever; I enjoyed following it. You are saying, I think, "Let's suppose that it were possible to get data so good that the scatter plot would be a straight line. Then, if the data were that good, would the physical world allow the data to demonstrate that the theory is correct? If such good data cannot support the data, then we should certainly not expect poorer data ever to support it, should we?" Congratulations. I don't think I would ever have thought of that strategy.

Page 12, end of first paragraph. To those steeped in the method of relative frequencies, "principle" often means little more than a relation--a statistical relation, one of frequencies. To them, a causal relation means either (a) a relation postulated top be causal without direct proof or (b) a comparison of an "experimental" and "control" outcome. But I think I have demonstrated that the meth of rel freq can only very rarely provide data for convincing demonstration of causality.

So what do you want the reader to think of when you speak of a "principle"? When I try to put into words what else a "causal principle" could be, my tongue stumbles. I can pick out a few assertions that I think are principles (causal or not) and are not merely statistical relations. For example, it seems to me a principle that in the living creature, causation acts in a continuous feedback loop. That principle or assumption can be implied in the equations of a model (it can be used to shape the equations), so the principle exists as a more specific and quantifiable thing than merely that string of words. But how do I put in one sentence what we mean by "principle"?

Page 13, middle. "Meaningful" is not meaningful to me. Do you mean having a standard deviation of y on x of no more than five percent of something-or-other?

Page 21, your argument that a test of a model in a simple situation must be successful if the model is to be successful in a complex situation. I agree, but I think few people who have worked always with the meth of rel freq will understand what you are saying. Most of them think of a "complex situation" as one they would want to slice by many variables to "understand" it or succeed in predicting action within it. (Here I am almost repeating what I said in chapter 7 on fine slicing in Casting Nets and Testing Specimens.) Those who think the "model" might "work" in a complex situation are thinking something like this: "In the complex, more real situation, there are variables I don't know about operating in ways I do not know about. But I think I have a good hunch about what often shows up in the complex world. I think the few variables I have seen working are confounded with those unknown variables in such a way that when the variables I know about are at the values I specify, they are accompanied by the right values of those unknown variables. And those values of the unknown variables are necessary, too, to get the right value of the dependent variable. But in this simplified situation in the laboratory, those unknown variables are not free to act, and therefore they are not lending their strength to the variables I do know about."

You, in contrast, are saying that a principle that is going to be a reliable one must be more than an empirical preponderance of frequencies; it must be an organization of action that is the same whatever variables show up to impinge on the action. Or something like that? Anyway, I think the important or key sentence in this section is the last one, which simply makes the assertion itself.

Then, too, some people will say (I have sent Bill a couple of articles of this sort, and maybe the article by Shimp is of this sort) that your simple modeling of SR theory shows you don't know the current sophistication of the theory, and you are not testing today's theory. Should we be spending time testing the theory of phlogiston? I guess all you can do with those people is what you have done here--invite them to do a computer simulation of their current model.

Page 25. The figures have no figure numbers on them! I hope some editor is not sending the figures on to some printer with no numbers! Also, I hope you will have time to take a pen and make either the target trace or the cursor trace clearly different from the other.

Page 26, bottom. But in figure 3d, the H does not seem to begin at the same value as T and C.

Page 32 near bottom: "error." I am tempted to warn readers accustomed to inferential statistics that this is not error of measurement nor error of the experimenter in making a prediction, but discrepancy between what the subject wants and sees. Put such a warning in a footnote? I don't think so. I don't suppose you can put in a warning for every possible lapse of a reader's attention.

So that's all I have to say. I think it a superbly argued paper and very attractively written. It ought to capture a few imaginations.

ACTION MEMO: Please send me a copy of the published version or, if you revise it still another time, send me a copy of that.

ACTION MEMO: Please send me a copy of Bourbon, Copeland, Dyer, Harman, and Mosley, 1990. I don't have a copy in my files. Maybe I have a copy in a pile of stuff not yet filed, but send me a copy anyway, if you will. *

You asked me to tell you the reference for "interaction goal." Interesting what one remembers from long ago. Well, that's one. And I remembered it well enough so that I was able to go to the library's computerized catalog and find it at first try. You will find enclosed some print-outs with red circled numbers on them. No. 1 is the reference for which you asked: the chapter by Jones and Thibaut in the book edited by Tagiuri and Petrullo in 1958.

Since then, however, a lot more thinking and empirical work has been done about the constraints on face-to-face communication. The writing of Goffman, for example, are well known. I don't know how much further comment about this you want, but I'll throw a few more items in for good measure, and you can feel free to ignore what you please.

I think everybody ought to know about the concept of the "behavior setting" invented by Roger Barker. Nos. 2 through 5 are about behavior settings. If you want to dip into the concept, probably an early chapter in No. 2 will serve. (Schoggen was a member of the psychology department here 20 years ago. I enjoyed talking with him.) I am sending along the citations for Nos. 3, 4, and 5 mainly so that you can see what will be behind what Schoggen says.

Jones and Thibaut say that to carry on a conversation, you must agree with the other person on some features of carrying it on--I forget the details (1958!). More recent writers (including Goffman) have investigated the norms or agreements made implicitly or explicitly that enable a conversation to be maintained and to go one direction or another. I just happened upon a book review the other day of a book that apparently deals with this sort of thing. I enclose a copy of it.

I hope you are well and happy. Write again.

Sincerely yours,

Philip J. Runkel

* Bourbon, W.T., Copeland, K. C., Dyer, V.R., Harman, W.K., & Mosley, B. L. (1990). On the accuracy and reliability of predictions by control-system theory. *Perceptual and Motor Skills,* 71, 1331–1338.

5070 Fox Hollow Road
Eugene OR 97405-4008
16 June 1992

Dr. Terence R. Mitchell
University of Washington Dept of Mgt. Seattle 98105

Dear Dr. Mitchell:

I was making my way today through the issue of 1989, volume 14, number 3 of the Academy of Management Review and read the article by you and James entitled "Conclusions and Future Directions." In your three sections headed "Epistemology," "Control Theory," and "Cognitive Theory," I was happy to see that the theories you mentioned all made the purpose of the individual a necessary feature. It seemed to me, too, that there was at least a hint in your descriptions that most of the theories recognized the fact that living creatures use varying means or acts to reach or maintain stable ends (that is, goals or purposes). I was glad to see that, too, since that recognition is the first step toward relinquishing the mistaken research strategy of trying to find high correlations between actions and environmental conditions.

I was glad to see that you made reference to the 1973 book by Wm. T. Powers. I wish, however, that you had used Powers's specifications in your brief description of control theory instead of the conceptions of Lord and Hanges. The thing that is distinctive about the theory and experimentation of Powers is that he actually builds physical models (not just conceptions or words on paper, but actual physical models in computers) that reproduce the behavior of individual humans in randomly varying environments to correlations touching .98 and even .99 between the quantities of motion of the model and the human. One of the followers of Powers (namely W. Thomas Bourbon) constructed a model of a human that predicted to that exactitude the behavior of the human an entire year later--in an unpredictable environment. The versions of control theory put forward by Lord and Hanges and by Carver and Scheier will not produce specifications for working models of the behaving human. In fact, I do not know any psychological theory other than that of Powers that has produced tangible, accurately functioning models, not to speak of producing behavior by the model that mimics human behavior so exactly, and does so in environments that vary unpredictably, ~~and~~ as real environments do.

It may be that you have not read Powers closely enough to see that he is doing that. So, now, are several of his followers. I have seen it happen, as anyone can who attends the annual meetings of the Control Systems Group.

Some easily accessible literature has come out since you wrote your 1989 article; for example:

Wayne A. Hershberger (Ed.) (1989). Volitional action: Conation and control. Amsterdam: North-Holland (copyright Elsevier Science Publishers B.V.). Contains papers by Powers, Bourbon, Marken, Plooij, and others.

Marken, Richard S. (Ed.) (1990). Special issue on control theory. American Behavioral Scientist, 34(1). Contains papers by Bourbon, Marken, Powers, Runkel, and others.

Richard S. Marken (1992). Mind readings: Experimental studies of purpose. Gravel Switch KY: Control Systems Group.

Powers, Wm. T. (1989). Living control systems: Selected papers of William T. Powers. Gravel Switch KY: Control Systems Group.

Powers, Wm. T. (1992). Living control systems II: Selected papers of William T. Powers. Gravel Switch KY: Control Systems Group.

The paper describing the year-ahead prediction that I mentioned is W. Thomas Bourbon, Kimberly E. Copeland, Vick R. Dyer, Wade K. Harman, and Barbara L. Mosley (1990). On the accuracy and reliability of predictions by control-system theory. Perceptual and Motor Skills, 71, 1331-1338.

Well, I was glad to see that you were keeping track of control theory. I hope you will continue to do so. And if you think of a way to build a model of the behavior of an organizational member, please let me know how it goes.

Sincerely yours,

Philip J. Runkel
Professor Emeritus of Education
and of Psychology
University of Oregon

cc: Dr. Lawrence R. James
University of Tennessee
413 Stokely Mgt Center
Knoxville 37996

Dexr Bill:

Just a small amusement.

Your wrote in the Amer Beh Sci about "Control Theory and Statistical Method." I wrote about an analogous pattern in the methods book I wrote with McGrath, 1972. Enclosed is the excerpt; the analogous pattern is described in the part I have marked in red. *

Phil R
10 Sept 92

* 920910_RunkelMcGrath.pdf —enclosure at this volume's web page.

5070 Fox Hollow Road
Eugene OR 97405-4008
24 September 1992

Dear Joel:

Thanks for sending me the paper and the copies of the remarks by the reviewers. Your reviewers are similar to some of those who reviewed my book manuscript.

The rantings of critics make about whether one is doing things right often sound to me like those from clergy. One has violated long-standing doctrine. As to assumptions about how the universe is put together, everybody knows the true nature of God, and those who don't should simply sign up for another semester of instruction.

Yes, the remark about "controlling" data with statistics is choice. Then Reader B follows that remark a sentence or two later with the statement that "humans are likely to give you fuzzy data." That is the recurring plaint that humans are naturally and unpredictably variable--that it is impossible in the nature of things ever to expect a person to behave twice in the same way. That comes, as Powers keeps saying, from looking at the varying and unpredictable particular acts with which humans maintain constant and predictable internal standards. Many psychologists do have a glimmering about internal standards--perhaps particularly the personologists--but they think of them as qualities of individuals, or densities of ingredients, or degrees of proclivities that are made salient by circumstances, instead of as goals.

I was struck by another of Reader B's remarks: that most SLA people care about cognitive mechanisms, not cerebral mechanisms. Presumably cognition goes on somewhere outside the head.

Readers C and D made remarks about the difference between studying the functioning of an individual, on the one hand, and seeking general laws of behavior, on the other. The conception of "generalizing," as it has developed within statistical theory, is an astonishing absurdity. Reader D says that you should not complain about research that

> does not explain the functioning of an individual learner. OF COURSE, [it does] NOT! It does not have that purpose. The goal ... is to find general laws--"on-the-average" phenomena. It seeks to find laws or principles that have likelihood of some applicability to most individuals.... Understanding individual learners' behavior is also a goal of a teacher; it simply is not the goal of the researcher seeking general laws.

And Reader C does not think

> that it is the basic aim of ... research to study how a particular person learns a language.... I rather think it is the aim of our field to specify the general laws or principles or regularities.... What a particular learner does is just an instance of these principles. But it may well be that we know the general principles without being able to predict their application in the specific case, just as we may well know the laws of mechanics without being able to predict the fall of a particular stone.

It astounds me that those writers do not see the contradictions and absurdities in what they write. How would you like to drive around in an automobile without being able to predict the application of the principles of mechanics to its particular case? And what does it mean to "know the general principles" without being able to predict whether turning to the right will rock the car to the right or to the left?

As I pointed out in my book (and as anyone familiar with the arithmetic of averaging ought to know) an average can be calculated from a string of data some of which actually go opposite to the direction of the final average. The average of +10, +7, -1, -2, -3, and -4 is +7. Twice as many cases go contrary to the direction of the average as go with it. But innumerable researchers will claim, given a distribution like that, that the subjects "generally" or "on the average" moved in the positive direction. If your automobile, when you steer to the right, now and then takes you sharply to the right, but more often takes you some degree to the left, will you be pleased with it because it "generally" or "on the average" goes in the direction you steer it?

Reader D seems to think that a teacher can indeed understand an individual learner's behavior. But Reader D argues that the teacher's goal and the researcher's goal do not overlap. Reader D seems to be saying that if a teacher succeeds in understanding the behavior of an individual, the researcher will find that uninteresting. And if the teacher succeeds in understanding another individual, and another, and another, will the researcher find them all uninteresting? I suppose if the teacher were to calculate the <u>average</u> success, the researcher would prick up his ears. But if the teacher merely has some individual successes, I suppose the researcher would find the account of the "case studies" unutterably boring.

I am reminded of a doctoral student at the University of Illinois many years ago. The university had (maybe still has) a world-class department of gymnastics. The student had compared three or four dozen Olympic gymnasts with 3000 run-of-the-mill sophomores. On half a dozen physiological indices, <u>all</u> the Olympic gymnasts fell well above the average of the 3000 sophomores. The student came to me for statistical help, because his two groups were so different in size, and the t-test doesn't make a good estimate when the two groups are too different in size. So how was he going to show that the Olympic gymnasts were better than the sophomores on those indices? Both he and his adviser (who was a nationally-known expert on the t-test) actually took that question seriously.

And how does the teacher succeed in understanding individuals? Is the teacher to reach his or her understanding without any help from general laws? Are the general laws to have no use to the teacher or counselor?

Reader C seems to think that researchers do not care how a particular individual learns a language. Rather, they seek general laws. So presumably researchers do not care how this next person learns a language, either. Or this one, or this one--indeed, investigators of language learning do not care how anybody learns a language. They are above all that; they are seeking general laws.

Doesn't that sound like Jonathan Swift?

Sincerely yours,

Philip J. Runkel

April 8, 1993

Hi, Phil.

Haven't written to you for years; I spend about six hours a day reading and writing email for CSGnet, and have almost forgotten how to write an ordinary letter.

In the diagram, the "T" stands for "Muscle Tone," but if you wish you can change it to "Z" for "Muscle Zone," although that wouldn't make a lot of sense.

The three reference signals at the top set (x) the muscle force component in the x direction, (y) the muscle force component in the y direction, and (t) the sum of the three muscle forces. This diagram, in the Byte article, was accompanied by a program that actually simulated the system shown. The point was that by SENSING the muscle forces with suitable weighting, the two-level system could achieve independent control of force in the x and y directions even though all three muscles contributed to both x and y forces. The muscle angles could be varied so the user could see how the control systems compensated. The first level just made the tension in each muscle be what it was told to be by the reference signals. The second level systems sensed the combinations of forces and adjusted the reference signals for all the muscles, as required for each system.

The "tone" control simply made sure that no muscle ever had to produce a negative force, by making all the muscles have some level of tension at all times.

This is good news, that you're launching a new book. Your Casting and Testing gets mentioned frequently on CSGnet -- whenever some new character comes aboard and says "What's wrong with using statistics?" I don't know how widely your book has spread, but wherever I have heard of its being adopted, people seem to consider it the definitive work on the subject.

You're probably wise to stay off CSGnet, although your wise words are missed. The traffic is incredible -- over a megabyte a month -- and it involves conversations among people all over the world. As you might guess, I'm sort of in the middle, with comments expected on just about everything. A lot of it is very difficult for me, because the people on the net tend to be heavily invested in some other approach, and are willing to consider just about anything except a change in their understanding. We spent a long time on linguistics, then on social control, and now are deep in a discussion about information theory. You'll probably be seeing the latter down the line in a Closed Loop (I actually wrote "Open Loop" first! Mind rot).

One bit of bright news I haven't announced on the net yet: a convert * in Wales has agreed to set up, through his university, the first meeting of the European Control Systems Group a year from now. Mary and I plan to take a vacation in Great Britain then and help with the inauguration. We should be able to get people from Wales (of course), Scotland, England, Holland, Germany, Luxembourg, France, Switzerland, and Spain. At least there are people on the net from all those countries who have been participating for considerable lengths of time. It's possible that some other CSG people might be able to get support to attend, also.

Hope all is well with you and Claire.

Bill

* Credit where credit is due: That's Marcos Rodrigues of the University of Wales at Aberystwyth.

Todd & Morris

8-17-93

Dear Phil,

I'm sorry you had to prod me to return the journal. It spent most of the time languishing under a pile of other stuff. The most interesting article in it at the moment is the one on academic folklore about behaviorism - control theory is also known mostly by what other people think it is - there is an article in Current Directions in Psychological Research 2(4) (the current issue) by Todd Nelson which is full of misunderstandings. The dismaying part is that he has been on CSGnet for some time. Some of us are working on a counter-article, and I hope this amounts to something.

Tom Bourbon seems quite pleased with where he is now, at the U. TX medical school in Houston. His tour-de-force at the conference was to use a model based on his tracking performance 5 years ago to compare with his performance on a new tracking task at the meeting. The correlations on two runs were .998 and .997. Pretty nice.

The conference went well. We had 31 participants, plus four guests. Several excellent presentations - especially in sociology. A number of people were pretty uncomfortable about the frequent, long-winded, pedantic comments of one person, but no one could hit upon an effective way to shut him up. This hasn't happened before, and did detract from the pleasant atmosphere. We also had a problem with hordes of ravenous and aggressive elderly people from another group descending on our coffee break goodies like they hadn't had a meal in weeks. On a personal note, my invitation to all who came early to come have dinner with us resulted in fixing a meal for 18 people, which was a bit of a stretch. But all went well, and I'll do it again.

Thanks again for lending me the journal,

Best to you and Claire,

Mary

MARY A. POWERS
73 RIDGE PL. CR 510
DURANGO, CO 81301

Hi, Phil -- March 6, 1994

It's good news that you're thinking of getting on the network with a new computer. The new computer will let you run all our software at full speed. And getting on the net will give us the benefit of your sensible ideas and attitudes, which are often sorely needed. Your work is mentioned every now and then when statistical matters come up -- it will be good to get it from the author.

Don't concern yourself about hackers breaking into your personal computer. They can't do it. They break into mainframes because the mainframes give some users, with sufficient levels of authority, the ability to operate the computer as if they were sitting at a console. What hackers do mainly is to figure out bugs in the mainframe programs that let them assume those authority levels, or else to guess the passwords by repeated trial and error so they can log on as a person with complete authority to do anything. And then they do anything.

They can't do this with your home computer because the only link between your computer and the mainframe "host" is a phone line that carries data only, not commands that run your computer. You might be given some authority to run programs on the main frame, but there's no way it can work the other way around. Your keyboard is the only way to put commands into your computer; the mainframe can't do it.

I recommend that you buy an IBM PC compatible computer with a 80486-DX (NOT SX) 33 MHz central processor, a hard disk with at least 120 megabyte capacity, one 3-1/2 inch floppy disk drive (capable of high-density operation, 1.44 megabyte capacity), one 5-1/4 inch floppy capable of 1.2 megabyte storage (if you need it to read your old disks), one 120 megabyte backup tape drive with 4 tape cartridges, a mouse (not the kind that senses acceleration, just one that senses motion linearly), a 14-inch or larger SuperVGA color monitor with .28 millimeter dot pitch or smaller, with a superVGA driver card, one SoundBlaster-16 basic sound card with a small pair of loudspeakers and a microphone, a fax-modem with the modem capable of all baud rates from 14400 down to 2400 or 1200, and a box that will automatically detect fax, modem, and voice connections (an optional convenience). The fax feature is optional, but will allow you to receive faxes as images stored in your computer (and view them on the screen or print them on your printer), and send text and computer-generated drawings to other faxes.

For software, I recommend the latest version of DOS (version 6.2), Windows 3.1, whatever word processor you're used to, and Procomm for a communications package if your University doesn't supply PC software.

All this no doubt sounds prepossessing, but when all this stuff is in the case and hooked up it will just be a computer. The best way to buy it is through Dell or Gateway (mail order houses with low prices and excellent support programs, including a first year of free in-your-house maintenance) or through your local computer store which will probably be able to assemble a system from components to your specifications. You may pay a little more from the local store, but you get quick support if something goes wrong, and I prefer this route. Your local store will make sure everything is running properly, and may well do this in your home.

Make sure the case you get has room for 3 or 4 future circuit cards.

The total cost will be under $2500, and perhaps considerably under it. My computer store gave me $500 for my old computer and $100 for my old printer (If you don't have one, I recommend getting a 24-pin dot-matrix printer which will be fast in draft mode yet which will type very nice-looking letters).

With this setup you will be able to run the most complex programs we have written at reasonable speeds, and do all the experiments we have done or are likely to do for some years, including experiments with sound and voice which we are only starting to think about. The color monitor is not a frill; we often use color for multiple-trace plots and modern color monitors are just as sharp as black and white ones. The large capacity hard disk is not overkill; when you get on the internet you will be able to download onto your hard disk from an immense variety of sources -- and saving the messages from the CSG net for 1993 alone used 16 megabytes of space (9 megabytes compressed). The tape backup unit is a must for such a large disk -- you would not want to back up your data on 50 floppies!

Windows 3.1 will give you a MAC-like user interface, which you may or may not like. I don't use it much, but there are some programs (like GEPASI, for simulating biochemical systems) that run only under Windows. DOS 6.2 contains an on-the-fly compressor which stores all data in compressed form -- for text files, the compression ratio is about 2:1. This increases the effective capacity of your hard disk at essentially no cost.

It will take you quite a while to explore all the capabilities of your new computer, but you will be able to start using it right away just as you used your old one. I will help you get connected to the internet if your university computing center won't help. I use the communications program Procomm, which allows "scripts" to be written that can automatically dial up the university mainframe, log on with a password, pack your mail messages into a single file, transmit the resulting file to a file with the same name on your hard disk, log off and hang up -- all with a single keyboard command. A similar script allows sending files from your hard disk to an individual or to CSGnet as a broadcast to all subscribers. So you can compose messages and read your mail using your normal word processor, off-line. I find that downloading a day's mail takes about 3 minutes.

I will be GREAT to have you on the net!

Love to you and Claire,

Bill (and Mary)

5070 Fox Hollow Road
Eugene OR 97405-4008
18 April 1994

Dear Bill:

Here are a couple of papers I have written mostly for myself. They have come about in the course of my thinking about the new book, but I do not now know where a good place in the book might be for them, nor even whether they will belong there at all. *

At this point, I yearn for some criticism. I am wondering whether I have made any indefensible statements: have I set down an idiocy someplace? Have I written 10 pages where 5 would do? Can some sentences or words be crossed out? Is the whole idea passe? I am trying to write here for people who know nothing about control theory and little about psychology.

So I am sending copies to a few people who might care to read one or both. If you can answer one or more of the questions above--or some other question you like better--I'll be very grateful. Please do not return these copies--unless you want to use them for stationery.

I hope you and Mary are well and happy.

Sincerely,

Phil

Philip J. Runkel

* 940418_Probabilistic.pdf By Runkel 5 pages —enclosure at this volume's web page.
940418_Replication.pdf By Runkel 9 pages —enclosure at this volume's web page.

April 22, 1994

Hi Phil --

I'm sending this to your email address -- don't know if any previous attempts got through. I'll also send it ground mail.

I do like both of your new articles/essays/chapters. Since you say that this is aimed at people who don't necessarily know anything about psychology or statistics, you may want to expand the explanations of how statistics is customarily used. Most of the reports on scientific research that the public sees are summaries of statistical results, generalized so they sound like universal truths.

Just think: aspirin protects against heart attacks. Fresh fruit protects against colon cancer. Vitamin C protects against colds. Just lately, "... nicotine may have medicinal value for people afflicted with a debilitating digestive disorder" (Science News, _145_, p. 199, March 26, 1994). The whole drug industry is based on people assuming that "drug y helps people who have condition z" means that drug y helps people who have condition z.

People think that the larger the scale of a study, like the cholesterol studies that involved huge numbers of subjects, the more certain the effects that are found. The exact opposite is true: the larger the study, the smaller the effect can be and still reach statistical significance. The only reason for doing a very large study is that the hypothesized effect is so small it can't even be seen in a normal-sized study. So when you hear of a giant project to determine the effect of something on hundreds of thousands of people, you are safe in ignoring it; the chances that it will pertain to you are negligible.

Tom Moore has written two wonderful books on medicine, including some most interesting observations about statistics. The first was _Heart Failure_, in which he does a great job with the cholesterol studies. The second is _Lifespan: who lives longer and why_. In the second one he discovered some extraordinary facts about published figures on risks. For example, the NIH (I believe) published a list of the number of people at risk of dying from heart disease every year due to series of hazards; the total number was something like 1,000,000. This number was several times the number of people who actually die of heart disease per year. By the time Moore had finished sifting the figures, the actual independent risks reduced the number to about 50,000 -- 1/20 the published number!

I hope you will take the opportunity to educate people a little about how to interpret research results based on statistics, in cases where a person might be tempted to take the results personally.

In the "replication" paper, there is a lot of good stuff. You do leave the impression, however, that it's impossible to replicate experiments with behavior. In fact, this depends on how you do the replication.

In a tracking experiment, we don't ordinarily get exactly the same behavior from any two people, but that is because different people have different control parameters. When you determine the control parameters for one person, you will find the same parameters, within a percent or two, in future experiments related to the first one, even using randomly different disturbances. What we replicate is not the detailed movements, but mathematical characteristics of the behavior that are one level more general than the movements. We can say that for _any_ pattern of disturbances with a certain frequency distribution, a given person will show a tracking error of x%, and a typical integration factor of x.xxx +/- 5% in units of inverse seconds. So the results are replicable.

You emphasize the unknown variables to such an extent that a person could conclude that behavioral research is futile. but the main unpredictable factor is not how a person will control, but what disturbances will occur. Through protracted study, it is possible to get some idea of how one person controls; also, it's possible to ask people to control in certain ways, and verify that the model fits what they do. But that doesn't amount to predictions _actions_, because actions depend on disturbances as well as reference signals, and to predict disturbances you'd have to predict the world. You might be able to characterize the way a person opens and manipulates an umbrella, but you can't predict when he will do that unless you learn to predict the weather.

Well -- keep it up. I think that a series of essays like these could make a book without necessarily having to be connected into one story. "Observations on human behavior."

Dinner time. Nice to hear from you.

Bill

* Date: Sun, 16 Oct 1994 22:02:24 -0700
From: RUNKEL Philip <RUNK@OREGON.UOREGON.EDU>
Subject: Hello

I, Phil Runkel, am now a subscriber to CSGnet. I like to write about psychological and social-psychological matters. As my years as a social psychologist went by, I got more and more dissatisfied (even incensed) with what academics call psychology. In 1985, I began reading CST and conversing about it. In 1990, I published the book CASTING NETS AND TESTING SPECIMENS, which explains the purposes inferential statistics can and cannot serve in research; it explains how the methods of CST can show the invariants in animal functioning. I have done no experimentation within CST. More than to experiment, I want to write, to show how social psychology can make more sense and be more helpful to actual living if it starts with CST. Anyway, I last studied the calculus in 1939, except for a brief refresher in 1954, and right now I am too lazy to start that again. I am retired from the University of Oregon.

Greetings to all.

* This is Phil Runkel's first post to CSGnet, the Control System Group network. Phil received several welcome notes in reply.

To follow the conversation as it proceeds on CSGnet, go to www.pctresources.com, where a complete archive is available.

[From Bill Powers (941024.0945 MDT)]

Hello, Phil --

Note hand-entered time-date stamp identifying sender of this post, so you can see who it's from without scrolling to the end or deciphering internet header information. This is entered by hand as the first line in a message. The format many of us use is

 [From NAME (yymmdd.hhmm ZONE)]

When replying to other persons' posts, we refer to the time-date stamp of each post this way:

NAME (yymmdd.hhmm ZONE) --

comments on first person's post

NAME (yymmdd.hhmm ZONE) --

comments on second person's post ...

etc.

then, usually, we put our signature at the end.

>... often people seem to be quoting passages by having each line be-
gin
>with ">". I don't know how to do that except "by hand." Is there a
>trick?

On some mainframe mail systems you can mark part of a post for inclusion in a message you're sending, and the ">" marks are inserted automatically. Since I'm dialing from home I can't do that. I download all my mail as one big file, hang up, then read it and compose replies with my own word-processor. Then I dial up again and send my reply.

My word processor lets me have the post I'm replying to on an alternate screen so I can switch back and forth between what I'm writing and what I'm replying to, and mark text for copying from one screen to the other. Then I eliminate hard carriage returns (substituting a space) so my word processor will put in its own margins, then add the ">" marks by hand. It sounds tedious but I have got used to it.

This being a network of PCT aficionados, the format you use is unimportant; what is important is the effect created by the format:

1. You can see who the whole post is from before you start reading it (and without having to decipher the internet header information).

2. You can see whose message is being replied to before you start reading the reply

3. You can tell which parts of the text were written by whom.

The funny way of writing dates is a computer-world custom, I think. It has the advantage that if you subtract one date from another you can tell immediately which is earlier. And it uses fewer keystrokes than most other methods, and no shifting.

>I am impressed by how quickly many members respond to a message.
>Don't they ever help with the dishes or go to a movie?

No, and many of them don't do their jobs, either. You will often see "Well, you won't be hearing much from me for a while because I am behind on six projects and have to make a living," followed within an hour or so by another even longer post. Watch out.

>I think you made a very wise judgment that if you want to influence
>people to dig into CST, the CSGnet is a much better way to do it than
>a book.

Good thing, too, because the three times I've started to write another book, I've had to stop because I was just rewriting B:CP. I respond well to stimuli in the form of comments and questions, but don't seem able to generate a whole book from scratch any more. I fear that control theory doesn't apply to me any more.

Credit Gary Cziko with starting CSG-L, by the way.

Our biggest problem on the net has been people who profess a strong interest in PCT but spend most of their time trying to justify whatever they had been doing before. I have begun putting up a strong resistance to that -- that's why the big push to get back to modeling. Of course anybody can still talk about anything they want to, but unless what they say furthers the development or understanding of PCT in some way I'm going to keep my comments short, if I make any at all. I'm sure you will sympathize with my feeling that we do not have all the time in the world to diddle around with philosophy and "Yeah, but how do you explain ...?"

My answer to the latter sort of question, if I can maintain my present attitude, will be "OK, you brought it up, how would YOU explain it with PCT?"

Hello to Claire, too --

Bill

Hi, Phil -- July 14, 1995

The [Susan] Motheral curve is a plot of behavior rate (vertical) against reinforcement rate (horizontal) for a variety of schedules, ranging from fixed-ratio 1 to fixed-ratio 160 (estimated from a figure). Starting with the highest ratio, we have a low rate of reinforcement (8 per session) and a medium rate of behavior (1300 per session). As the ratio declines to 80 and then 40, we get more behavior and more reinforcement. A peak is reached at about FR-40, at about 3000 behaviors per session and about 90 reinforcements per session.

As the ratio declines even further, the behavior rate begins to fall off again as the reinforcements increase, until at FR-1 we are getting 210 behaviors per session and 210 reinforcements per session. Extrapolating the curve on the right (which is nearly a straight line), we find an intercept at about 220 reinforcements per session; at that point the behavior rate would be zero. That is the formal definition of a reference level: that level of input at which the output just becomes zero.

The curve in question can be found on p. 214 in

Staddon, J.E.R.(1983); _Adaptive Behavior and Learning_. Cambridge: Cambridge University Press.

Staddon actually started developing control-system equations earlier in the book, but didn't know where to go from there.

I'll miss you at this year's meeting!

Bill

From: RUNK@OREGON.UOREGON.EDU 1-SEP-1995 18:26:05.87
To: POWERS_W@FORTLEWIS.EDU "William T. Powers"
Subj: Method of levels

Do you or Mary know about "Client-Centered Therapy" by Carl Rogers?

==

Hi, Phil -- (950901)

> Do you or Mary know about "Client-Centered Therapy" by Carl Rogers?

Yes. When I met Mary, she was an intern at Rogers' counseling center at the U. of Chicago. Rogers wrote one of the blurbs for B:CP on the back cover. There are obvious resemblances between the method of levels and Rogers' approach. One important differences is that Rogers was famous, while I am not.

Hi to Claire.

Bill

Date: Wed, 13 Oct 1999 19:49:26 -0700
From Phil Runkel on 13 October 1999:
Subject: Powers
To: CSGNET@POSTOFFICE.CSO.UIUC.EDU

Dear Bill:

In a moment of musing on the fragility of life, it occurred to me that I had set down my admiration, respect, and affection for you in only two published places, both of which were constrained by narrow purposes. And I do not want one of us to expire before I have set down in some public place some further testimonial. Therefore this.

As you know, I have been reading your writings and those of your followers since 1985. I have told you before how, as I strove to understand your view of perception and action, I found my own accustomed views undergoing wrenching, unsettling, unhinging, even frightening changes. I found myself having to disown hundreds, maybe thousands of pages which at one time I had broadcast to my peers with pride. I found, too, that as my new understanding grew, my previous confusions about psychological method, previously a gallimaufry of embarrassments, began to take on an orderliness. Some simply vanished, as chimeras are wont to do. Others lost their crippling effects when I saw how the various methods could be assigned their proper uses -- this is what I wrote about in "Casting Nets." For me, the sword that cut the Gordian knot -- my tangle of methodological embarrassments -- was the distinction between counting instances of acts, on the one hand, and making a tangible, working model of individual functioning, on the other. That idea, which in retrospect seems a simple one, was enough to dissipate (after some months of emotion-fraught reorganization of some cherished principles and system concepts) about 30 years of daily dissatisfaction with mainstream methods of psychological research.

The idea that permits making tangible, working models is, of course, the negative feedback loop. And that, in turn, requires abandoning the almost universally unquestioned assumption by most people, including psychologists, of straight-line causation -- which, in turn, includes the conceptions of beginning and ending. Displacing that theoretical baggage, the negative feedback loop requires circular causation, with every function in the loop performing as both cause and effect. That, in turn, implies continuous functioning (beginnings and endings are relegated to the convenience of perception at the fifth level). One cannot have it both ways. Living creatures do not loop on Mondays and straight-line on Tuesdays. They do not turn the page with loops while reading the print in linear cause-to-effect episodes. William of Occam would not approve.

The loop, too, is a simple idea. I don't say it is easy to grasp. I remember the difficulty I had with it in 1985. I mean it is a simple idea once you can feel the simultaneity of its functioning.

You did not invent the loop. It existed in a few mechanical devices in antiquity, and came to engineering fruition when electrical devices became common. Some psychologists even wrote about "feedback." But the manner in which living organisms make use of the feedback loop -- or I could say the manner in which the feedback loop enabled living creatures to come into being -- that insight is yours alone. That insight by itself should be sufficient to put you down on the pages of the history books as the founder of the science of psychology. I am sure you know that I am not, in that sentence, speaking in hyperbole, but in the straightforward, common meanings of the words. In a decade or two, I think, historians of psychology will be naming the year 1960 (when your two articles appeared in _Perceptual and Motor Skills_) as the beginning of the modern era. Maybe the historians will call it the Great Divide. The period before 1960 will be treated much as historians of chemistry treat the period before Lavoisier brought quantification to that science.

Using the negative feedback loop as the building-block of your theory also enabled you to show how mathematics could be used in psychological theorizing. (I spent a few years, long ago, reading here and there in the journals of mathematical psychology. I found that most articles were actually dealing with statistics.) Your true use of numbers has made it possible at last to test theory by the quantitative degree of approach, in the behavior of each individual, to the limits of measurement error, as in other sciences. This incorporation of mathematical theorizing was another of your contributions to the discipline.

But even making a science possible was not enough to fill the compass of your vision. You saw the unity of all aspects of human perception and action. You saw that there was not a sensory psychology over here, a cognitive over there, a personality in this direction, a social in that, and so on, but simply a psychology. You gathered every previous fragment into one grand theoretical structure -- the neural hierarchy. As you say, the nature of the particular levels is not crucial. What is crucial is the enabling effect of organization by levels -- the enabling of coordination among actions of all kinds. Previously disparate psychologies with disparate theories can now all begin with the same core of theoretical assumptions. Though it will take a long time to invent ways of testing the functioning of the hierarchy at the higher levels, I find it exhilarating to realize that you and others have already built models having two or three levels organized in the manner of hierarchical control and that the models actually work.

The neural hierarchy is far more than a listing of nice-sounding categories. The theory itself tells how we can recognize the relatively higher and lower placements of levels. It tells us, too, some of the kinds of difficulties to be anticipated in doing research at the higher levels. That kind of help from early theory is a remarkable achievement.

For any one of those three momentous insights, I think you deserve a bronze statue in the town square. To put all three together in one grand system concept is the kind of thing that happens in a scientific field once in a century or so. I am lucky to be alive when it is happening. How lucky I was in 1978 to have my hands on the _Psychological Review_, volume 85, number 5!

I do not want to give the impression that I think I have acquired a deep understanding of PCT. After 15 years of reading, conversing, writing, and thinking about PCT almost every day, I still feel the way Lewis and Clark must have felt when they began rowing their boats up the Missouri River. I know the general nature of the territory, but I know that much of what I will come upon will be astonishing and baffling, and I know that every mile of the journey will be hard going. As I work on the book I am writing, much of which will be elaborations of the three simple ideas I set out above, I find time and again that I must take an hour or a day to struggle with ways of keeping the words as simple as the idea. The ramifications of those simple ideas are multifarious, intertwined, and subtle. As I set forth to describe a complication in the way those ideas work together, I find now and again that I have opened further regions of complexity for which I am wholly unprepared. Then I must take an hour or a day or a week to find my way back to firm footing. I do not feel that I am trudging along a prescribed path. I feel that I am taking every step with caution, but also with awe and exhilaration as I wonder what I might come to understand. But I am sure I have only an inkling of the exploratory feelings you have had; you have guided your footfalls by experimentation, and I have guided mine only with thinking.

To those who know you, Bill, you are a treasure not only as a theorist and researcher, but also as a person. In our very first conversation by letter in 1985, I learned about your generosity. Without any hesitation, you spent eight single-spaced pages answering my ten questions of 23 July of that year about your 1978 article in the Psychological Review and four more single-spaced pages answering my letter of 9 September. In my experience with academic social scientists, my questions have usually been ignored or sometimes answered in three or four lines or by a reprint or two -- or sometimes just a reference to a publication -- without any personal words at all. I don't mean all my letters have drawn that sort of disappointing response; I have formed several happy professional friendships by letter. But you were more generous with thought, time, and paper than any.

You have bestowed thought, time, paper, and computer screens, not to speak of hospitality, on everyone who has evinced the slightest interest in PCT. You have understood the internal upheavals suffered by those of us who try to comprehend this strange new world -- our intellectual foot-dragging and our anguished obsequies muttered at the graves of our long-cherished beliefs. You have been patient with misunderstanding, persevering in the face of disdain, forbearing of invective, and modest under praise.

In all of this, you have been aided immeasurably by the intelligence, stamina, and love of Mary.

I owe you, for your help to me, a great debt. You have given me a way, after all these years, of laying hold of a system concept, a psychology, that is more than a grab-bag and a tallying. You have given me a way to set down thoughts that will come to more than a mere rearrangement of what every other psychologist would say. To join you and your other followers in the effort to make PCT available to others is, for me, here in my last years, a joy, a privilege, and a comfort.

Thanks, brother.

Date: Thu, 14 Oct 1999 09:53:36 -0600
From: Bill Powers <powers_w@FRONTIER.NET>
Subject: Re: Powers
To: CSGNET@POSTOFFICE.CSO.UIUC.EDU

[From Bill Powers (991014.0946 MDT)]

Phil Runkel on 13 October 1999--

Your post left me in tears, Phil, my brother in this adventure. How petty you make all our squabbling look! You and I have no time left to waste on that. Would that the young realized how little time they have left.

With greatest affection,

Bill P.

June 8, 2007

Dear Dag and Christine,

As Philip Runkel's wife, I sadly send you notice that Philip's cancer has gotten the better of him, and he died quietly at home at 4 a.m. Thursday morning, the seventh of June.

His stepson Pierre held a hand on one side of his narrow bed, and Pierre's wife Linda held a hand at the other side, assuring him of their love and concern. With closed eyes and very light breathing that seemed to signal a man beyond reach, he gradually slowed to total relief from the merciless pain of the disease. (Unfortunately, I had been advised to accept a heart pacemaker, so was in a hospital bed a few miles away from Phil's deathbed.)

The aging effects of the cancer, and its accompanying heavy pain, came on with gradual though swift effect, so that no more than four or five weeks called for heavy medication. This, of course, left him groggy, incapable of expressing himself and vaguely miserable until he died. Whether or not you wished to know the details, there they are.

I'm sure you can imagine the disappointment of me and my family in having to make this report ... Phil had told us he was looking forward to his 90th birthday (25 June) to celebrate some of the good things of life, but he didn't quite make it.

My family and I plan to rally with Phil's wellwishers sometime in the months ahead to celebrate his life, and will be inviting you to be among them.

With my best wishes and good thoughts of Phil,

Claire

Claire Runkel

P.S. We were so glad to have you with us so shortly before Phil died. It was good for him.

Date: Mon, 11 Jun 2007 15:43:19 -0700
From: Dag Forssell <dag@livingcontrolsystems.com>
Subject: Philip Runkel
To: CSGNET@LISTSERV.UIUC.EDU

[From Dag Forssell (2007.06.11.1530)]

I choke as I sit down to write.

A letter from Claire Runkel just arrived telling me that Philip's cancer got the best of him and he died quietly at home at 4 a.m. Thursday morning the seventh of June. Phil had been looking forward to his 90th birthday (25 June) to celebrate some of the good things in life, but he didn't quite make it.

I sure am glad I drove up to Eugene in early May to see Phil and Claire and to tell them what they have meant to me.

Best, Dag

Date: Wed, 13 Jun 2007 11:12:50 -0600
From: Bill Powers <powers_w@FRONTIER.NET>
Subject: Phil Runkel
To: CSGNET@LISTSERV.UIUC.EDU

[From Bill Powers (2007.06.13.1035 MDT)]

I returned from a visit (without email) to Ed Ford and his RTP group last night, and learned only this morning, in a phone call with David Goldstein, that Phil Runkel had died a week ago. My first thought was how lucky it was that I picked May to go see Phil.

Neither of us realized that he had only four weeks left, though we knew it wouldn't be twice that -- Phil hoped to make it to his 90th birthday on, I think, June 25. He and I both knew that this was our last meeting. We talked about his recalcitrant computer, about PCT, about our long friendship and its beginnings, about his wonderful life with Claire and his delight in her character and mind. Phil, Claire, and I talked a lot about death and dying, and Phil was glad I had come when I did instead of waiting for his funeral when he would not be there any more, and so was I. That's exactly why I made the trip. Getting to know Claire better and seeing what Phil saw in her was a delight.

Others offered tributes on CSGnet to Phil Runkel in the same time frame.
See CSGnet archives at www.pctresources.com.
For more on Phil Runkel and the family's celebration of his life, see About Phil Runkel at the site.
—Link at the web page dedicated to this volume.

Phil was under hospice care then and was keeping pain more or less in abeyance with medications that, to his impatience, dulled his senses and made him sleep a lot. But in the afternoons he was clear and sharp and funny as always. We went out to dinner at restaurants twice and he ate a bite or two, but clearly he enjoyed the company and being out in the world a lot more than the eating.

We spoke of love now and then; long ago we decided that Phil was the brother I never had. Our e-mail posts were often signed Brother Bill or Brother Phil, like a couple of old monks, or even almost-real-brothers.

I admired Phil for all the time I knew him, from the first letters in the early 1980s to the last brief morning of May 15, 2007 when I drove back to Eugene with a fix on one of my demos he couldn't run on his machine. He and Claire came out to the car to see me off, and we were all so sad we could hardly speak, or see. One last hug on the side that didn't hurt, and that would be my last memory of Phil Runkel. How I will miss him. How glad I am that he doesn't have any more pain.

We've lost a great man, as I don't need to tell anyone here.

Best,

Bill P.

2

There is no fulcrum on which we can rest the lever to move the world. There is no place for the observer to stand from which he or she can see the true nature of the universe. I've had that notion for a long time, though it means a great deal more to me now that I have read W.T. Powers. I had that notion in mind when I wrote in 1983 about the "operational definition" (enclosed). I was wrong, but I was not as wrong as a lot of other people.

It is indeed important to keep in mind that people do fly airplanes, and build airplanes, and write letters and get answers, and meet one another at the Biltmore. That is also what I mean by dependable bundles of perceptions. And it is important to keep in mind that people can conceive of senses and of ways of converting energies we cannot sense to energies we can sense, so we have microscopes and telescopes and radio receivers and cloud chambers and oscilloscopes (sp?) and devices to convert frequencies we can't hear to frequencies we can, and devices to convert light to sound and sound to light, and so on. And that helps us, I think, to estimate more of the degrees of freedom in the belfry. Or at the very least to be more confident that they are there.

And I guess we manufacture our own multiplying degrees of freedom. Don't we add to them with every level of control system? And the only way we can find out about them is to estimate those __this__ person has and then __that__ person. including ourselves. There is no virtue, of course, in setting out to make an exhaustive catalog of the estimated internal standards (reference signals) of every living person (or dead, either, as some historians and psychoanalysts like to do). But it would be worth while to undertake a catalog of a hundred or so persons just as an example of the kind of thing one can expect to find. And certainly social scientists should stop trying to find the one sock that fits all. They should give up the search for a few "principles" that explain all of human behavior. Sounds silly, but I am sure most of them do dream of such a thing. When I say "all of human behavior," I mean, of course, all human __acts__, because that is what they mean. Trying to find out __when acting__ will occur (your quest) is much more feasible.

Then, going back to the method of frequencies, making a tally of the kinds of acts that will be frequent in the presence of particular environmental resources can also be very useful. It can lead to ways (education, for example, or altering the environmental opportunities) of enabling people to make better use of the environments.

So there.

Phil R.

Opinions/Columns

```
I may be an incorrigible
romantic or sentimentalist,
but this sort of thing hits me
where I live.  It makes me
want to cry with yearning.
```

The Kansas City Times A-9

All I ever really needed to know I learned in kindergarten

By Robert Fulghum

Most of what I really need to know about how to live, and what to do, and how to be, I learned in kindergarten. Wisdom was not at the top of the graduate school mountain, but there in the sandbox at nursery school.

These are the things I learned: Share everything. Play fair. Don't hit people. Put things back where you found them. Clean up your own mess. Don't take things that aren't yours. Say you're sorry when you hurt somebody. Wash your hands before you eat. Flush. Warm cookies and cold milk are good for you. Live a balanced life. Learn some and think some and draw and paint and sing and dance and play and work every day some.

Take a nap every afternoon. When you go out into the world, watch for traffic, hold hands and stick together. Be aware of wonder. Remember the little seed in the plastic cup. The roots go down and the plant goes up and nobody really knows how or why, but we are all like that.

Goldfish and hamsters and white mice and even the little seed in the plastic cup — they all die. So do we.

And then remember the book about Dick and Jane and the first word you learned, the biggest word of all: LOOK. Everything you need to know is in there somewhere. The Golden Rule and love and basic sanitation. Ecology and politics and sane living.

Think of what a better world it would be if we all — the whole world — had cookies and milk about 3 o'clock every afternoon and then lay down with our blankets for a nap. Or if we had a basic policy in our nation and other nations to always put things back where we found them and cleaned up our own messes. And it is still true, no matter how old you are, when you go out into the world, it is best to hold hands and stick together.

Robert Fulghum is minister emeritus of the Edmonds, Wash., Unitarian Church. This piece, reprinted by permission, appeared in Church and Public Education.

Thou shalt choose thine own reference signals
so that they stealeth not from one another.

That is the first and great commandment.
And the second is like unto it:

Thou shalt choose thy feedback functions
so that they stealeth not
from those of thy neighbor.

On those two commandments hang all the law
and the prophets.

Single page found among enclosures. Date and context unknown. Style Runkel.

Part II

PCT, Science and Revolutions

See also:

A Look At Where We Are —Chapter Two in *The Method of Levels: How to do Psychotherapy Without Getting in the Way* by Timothy A. Carey (2006)
included in

Perceptual Control Theory; Science & Applications —A Book of Readings
Available printed from bookstores and as a free pdf download from www.livingcontrolsystems.com, featuring another 15 articles and complete chapters from most books on PCT.

Perceptual Control Theory—A Business Proposal at www.livingcontrolsystems.com

Why Study Perceptual Control Theory? at www.livingcontrolsystems.com

PCT is Reverse Engineering at www.livingcontrolsystems.com

PCT in 11 Steps

By William T. Powers

1 Behavior as Control

Control is a process of acting on the world we perceive to make it the way we want it to be, and to keep it that way. Examples of control: standing upright; walking; steering a car; scrambling eggs; scratching an itch; knitting socks; singing a tune. Extruding a pseudopod to absorb a nanospeck of food (all organisms control, not only human beings).

The smallest organisms control by biochemical means, bigger ones by means of a nervous system. Whole organisms control; the larger ones have brains that control; most have organs that control; if they are composed of many cells, their cells control; the DNA which directs their forms and functions controls; even some molecules, certain enzymes, control by acting on the DNA to repair it when it's damaged. Control is the most basic principle of life and can be seen at every level of organization once you know what to look for.

In this series[1] we will examine the process of control to see how it works, how it explains the behavior of organisms, how we can recognize it when we see it, and how understanding it can change our theories. In the first 11 mini-chapters we will see how PCT, Perceptual Control Theory, grows out of and replaces its main theoretical predecessors.

We will start by seeing how the mainstream of behavioral science found itself in channels that led to confusions and impossibilities, and how engineers who had no interest in psychology at all managed to discover the one basic principle that could have saved the sciences of life from a 300-year search down one blind alley after another. The problem is not that the life sciences got everything wrong; it's just that they got the most important things wrong: what behavior is, how behavior works, and what behavior accomplishes.

1 Bill Powers wrote this compact series of 11 brief statements to serve as an outline for a proposed TV program. The program did not come to pass, but this is an excellent summary of PCT.

2 Behavioral Science I

Before PCT, there was behavioral science. The "behavioral" part indicates that if we're behaviorists, we're interested in what we can see organisms doing, not in what we might guess goes on inside their minds, or brains, or other insides. Others have tried guessing, but without much success.

When a person accidentally moves a bare foot too close to a fire, an observer can see the foot pull away from it. In Descartes' *Treatise on Man* (1631) he says "If the fire A is close to the foot B, the small parts of this fire, which, as you know, move very quickly, have the force to move the part of the skin of the foot that they touch, and by this means pull the small thread C, [running up the back to the brain] ... simultaneously opening the entrance of the pore d, e, where this small thread ends... the entrance of the pore or small passage d, e, being thus opened, the animal spirits in the concavity F enter the thread and are carried by it to the muscles that are used to withdraw the foot from the fire."

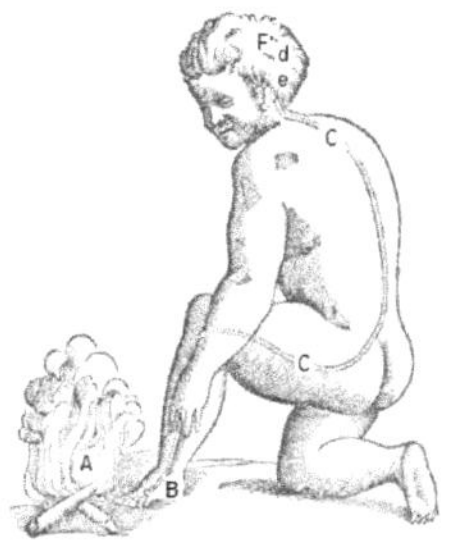

This sounds like an attempt to understand responses to stimuli, but 380 years later we can understand it as a description of a negative feedback control system, which we will get to before long.

If the observer happens to be the organism with the overheated foot, one more effect can be observed: it hurts. This leads to noticing that the foot is generally moved according to whether the sensed warmth is too little, too much, or just enough. The fire affects the sensed temperature of the foot in one direction; the response affects the same sensed temperature in the opposite direction. This turns out to be an exceptionally important observation. It's a pity that nobody could have analyzed it in 1631, but Newton's calculus then lay 73 years in the future. A differential equation would have explained this circle of causation that baffled philosophers of science until, 400 years later, control system engineering appeared.

3 Behavioral Science II

Just as PCT began to get organized, a new branch of behavioral science appeared: cognitive science. The emphasis moved from externally visible variables to those experienced by each individual. Now it was permissible to explore processes inside the brain and try to analyze them, but the phenomena to be explained scientifically were still basically the way stimuli cause responses. Theoretically, stimuli from the environment were now analyzed by cognitive processes in the brain, which then would formulate plans for generating responses appropriate to the stimuli.

The main task for the brain was now to figure out what commands should be sent to the muscles to generate appropriate results, given all the information coming into the brain from outside. This required the brain to have knowledge of neural and physiological processes as well as physical processes in the external world, and entailed rapid computation of the "inverse kinematics and dynamics" of body and environment ("kinematics" = properties of linkages, " dynamics" = movements of masses). Once this plan of action was turned into the set of necessary commands, it could be executed to produce the actions and their anticipated results.

There is something wrong with this picture. Rabbie Burns observed that the best-laid plans of mice and men gang aft agley, which is true not because we are bad at analyzing and planning but because plans of *action* are always close to their expiration dates. A planned action such as turning a steering wheel might produce exactly the wrong result if another car, a second later, changes direction by only a small amount. Planning all the turns of the steering wheel needed to drive from home to work couldn't conceivably get you to work the next day, no matter how precisely executed, even if exactly the same movements worked perfectly the day before. Think about other cars, traffic lights, pedestrians, weather, road repairs.

While planning clearly does take place, it can't operate by planning actions. We plan results, not actions, and that requires a new model of behavior. Even before cognitive science appeared, that new model was under construction.

4 Understanding Purpose

The new model was born in a parallel universe. Electronics engineers of the 1930s were using their new skills at designing electromechanical systems to automate tasks formerly done only by human beings. These tasks entailed a specification for some external condition to be brought about and maintained, even though it was impossible to predict or even detect all the events that might disturb that condition. The tasks included such things as aiming guns from the deck of a rolling ship; stabilizing the temperature of a room subject to opening and closing of doors and windows at unpredictable intervals on cool or cold days; adjusting the course of a torpedo to arrive at a moving target that made propeller-noises; keeping an airplane flying through rough air at constant altitude and speed, and on course.

To build such devices the engineers had to solve some basic problems. How could a (preferably) simple electromechanical device be given a specification for some effect that didn't yet exist, to be caused by a behavior that was not yet being carried out? How could this future state be made to cause an action in present time that would lead to that state? What if the effect of the action were disturbed *while* the device was producing the action? The engineers of the 1920s and 1930s, not knowing that the behavioral sciences had declared a device of this sort to be impossible (because future effects can't bring about their own causes), kept working at this problem until they solved it. The result was a new occupation called control system engineering, and (accidentally) a new theory of just about everything that lives.

These engineers had inadvertently discovered how purposive systems work. This discovery re-opened the door to the concept of intentional behavior directed by internal mental goals (which Watson, the founder of behaviorism, called a primitive superstition). The next logical step would have been to introduce this new understanding to the behavioral sciences. However, the sciences of life already had dozens of theories, all based on the idea that purpose is just causation misunderstood. They resisted mightily and that giant leap for mankind didn't happen.

5 Cybernetics *en Passant*

The Mexican physiologist Arturo Rosenblueth did notice the new ideas. He had been primed by studying under Walter B. Cannon, who worked to understand homeostasis, a process inside organisms that stabilizes critical variables such as nutritional state, body temperature, CO_2 level in the bloodstream, and other details of the life-support systems. Rosenblueth noticed that in the human body were many systems, behavioral systems, that appeared to work almost exactly in the way that the new artificial servomechanisms work. He communicated this discovery to Norbert Wiener, a mathematician at MIT where control engineering was rampant, and cybernetics was born.

Unfortunately, the main founders of cybernetics were not control-system engineers. They learned just enough about control systems to pattern cybernetic thinking around concepts like circular causation, but were more interested in subjects like communication, information theory, and (later) artificial intelligence and failed to carry the transformation to its ultimate conclusion.

That last step was not begun until the 1950s. That was when I learned of a recent school of thought called engineering psychology, and also started following the lead of W. Ross Ashby, a psychiatrist in the cybernetics movement who did have an understanding of control systems. With the help of R. K. Clark and R. L. MacFarland, I began to explore control systems with the idea of joining the cybernetics movement. After our first paper was published in 1960, we made overtures to psychology and cybernetics, but were put off by a general lack of interest. Clark and MacFarland went on to other things, and I kept working on PCT on my own. This led to my first book in 1973, then eventually to the formation of the interdisciplinary Control Systems Group in 1985, which in 1994 started a move toward becoming international by holding a meeting in Wales, and a few years later two meetings in Germany. The 22nd annual meeting of the CSG took place in 2006 at South China Normal University in Guangzhou, PRC, in collaboration with the Systems Society of China. PCT is part of the mainstream now. Almost.

6 A Scientific Revolution

The nature of a control system was almost understood by those who adopted behaviorism and cognitive science. There is something of each one in a control system.

The behaviorists realized, correctly, that behavior is based on perceptions that are caused by the physical events called stimuli. A driver can't keep a car on the road with both eyes closed. The kind of problem unsolved by behaviorism was how the stimuli could affect the driver's steering-responses in exactly the quantitative way needed to keep the car in its lane or steer it onto the correct exit ramp. This problem becomes worse when we realize that the driver also has to respond to *invisible* stimuli such as a crosswind. If the driver doesn't steer slightly into the wind by exactly the right amount, the car will drift into a ditch or into oncoming traffic. In general, stimuli as classes of happenings given names like "oncoming traffic" might lead to the right *consequences* of behavior ("avoiding collisions"), but are simply not the sort of thing that can produce the *quantitative amount and direction* of behavior needed.

Cognitive scientists realized, correctly, that behavior is the means an organism uses for achieving goals. An organism with a goal, they thought, must somehow figure out how to behave to achieve it. They noted, correctly, that the required behavior is not just a qualitative class of actions, but the quantitatively correct amount of action in exactly the right direction. The driver needs to perceive the environment to steer a car; the perceptions are supposedly the basis for the computations by which the organism calculates the actions that will achieve the goal. But it seems unbelievable that the driver could carry out all the repeated mental calculations required in the short time available, based on rather imprecise perceptions of what is going on out there.

In fact, neither behaviorism nor cognitive science hit on what now seems like the right explanation of behavior, though both hovered near it. The main mistake of both was to assume that the final product of brains was behavior, overcomplicated by the idea that behaviors must be exactly calculated.

7 The Solution: PCT

Here are the main questions unanswered by previous theories. How can stimuli produce not just responses, but *specifically appropriate* responses? What is a goal, that it can lead to just the behavior that will achieve it?

To answer these questions we have to look at things like perception and action a little differently. When someone steers a car, the perception that matters is the relationship of the car to the road as seen through the windshield. All the steering behavior has to be based on that perception—but not that perception alone.

It is also necessary for the driver to know, somehow, how that picture framed by the windshield *should* look if the car is to be properly located. This picture has to exist in the same place that the perception exists: in the brain. Without getting too neurological about this, we can say that whatever form the perception takes in the brain, the image of how the car and road *should* look must be in that same form, because the perception has to be compared with that image, the *reference image* ("goal:" goals are In Here, not Out There).

The difference between the imagined reference image and the real perception tells the driver how much steering error there is. "Error" just means the difference between reference and real. If the two coincide exactly, there is no error. If there is a mismatch in one direction, the driver should steer to the right. If in the other direction, to the left. That is basic control theory.

Now the cognitive scientist wakes up and says, "Yes, but exactly how much left or right? The brain has to calculate that, doesn't it?" The answer is yes, but. Yes, if there's a big error the brain should cause the steering wheel to turn a lot or if a small error, a little. But (and now we see the beauty of classical negative feedback control theory) the brain doesn't have to compute the exact amount because it can continuously adjust the action as the error changes, making smaller and smaller approximate adjustments as the error gets smaller until there is no error. Then no more changes in steering effort occur and the car is where it belongs in the lane. No complex computations. No planning. Just one swift simple process that converges smoothly to a final condition.

8 Behavior in the Real World

A driver traveling along a straight level road sees the picture in the windshield as exactly right; he steers neither to the right nor to the left. But is that true in the real world? Riding with a driver, we see endless little movements of the steering wheel, yet we don't feel or see the car moving left or right in its lane. The driver's steering efforts seem to be having no effect.

The reason is simple once you work it out. When the car starts drifting a little to either side for any reason, the driver immediately turns the wheel the other way as much as needed to keep the drift from getting larger, then a tweak more to eliminate it. If the driver can detect changes of the car's position as small as we can detect, or smaller, then we will never see or feel anything but tiny, barely-detectable, changes in position—if any at all. But the steering efforts can be quite large, in a gusty crosswind. It really looks as if the driver is responding directly to the crosswind, but of course in a closed comfortable car there is no way to detect the crosswind, except through effects on the car that the driver is mostly preventing. The result is that the deviations of the car are kept very small, especially in comparison to what would happen if the driver *didn't* make those steering movements. This is called negative feedback control—the same thing Descartes described.

So it seems that control means keeping disturbances from having much effect. But now, suddenly, the driver is turning the wheel so the car veers entirely out of its lane, a huge steering error. We immediately see why: it's an exit ramp. But why doesn't that steering control system act immediately to counteract the error? Because the reference image has been changed (one more time: reference image, reference perception, reference condition = GOAL). In fact, the driver's brain has smoothly changed the reference image from that of a car going straight in its lane to that of a car curving off to the right and up the ramp. The control system, still keeping the perception of the car's position matching the reference image, automatically alters the steering actions so as to keep the steering error close to zero. We see that simply by smoothly altering the goal of the behavior, the driver accomplishes the required change in behavior in an extraordinarily simple way, with no complex calculations.

9 Behavior: The Control of Perception

Behavior is the externally visible part of a process by which perceptions of various aspects of the experienced world are controlled. It is not the end-product of either the effects of stimuli or the goals sought by the organism. Behavior is simply the adjustable means by which an organism can keep its perceptions matching reference conditions. As disturbances come and go, behavior changes to have equal and opposite effects. As reference conditions vary, behavior changes to cause perceptions to vary in a matching way.

Behavior changes to cancel the effects of the disturbances on whatever the organism is controlling. The appearance is that the disturbances cause the actions, the observable behavior. But the real story is that the actions prevent the disturbances from significantly altering what the organism is concerned with: the perceptions it is controlling. This is how PCT explains the appearances that led to behaviorism.

When we make plans, the appearance is that we plan what behaviors will be needed to achieve what we want. But we can't predict what disturbances and changes in properties the environment is going to throw at us. What we can do is plan the perceived consequences we want to happen. We don't plan actions; planning successfully means planning perceptions. Higher levels in us tell lower control systems what perceptions to experience. The lower control systems adjust their actions to make their perceptions match the reference conditions they are given, and (without being told) enough more to cancel the effects of any disturbances that might be happening. This is how PCT explains the appearances that led to cognitive science. PCT does not require the brain to perform miracles of prediction and impossibly fast, complex, and accurate computations.

PCT thus encompasses the concepts of behaviorism and cognitive science, providing a single framework in which the observations of both can be understood. With one more added concept—levels of control—it expands to encompass all that human beings and perhaps all organisms experience.

10 Emotion

The control hierarchy can control perception at many levels by using actions from mild to strong, but there is something missing: feelings. This model doesn't suggest the physical feelings that accompany emotions, but one modification of the model can put feelings into relationship with the goals that go with them, to cover both the cognitive and feeling sides of emotion.

Disturbing higher control systems or changing goals causes errors that generate a cascade of changes in the reference signals passed down the hierarchy of control. We now divide this cascade into two branches. A behavioral branch goes to systems, mostly learned, that control using muscles. A somatic branch, primarily a product of evolution, goes into the amygdala, then the hypothalamus, and then the pituitary gland and autonomic nervous system which control the state of the body. This branch is where emotions supposedly originate, but in the PCT theory emotional feelings are effects, not causes.

Some control systems are inherited; most are learned. All act to adjust both the somatic systems and the action in the behavioral branch. The somatic branch adds sensations that generate the feeling component of the configurations we call emotions. Example: Either learned or innate systems can specify goals like escaping or attacking. If the perception differs from the reference, a "motivating" error signal is sent to multiple lower behavioral systems as reference signals. The effect of the error signal on the somatic branch provides the feeling part of the experience, the so-called fight-or-flight syndrome. The goals of attacking or fleeing distinguish fear from anger; the physiological states have been found to be identical in both emotions.

The feeling part of emotions often arises without any consciousness of the cause. This can happen if awareness is engaged at higher levels, and a disturbance occurs that affects lower-level control systems not currently in awareness. Those systems will react automatically by using the muscles and, according to this theory of emotion, will also adjust the physiological state of the body. The sensations arising from the physiological states will be processed level by level up the hierarchy, and when the perceptions reach a level accessible to awareness, will attract attention exactly as if they had occurred spontaneously, or had been caused from outside the body. An injection of adrenaline can be interpreted and experienced as fear *or* anger.

11 The Hierarchy of Control

The driver keeps the car in its lane, yes. But why? To stay alive, surely, but there are more immediate reasons. The driver has a destination in mind, and wants to get there. The reference perception: *I am at the entrance to the parking lot at the mall.* The actual perception: I am on 55th street a mile from the parking lot. So keep the car moving along in its lane. When the entrance appears, change the reference: *the car is following **this** path into the lot.*

The higher system is not telling the lower one what to do but showing it what to perceive. It does so by continuously varying the reference image, not by commanding steering wheel movements. The lower system automatically corrects the effects of disturbances and little steering errors on the car's path without having to be told to do it. The higher system needs only to alter the images that the lower system is to reproduce by turning the wheel. The lower system determines when, how much, and which way to turn the wheel.

The reason for going to the mall is to buy a dress shirt. The reason for buying the dress shirt is to look good at a wedding. The reason for looking good is to please the woman you're going to marry. The reason for pleasing her is that you want to show respect for her opinions. The reason you show respect for her opinions is that you want to make the marriage as ideal as you can, and see respect as an essential principle for making a good marriage.

Each level of control sets multiple goals for the next level down to perceive; that's how any higher system controls its own perceptions. The higher system's perception is built out of the perceptions that exist, some being controlled, at lower levels. There are many control systems at each level, and more than a few levels. The only systems that act on the environment directly are those at the first level. All the rest act by adjusting the perceptual goals for lower systems. All control their own perceptions, not their actions.

Now you know the essence of Perceptual Control Theory, which replaces the basic concepts of behavior in both behaviorism and cognitive science. A revolution, in progress.

Bill Powers,
Lafayette, Colorado, October 2009

This series continues with ***Reorganization and MOL***, an overview of how control systems may come into being, change, cause internal conflict, and ways to resolve internal conflict.

(Posted at the website)

Foreword to Living Control Systems II

Living Control Systems I & II
are collections of selected papers by
William T. Powers
published in 1989 and 1992, respectively

Foreword by W. Thomas Bourbon

In 1979, Bill Powers wrote a prophecy: "A scientific revolution is just around the corner, and anyone with a personal computer can participate in it.... [T]he particular subject matter is human nature and in a broader scope, the nature of all living systems. Some ancient and thoroughly accepted principles are going to be overturned, and the whole direction of scientific investigation of life processes will change." (William T. Powers, "The Nature of Robots: Part 1: Defining Behavior," *BYTE* 4(6), June, p. 132) Powers foresaw the overthrow of the idea that either stimuli from the environment, or commands from the mind or brain, are sole causes of behavior. In its place, he offered the concept that people (and in their own ways all other organisms) intend that they will experience certain perceptions and behave to cause the perceptions they intend. The social, behavioral, and life sciences had simply missed the fact that living things control many features of their environments. Powers acknowledged that fact, and he realized that to an organism the environment exists only as perceptions, hence his insight that organisms act to control their own perceptions. His formal statement of the new concept was control theory, and he said amateurs, working with personal computers on their tables at home, would be major players in the revolution. Thirteen years later, the revolution is not accomplished, but it is underway.

Powers' perceptual control theory is new, but he is not the first to describe many of the key ideas in the theory. Over 2200 years ago, Aristotle wrote about intention—"that for the sake of which," the desire or wish that causes actions that result in a particular end. Aristotle used many examples in which a person acts to produce an intended object, such as a bed, statue, tray, or house. The person's intention to create the object is the "final cause" of the actions that produce the object. Aristotle wrote that, depending on the condition of the world and the intention of the person, the same actions sometimes produce different ends, and different actions sometimes produce the same end. All of that sounds like good control theory, so why are those ideas considered revolutionary today?

For many centuries, Aristotle's ideas disappeared from Europe and were preserved by scholars in the Arab world. They returned, in altered form, to a Europe dominated by Christian theology. Theologians changed "final cause," which to Aristotle often meant only a person's intention to manufacture a bed out of wood, into God's original plan for the linear unfolding of history, from creation, to Calvary, to Apocalypse, to the end of time. Aristotle's original idea was unrecognizable.

Most early European scientists worked within Christian theology, embracing its notion of linear time and its implication of linear cause and effect. Many of these scientists mistakenly assumed that the original concept, that a final cause is a goal, implied that the future influences the present—a clear violation of the assumed linear flow of cause and effect. Eventually, potentates of The Church and potentates of Science came to a falling out over dogma. Those who established the canon for Science had yet another mistaken reason to reject final cause: they said it represented an appeal to the supernatural, in the form of God as agent. The idea that there is purpose or intention in the behavior of any living thing was pronounced "unscientific." Most aspiring behavioral and biological scientists still affirm that credo.

When William James wrote one hundred years ago, the ideas of purpose and intention were popular again. James said purposive behavior is the distinguishing feature of intelligence—of life. He said that in a variable world an organism's behavior necessarily varies to produce unvarying intended results. James wrote that people do not intend their specific actions; they intend to experience perceived consequences of their actions, then they vary their actions any way necessary to produce those perceptions. For a while, it looked as though the idea of intention might take hold, but once more the idea was purged from the sciences of behavior and life. Orthodox scientists asserted that intention implies final cause, which necessarily implies an appeal to supernatural forces and to a temporal reversal of causality. Purposive behavior was banished, on the one hand by behaviorists, environmentalists, and reflexologists who claimed that events in the environment determine behavior, and on the other by those who claimed that instincts acting as internal stimuli cause behavior. People on either extreme believed their positions were dramatically different, but they all portrayed behavior as the end result of a linear chain of cause and effect.

Powers writes at a time when purpose and intention remain unacceptable to most scientists. Behaviorists still believe environmental "stimuli" have the "power" to control behavior; and most cognitive scientists and neuroscientists say the mind-brain issues "commands" that cause muscles to produce appropriate behavior. Cognitive-neuroscientists frequently claim behaviorism is dead and a cognitive revolution has swept the behavioral and life sciences; in return, behaviorists pronounce themselves very much alive, and some portray cognitive theorists as "creation scientists," bent on keeping alive the concept of soul-as-mind. Once again, each camp believes its views differ markedly from those of the other, but both embrace the wearisome model of linear cause and effect—a model that was necessary a few hundred years ago to establish the physical sciences, but a model that mistakenly rejects what Powers recognizes as the defining properties of life. Neither wing of the cause-effect orthodoxy recognizes the abundant evidence that organisms control many parts of their world. But revolutions have a way of changing the minds of the orthodox.

Powers turned the millennia-old idea that living systems act to produce intended perceptions into a formal theory of behavior: perceptual control theory. Perceptual control theory identifies behavior as the necessarily variable means by which organisms control their perceptions of the world. Working first on a build-it-yourself computer (the one he used when he wrote his prophecy), then on a first-generation IBM personal computer, Powers created elegant demonstrations in which the simple-idea-turned-formal-model generates remarkably accurate quantitative simulations and predictions of behavior and its consequences. He identified a first principle for behavioral, social, and life sciences and showed the way to a new foundation of theory and method.

For several years, only a few people followed Powers' lead, and even fewer gathered the data and performed the modeling that could establish control theory as an alternative to traditional science. But interest in the theory grew—a tribute to the dogged efforts of William and Mary Powers, over three decades, to maintain the visibility of the theory. During most of that time, Powers published only one book and a few papers. More recently, information about control theory burst into wider circulation through two functions of personal computers that no one predicted in 1979: desktop publishing and electronic-mail networks. Those applications freed perceptual control theory from the heavy hands of editors and reviewers who routinely rejected manuscripts on the theory. They were true defenders of cause-effect orthodoxy, rejecting control theory as uninteresting and unnecessary, or as merely another way to describe things that were already understood. The new media let many people see control theory, then judge it on its own merits. The once-small circle of people aware of the theory grew into a network spanning the world, including people from many disciplines, specialties, and professions. And the demand for Powers' writings grows.

In the Foreword to the first volume of Living Control Systems, Richard Marken wrote about the difficulty he experienced several years ago when he tried to locate published material by Powers. Volume I was a collection of Powers' published work But Powers has written far more than he has published. When he writes, Bill does not revise his drafts. If he encounters a block or is dissatisfied, he starts over. He has cast aside several beginnings of books and many drafts of chapters and papers that he never submitted, or that were rejected by editors and reviewers. Most of us would be happy if any of our publications equalled the quality of the work Bill put away in drawers and boxes and, more recently, on disks.

Over the years, only a few people have had a chance to read parts of Bill Powers' unpublished work. The opportunity to rummage about, looking for those gems, was at least part of "that for the sake of which" some of us travelled to his "laboratory" in the back room of his home in Northbrook, Illinois. When Mary and Bill decided to move to Colorado, Edward Ford, a counselor in Arizona, suggested that the mandatory gathering of possessions into boxes provided an excellent chance to select part of Bill's unpublished work for an edited volume. Greg Williams, a frequent visitor to Northbrook, journeyed there from Kentucky for the last time to gather the pages and disks and take them away so he could select the pieces in this volume.

This volume contains a small sample of the previously unpublished material from the years when Bill and Mary Powers were in Northbrook. If you want to rummage through the next accumulation, you must travel to the new site of The Laboratory of William T. Powers. That is the locus of many of today's clearest insights into purposive behavior. Over the millennia, that locus has moved from Aristotle's Lyceum, to James' Harvard, to Northbrook, and now to a house atop a ridge near Durango, with a view of the San Juan Mountains, located only a few miles away, across a broad valley—a view that, years ago in Illinois, Mary and Bill Powers said they intended to see out their back door. Stated intention, actions, and perceived consequences that match the intention. It looks like control to me!

W. Thomas Bourbon
Nacogdoches, Texas
February 1992

On the Phenomenon of Control. In the foreword above, I sketched a history of the often-rejected idea that living things act to control their own experiences. There is also a long history of devices that mimic control by a person. In classical times, observers of manufactured control devices often identified them as "mysterious" or "miraculous." There were lighted lamps in which the wicks and oil were never consumed, and vessels in which, no matter how much was consumed, the levels and flows of water or wine never changed, and statues that seemed to move of their own accord. The "miraculous" phenomenon of control was there for all to see, but the ingenious devices that actually controlled were hidden from view and the principles of control went unrecognized.

Centuries later, the metaphor of the machine was dominant in European thought. People were compared to lineal machines, embodying discrete, sequential cause and effect. The idea that people *resemble* machines soon gave way to the still-popular assertion that people *are* lineal cause-effect machines. Overextended metaphors aside, the design, and eventually the theory, of control devices moved on, from a variety of hydraulic and mechanical governors and regulators in the 1600s and 1700s, to electronic controllers in the 1920s and 1930s. Today, control devices are ubiquitous, yet most people who say a person is a machine (probably a computing machine), mean people *are* lineal cause-effect machines, not controllers or regulators.

To most people, the phenomenon of control typically goes unnoticed or unacknowledged, whether the controller is a living system, or an ingenious device. Control: it is everywhere, and everywhere it is denied.

December, 1994. W. Thomas Bourbon
University of Texas Medical School-Houston

PCT—An Engineering Science

by Dag Forssell, October 1994

Basic PCT offers a clear explanation for the pervasive phenomenon of simple purposeful behavior or control. Hierarchical PCT (HPCT) outlines a hierarchical arrangement as the likely organization of multiple control systems in humans.

The kind of explanation PCT offers for human behavior is the kind of explanation responsible for the successes of modern engineering.

Just hold up a finger in front of you and bend it. Notice that just before it bends, you will it to bend. The willing and the bending are facts we experience. How can you explain this phenomenon of behavior?

A "popular theory" approach has been to describe appearances in terms of themselves. Life scientists think and talk in terms of reflex, stimulus and response, affordances, conditioning, reinforcement, and cognition—terms which give apparent phenomena names without actually explaining them. Much research in the life sciences is focused on accumulating descriptions where weak statistical correlations suggest mysterious causal relationships.

An "engineering theory" approach is to suggest and describe the *properties* and *organization* of elements which when they *interact* with each other and their environment *produce* the kind of behavior we observe. Thus an engineering theory approach proposes a *model* or *simulation* of an underlying set of properties and causal relationships which are invisible and cannot be experienced directly, but where we gain confidence through repeated successful experimentation. Engineers learn to visualize and think in terms of models and simulations in the course of their training as they repeat the basic experiments which define the many invisible "laws of nature" or "first principles" of engineering science. In practice, engineers deduce properties of new designs from these first principles and the behavior of the designs from the properties. Engineers predict the performance of a design or model in various environments and circumstances. Thus they predict experiences they have not yet had, and with confidence. The in-depth understanding fostered by the approach of modern engineering theory is the reason for spectacular progress in the engineering sciences in the last several centuries.

Your bending of the finger (converting your thought into action) is an example of control with a changing reference signal. Behavior "emerges" from the natural properties of control systems as they interact with their environment. In engineering, control has been well explained only since the 1930s. In the life sciences of today, control is not yet part of the explanation for behavior. Thus life scientists attempting to explain "finger-bending behavior" do so without recognizing or understanding the organization and properties of the basic organizing principle of behavior.

HPCT offers a new explanation for human experience. It is technically elegant, conceptually simple, testable, and better than "common sense." The principles of HPCT are readily understood by any attentive person. In practice, a person who has learned HPCT can deduce properties of organisms and people from the principles of HPCT and see how the behavior and interactions of people "emerge" from those properties in different circumstances.

When you learn the explanations of HPCT, you can apply them to explain past experience as well as think ahead. Your own experiences suddenly make more sense to you, and you can manage and lead better in the future.

An Essay on the Obvious

William T. Powers January 1991
Post to CSGnet

. . . I'm reminded of a lot of the "new physics" stuff that's been going around—The Emperor's New Mind, The Quantum Self, chaos in the brain, and so on. I'd like to say this about that:

AN ESSAY ON THE OBVIOUS

I think that all attempts to apply abstract physical principles and advanced mathematical trickery to human behavior are aimed at solving a nonexistent problem. They all seem to be founded on the old idea that behavior is unpredictable, disorderly, mysterious, statistical, and mostly random. That idea has been sold by behavioral scientists to the rest of the scientific community as an excuse for their failure to find an adequate model that explains even the simplest of behaviors. As a result of buying this excuse, other scientists have spent a lot of time looking for generalizations that don't depend on orderliness in behavior; hence information theory, various other stochastic approaches, applications of thermodynamic principles, and the recent search for chaos and quantum phenomena in the workings of the brain. The general idea is that it is very hard to find any regularity or order in the behavior of organisms, so we must look beyond the obvious and search for hidden patterns and subtle principles.

But behavior IS orderly and it is orderly in obvious ways. It is orderly, however, in a way that conventional behavioral scientists have barely noticed. It is not orderly in the sense that the output forces generated by an organism follow regularly from sensory inputs or past experience. It is orderly in the sense that the CONSEQUENCES of those output forces are shaped by the organism into highly regular and reliably repeatable states and patterns. The Skinnerians came the closest to seeing this kind of order in their concept of the "operant" but they failed to see how operant behavior works; they used the wrong model.

Because of a legacy of belief in the variability of behavior, scientists have ignored the obvious and tried to look beneath the surface irregularities for hidden regularities. But we can't develop a science of life by ignoring the obvious. The regular phenomena of behavior aren't to be found in subtleties that can be uncovered only by statistical analysis or encompassed only by grand generalizations. The pay dirt is right on the surface.

The simplest regularities are visible only if you know something about elementary physics—and apply it. Think of a person standing erect. This looks like "no behavior." But the erect position is an unstable equilibrium, because the whole skeleton is balancing on ball-and-socket joints piled up one above the other. There is a highly regular relationship between deviations from the vertical and the amount of muscle force being applied to the skeleton across each joint. There is nothing statistical, chaotic, or cyclical about the operation of the control systems that keep the body vertical. They simply keep it vertical.

The same is true of every other aspect of posture control and movement control, and all the controlled consequences of those kinds of control. Just watch an ice-skater going through the school figures in competition. Watch and listen to any instrumentalist or vocalist. Watch a ballet dancer. Watch a stock-car racer. Watch a diver coming off the 10-meter platform. Watch a programmer keying in a program.

It's true that when you see certain kinds of human activity, they seem disorganized. But that is only a matter of how much you know about the outcomes that are under control. The floor of a commodities exchange looks like complete disorder to a casual bystander, but each trader is sending and receiving signals according to well-understood patterns and has a clear objective in mind—buy low, sell high. The confusion is all in the eye of the beholder. The beholder is bewitched by the interactions and fails to see the order in the individual actions. When you understand what the apparently chaotic gestures and shouts ACCOMPLISH for each participant, it all makes sense.

Of course we don't understand everything we see every person doing. It's easy to understand that a person is standing erect, but WHY is the person standing erect? What does that accomplish other than the result itself? We have to understand higher levels of organization to make sense of when the person stands erect and when not. We have to understand this particular person as operating under rules of military etiquette, for example, to know why this person is standing erect and another is sitting in a chair. But once we see that the erectness is being controlled as a means of preserving a higher-level form, also under control, we find order where we had seen something inexplicable. We see that an understanding of social ranking, as perceived by each person present, results in one person standing at attention while another sits at ease. Each person controls one contribution to the pattern that all perceive, in such a way as to preserve the higher-level pattern as each person desires to see it.

It seems reasonable that once we have understood the orderliness of simple acts and their immediate consequences, we should be able to go on and understand more general patterns that are preserved by the variations that remain unexplained. As we are exploring a very large and complex system, we can't expect to arrive at complete understanding just through grasping a few basic principles. We must make and test hypotheses. But if we are convinced that the right hypothesis will reveal a highly-ordered system, we will not stop until we have found it. If, on the other hand, we are convinced that such a search is futile, that chaos reigns, we will give up the moment there is the slightest difficulty and turn to statistics.

I claim that human behavior is understandable as the operation of a highly systematic and orderly system—at least up to a point. I say that it is the duty of any life scientist to find that orderliness at all discoverable levels of organization, and to keep looking for it despite all difficulties. We must explore all levels, not just the highest and not just the lowest; what we find at each level makes sense only in the context of the others.

We have a very long way to go in understanding the obvious before it will be appropriate to look for subtleties. I have no doubt that we will come across mysteries eventually, but I'm convinced that unless we first exhaust the possibilities of finding order and predictability in ordinary human behavior, we won't even recognize those mysteries when they stare us in the face. I don't think that anyone is prepared, now, to assimilate the astonishments that are in store for us once we have understood how all the levels of orderly control work in the human system.

We won't get anywhere by looking for shortcuts to the ultimate illuminations that await. Most of the esoteric phenomena of physics that are taught in school today were occurring in the 19th Century, as they always have. But who, in that century, would have recognized tunneling, or coherent radiation, or time dilatation, or shot noise? If we want to see a Second Foundation of the sciences of life, we have to begin where we are and build carefully for those who will follow us. If we succeed in trying to understand the obvious, the result will be to change what is obvious. As the nature of the obvious changes, so does science progress.

Teaching Dogma in Psychology

By Richard Marken

Dr. Richard Marken, Associate Professor of Psychology, joined the Augsburg College faculty in 1974.
Lecture date: May 15, 1985.

I HAVE COME HERE to come out of the closet—I am not a straight psychologist. I have been convinced for at least five years now that the foundations of my discipline are wrong. I feel like the little boy who noticed that the emperor was not wearing any clothes. All the people who would like to be considered smart are saying that behavior is controlled by environmental events.

This is the central dogma of scientific psychology and of the social sciences in general. It is the basis on which all research is conducted in these disciplines.

Things look quite different to me. It looks to me as if behavior controls the environment—not vice versa. Behavior is the process by which we control the things that matter to us—to behave is to control.

The difference between the conventional view of behavior and my own is fundamental. From my point of view the introductory psychology texts are wrong from the preface on. There are irreconcilable differences which I will try to make clear. As you can imagine, given what I have just said. It has been terribly difficult to teach some of the standard psychology courses, notably the intro course and the research methods course. It is not a problem that can be cured by putting a little section on "my point of view" in these courses. It would be like having to teach a whole course on creationism and then having a "by the way, this is the evolutionary perspective" section. Why waste time on non-science? From my point of view, most of what is done in the social sciences is scientific posturing and verbalizing.

First, let me tell you a little about how I came to this revolutionary position. I did not set out to be in this boat; I am not a revolutionary by temperament, and I have not been brainwashed by some weird cult.

I was trained as a standard experimental psychologist. My specialty was auditory perception. I did my thesis research on an esoteric but conventional topic—auditory signal detection. I knew my stuff—I became an expert in experimental design and some of the more powerful aspects of statistical analysis.

Shortly before coming to Augsburg, in 1974, I was browsing through the library at UCSB and noticed a new book with the intriguing title: *Behavior: The Control of Perception*, by William T. Powers. I was curious, because I was a student of perception and interested in behavior. But I couldn't imagine what this book might be about. I looked through it briefly. My impression was that the author knew what he was talking about. I, however, did not. The book, it turns out, was about control theory as a model of behavior. I had no idea, at the time, that control theory would eventually turn my professional life into agony and my intellectual life into bliss.

During my second year here I discovered that Powers' book was in our library. I went back to take a look at it. I had an idea that it might help me in a talk I was preparing, at the time, on the control of behavior. This talk was to be sort of a rebuttal to one given earlier by Dr. Ferguson on the glories of behavior control. I was trained at a school that was very oriented toward cognitive psychology, bristling with the then new computer-oriented approach to behavior. I thought Skinnerian behaviorism a dinosaur that had been comfortably interred so I was surprised to find so many people here who not only admitted but were proud of their adherence to Skinnerism. I was going to present the enlightened cognitive view. I know now that the differences between cognitive, behaviorist, and other approaches to psychology are matters of form more than substance—different verbalisms for the same basic model.

I tried formulating the talk on the basis of concepts from cognitive psychology—along with some of the stuff I was learning from Powers' book. But as I read and re-read Powers, he seemed to make more sense than anything I was reading in the cognition texts. Powers spoke directly and clearly to the fundamental problems that I had only intuitions about. I realized that cognitive psychology was trying to differ from behaviorism by talking bravely about mind, but the basic approach was the same: behavior is caused by inputs into the system; the inputs just swirl around more inside the system before coming out as behavior. I eventually based the entire talk on Powers' book, which I really didn't fully understand at the time.

After the talk, my interest in challenging Skinner diminished, but my interest in control theory continued to grow. I was still a conventional psychologist. I was even trying to do some perceptual research—based on the standard model. But control theory kept bugging me. I wanted to do research based on control theory. I tried to graft control theory into some of my research projects. This really didn't work; Control theory implies such a fundamentally different orientation to behavior that attempts to apply control theory to the results of most conventional research will be fruitless—I will explain why in a moment.

This was about 1978, and I was starting to see the beauty of control theory. My faith in conventional psychology was waning, and this was very troubling. I read all I could find on control theory. I started to realize that much of what was said about control theory or feedback theory in the behavioral science literature was wrong.

In 1978, Powers came out with an excellent article in *Psychological Review*. This was a significant event, because it was the first new publication I knew of, since his book, and it described some actual experiments demonstrating some of the basic principles of control theory. The article was rough going—mathematically and conceptually. But I set up the experiments on my computer and started really to understand what was going on—and what was going on was downright amazing. The process of behaving is a truly remarkable phenomenon; I began to understand what the title of Powers' book meant: To behave is to control, and what control systems control is not their actions but the perceptual consequences of their actions.

My understanding was further expanded by a series of four articles Powers published in *Byte* magazine in 1979. The experiments I was doing (and still do) look pretty simple. They involve controlling events on a computer screen. Though simple, the experiments demonstrate the way control systems work—and the results are completely inconsistent with all current models in psychology. Control systems behave in ways that are quite counter-intuitive. The experiments are simple for the same reason that the experiments in physics labs are simple—we know what results we're going to get. The results are perfectly repeatable. They show how control works. Once you know the principles and can repeatedly demonstrate them, you have a solid foundation for going on to more complex phenomena. The experiments I do are of a type completely alien to conventional "Psychology Today" mentality, so they are sometimes dismissed as trivial. To my mind, one quality fact is worth all the statistical generalities in all the social sciences.

In 1980 I began my own little research program on control theory. I designed a number of studies that were aimed at showing how the behavior of a control system (like a person) differs from that of the kind of system that psychology currently imagines people to be. I have had little difficulty publishing these reports, and the reception of my work at meetings has been positive—probably because no one really understood what I was talking about.

By 1981 I had become a complete prodigal. I now understood control theory rather well and knew precisely why it was usually a waste of time to try to interpret existing research findings in terms of control theory. This is the usual challenge I get—how does control theory explain this or that "fact"? My first answer is that the statistical results you find in the social sciences do not, for me, constitute meaningful facts. But the real problem is that facts obtained in the context of the wrong model are simply misleading and worthless.

Once you get to a certain point in your understanding of control theory, you realize that almost all of traditional psychology can be ignored. This is a rather sickening experience at first, and everyone I know who gets excited about control theory eventually encounters the problem. A clinician friend of mine in New Jersey, an avid control theorist, just isn't willing to cross the line and ignore what deserves to be ignored—yet. I sympathize. It's not easy to ignore everything you were once taught to take very seriously. But this is what had to be done in physics after Galileo. You just have to take off in the right direction. Physics doesn't need to spend a lot of time explaining why pre-Galilean physics is wrong. Revolutions are revolutionary—you don't gain anything by clinging to old ideas that are wrong, no matter how much you used to love them.

Current approaches to psychology and the social sciences are based on an input-output model of behavior. In every methods class you learn that the proper way to study behavior is to manipulate independent variables (environmental input, such as room temperature or reinforcement schedule) to determine their effects on dependent variables (behavioral outputs that you have carefully operationally defined so as to be measureable). This should all be done under controlled conditions, so that you can correctly infer causality—that is, if there is a change in behavior, this change can be attributed to variation of the independent variable.

In some social sciences manipulation and control is impossible, but the approach is the same: look for correlations between input and output variables, between environment and behavior. This is bread-and-butter psychology and sociology and economics and political science. It's easy to do once you get used to it.

This method of doing research will give you good results only if the objects of study are input-output devices. Whatever the verbalisms used to describe different theories, the model of research in the social sciences assumes that organisms are some type of input-output device—arguments concern only what type (computer, conditioning machine, etc.).

The social sciences have persisted in using this model in spite of the fact that it *clearly does not work*. The results of research in the social sciences are a mess by any reasonable scientific standard. They are extremely noisy. Statistics must be used to determine whether anything happened at all in most studies. The reason for all this variability in the data is usually attributed to random stimuli flying around in the environment. But after 100 years of doing this kind of research, using more and more sophisticated apparatus and control, the variability is still there and it is still large.

Nowadays the variability of data in the social sciences is attributed to the inherent variability of behavior. Besides being unscientific by blaming the failure to understand a phenomenon on the objects of study, this posture can be seen as ridiculous just by looking around. If the behavior of the architects, engineers and workers who built the buildings in this city were as variable as social scientists imagine it to be, few of these structures would still be standing.

In fact, behavior is variable only when looked at from the wrong point of view—the point of view of the input-output model. What's wrong with the model can be seen by considering the output side of the model in more detail: Just what is behavior? The textbooks say that it is anything that organisms do—but we know that's not so. Psychologists don't study the acceleration of animals as they are accelerated to earth by the force of gravity, but the animal is behaving.

The behavior we are interested in is the kind that is generated by the organism itself—not only generated by the organism itself, but consistently so. If organisms never did anything more than once, we would see chaos. Instead, we see regularity—pressing a bar, getting dressed, having a conversation, making love.

The events that we recognize as behavior are named for the uniform results produced by organism actions, not for any particular pattern of the actions themselves. Thus we see an animal pressing a bar, but fail to note that the result (the lever going down) is always produced by a different pattern of actions. In fact, the detailed actions that produce any behavior are always different and must be different if the result is to repeat. The appropriateness of this variability cannot be understood in terms of the input-output model, so it is ignored.

Students of behavior have noticed that organisms use variable acts to produce consistent results, but few have noticed that these variations are necessary. Skinner, for example, considered the different ways the rat gets the lever down to be arbitrary—one way is just as good as another. In fact, if the rat pressed in the same way each time, the lever would not go down on each occasion. The apparently random variability is really not random at all. But this causes a problem, because it then appears that the organism is varying its actions in just the right way to produce a consistent result. It looks like the animal is trying to get the lever down. This implies internal purposes, and there is no room for such things in an input-output model.

E.C. Tolman was on the right track. He showed that rats who could run a maze to a goal could still get to the goal when the maze was filled with water. Tolman correctly concluded that the rat had the purpose of getting to the goal and was using whatever means necessary to produce that result. But this was in the 1930s, before control theory and hence the tools to explain how purpose could be carried out. So everyone said, "response generalization" and went back to the labs with the input-output model intact (in their heads, if not in reality).

However, if one thinks about it for a moment, it is clear that Tolman's phenomenon—together with many everyday examples of the same thing—is completely inconsistent with the notion that behavior is the last step in a causal chain, as the input-output model implies. There is no way for any input-output system, however smart, to produce actions that will always have the same result in an unpredictably changing world. The straight-through causal model breaks down completely.

When we do anything we are adjusting our actions, usually without even being aware of it, to produce the intended result, regardless of the prevailing environmental circumstances. The rat pressing a bar is not just emitting this result—it is producing forces which, when combined with all other forces acting on the bar, produces the result "lever press." These "other forces," which I call disturbances, are always present when we do anything. We usually don't notice their contribution to behavior because their effects are usually precisely canceled by the actions of the organism. If I pressed a bit on the other end of the rat's lever, the lever would still go down because the rat would increase the forces it exerts in just the right way to produce the intended result. If I block a route you usually take to get to the store, you will get there by another route: the same result produced by different means. Thus, the effects of disturbances are not noticed, and behavior seems to just pop out of animals.

The process of producing consistent results in an unpredictable environment is called control. To behave is to control. The only system known that can do what organisms do every instant of the day is the negative feedback control system. A control system produces the consistent results we call behavior by producing pre-selected perceptions, not outputs. Control theory consists of the equations describing how closed loop control works. Control is not explained by muttering words like "feedback" and "error correction." I have never seen a correct treatment of control in the behavioral literature.

To the extent that behavioral scientists have dealt with it at all (and they have really tried), control theory has been twisted into what is really a disguised version of the old input-output model. This is usually done by imagining that closed loop control systems can be broken up into an alternating sequence of inputs and outputs. What you get is a sequential model where a person makes a response which produces a new input, which produces a new response. Input and output are preserved, alternating in time. In fact, such a system would not control anything. Real control systems work much more beautifully—there is no alternation in time. Input and output are joined in a continuous wheel of causation. The system is a wholly different thing from that which psychologists imagine it to be.

One reason psychologists have not learned control theory is that they think that they already know it. They don't—they just know terminology. When they get close to understanding it, they realize that it is completely different from their beliefs—so they redesign it to be consistent with their preconceptions.

Now I can try to explain why the results of behavioral research based on an input-output model is bound to be largely useless. According to control theory, when we are watching behavior we are watching a control system from the outside. This system will be controlling many different results of its actions (actually the perception of those results), some of which will correspond to very complex functions of the events that are part of the observer's perceptual experience. To control these results, which are almost certainly going to be quite abstract and, thus, hard for an outsider to notice, we will see the system doing many things in the process of protecting these results from the effects of disturbance. We might want to find the "cause" of one of these actions. So we do an experiment in which we manipulate stimuli to see if there is some effect on the action. Some effect is almost certain, although it will be only statistical. Almost anything you do is bound to disturb, in some way, some controlled result of actions. The behavior you are studying may be only incidentally related to the means used to protect against the disturbance you have created. Hence we get statistical relationships —usually by averaging over several subjects.

If you had a better idea of what the subject was trying to control, you could get more precise results. This is what happens in operant conditioning experiments. Of course, the experimenters would never consider reinforcement a controlled result of actions, but it is. In operant situations you create disturbances to the rat's ability to control the reinforcement rate. This leads to precise and dramatic corrective actions by the rat. For example, if you require more bar presses per reinforcer, the rat presses faster, preserving the rate of reinforcement. Of course, to the experimenter it appears that the change in reinforcement schedule is controlling the rat's bar pressing. But this is an unfortunate illusion that has prevented psychology from progressing beyond the input-output conception. This illusion of stimulus control (a well understood property of control system behavior) is just as compelling as the illusion that the sun goes around a stationary earth—just as wrong and just as difficult to dispel.

What you get by studying control systems as input-output systems is exactly what you have in the social sciences—a confusing and often inconsistent array of findings, only weakly reproducible and little more than verbal models to account for them, models with virtually no predictive or explanatory power. If you knew what the subject was controlling, you would not have to do such experiments any more. You would know how the system would respond to any disturbance. This is one goal of research based on control theory: to discover the kinds of things that can be or are controlled. Then you can ask how they are controlled, and why. The "how" question will take you to lower-order control systems (What results are controlled in order to control this result?). The "why" question will take you to higher-order control systems (What higher-order result is being controlled by controlling this result?).

Control theory is revolutionary, and the revolution is going to be tough. One reason is that most social scientists see no problem with the status quo. People will continue to do bread-and-butter social science because it's what they know how to do—they know what kinds of questions to ask and what kind of results to expert. Social scientists are experts at having an explanation for the results, no matter how they come out, so long as they are statistically significant. It is easy to turn the statistical crank. With sufficiently powerful statistical tools, you can find a significant statistical relationship between just about anything and anything else.

Psychologists see no real problem with the current dogma. They are used to getting messy results that can be dealt with only by statistics. In fact, I have now detected a positive suspicion of quality results amongst psychologists. In my experiments I get relationships between variables that are predictable to within 1 percent accuracy. The response to this level of perfection has been that the results must be trivial! It was even suggested to me that I use procedures that would reduce the quality of the results, the implication being that noisier data would mean more.

After some recovery period I realized that this attitude is to be expected from anyone trying to see the failure of the input-output model as a success. Social scientists are used to accounting for perhaps 80% (at most) of the variance in their data. They then look for other variables that will account for more variance. This is what gives them future research studies. The premise is that behavior is caused by many variables. If I account for all the variance with just one variable, it's no fun and seems trivial.

If psychologists had been around at the time that physics was getting started, we'd still be Aristotelian, or worse. There would be many studies looking for relationships between one physical variable and another—e.g., between ball color and rate of fall, or between type of surface and the amount of snow in the driveway. Some of these relationships would prove statistically significant. Then when some guy comes along and shows that there is a nearly perfect linear relationship between distance traveled and acceleration, there would be a big heave of "trivial" or "too limited"—what does this have to do with the problems we have keeping snow out of the driveway?

Few psychologists recognize that, whatever their theory, it is based on the open-loop input-output model. There is no realization that the very methods by which data are collected imply that you are dealing with an open-loop system. To most psychologists, the methods of doing research are simply the scientific method—the only alternative is superstition. There is certainly no realization that the input-output model is testable and could be shown to be false. In fact, the methods are borrowed, in caricature, from the natural sciences, where the open-loop model works very well, thank you. Progress in the natural sciences began dramatically when it was realized that the inanimate world is not purposive.

Psychologists have mistakenly applied this model of the inanimate world to the animate world, where it simply does not apply.

This was a forgivable mistake in the days before control theory, because before 1948 there was no understanding of how purposive behavior could work. Now we know, but the social sciences have their feet sunk in conceptual concrete. They simply won't give up what, to them, simply means science.

It is not, however, science, and the input-output framework is not the way to study closed-loop systems. There is a methodology for studying purposive systems; I have written a little about this. It is quite objective and experimental, and it gives results that are completely precise—and without statistics. But it is based on the rigorous laws of control, not on loose verbal, or mistaken quantitative, treatments of behavior.

I am not here seeking converts. I do not expect a social scientist to become a control theorist. Control theory requires a great deal of work; it is a lonely enterprise, and involves a painful change. But I hope that you can see why I can no longer teach the dogma.

I love psychology, and I consider it potentially the most exciting field left to explore. That is because it is basically virgin territory. All the attempts to understand behavior up to this point have been well-intentioned stabs in the dark. They have been based on the only tools available and on an allergic fear of committing metaphysics.

One might well ask. "Why should I believe you?" Well, you shouldn't. Understanding human nature is not a matter of finding the right words to use to describe a phenomenon, although one might easily get that idea by spending enough time in the social sciences. The only way to become convinced about the value of control theory is to learn it, to test it, to try to understand it. And then see if you can still buy the old approach. But learning control theory takes time, in my case at least two years—really four years before I was really comfortable with it.

I don't have a private pipeline to truth, and control theory is the beginning of a search, not the end. It won't solve all your problems. But it will, once you really begin to understand it, give you the extremely satisfying experience of finally knowing a little part of one of nature's secrets: the secret of purposive behavior. Then you can start looking at how learning, memory, consciousness, individual differences, and so on, enter the picture. But at least you will know that you are on the right track, proceeding from a solid foundation of replicable facts rather than from a trembling network of unreliable statistical generalizations.

Control theory has made me a revolutionary, not against psychology, but against the current dogma that passes for scientific psychology. If you are happy with the dogma, then go with it. If you want to understand human nature, then try control theory.

So my problem is what I, as a teacher, should do. I consider myself a highly qualified psychology professor. I want to teach psychology. But I don't want to teach the dogma, which, as I have argued, is a waste of time. So, do I leave teaching and wait for the revolution to happen? I'm sure that won't be for several decades. Thus I have a dilemma—the best thing for me to do is to teach, but I can't, because what I teach doesn't fit the dogma. Any suggestions?

PCT and Scientific Revolutions

William T. Powers,
Post to CSGnet February 2010

... The same question keeps coming up: are we going to have a revolution or aren't we?

PCT is the present state of a process that began with learning about, accepting, and starting to work out the implications of a revolutionary scientific concept, the idea of negative feedback control. The revolution started in the mind of H. S. Black on the morning of August 27th, 1927 as he was on the Lackawanna Ferry going to work at the Bell Labs, and spread rapidly over the next 20 years. When it started to leak out of the engineering world into the life sciences, however, it ran into resistance. The resistance arose because all of the life sciences with only a few minor exceptions had been developing for many decades in total ignorance of this new concept, and had created a huge network of concepts, terminology, and classifications based on other—and completely spurious—ideas of what makes behavior work. So not only did the revolution have to spread into the life sciences, it had to displace the ideas that were already there. And that aroused fierce defenses.

Arthur C. Clark gave us Clark's Theorem: the products of any highly advanced civilization will appear to us to work by magic. To this I want to add Powers' Corollary: to the inhabitants of any sufficiently retarded civilization, *everything* will appear to work by magic. Civilizations begin in ignorance and strive toward knowledge; they move from magic to science,

Magic is causation without mechanism. The mere fact that event A occurred is enough to cause event B to occur, with no intermediate processes to explain how A was transformed into B. The mere wave of a wand at Hogwarts causes someone on the other side of the quadrangle to fall flat on his back. What science does is to provide connections from A to B in the form of smaller magics. These smaller magics are called mechanisms, and while they still involve causation without mechanism, they also provide stepping stones from A to B that are useful and in fact are the source of immense increases in understanding. Having seen these new mechanisms, we can now see how combining them differently can lead not only from A to B, from A to C, D, E, and so on.

The structure of the behavioral sciences has been mostly magical, which is to say, empirical. I recently attended a seminar on motor behavior. What we have learned in the last hundred years, apparently, is how moving one hand or two hands to a target or to a target and back again, slowly or rapidly, with the same or different distances to the target, alternating hands or repeating with one hand, with pauses between the trials or no pauses, and with spaced or continuous learning sessions, affects the accuracy of pointing. Some conjectures were offered about what the subjects were thinking by way of strategy, but nothing organized or systematic. So this was pure magic: these changes of conditions affected accuracy just because they existed, not because of any intervening processes. Afterward I said to the presenter, "This is good old-fashioned experimental psychology, isn't it?" He was quite pleased that I put it that way. He would not, I presume, have been so pleased if I had said he was studying magic. But he was. All science begins with studying magic and formulating beliefs. But after 100 years of studying, you'd think it would have gone a little way toward knowledge, wouldn't you?

Anyway, one has to admire the presenter's skill, persistence, and patience to have spent 20 years meticulously studying pointing behavior.

So the question is, are we going to have a revolution or not? I think there is only one way to do that. Scrap everything and start over. If you don't go all the way, if you aren't willing to give up everything you think you know about behavior, it will simply be too hard to make the transition. You won't be free to explore any part of the new approach any way you please; you'll always have to be careful not to upset any of your favorite apple carts. That will inhibit your thinking and generate blind spots, like continuing to believe that the way to create repeatable results is to create repeatable behaviors.

Maybe—in fact quite likely, though we shouldn't start out by thinking this way—we may discover some things about behaving organisms that the old-time psychologists also discovered, even though they had the wrong explanations for them. Even after Lavoisier put an end to 150 years of phlogiston, it remained true that if you put mice into dephlogisticated air, they will die. Only now we know that there never was any such thing as phlogiston; the oxygen had merely combined with carbon and become unbreathable. Lavoisier had the role of H. S. Black, and the result of his finding the role of oxygen in combustion was a scientific revolution that ended up replacing alchemy with chemistry. So PCT is the start of a revolution that will replace psychology and many other allied disciplines with something entirely new. As Kuhn observed, the new science will not be built on the old science; it will replace the old science.

In *Living Control Systems III,* chapter three, a "Live Block Diagram" is discussed; the program comes with the book and can run on a Windows-based PC or an Intel-based Mac with a suitable virtual-machine program in it. In this diagram you will find all the basic features of the revolutionary idea behind PCT. You will see that despite time-delays in the control loop, the loop gain is high and the control is highly accurate, and *the control system is not unstable* as so many behavioral scientists seem to believe it must be. The time-constant of the output function, out of the box, is 30 seconds (that is, after a step-change in the error signal to a new constant value, the output will take 30 seconds to change 2/3 of the way to its final new value, 30 more seconds to change 2/3 of the remaining way, and so on). Despite that very sluggish response, the time constant of the overall control process is 0.3 seconds. The gain of the output function is 100: that is, the output is 100 times the magnitude of the error signal, after it comes to equilibrium. Reducing the output gain to 50—cutting it in half—reduces the output by 2%.

In other words, a negative feedback control system doesn't behave in accord with ordinary causal logic or common sense. Our common sense has been trained to fit a different model, the cause-effect model that underlies all conventional theories of behavior. If you want to be part of the PCT revolution, you have to retrain your common sense, which is exactly why you must simply give up every previous thing you learned about behavior that was based on the old common sense—that is, you have to give it all up. It is entirely wrong at its foundations.

Study the Live Block Diagram. Experiment with it any way you can think of, until it begins to make sense to you, until it starts to be part of your common sense about behavior, about control systems, about organisms. Behind it is a running model of a real control system, the same model that's used in Chapter 4 to match your own behavior in a real tracking experiment. There's nothing hypothetical about it any more; it really fits actual human behavior very closely. The more sense this block diagram makes to you, the less sense any other psychological theory will make. Do that enough and you will become part of the revolution whether you like it or not. You can't un-understand PCT once you have understood it.

Best, Bill P.

Three "Dangerous" Words

By W. Thomas Bourbon

This very personal essay was composed for consideration as a foreword for a book with the apt working title

Starting Over—Psychology for the 21st century

which became

Making Sense of Behavior—The Meaning of Control

"Behavior controls perception." Three simple words that summarize the subject of this little book. They don't look very dangerous, do they? But they are. What could possibly be dangerous about that little phrase? Many things, if you really understand it. Let me tell you about some of the "dangers" that I have seen during the 24 years since I first read the phrase. Remember that I am describing things I saw during a quarter of a century—everything did not happen all at once.

For one thing, many people don't perceive the words the way they are written, or spoken. Instead, they believe the phrase says "perception controls behavior." How could that be? How could people, including widely-respected behavioral scientists, influential editors of scientific journals, and respected educators all believe the phrase says something that means the opposite of what it really says? Ah, that's the danger! The phrase says that the relationship between behavior and perception is *exactly the opposite* of what most scientists believe it to be. Nearly everyone in behavioral science believes *perceptions cause behavior*, whether directly, or as a step in between stimuli from the environment as the cause, and behavior as the effect. When those scientists see or hear the phrase "behavior controls perception," they experience a feeling of *error*, between the way they *think things are*, and the way the phrase *says they are*; immediately, they say something to correct the error they perceive in the statement, so that they can hear themselves saying what they believe *should* be said. Those scientists behave to make their perceptions be the way they want them to be. They behave to control their perceptions.

This book is about those three simple words, and about what they imply for all of the sciences of behavior and for all of the practical applications that grow out of those sciences. When he first wrote those words, back in the 1950s, Bill Powers created an entirely new theory of behavior—an entirely new science of life itself. Bill's theory is called Perceptual Control Theory (PCT), and it is different from every other kind of theory I know in behavioral science, social science, or the life sciences. "Behavior controls perception." I can tell you, for certain, that if enough people ever understand that simple phrase, the world will be a different place—a better place. In this little book, Bill Powers gives you some clues about why that will be so, and he invites you to join in the excitement, and the challenge, of behaving to make it happen. I can tell you another thing for certain: the challenge in teaching people about PCT is great, and that brings me back to the "dangers." You need to know something about them, in case you decide to join in the PCT project. Let me describe just a little of what has happened to me, and to people I know, during the 24 years after I first read and understood Bill's little phrase. Let me tell you about some of the dangers, while we follow my path from the university to medical schools. Remember that nothing I describe here even came close to discouraging me, or any of others who are most closely associated with PCT. It is a unique and powerful theory. I simply want to tell you a few of the ways that some people misunderstand it, and the ways that others are threatened by it.

My first encounter with PCT came in 1973, when I read a journal article by Bill (William T. Powers, 1973, Feedback: Beyond Behaviorism, *Science, 179*, 351-356) [Reprinted in *Living Control Systems* (1989) p. 61-78.] I knew, immediately, that Powers had created a new theory that explained a festering mess in my own mind, he had found one clear principle that explained many seemingly unrelated facts in the behavioral and life sciences. The principle? You know it by now: behavior controls perception. That same day, I ordered Bill's book, *Behavior:*

The Control of Perception. The danger? I read it, and knew my life would never be the same. For one thing, I knew in a flash that my career as a traditional research psychologist was over. I could never go back to accepting all of the "theories" and research methods that I had learned were "true," and that I was teaching to innocent university students. It took many years for me to absorb some of the big implications of PCT and the process is not complete.

Immediately after I read Bill's book, the danger began to spread from me, to my students. I changed what I taught in all of my psychology courses, for undergraduates, and graduate students alike. For the sake of my students, who had to survive in traditional psychology, I still taught the "essentials," but I put them in the context of PCT—the comprehensive theory that explains how behavior controls perception. Over the next nineteen years, in practically every class, the time came for "The Declaration and The Question." A peer-selected class member raised a hand and declared (often with an appearance resembling fear and trembling), "What you are teaching us is different from what we learn in all of our other psychology courses." An accurate declaration, to which my reply was always "Yes, it is!" Then came the question, with unmistakable fear and trembling, "What are we supposed to do?" And my reply was always, "Each one of you will decide what to do."

My students accurately identified the danger of what they learned in my courses: behavior controls perception. Most of them did whatever was necessary to finish my class, and then they vanished back into the world of traditional psychology. However, during most semesters, at least a few students decided that PCT was a better scientific basis for psychology than the traditional ideas taught to them by my colleagues. Those students began to share in the rejection, and sometimes ridicule, that some of my colleagues had directed at me. Some of those students gave up trying to learn more about PCT, but others persisted. I shall always admire my imaginative and daring students who found ways to use ideas from PCT in clinical activities that were always closely monitored and regulated by members of the clinical faculty, some of whom were strongly opposed to anything having to do with PCT. Along with me, several students experienced the frequent rejection of research articles we submitted to scientific journals. Often, the editors and reviewers said bluntly that our papers were about a subject they were not familiar with, and they did not want to read anything about it. Bill Powers, Rick Marken, and anyone else who has tried to publish about PCT research, have all encountered similar rejections. So much for the myth that scientists are an objective and inquisitive lot! In spite of the obstacles in their paths, several of my students maintained their interest in PCT and they use it today, in their clinical practices and their research.

From time to time, one of my faculty colleagues would examine PCT, even if only a little bit. One day, a bright new faculty member, with a shiny new Ph.D. in experimental and theoretical psychology from a major university, came to my lab to learn a little about PCT. One of my thesis students had asked the fellow to serve on his thesis committee. I ran a few simple PCT demonstrations. One product of those demonstrations is a set of statistics that describe what happened during the session. Some of those statistics reveal, unambiguously, the inadequacy of traditional methods in experimental psychology. After one demonstration, my young colleague sat quietly for a while, staring at the computer screen. Then he turned slowly, looked at me, and said, "You know, of course, what this implies about the past three hundred years of research on behavior." Perhaps he expected me to realize the folly of my PCT ways and retract the point of the demonstration. Instead, I paused, then said, "Of course." He sat a while, quietly. He was a bright and energetic fellow, with a brand new doctoral degree. To earn that degree, he had to demonstrate that he knew all of the traditional theories and methods in psychology. Here he was, at the beginning of his professional career, staring directly in the face of something he knew refuted what he had just learned. I ran a few more demonstrations, with their inescapable evidence that most of the traditional statistical analyses in psychology are worthless. Once again, my colleague looked up slowly and said, "You know what this means about the things we teach in statistics and research methods." (In our department, he taught those courses. Back then, all psychology majors took them.) I replied, "Yes." My young "colleague" understood, perfectly, what he had seen, and the danger in it was as clear to him as it could possibly be: he had witnessed compelling evidence that traditional behavioral science was indefensible. How did he handle the danger? He became one of the faculty members who was the most critical of my students when they expressed an interest in PCT.

Nineteen years after I first read the phrase, "behavior controls perception," I decided I would never convert my faculty colleagues, or the community of research psychologists, to an understanding of PCT. I left the university for a new career of research in medical schools. Perhaps there I would find people who were more interested in understanding this exciting little phrase. How could there be any danger in a move to a place where there are "real scientists," rather than just a crowd of traditional psychologists? Three years later, I left the medical schools. My interest in PCT, and my work related to the theory, did not fit there, any more than they had in the university.

Most of the scientists were intent on discovering something in the environment, or in the brain, especially in the brain, which controls behavior. Their reputations, and their funding, were firmly rooted in one or the other of those two ideas about where to look for what causes behavior. Even a passing glance at the idea that behavior controls perception could prove dangerous, in the extreme, to a respectable scientist's professional well being! Four or five brave souls did look, briefly, at our simple demonstrations of control, and at the precision with which the model from PCT explains how behavior controls perception. Each of them described the demonstrations and the model with terms like, "interesting," or "intriguing," and then they went their traditional (safe) ways.

On the clinical side, I made a modest proposal, and a couple of clinical neuropsychologists agreed that we should test it. I suggested that some of the performance tasks and research methods used in PCT yield behavioral data and modeling coefficients that might help assess the functional status of various clinical patients. (Most of the patients had a history of stroke, or of injury to the head or spine.) I survived long enough at the medical school to make a start on testing that proposal. It looked like we might be able to identify effective levels of control in some patients who were classified as, "nonfunctioning," after conventional diagnostic procedures in neurology, and clinical neuropsychology. (In those clinical areas, practically all of the diagnostic procedures grow out of research and theorizing about environment, or brain, as the locus of whatever it is that allegedly controls behavior.) It looked like we could also identify a range of ability to control, in patents who were all lumped into single categories of functioning, or non-functioning, by conventional diagnostic procedures.

I vividly recall several patients who expressed thanks, and appreciation, that someone finally tested them in a way that allowed them to show what they can do, rather than in ways that always show how they fail.

Some of the clinicians described our early results with terms like, "fascinating," and "interesting, but... You knew it was coming! ...there was no way to use results like those. The numbers did not fit into existing diagnostic protocols or categories, and... Purely incidentally, of course! ...there was no way to bill an insurance provider for procedures like those. Now *that* is real danger! And so it goes.

The simple idea described in this little book is unique in behavioral and life science, therefore it is viewed as a threat by many people in those fields. That's too bad. They are missing out on a chance to participate in the creation of a new science of life, an experience I would not miss for the world!

Well, there you have a quick tour of some of the dangers I have seen for people who understand the simple phrase of Bill's that I first read in 1973. Bill Powers, and his wife Mary, have lived with those dangers since the 1950s. Many others have lived with them over the past few decades. Most of us have "survived," although a few former colleagues have dropped by the wayside, professionally and intellectually. For all of us who remain, and for the many others who have joined us, we would not miss a minute of the adventure. When it comes to developing the science and the applications that grow from the idea that "behavior controls perception," nothing I have described is really a danger, after all. At the worst, they are annoyances and nuisances. If "dangers" like the ones I described don't frighten you, and if you want to become part of the revolution that PCT *will* bring to the behavioral and life sciences, and to all of human kind, then I urge you to read this little book. There is no better place for you to begin your adventure!

Tom Bourbon
Houston, Texas
July, 1997
Revised January, 2008

Things I'd like to say if they wouldn't think I'm a nut

Or — Overgeneralizations that aren't that far over.

William T. Powers, 1989

When you study human beings, remember that you are a human being. You can't do anything that they can't do. You think with a human brain, experience with human senses, act on the world as human beings experience a world. Whatever you say about them is true about you. Whatever you can do, they can do.

Understanding human nature means more than having a large vocabulary. You experience the world at many levels, some lower than symbols and some higher. If you try to understand by using nothing but words, you'll miss most of the picture. What most people call "intellectual" is really just "verbal." If you always use the same terms to refer to the same idea, it's not an idea but a verbal pattern. Most important words don't mean much. Words that "everybody knows" don't mean anything. Words that are used to describe psychological phenomena are almost all informal laymen's terms that have negative scientific meaning: they imply the existence of things that don't exist, like "intelligence" or "aggressiveness" or "altruism." Or "conditioning" or "habits" or "aptitudes" or—see the literature.

Knowledge isn't what you can remember or name: it's what you can work out from scratch any time you need to, from basic principles. The behavioral sciences don't have any basic principles. None, that is, that would survive scientific testing.

Statistical findings are worse than useless. They give the illusion of knowledge. Even when they're true for a population, they're false when applied to any given person. To rely on statistics as a way of understanding how people work is to take up superstition in the name of science. It's to formalize prejudice.

When you propose an explanation of human behavior, you ought to make sure that the explanation works in its own terms: what exactly does it predict? Most explanations in the behavioral sciences consist of describing a phenomenon, saying "because," and then describing it again in slightly different words.

Perceptual control theory may have a long way to go as a theory of human nature, but it's the only theory that deals with individuals and accepts them as autonomous, thinking, aware entities. You might say that thinking about them that way is what makes control theory possible to understand. Using control theory, you don't have to ignore individuals who deviate from the average. Using control theory you can propose explanations that you can test. Using control theory you can learn that scientific understanding isn't any different from ordinary understanding. A scientist would judge that a cooling device used in regions of very low ambient temperatures would be inefficient, and you can't sell a refrigerator to an Eskimo, either.

But never forget that science bought Phlogiston for 150 years, and stimulus-response theory—so far—for 350 years. We're still crawling our way out of one system of faith into the next, still looking for dry land and solid ground. Is control theory the new faith? Not as long as you can forget everything you've memorized and reason it out for yourself.

The behavioral sciences don't have any basic principles. None, that is, that would survive scientific testing.

PCT, Biology and Neurology

William T. Powers,
Posts to CSGnet, February 2010

Bruce Gregory, BG:

You seem to be ignoring the possibility that inhibitory connections can be strengthened. If you are at all interested in learning something about this topic you might look at *Neural Networks and Animal Behavior* by Magnus Enquist and Stefano Ghirlandia (Princeton, 2005).

Richard Kennaway, JRK:

What an extraordinary book that is. I've only looked at the pages available on Google Books and Amazon, but I think that's enough.

Models? Here's a model from chapter 1:

$$r = m(x)$$

x is the animal's state and r is its response—its behaviour. The author calls m the behaviour map, and says it is common to all theories of behaviour.

The first few pages of Chapter 1 make it clear that the book is solidly grounded in the assumptions of stimulus-response operation and behaviour controlled by perception.

This brief look does not suggest to me that my time would be well spent in getting hold of a copy on interlibrary loan.

Bill Powers, BP:

That's the impression I'm getting so far from neuroscience in general. It really does look as if S-R has simply migrated from psychology into neuroscience—though in truth, it probably began with neurology (and biology) and has simply stayed there. Watson's initial formulation of behaviorism is said to have originated in biology. This makes me wonder about the number of departments of psychology I have seen which are now called departments of psychology and neuroscience, or neuropsychology, or some other term with a "neuro" prefix used somewhere. The "neuro" part looks like a signal that says "Stimulus-response spoken here." But I don't really know if that's fair. My sample is very small—though what psychology does get into Science and Nature is pretty solidly on the side of SR, and that indicates a pretty wide penetration, or chronic infection, by SR ideas.

If you look at the basic control-system diagram we use in PCT, you can see that it's possible to trace signals from sensory receptors to more central structures and back out again to the muscles. If that's all you consider, the only possible model is the SR model. A neurologist using this concept as a guide for exploring the nervous system can obviously trace cause and effect, progressing synapse by synapse, from input to output. Why look for anything else?

The problem is that this sort of experimental investigation involves applying inputs under control by the experimenter, which prevents any effects by the organism's actions on the stimuli while they're being applied. This guarantees that no concurrent feedback can occur. Because of that, it's impossible to see any sign that the actions are organized to have specific effects on the inputs. And this means that it's impossible to see that these feedback effects tend to maintain input variables in specific states that we would call reference levels, which means that nobody will ask how those reference levels are specified. Nobody will discover the existence of reference signals in the brain. That means that nobody will discover the hierarchy of control. All of which is what happened.

There are some current opportunities for me to sneak PCT into some departments that are joined with neuroscience. It seems clear that the main thing to do, to make any sort of worthy contribution, is to get people to see and understand the feedback loop, the control system idea. It's probably unnecessary to do much more than that, other than thinking of more demonstrations and experiments to make the understanding more solid. Once neuroscientists understand how feedback control works (I haven't yet met more than a couple who do), they will see that there are a lot of new phenomena to explain, and a lot more kinds of neural functions to account for. Suddenly a lot of new pathways will be found in the nervous system, not because they suddenly grew, but because the investigators were not equipped to recognize what they were looking at.

It's fortunate that my new book is subtitled "the fact of control." I wasn't thinking quite this way while I was writing it; I just wanted to get the demonstrations out there for people to look at. But now it's clear that the main thing we have to do to bring PCT into mainstream science is to convince neuroscientists and the rest of them that control is a fact of nature that has to be taken into account. They will know what to do about that fact once they decide to accept it, and once they realize that a lot of what they know about feedback ain't so. If we can just accomplish that much, I think the revolution will become self-sustaining.

What I have learned about modern neuroscience so far shows that it has made a lot of progress in exploring neurological phenomena. It's obvious that I still have a lot to learn about the state of this field today, as opposed to what I could find out before publishing B:CP *[Behavior: The Control of Perception]*. But it's also obvious that the weakest part of neuroscience is in its idea of what behavior is and how it works. What we have now is a picture of very sophisticated experimental methods being applied to an outdated and inadequate concept of the behavioral phenomena to which neuroscientists are trying to link the neurological and biochemical findings.

Best, Bill P.

Are All Sciences Created Equal?

By Dag Forssell 1994

ABSTRACT:

Sciences of today are not created equal. The physical sciences we depend on today were not always dependable. The life sciences we cannot and should not depend on today may become dependable in the future. While *Perceptual Control Theory (PCT)* deals with a "fuzzy" subject, it differs from contemporary life science in the kind and quality of explanations offered. To clarify this difference, categories of experience and explanation are defined and illustrated. PCT is not explained in this paper, but perspectives, basics and some explanations are discussed.

Introduction

I have long been interested in "what makes people tick." When I read *Behavior: The Control of Perception* by William T. (Bill) Powers, the detailed, in-depth explanations made perfect sense to the engineer in me. Demonstrations were compelling in their universal application and validity. I found the book very different from seminars, books and tape programs I had studied before. Powers provides a lucid synthesis, showing how neurons interacting in a hierarchy of control systems can account for most of the phenomena we experience.

I found that applying my understanding of PCT can help me develop and maintain pleasant, productive personal relationships on and off the job. PCT shows me that I am an autonomous living control system, and I value my ability to control my perceptions freely. In my roles as father, husband, friend, teacher and manager, I now strive to support others, especially those close to me, to control their perceptions in a way that is satisfying to them. This motivates me to teach PCT.

I have become acutely aware that PCT has been distorted, misunderstood, oversimplified and dismissed by scientists who deal with the descriptions and explanations PCT improves upon—or replaces. I have wondered why some people grasp and appreciate PCT with ease while others find it difficult to understand, accept or both. There appear to be two reasons for this.

The first reason, well explained by PCT, is that once a person has been taught an idea and decided to believe in it, that idea becomes part of the person's control hierarchy and any suggestion that the idea is false is resisted. Kuhn (1970), shows how this has been true for many scientific revolutions. Any adult has woven a personal web of ideas of how people "work." Suggestions that don't fit this web of principles and systems concepts are quite naturally resisted—or misinterpreted or distorted so they do fit.

A second reason may be that there are significant differences between the kind of theory and explanation scientists are used to in different fields, and that these differences make comprehension difficult. Scientists who are used to deal with descriptions alone may fail to understand the kind of explanation PCT offers. In this paper I address this second reason by discussing theories and explanations. "Theory" can mean anything from a hunch to a law of nature. I propose the categories *Experience, Description, Descriptive Non-Explanation*, and *Causal Mechanism*, and shall point out the advantages of causal mechanisms.

Language and expectations

We like to say that we live in a scientific age. Every day newspapers and TV programs announce new findings by scientific researchers. Scientific research done by a scientific method suggests definitive information, double-checked by researchers and 100% valid. This interpretation may be overly generous. All sciences are *not* created equal. Some very important fields of science are not very scientific at all, lacking explanations that have proven valid.

Theory and science go together, but in popular usage the word *theory* can mean anything from a casual hunch based on personal experience (which is hard to articulate) to a law of nature which has been confirmed in innumerable rigorous experiments. A *paradigm* means any personal way of looking at the world. A *science* means a field of study. A *scientist* means anyone doing *research,* no matter how. The new theories and scientific research we hear about on the evening news vary all the way from conjecture and questionable statistical "facts" to newly discovered, experimentally confirmed laws of nature.

Bill Powers writes about different interpretations of theory on an E-mail network:

> Theory, as I see it, purports to be about what we can't experience but can only imagine [with respect to PCT:] (neural signals, functions like input, comparison, output and mathematical properties of closed feedback loops), while evidence is about what we can experience. Both theory and evidence are perceptions, but the way we use these perceptions in relation to each other puts them in different roles.
>
> In the behavioral/social sciences, the word "theory" seems to mean something else: a theory is a proposition to the effect that if we look carefully, we will be able to experience something. A social scientist can say "I have a theory that people over 40 tend to suffer anxiety about their careers more than people under 20 do." The theory itself describes a potentially observable phenomenon. The test is conducted by using measures of anxiety and applying them to populations of the appropriate ages. If we observe that indeed the older population measures higher on the anxiety scale than the younger, we say that the theory is supported—or, as some would put it, the hypothesis can now be granted the status of a theory that is consistent with observation.
>
> This meaning of theory leads to the popular statement that a theory is simply a concise summary of, or generalization from, observations. That definition has been offered by quite a few scientists past and present. I think it misses an essential aspect of science, the creative part that proposes unseen worlds underlying experience. Before the "unseen worlds" definition can make any sense, however, it is necessary to understand, or be willing to admit, that there is more to reality than we can experience. . .

Scientific perspectives

A traditional scientific perspective. It is my impression that most adults take the world for granted. I do. As adults discussing the world, we all have a sense of what some call *objective reality.* We see it in living color, touch it, hear it, smell it, chew it, walk on it, and swim in it. Sometimes we hit it, or it hits us, and it hurts.

Most of us agree that some mental constructs have no equivalent in the physical world we live in. They are what we call *subjective* or *personal.* There is no way to definitively compare one person's subjective impression of things like beauty, marriage, courage, friendship, loyalty, ownership or self-esteem with that of another. What is unclear is where to draw the line between the objective and the subjective.

In electronics, engineers sometimes talk about *black boxes*—electronic assemblies or mechanisms with secret insides but observable and most often very dependable functions. One could say that the function of science is to uncover the secrets of the black boxes we find in nature. In management or behavioral science, the black box is the human being.

An alternative scientific perspective. Instead of taking the world for granted and studying the brain as a black box, we can take the brain for granted and look at the world outside the nervous system as the black box. The challenge now is making sense of that world, starting from the time of emerging awareness in the nervous system of a fetus still in the womb. To see how the nervous system can possibly make sense of its environment, we will need to consider the best available information about neurology, mechanics, physics, chemistry, and biology. We may learn more about the brain looking out from the inside than in from the outside.

Some observations about nerves. Nerves interact with our physiology and the world around us to create the high level human experience we take for granted. Researchers in the fields of biology and neurology tell us that:

1) Nerves are capable of sending streams of pulses through their fibers. Frequencies range from zero to about 1,000 pulses per second. Propagation velocity ranges from 50 to 300 meters per second, which approaches the speed of sound in air.
2) The rate at which pulses are sent appears to be caused by a variety of influences, singly or in combination. Pulses may be originated by

the neuron itself (some continue throughout life), or result from light, vibration, chemicals (hormones), pressure, stretch, temperature, and electricity. Pulses from connecting nerves are another typical source, causing pulses to propagate from nerve cell to nerve cell. A stream of pulses can be called a neural current. Depending on how the neurons are arranged and connected, currents can be added, subtracted, branched, multiplied, integrated, etc., making almost any logical manipulation possible.

We can never know REALITY. Philosophers have argued about what really exists. I accept that the physical world exists, and that we as physical entities are part of and exist in this physical world. The physical world as it exists, I call physical REALITY. I recognize that we are limited in what we can know about the REALITY we are part of. I call the representation we develop in our minds perceptual *reality*.

The complexities of nerves and nerve function are interesting in their own right and will be the subject of detailed research for centuries to come. The intent here is simply to note that all the nervous system can possibly know about its environment (REALITY) are the neural currents travelling in nerve fibers *(reality)*. No organism can possibly have direct knowledge of the world around the brain (REALITY), even though it sure looks that way and many scientists who have not considered this, take for granted that we do. Exhibit 21.

With this realization, it is no longer useful to draw a line between the objective and the subjective. All anyone can know is subjective *reality*. But the dependability—the effectiveness—of a person's personal *reality* varies greatly. Most of us experience it as 100% dependable when dealing with simple physical phenomena. At the same time, we experience it as uncertain when we deal with high-level mental constructs, both in ourselves and in other living beings. Good theory serves to improve the quality of this uncertain *reality* so that we can deal more confidently with the REAL world we live in.

Good theory serves to improve the quality of this uncertain *reality* so that we can deal more confidently with the REAL world we live in.

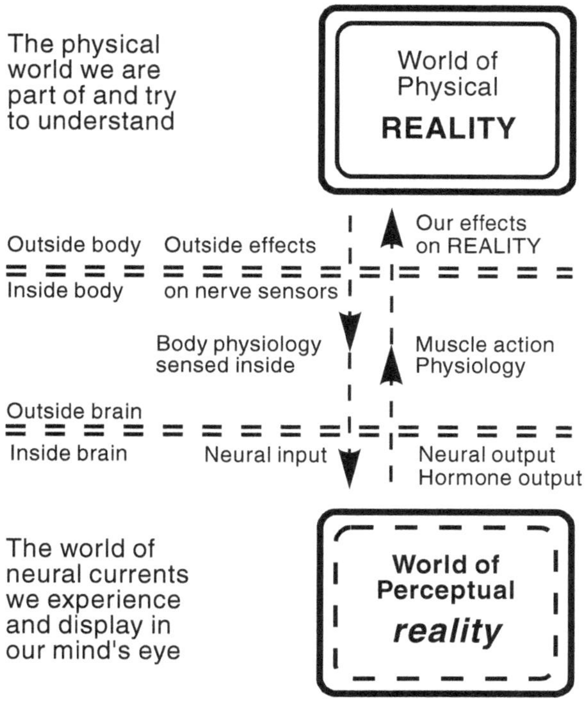

Exhibit 21. REALITY outside. *reality* inside.

Infant perspective: The world as a black box. The challenge for the developing infant is making sense of the currents in its nervous system as signals representing the world outside the brain. The currents originate in a variety of nerve-cell sensors inside the body: in organs and muscles, in eyes and ears, in the nose, mouth and in the skin.

Adult perspective: The brain as a black box. A challenge for life science is to determine the organization of our nerve cells. Taken together, nerve cells make sense of all these currents and develop into a human brain. The adult experiences the world in living color with stereophonic sound—then turns around, takes the world the infant brain has made sense of for granted, ***as if*** it is experienced directly, and asks questions about the mysterious brain.

Making sense of the black boxes. I certainly don't remember when I became aware of my existence. Adults don't remember much of their early development, but as adults we can observe that the development of infants is rapid. Fetuses still in the womb move about, kick, probably listen and may suck their thumb. A newborn is clumsy at first, but by trial and error finds out what works. When nerves sensing hunger, thirst, heat or cold send signals, other signals are created in the brain, perhaps at random in the beginning, causing the little body to act. If a particular act alleviates the problem, the signals that caused it become part of the brain's specifications to keep itself satisfied; to minimize those hunger, thirst, heat or cold signals. For example, many babies try crying and discover that—as if by magic—crying helps eliminate problems.

As the infant and its brain develop, the brain receives perceptual signals from organs deep inside the body as well as at the surface and sends out neural and chemical signals, causing the muscles to contract, organs to change, and the body to act on the world. The brain senses the new condition. Over time it develops a structure and memories that allow it to effectively act on the world so that the perceptions it experiences are the ones it wants to experience. *The brain acts* (sends neural and chemical signals to muscles and organs) *in order to affect what it experiences.* As time progresses, the baby learns to control its perceptions in ever more sophisticated ways.

As the baby focuses its eyes, coordinates its limbs, enjoys stroking, recognizes sounds and tastes everything it can bring into its mouth, the brain develops a *reality*, an interpretation of the world around the brain. We might say that the baby does scientific research and develops paradigms about the world. In this sense there is no difference between Nobel Prize science and an infant exploring its world. We are all scientists from the beginning of our awareness. But just as Eskimos have many words for different shades of white, we need several words for different shades of theory.

Tools for explanation

Before I discuss theory and explanation, I will review tools we use to describe and explain.

Language: Categorization and generalization. As humans, we benefit from a well developed capability to hear and utter sounds. The infant soon learns to associate sounds with experiences. While some sounds are associated with singular experiences, many words soon come to represent a whole class of experience. The meaning of food, chair, tired, hurt, shoe, walk, sit, and high include several possible configurations of objects, feelings, posture and physical relationships. Language facilitates generalization. Instead of having to duplicate experience, we can describe and categorize experiences.

Logic and Reasoning. Logical reasoning, mathematics and geometry are in a class by themselves. Based on idealized hypothetical postulates, they are logically rigorous. They do not represent physical experience. Therefore, they are not physical sciences, but are valuable as supplements to our descriptive language—tools to manipulate and give precise meaning to descriptions and mechanisms of all kinds, at all levels of sophistication, in all the physical sciences.

Measurement. Measurement is a different kind of tool, linking physical experience with description. Careful measurement has been very important to the development of modern physical science, as exemplified by Galileo's measurements of acceleration.

Statistical Analysis. A special branch of mathematics, statistics is widely used as a diagnostic tool. High correlations between observed variables can prompt guesses about underlying causal mechanisms, which can then be tested to see if the guess is valid. But it is important to recognize the strength as well as the limitation of statistics. In his book *Scientific Explanation and the Causal Structure of the World*, Wesley Salmon (1984) writes:

> Even if a person were perfectly content with an "explanation" of the occurrence of storms in terms of falling barometric readings, we should still say that the behavior of the barometer fails objectively to explain such facts. We must, instead, appeal to meteorological conditions. ... Statistical analyses have important uses, but they fall short of providing genuine scientific understanding .... A rapidly falling barometric reading is a sign of an imminent storm, and it is *highly correlated* with the onset of storms, but it certainly does not *explain* the occurrence of a storm.

Statistical descriptions are useful in terms of populations, whether of people or products, and can be used for prediction in terms of populations. But making decisions about individuals based on statistical prediction amounts to abuse. We call it prejudice. For a discussion of strengths, limitations (why statistical methods are incapable of delivering the secrets of human nature) and misapplication of statistics, please read *Casting Nets and Testing Specimens: Two Grand Methods of Psychology* (Runkel, 1990, 2007).

> The time has come... to put the "cause" back into "because."

Theory, explanation and prediction

Experience. PCT shows that organisms control perceptions, not actions. This explains why organisms do not need to "understand" their environment and why faulty explanations discussed among humans are simply ignored in practice. All an organism needs to do is to pay attention to a perception it wants to control while it acts and remember which way actions influence the variable. An infant lying in the crib reaches for an object hanging overhead. At first the image may be fuzzy and the hand miss the object, but the infant does not give up. It persists and over weeks, months and years learns by trial and error to act on its world so that it can experience it the way it wants to. As adults we have accumulated a large "world" of perceptions which make up and help us function in our individual *perceptual reality*. We call it *experience*. Predators teach their young to hunt through play, demonstration and practice. Consider the tradition in many arts of the master showing the apprentice what and how to perceive: what to look for, how it should feel, sound, smell and taste. We describe only a fraction of our perceptual reality in words. Exhibit 21 and 22.

Predicting from experience. The word hunch captures the idea of theory and prediction in the nonverbal world of *experience* at a very simple level. When we express a hunch we use a few words to summarize a vague or complex notion that we sense, visualize or imagine in the world of perceptual reality, but cannot put into words.

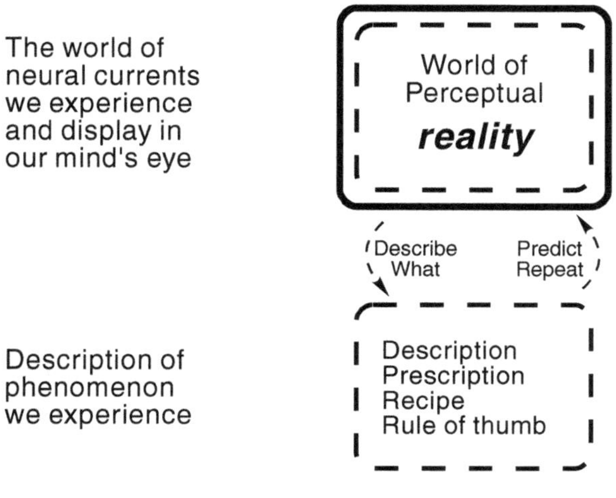

Exhibit 22. Experience and description

Verification and dependability of experience. The words hunch, gut feel, wisdom, and mastery suggest degrees of confidence in the predictions we make from experience.

Description. Language allows us to describe our experiences. It becomes possible to learn from experiences of others without having to take the time or suffer the risk of duplicating the experience itself. Our infant becomes a toddler and begins to express experiences in words. Lemons taste sour. Fire burns your skin. Objects fall when you release them. These are simple *descriptions* of phenomena we experience. Exhibit 22.

Prediction from Description. I can now predict that if I bite into another lemon, I will experience sour taste. If I touch fire, I get burned. I predict that when I release an object, it will fall. Predictions are based on regularities; things that usually happen "other things being equal." We use *Rules of thumb*, *Prescriptions* and *Recipes*. Exhibit 22.

Verification and dependability of descriptions. Since descriptions can be shared, they can be compared and the rules can be tried by many people, under different circumstances. We find some rules to be very dependable, while others are uncertain.

Descriptive Non-Explanation. Our preschooler pesters mother with questions. Why, Mother? Why is the Dandelion yellow? Why doesn't the rope break? Because! Because it is strong. Our little scientist is asking questions to make sense of the black box that still holds secrets everywhere you look. Some of mother's answers fit the category of theory I call *Descriptive non-explanation:* The Dandelion is yellow because all Dandelions are yellow. The rope does not break because it is strong, but strong is defined by "does not break." We notice that these are not explanations at all, but restatements or further descriptions of the same experience.

We often explain a phenomenon by using its description, somewhat transformed, as its explanation: You have trouble reading because you are dyslexic. By switching from the English "read" to the Greek "lexia" you make it sound as though you are naming a cause, whereas in fact you are simply repeating the description in a sentence that has the form of an explanation. In one of Molière's plays, a physician explained to a patient that sleeping medication worked because it contained "dormitive principles," where dormir is French for sleep. This term has been used to signify descriptive non-explanation. This is a popular mode of explanation in any field where people keep pestering you for explanations and you find it embarrassing or impolitic to keep saying "I don't know." Exhibit 23.

Exhibit 22 continued:

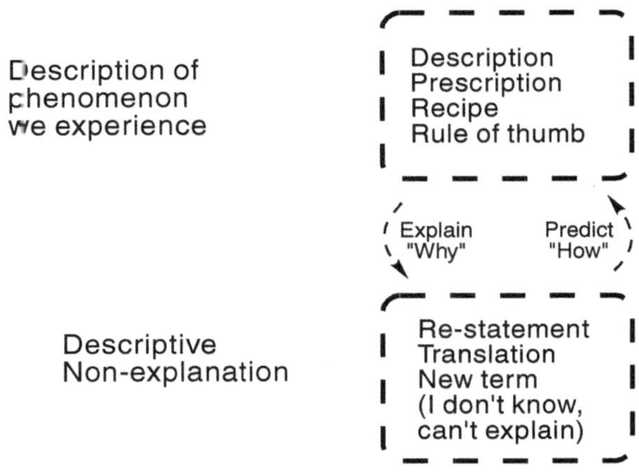

Exhibit 23. Descriptive Non-explanation

There really isn't any difference between descriptions and descriptive non-explanations except for the pretense of explanation and the introduction of a new term. The new term is incorporated in our language.

Causal mechanism. Wesley Salmon (1984) advocates causal mechanisms:

> The time has come, it seems to me, to put the "cause" back into "because." ...The relationships that exist in the world and provide the basis for scientific explanations are causal relations. ...To understand the world and what goes on in it, we must expose its inner workings. To the extent that causal mechanisms operate, they explain how the world works. ...A detailed knowledge of the mechanisms may not be required for successful prediction; it is indispensable to the attainment of genuine scientific understanding.... Explanatory knowledge involves something over and above merely descriptive and/or predictive knowledge, namely, knowledge of underlying mechanisms. ...To untutored common sense, and to many scientists uncorrupted by philosophical training, it is evident that causality plays a central role in scientific explanation. An appropriate answer to an explanation-seeking why-question normally begins with the word "because," and the causal involvements of the answer are usually not hard to find.

Causal mechanisms suggest the property, structure or functional relationships and interactions of elements below the level of described phenomena. Initially made up in one person's creative imagination, causal mechanisms offer explanations of why and how things happen. The physical sciences, based on causal mechanisms, have progressed far. Exhibit 24 illustrates a series of causal explanations in principle, reaching deep below the surface of the experienced phenomenon and its description.

A detailed knowledge of the mechanisms may not be required for successful prediction; it is indispensable to the attainment of genuine scientific understanding

Prediction from causal mechanisms. Visualizing the operation of the mechanism in different circumstances, we can predict what effects will emerge. We gain a deeper understanding of what is meant in any given instance when we make a prediction based on some regularity; things that (with high confidence this time) happen "other things being equal." What must be equal? In what way must it be equal? (Ways that allow the mechanism to operate). What does not have to be equal? (Things that do not affect the mechanism). Even a single level of causal mechanism below the level of the phenomenon allows much better prediction.

Verification and dependability of causal mechanisms. We can predict how the mechanism will perform in a multitude of circumstances, even ones we have never experienced before. Experimentation allows us to either reject the proposed mechanism as false and therefore unable to improve our predictions of what will happen, or as 100% dependable. With several levels of such dependable causal mechanisms in the physical sciences, one explaining the other, we have been able to travel to the moon and beyond.

Applications of theory

Causal mechanisms, descriptions and personal non-verbal experience mix when applied. Physical science, rich in causal mechanisms, depends on descriptive empirical data at several levels. A largely descriptive science may have pockets of insight that are of a causal mechanistic nature, whether formalized or not.

To illustrate, I'll share my perspective on applied sciences:

Medicine. Much of medicine is unexplained, and descriptions of symptoms (syndromes) abounds. Much drug research is done by systematic trial and error, just like Edison developed the light bulb. Practicing physicians know that a large part of their job is to comfort and support their patient while nature takes care of healing. Descriptive non-explanations are popular: you have red itchy eyes because of conjunctivitis[1], a red itchy nose because of rhinitis[2], and are cross-eyed because of strabismus[3].

Medicine has made great strides in the last century thanks to the discovery of some causal mechanisms explaining what happens in the body. One example is the discovery of the mechanism of bacterial growth causing the phenomenon of infection. People have learned to avoid harmful bacterial growth through hygiene. Scientists have learned to interfere with bacteria through vaccination and antibiotics, reducing infectious disease. We know that you get other diseases through the mechanisms of virus growth, but have had limited success in interfering with these mechanisms.

When repairing mechanisms of the body, surgeons successfully employ many different causal mechanism explanations derived from the physical sciences.

Mechanical Engineering. Ancient feats of engineering are still admired today: sophisticated compound bows and arrows, ocean crossing canoes, aqueducts, large bridges.

We have few records of exactly how these things were designed and built, but I think it is fair to say that they were based on experience and description, along with some causal mechanism explanations.

1 **çon·juñc·tī'vȧ,** n, the mucus membrane lining the inner surface of the eyelids, covering the front of the eyeball.

2 **rhĭ•nī'tis,** n. [*rhino-* and *-itis.*] inflammation of the mucous membrane of the nose.

3 **strä•bis'mus,** n. [from Gr. strabismos; *strabizein,* to squint; *strabos,* twisted.] a disorder of the eyes, as cross-eye, in which both eyes cannot be focused on the same point at the same time; squint.

Exhibit 21 and 22 continued:

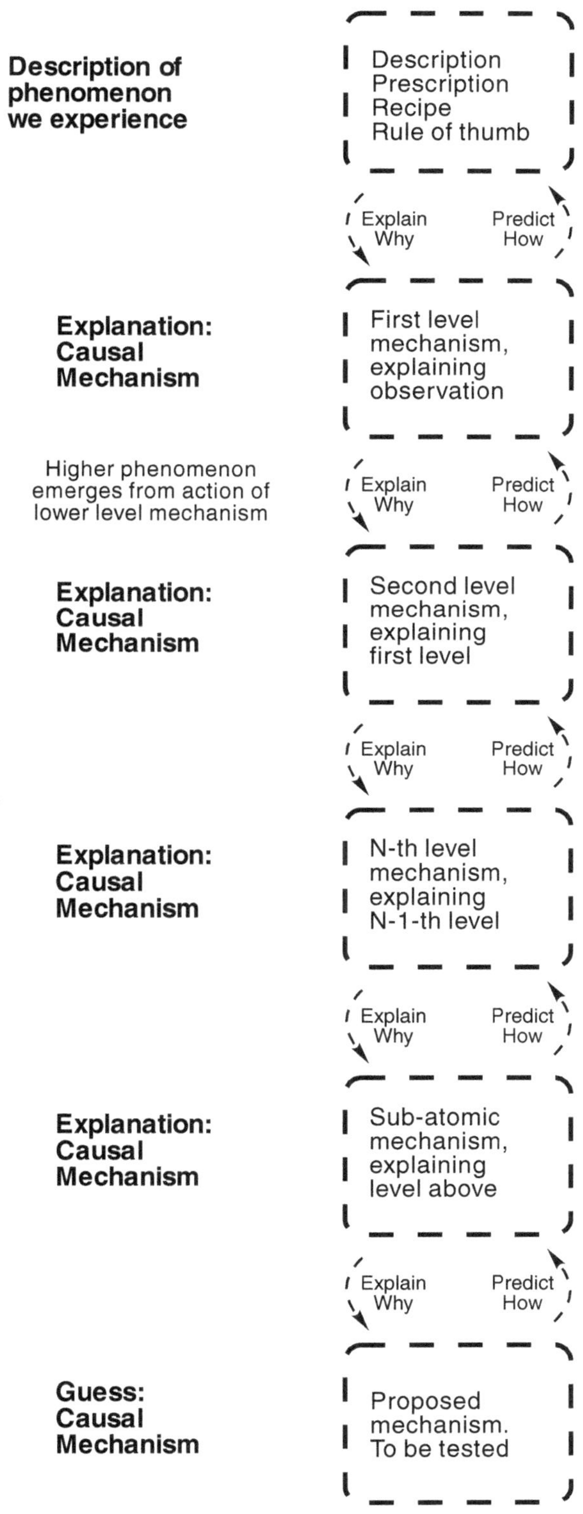

Exhibit 24. Causal mechanisms in depth.

With the advent of the Newtonian laws of nature in the late 1600's and the new rigor of measurement and test of theoretical models, the physical sciences began a development that is qualitatively different from what went before. The new causal mechanisms are much more consistent with observation and provide explanations in much greater depth than had been possible. The last few centuries have seen unprecedented progress in engineering. Causal mechanism explanations, coupled with descriptive data (about such things as material properties) have allowed us to extrapolate from small experiments and design complex machines. But we still have no idea what is causing gravity; we can only measure it and speculate about possible causal mechanisms to account for it.

Railroads, the Golden Gate bridge, aircraft, spacecraft, television, computers... The list of modern engineering accomplishments is long. They depend on the development and verification of in-depth causal mechanisms.

Chemistry. I am aware of three phases of chemistry. To describe the first, let me quote from *Alchemy: Ancient and Modern*, by H. Stanley Redgrove (1911):

> ... we find a school of Arabic alchemy arising in the eighth century A.D. Its inspiration was primarily Hellenistic, and from the contents of many of the texts, much of its theory and practice derived from Egypt. ... The basic idea permeating all the alchemistic theories appears to have been this: All the metals (and, indeed, all forms of matter) are one in origin, and are produced by an evolutionary process. The Soul of them all is one and the same; it is only the Soul that is permanent; the body or outward form, i.e., the mode of manifestation of the Soul, is transitory, and one form may be transmuted into another. ...The old alchemists reached the above conclusion by a theoretical method, and attempted to demonstrate the validity of their theory by means of experiment; in which, it appears, they failed. ...The alchemists cast their theories in a mould entirely fantastic, even ridiculous—they drew unwarrantable analogies—and hence their views cannot be accepted in these days of modern science.

Alchemy in its long history produced products of many kinds—metals, plating, medicine. Alchemy was a descriptive science, a body of prescriptions and recipes based on accumulated experience. The causal mechanism explanations it suggested were failures.

The next phase was dominated by *Phlogiston Theory*. This was an explanation for combustion proposed by Johann Becher (1635-82). It postulated that combustible materials contained an odorless, colorless, weightless (it would rise when released) material called *Phlogiston*. The search for Phlogiston gave direction to much experimentation and by 1775 resulted in the isolation of what was thought to be dephlogisticated air. Today we call it *Oxygen*.

Thus the causal mechanism of Phlogiston failed but was replaced by new explanations for combustion, which we are confident of today. Since the discovery of Oxygen, the science of chemistry has made rapid progress, and is now supported by many additional in-depth mechanisms such as the periodic table of the elements, atomic structure and chemical bonds.

Astronomy. To say that the Sun travels across the heavens in a chariot is indeed to propose a causal mechanism. This and other explanations of celestial phenomena were supplanted by Ptolemy's *Earth centered* model of the universe (c:a AD 140), which placed the Earth at the center of the universe with the heavenly bodies in circular orbits around it. It was apparent that some bodies traveled in reverse periodically, so epicycles, small circular motions, were superimposed on the major circular motion, to describe the apparent paths of individual planets. Over time, this model grew increasingly complex.

Copernicus published an alternate, *Sun centered*, causal mechanism in 1543. This model actually provided predictions which fit observations worse than the existing model. Galileo (1564-1642) developed and published much physical evidence in support of this model. Johannes Kepler (1571-1630) inherited Tycho Brahe's (1546-1601) twenty years of meticulous, descriptive astronomical records, spent additional decades analyzing them, concluding that the planets moved in *ellipses*, not circles. The fit between prediction and data improved. The fit became perfect when Isaac Newton (1642-1727) placed the sun not in the center of the ellipses, as Kepler had done, but in one of two ellipse *focal points*. Newton suggested causal mechanisms to explain how the elliptical motion is created by the heavenly bodies in motion, tugging on each other with (the still unexplained phenomenon of) gravity.

This sequence is interesting as it moves us from an elaborate causal mechanism that appears to work but is fundamentally mistaken, to a fundamentally sound mechanism that appears to work worse, through refinements in several stages to a 100% dependable causal mechanism that today gives us precise results as we continue to map the universe and send spacecraft to the far ends of our solar system.

Psychology: Professional insecurity. Several psychological theories compete for acceptance, with many methods competing for practical use. Many psychologists say that their psychological theories and clinical practice have nothing to do with each other. Scientific psychologists and clinical psychologists have separate societies and professional journals. The diversity of opinion in this field is bewildering. To an electrical, mechanical, or chemical engineer, it would seem strange indeed to be told that there are several electrical, mechanical, or chemical theories, and that practical applications have little or nothing to do with any of the theories.

Psychology: Experience. We all develop an understanding or "feel" for how to deal with people. Most of this "feel" is very personal, intuitive and difficult to express. The style, personality and interpersonal effectiveness that develops from personal experience vary considerably.

Psychology: Description. The vast majority of research in psychology describes apparent phenomena and attempts to relate one description to another by statistical correlation, implying some underlying causal relationship. Such relationships (tendencies, propensities) often are reported despite correlations which sometimes approach pure chance. Over time, stripped of the original uncertainty, many such relationships attain the status of "fact," referred to by subsequent researchers and widely discussed in media. Hidden by statistical summaries are large numbers of counter-examples, where observations are the opposite of reported and popularized "facts." Given more stringent criteria for facts of the physical sciences, where a single counter-example disproves theory, a large number of accepted facts in psychology must be recognized as groundless and simply false. It is unfortunate that psychological descriptive theory is not discarded in the face of counter-examples which disprove it. Instead, uncertain tendencies are used for prediction and judgement of individual behavior. This does not help us resolve conflicts, develop personal relationships, educate capable parents or managers and understand the dynamics of leadership.

Psychology: Descriptive non-explanation. Many popular explanations in this field are descriptive non-explanations. To illustrate, let us take a look at emotion. William T. (Bill) Powers, the creator of PCT, wrote on an E-mail network:

> Emotions are hard to untangle because some people place great value on emotions and don't like to think that emotions might have a rather simple explanation. Emotions, traditionally, are treated as a separate branch of motivation, reaction, or experience, having a somewhat mysterious kind of existence that is neither physical nor mental. Scientists decry arguments that appeal to emotion rather than reason. Their opponents often sneer at emotionless scientists for their coldness or indifference to feelings. Both, when asked to explain what they mean, fall back on descriptive non-explanations.
>
> Consider the emotion called anger. How do you know when you're feeling anger? In one episode of the television series *Star Trek: The Next Generation*, the android Commander Data asks this question of Geordi, the blind Chief Engineering Officer. In an effort to learn, Data asked Geordi to describe anger without using the word "angry." Geordi (and presumably, the show's writers) are at a loss. "You just—you know—feel *angry*." If you don't know what anger is, how can you understand a description of it? Geordi refuses to fall back on a descriptive non-explanation, and admits that he can't describe anger.
>
> Well, what does happen when you feel angry? You feel a surge of sensations from your body, and an urge to do something energetic to something. If you have no "self-control" you may well lash out and do damage to something or somebody—anger most often has an object at which you're angry, and it's usually a person.
>
> The term anger refers to an experience of a surge of bodily feeling and an urge to do something extreme. Anger is just the short way of saying "bodily feeling and an urge to do something." "Anger" isn't an explanation: it's a word referring to a phenomenon that needs an explanation. You don't feel the sensations and the urge to act because of anger, or vice versa. You feel the sensations and the urge to act, or alternatively, you feel anger. The two ways of putting it say the same thing. The word "anger" and the phrase "a surge of bodily feeling and an urge to so something extreme" refer to the same experience. What passes for an explanation is actually a descriptive non-explanation.

Psychology: Failing causal mechanisms. Two major suggested causal mechanism dominate psychology today: behaviorism and cognitive psychology.

Behaviorism[2] suggests that organisms respond to stimuli: What people do depends on what happens to them. Behaviorism includes the ideas of operant conditioning, reinforcement, and affordances; properties of the environment that somehow stimulate us to do what we do. Behaviorism has had a major influence on the psychological understanding of today's teachers and managers. It lays the scientific foundation for our society's love affair with reward and punishment. Data from experiments has varied, so additional, unexplained and unidentified internal and external stimuli have been proposed to account for any mismatch. Critics point out that "behavior" and "stimuli" both are poorly defined.

A major problem with the causal mechanisms suggested by behaviorism is that organisms not only experience stimuli, they create their own. Their behavior obviously, immediately and continuously changes the stimuli that supposedly cause the behavior.

Cognitive psychology describes many phenomena of perception and suggests that behavior is the execution of plans created in our minds.

A major problem with the causal mechanisms suggested by cognitive psychology is that when the brain has to calculate the signals sent to muscle fibers, things will start to go wrong the moment the world around the organism changes. The world may not change in the laboratory, but it sure does in everyday life.

Another problem for contemporary psychological research can only be understood once basic PCT has been understood. The scientific method used in both physical science and psychology simply put is this: Push here and see what happens there. (Change the Independent Variable and observe the Dependent Variable). This method shows what happens naturally with inanimate physical objects, but *not* with animated, active control systems. Control systems resist disturbances! You can learn from the presence or absence of this resistance, but you must understand how a control system works and that you are in fact dealing with a control system. PCT shows that the scientific method has been used incorrectly in psychological research and that all such research must be questioned.

2 **bē·hav'iŏr·iṣm,** n, in psychology, the theory that all investigation of behavior must be objective or observed as [because] introspection is considered invalid.

Psychology: Present status. Great variation of psychological terminology and interpretation has made it very difficult to agree on consistent descriptions of results. Psychological research is often published despite poor correlations. Studies are rarely replicated to confirm results through independent experimentation, as is routinely done in basic research in the physical sciences. I was startled the first time I was told by a psychologist that psychological theory and practice have nothing to do with each other. Now I understand that this schism is necessary for wise practice based on accumulated experience, since the causal mechanisms offered have not proven valid. But I don't accept that this state of affairs is the nature of science, which the psychologist also claimed.

> Scientists must first understand the new explanation before they can see what is wrong with the old one.

Psychology of the future:
Successful causal mechanism.

Organisms live and behave in a world full of influences (disturbances), some of them invisible, (crosswind when you drive), which affect our world (direction of the car) just like our actions (steering) do. These influences should produce instability and failure since they affect outcomes of our actions, but do not. The reason is that our actions automatically compensate for invisible disturbances. The causal mechanisms of psychology discussed above fail because they do not recognize and cannot deal with disturbances in a changing world.

We overcome disturbances and achieve consistent ends by variable means in a changing world because we control. PCT offers a clear and compelling explanation for the phenomenon of control.

HPCT suggests an architecture—an organization in principle of the entire nervous system—suggesting how a system of control systems made up of neurons can develop in the infant and make sense of the world, the black box outside the system.

Neurologists have identified the structure and organization of the neurons surrounding muscle fibers as a control system called the *tendon reflex loop*. A tendon receptor senses tension and sends a perceptual signal (current) representing the tension. A *reference signal*, a signal specifying the momentarily desired tension, arrives through a string of neurons from a higher level in the nervous system. The last neuron in this chain is called the *spinal motor neuron.* The current conveyed through this cell stimulates the muscle fiber to contract, increasing tension at the tendon. A branch of the perceptual signal from the tendon receptor contacts the spinal motor neuron and inhibits its pulse rate. The result of this arrangement is a *comparison* (subtraction) of the stimulating current specifying tension and the inhibiting current reporting perceived tension. This difference is called an *error signal.* In this diagram, the error signal drives muscle contraction directly. In the PCT architecture, a high-level error signal works through other control systems and neural *output functions* to drive *action.* Exhibit 25.

This causal mechanism of neuron interaction explains the lowest level of muscle control we observe when we use muscles in our own bodies and when we experiment on the muscles and nerves of simple animals.

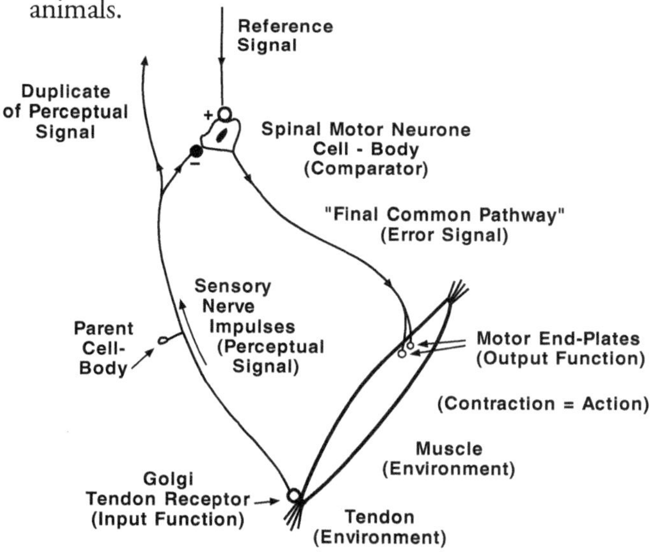

Exhibit 25. The basic first-order control system; the tendon reflex loop. (Powers, 1973).

PCT explains feelings. Bill Powers continues his discussion of emotion:

> How would we explain the experience of anger in terms of the PCT control architecture? Clearly, "a surge of bodily feeling" is a perception, and an "urge to do something extreme" implies a control system containing a large error signal. Why, we may ask, would the occurrence of a large error signal in a neural control system be accompanied by a surge of bodily feeling? One answer that seems reasonable is that the same output of the control system in question that would set reference levels calling for extreme action by the lower motor systems would also set reference levels calling for an altered state of the biochemical systems that support action. Thus we would expect blood sugar to rise, respiration to increase, heart-rate to increase, and so forth—the so-called "general adaptation syndrome." These sudden changes in somatic state can obviously be sensed; they are experienced as bodily feelings.
>
> So when a reference signal is suddenly changed to a relatively extreme value, or a large disturbance suddenly appears, the result is an error-signal-driven urge to change the state of the motor systems and the state of the biochemical systems by a large amount. There is thus a surge of sensation from the body as the biochemical systems are called upon to change to a significantly different state.
>
> Under normal circumstances and in a well-balanced system, the heightened state of preparation of the body is immediately "used up" by the accompanying motor action. There is a momentary sense of elevated somatic state that is simply part of the sensed action. The word "anger" would not be likely to be used to refer to the result.
>
> If, however, the person who experiences the large error has good "self-control," a conflict immediately ensues. One control system receives a reference signal implying an immediate change of state of the whole system, and at the same time a second control system says "No, a civilized person like me does not punch a boor in the nose, whatever the provocation." The "civilized" system cancels the reference signals going to the motor systems, and the punch does not take place.
>
> However, the control system gearing up for the punch is still there, and it is still telling the somatic systems to prepare for violent action. This state of preparedness is now not dissipated by the appropriate motor behavior and disappearance of the error signal; it is maintained by the same error signal that would throw the punch if lower systems were not receiving canceling reference signals from the "civilized" system. The reference signal calling for extreme action is not matched by the appropriate perception, so the urge to act continues and the sensation from the body persists, too. **Now** the person would say "I am angry!"

> Moreover, the person would say "I am angry **at him**." The person still wants to see and feel a fist mashing the other's nose, the other person crying out in pain, falling, becoming abject and apologetic and tearful and otherwise suffering all the embellishments of a thoroughly satisfying retribution. All these desires are the immediate source of the reference signal that suddenly changed so as to call for an energetic punch. As long as these desires are in effect, the "civilized" system will have to keep canceling the actual motor reference signals, and the anger and hatred and whatever else we call it will continue. The emotion will persist until the source of the reference signal is turned off. One ceases to be angry when one ceases to want retribution.
>
> This is a PCT explanation of anger that does not rely on a descriptive non-explanation. The same can be done for all the other experiences we label with emotion-names. The feeling component is the perception of a change in the biochemical state of the body, or more generally, somatic state. The goal-component is the reference signal that is calling for both motor action and the somatic state appropriate to the action. If the goal is to get the hell out of there, the same somatic changes take place as in anger, but now the combination of goal and feeling is called alarm, fear, fright, terror, panic, and so on. When the action is prevented from succeeding in achieving the goal, the emotion is felt the most strongly.

Powers concludes:

> True connoisseurs of emotion have as large a vocabulary for describing emotions as epicures have for describing tastes and smells. We can speak of feeling annoyed, offended, irritated, provoked, exasperated, angered, incensed, aroused, inflamed, infuriated, and enraged. I've just arranged the terms under "anger" from Roget's Thesaurus in order of increasing error signal and increasing shift in somatic state, as I understand them.
>
> Notice how those adjectives imply the passive voice. It isn't common to attribute emotions to one's own desires. Emotions—particularly the somatic feeling part—seem to arise as though they're being done to us by something else. "You make me angry!" We don't understand where they come from; that's why we need causal mechanisms. In this case, the PCT mechanism tells us we gambled on the wrong voice: we produce our own emotions, which arise from what we want. All these terms should be used in the active voice, which sounds really strange when you do it: "I'm angering at you!"

PCT offers detailed causal mechanisms, subject to refinement in coming decades and centuries. It is possible to generate predictions and effective practices from an in-depth understanding of these causal mechanisms.

Productive and satisfying relationships in the work place, non-manipulative buying and selling in business, loving family relationships, effective education, confident individuals, effective counseling, better understanding of biology, neurology and medicine. The list of improvements will be long. Just like the progress we have already enjoyed in the physical sciences, they will depend heavily on the development and verification of causal mechanisms.

Obstacles to new ideas. Scientific revolutions are not easy. Kuhn (1970) writes:

> Because it demands large-scale paradigm destruction and major shifts in the problems and techniques of normal science, the emergence of new theories is generally preceded by a period of pronounced professional insecurity. As one might expect, that insecurity is generated by the persistent failure of the puzzles of normal science to come out as they should. Failure of existing rules is the prelude to a search for new ones. .... Though [scientists] may begin to lose faith and then to consider alternatives, they do not renounce the paradigm that has led them into crisis. ....The decision to reject one paradigm is always simultaneously the decision to accept another, and the judgment leading to that decision involves the comparison of both paradigms with nature and with each other.

The comparison with nature that Kuhn writes about requires the kind of scientific rigor and understanding of causal mechanisms found mostly among those schooled in the physical sciences. Professional insecurity has been present for a long time in the social sciences. A new paradigm is available: The PCT revolution has begun.

PCT: Foundation for physical life science

Exhibit 26 illustrates layers of in-depth explanation in the format of exhibit 24.

At the level of description, PCT deals with familiar phenomena. This can create a problem when communicating about PCT, since some people (not used to causal explanations) look no further and conclude that PCT offers "nothing new."

At the first level of interaction, many lay people have a feel for how individual control (self-direction, freedom) manifests itself in autonomy, conflict and cooperation.

At the second level of explanation, PCT demonstrations of how people can control a single task, acting as an apparent single perceptual control system, are compelling. (Understanding to this level clarifies conflict resolution and personal interactions).

At the third level of explanation, Hierarchical PCT (HPCT) suggests an outline of a hierarchical arrangement of control systems as the organizing principle for the human nervous system. Demonstrations show the operation of such a hierarchy in humans, particularly at lower levels of perception and control. (Understanding to this level clarifies leadership issues).

At the fourth level of explanation, neurologists have identified control systems made up of a few neurons. See exhibit 25.

At the fifth level of explanation, researchers study the structure and interaction of neurons in terms of biology, chemistry and electronics.

PCT and HPCT offer no suggestions for mechanisms behind phenomena such as consciousness, awareness or attention. Understanding the operation of the human mind in greater detail will require research for many years to come, especially at the third through fifth levels of explanation outlined here, including biochemical control systems of several kinds.

It is not necessary to wait for additional research. Even a cursory understanding of the demonstrable concepts of PCT and HPCT offer immediate advantages, as this understanding leads to more effective and satisfying personal interactions.

Exhibit 21 and 22 continued with application to PCT:

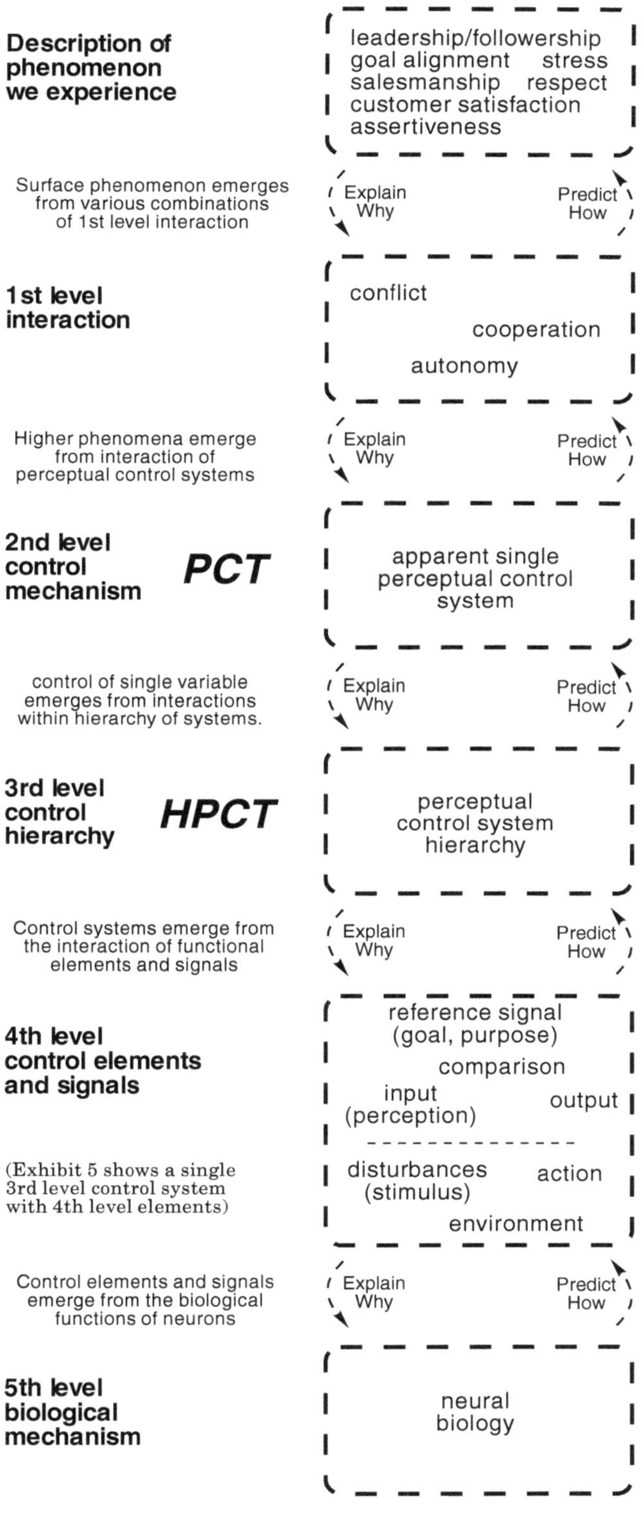

Exhibit 26. PCT psychology: Causal mechanisms in depth.

Conclusion

The point of this discussion of theory and explanation is this: **All sciences of today are *not* created equal. The physical sciences we depend on today were not always dependable. The life sciences we cannot and should not depend on today may become dependable in the future.** The difference lies in the kind and depth of theory and explanation a science is based on. Descriptions in the life sciences are often uncertain to the point of uselessness compared to in-depth explanations based on causal mechanisms in the physical sciences. Progress can best be made when we discover, validate and apply in-depth casual explanations in the life sciences, just like we do in the physical sciences.

References

Kuhn, Thomas S., *The Structure of Scientific Revolutions,* Chicago, IL: University of Chicago Press (1970).

Powers, William T. (1973). *Behavior: The control of perception*. Chicago: Aldine.
Second edition (2005), revised and expanded, Bloomfield, NJ: Benchmark Publications.

Redgrove, H. Stanley, *Alchemy: Ancient and Modern*, (1911) Reprint by Harper & Row (1973).

Runkel, Philip J., *Casting Nets and Testing Specimens: Two Grand Methods of Psychology* (1990, 2007) Hayward: Living Control Systems Publishing.

Salmon, Wesley C., *Scientific Explanation and the Causal Structure of the World*, Princeton, NJ: Princeton University Press (1984).

Descriptive vs. Generative Scientific Theories

by Dag Forssell 2004
Note on page 3 added 2010

The spectacular progress we have seen in the physical sciences in the last 400 years, compared to previous millennia, is largely due to a historic shift from descriptive science to generative science.

By a generative theory we mean a postulated organization of functional components with well defined, quantified interactions. Operating by itself as a model or in simulation, this organization generates action which validates or disproves the particular theory. Other terms used to describe the two kinds of theory are Empirical versus Fundamental, where empirical means derived from data using correlations or statistics (without any understanding of underlying reasons) and fundamental means derived from basic ideas, or laws of nature.

This comparison of descriptive and generative science in the fields of astronomy and psychology illustrates the well-known scientific revolution in astronomy and suggests that a similar upheaval is overdue in psychology and related fields.

The starting point for the modern era of physical science was the Copernican idea of a Sun-centered universe. Copernicus's model was adopted and promoted by Galileo, who among other things carried out meticulous studies of acceleration, thereby establishing the basic methods of modern physical science. The model of the solar system was later refined by Kepler and the laws of nature that govern it defined by Newton, completing the conversion of astronomy from descriptive to generative. Replacing the previous descriptive, "cut-and-try" approach to physical science, this sequence of developments laid the foundation for our contemporary, generative, physical and engineering sciences.

As new theories have been proposed and tested in the physical sciences, numerous scientific revolutions have followed, but as Thomas Kuhn explains in *The Structure of Scientific Revolutions*, textbooks don't usually explain or even mention previous concepts, so students are left with the impression that science is a matter of accumulating facts, where of course all new facts must fit previous facts. Not so. Numerous upheavals have taken place in physical science in the last 400 years.

2 *Descriptive Versus Generative Scientific Theories*

DESCRIPTIVE ASTRONOMY

Concept
Formalized by Greek astronomer Ptolemy (approx. 87–150 AD) in one of the world's oldest scientific works, the *Almagest*, the basic concept was that the Earth was an immovable object at the center of the universe. The idea that the Sun and all the other heavenly bodies rise in the East and revolve around the Earth seemed obvious and was accepted by scientists and lay people alike.

Study
You study the description of each heavenly path and master the tools of this science—the geometry and mathematics of circles and epicycles.

Description and interpretation
Descriptions assume that we experience reality directly through our exquisite senses—in living color and stereophonic sound. What we observe in the heavens is what is going on.

Prediction and testing
You predict future positions by projecting forward from current observations, using the descriptive mathematical tools. Because of the great regularity of the heavenly movements, such projections were very accurate. Lunar eclipses could be forecast years in advance. Ptolemy's descriptive model must be said to have been quite successful.

Limitations and complications
Ptolemy's descriptive mathematics provided no explanation for the phases of the moon or planets. About eighty epicycles (read fudge-factors) were defined by Ptolemy to make the basic geometric descriptions hang together.

Use
Heavenly constellations were noted, named and invested with significance by the Ancient Egyptians, from whom we have inherited Astrology. The model served as the basis for development of the calendar and was helpful for navigation at sea. The Catholic church accepted Ptolemy's circles and spheres and concluded that the planets are supported and carried by perfect crystal spheres as they revolve around the Earth.

To learn more
The University of St. Andrews web site:
http://www-gap.dcs.st-and.ac.uk/~history/Mathematicians/Ptolemy.html
is one good source of information on Ptolemy.

DESCRIPTIVE PSYCHOLOGY

Concepts
Basic concepts have included sequences of stimulus and response.

Behaviorists believe the environment determines what we do. Cognitive psychologists believe the brain issues commands for particular actions.

In both cases, explanations focus on output—on particular actions. Both these beliefs are at present almost universal among scientists and nonscientists alike.

Study
You study a vast number of theories put forth by a multitude of psychologists. You master the tools of statistics, which can provide an illusion of causal relationships and thus an illusion of understanding.

Description and interpretation
Descriptions assume that we experience reality directly through our exquisite senses—in living color and stereophonic sound. What we observe and describe is objective truth.

Prediction and testing
You predict future behavior basically by saying: "I've seen this before—I'll see it again." Due to the great variety of conditions and individuals, such predictions have an extremely poor track record.

Comparison with a working model has never been required. No psychological theories have ever been disproven or discredited.

Limitations and complications
The field of psychology is extraordinarily fragmented. The focus is on behaviors, which are classified and discussed, but no functional, physical explanations are offered for even the simplest phenomena.

Use
Descriptive psychological ideas of many different kinds are used throughout our culture. They are part of our language and pervade education, politics, management etc.

People have long used unverified concepts from these descriptive sciences to feel they are explaining events.

To learn more
We live in a culture dominated by descriptive sciences of psychology. Umpteen books on various psychologies are published every year. Findings are regularly reported on the evening news.

GENERATIVE ASTRONOMY

Origin

Polish astronomer Nicolaus Copernicus (1473–1543) proposed the Sun-centered alternative to the Earth-centered Ptolemaian model. Copernicus distributed a handwritten book called *Little Commentary* to other astronomers already in 1514. His major work *On the Revolution of the Heavenly Spheres* was published in 1543. Copernicus work (still descriptive, featuring some epicycles, but on the right track) was championed by Galileo Galilei (1564-1642), who found evidence supporting the concept, such as phases of Venus and moons of Jupiter using the newly invented telescope. Johannes Kepler (1571-1630), using observations collected by Tycho Brahe (1546-1601), found that if planetary paths were elliptical, not circular, they would fit the data—doing away with the need for epicycles. Finally, Isaac Newton (1642-1727), formulated the laws of motion and gravity, which, when operating on heavenly bodies interacting in the mechanism we call the Solar system, generate the elliptical motions observed in the heavens. The 200-year conversion of astronomy from a descriptive to a generative science was complete.

Postulates

Copernicus's *Little Commentary* states seven axioms, which suggest the structure of the universe:

1. There is no one center in the universe.
2. The Earth's center is not the center of the universe.
3. The center of the universe is near the sun.
4. The distance from the Earth to the sun is imperceptible compared with the distance to the stars.
5. The rotation of the Earth accounts for the apparent daily rotation of the stars.
6. The apparent annual cycle of movements of the sun is caused by the Earth revolving round it.
7. The apparent retrograde motion of the planets is caused by the motion of the Earth from which one observes.

GENERATIVE PSYCHOLOGY

Origin

Developed by William T. (Bill) Powers (1926–). Bill was trained by the U.S. Navy as an electronic technician to service control (servo) systems. After WW II, he obtained a B.S. degree in physics. An interest in the important subject of human affairs led him to enroll in a graduate program in psychology, but he left after one year because his proposed Masters Degree thesis, involving control by rats, was not acceptable to the Spencian psychologists then in charge. He began his development of Perceptual Control Theory (PCT) in the early 1950s by applying control engineering and natural science to the subject of psychology. His major work *Behavior: the Control of Perception* was published in 1973.

In this work, Powers proposes a structure of our nervous system, complete with mechanisms in some detail and, most important, functional interactions between the various elements and clusters of these mechanisms. The result is a coherent whole that can be tested to see if it functions in a way that rings true when compared to our observations of the real thing—human beings and animals. PCT lays a foundation for a new beginning, a new way to think about and perform research in psychology and related fields.

Postulates

Philip J. Runkel spells out postulates of Perceptual Control Theory (PCT) in *People as Living Things*, (page 57):

1. Causation in the human neural net is circular and simultaneous.
2. Action has the purpose of controlling perception. Controlling perception produces repeatable consequences by variable action.
3. A controlled perception is controlled so as to match an internal standard (reference signal). Every internal standard is unique to the individual, though two individuals can have very similar standards.

One of the deductions one can make from these postulates is that particular acts are not, in general, predictable.

Note:

As discussed in *Big Bang* (2004) by Simon Singh, page 22 ff, and *The Structure of Scientific Revolutions* (1970, 1996) by Thomas Kuhn, page 75, Aristarchus of Samos (circa 310-230 BC), proposed a heliocentric solar system.
On pages 34-35 and 68-69, *Big Bang* features informative overviews of the evidence for the earth-centered model and the sun-centered model in Aristarchus' era and as of 1610 AD, after Galileo's observations. I leave it to another student of PCT to present a similar overview of the evidence for descriptive versus generative psychology.

Generative astronomy, continued

Postulates, continued
Newton's three laws of motion and law of gravity suggest the dynamic physical states of and interactions between moving objects:

Motion:
1. Every body will remain at rest, or in a uniform state of motion, unless acted upon by a force.
2. When a force acts upon a body, it imparts an acceleration proportional to the force and inversely proportional to the mass of the body and in the direction of the force.
3. Every action has an equal and opposite reaction.

Gravity:
> Every particle attracts every other particle with a force that is proportional to the product of their masses and inversely proportional to the distance between them.

The structure and functional interactions allow the scientific model to generate action by itself. This can be compared to actual observations as well as used to predict future states of the heavens.

Study
You grasp the idea and generative model of the solar system by studying the mechanism and dynamic physical relationships between moving objects.
You realize that the concept of an Earth spinning around its axis while revolving around the Sun is counter-intuitive, but once the mechanism and the quantifiable physical interactions have been studied, it is not particularly difficult to visualize and understand.

Description and interpretation
You realize that appearances in the heavens can be very deceiving. What looks obvious to the intuitive observer may be better explained by a very different mechanism operating in ways that are not readily apparent and can only be inferred from various observations, interpreted through the framework of a proposed mechanism.

Generative psychology, continued

Postulates, continued
These postulates are summarized and amplified on page 129:

> Perceptual control theory claims that behavior controls perception—at every time, in every place, in every living thing. The theory postulates that control operates through a negative feedback loop—neurally, chemically, and both. The theory postulates the growth of layers of control both in the evolution of the species and in the development of individuals of the "higher" animals. Those are the crucial postulations of invariance in PCT. They are asserted to have been true for the single cells floating hither and thither a billion years ago, which might have had only two layers of control, and they are asserted to be true for you and me with our many layers. They are asserted for all races, nations, sexes, and indeed all categories of humans—and indeed all categories of creatures. Furthermore, if one creature is found reliably to violate any one of those postulations (and yet go on living), the theory will immediately be revised.

Study
You grasp the idea and generative model of PCT by reading the basic text, studying tutorials that explain control in detail, by experiencing physical control systems, and by studying informative simulations you can run on your own Windows computer.
You realize that the concept is counter-intuitive, but once the mechanism and the quantifiable physical interactions have been studied, it is not particularly difficult to visualize and understand.

Description and interpretation
You realize that our various sensors merely originate neural signals when "tickled" by various physical phenomena in a physical reality we as humans will never know, but certainly do our best to draw conclusions about. You realize that everything you see, hear, touch and smell is made up of neural signals in your nervous system. The sights and sounds you enjoy are fabricated by your nervous system and "displayed" in your mind. You never experience reality directly.

Generative astronomy, continued

Prediction and testing
You build a model of the Solar System, either a physical model or a simulation of the physics, implemented in a computer program. You make sure that you program functional interactions correctly with regard to the laws of nature, such as Newton's laws of motion. You predict by allowing the model to operate by itself, generating future positions. You test these predictions against the best possible observations of the motions of heavenly bodies. You expect agreement as closely as you can measure, or you modify your model.

Predictions based on contemporary astronomy routinely match observations to the limit of measurement. Rockets launched into space have found their targets.

Consequence
Copernicus's theory was not compatible with the existing, predominant Ptolemaian theory. It ultimately gave rise to a scientific revolution, which took a long time to play out. Once you understand the mechanism of the Solar system, Newton's laws of motion and gravity and accept the generative model, you reject all the explanations inherent in the old, descriptive astronomy, though not necessarily all of its observations. You may retain some of its language, such as "The Sun rises in the East." You realize that if you are interested in moving beyond the scope of simple observation, such as calculating trajectories and forces required for space travel, the old descriptive astronomy would have been utterly useless. You recognize that the physical model and mechanisms implied by the descriptive science, such as the stars revolving around the Earth in 24 hours, was not physically feasible. You recognize that accepted phenomena of the old science, such as the epicycles, planets moving in small circles as they move in big circles, were illusions.

Use
The transition from descriptive to generative physical science laid the foundation for the engineering progress we have enjoyed for the last 400 years.

Generative psychology, continued

Prediction and testing
You build a model of an organism, either a physical model or a simulation of the physics, implemented in a computer program. You make sure that you program functional interactions correctly with regard to the laws of nature, as known from physics, kinetics, neurology, etc. You predict by allowing the model to operate by itself, generating activity on its own. You test these predictions against observations of actual, living organisms operating by themselves. You expect very close agreement, or you modify your model.

Tests to date shows correlations above .95, often around .98, between the model and the actual person.

Consequence
Powers's theory is not compatible with existing, predominant psychological theories. It causes a scientific revolution, which will take a long time to play out. Once you understand the mechanism of perceptual control and recognize that control is the pervasive, defining quality of living things, you reject the basic concepts of descriptive psychologies, though not necessarily all of their observations. You have little choice but to continue using the languages of contemporary psychologies, such as "What are you doing," because that is part of our current culture and language. (PCTers might say "What are you controlling for.") You realize that if you are interested in moving beyond the scope of repeating observations, such as developing harmonious management programs or effective educational programs, descriptive psychologies have severe limitations. You realize that the physical mechanisms implied by descriptive science, such as super-computer brains issuing commands, are not feasible in a rapidly varying environment. You recognize that many widely held ideas, such as people controlling their behavior, or responding to stimuli, are illusions.

Use
PCT, seen as an overall organizing principle for living organisms, lays a foundation for a fresh review of the life sciences, promising great progress in the future.

Generative astronomy, continued

Limitations
By the time the transition from an Earth-centered to a Solar-centered astronomy was complete, the evidence for the Solar system was compelling to those who looked at the evidence. However, at that time there was much detail left to be worked out, such as detailed equations that portray the movement of the moon relative to the Earth, and astronomers are still uncovering wonders of the universe. Newtonian physics has been extraordinarily successful, but we still don't have any explanation that tells us how gravity works. But we have no doubt that it does.

Willingness and ability to understand
If you were raised at an age and in a society where everybody *knew* that the Earth rests at the center of the universe, and somebody suggested the idea of a Sun-centered universe. What would you make of it?

Would you have been willing and capable of making the effort to grasp the model? Might you have found the idea strange and obviously false?

Acceptance
The basic Sun-centered model of our local universe is widely accepted today. You most likely take it for granted because you learned the concept already in kindergarten. It was not intuitively obvious, was it?

To learn more
The Internet features numerous web sites about Copernicus, Galileo, Kepler and Newton. *On the Shoulders of Giants*, edited by Stephen Hawking, (2002) features the full text of *On the Revolution of the Heavenly Spheres* by Copernicus, *Dialogues Concerning Two Sciences* by Galileo , *Harmony of the World*, book five, by Kepler, and *Principia* by Newton.
For information on the numerous scientific revolutions in the natural sciences, see Thomas Kuhn's *The Structure of Scientific Revolutions*.

Generative psychology, continued

Limitations
PCT is a natural science in its infancy. Evidence that living organisms control their perceptions is compelling to those who examine it, and this makes all the difference for our understanding of behavior. Detailed simulations show how a hierarchy of control systems can work. Some levels of control in people can be clearly demonstrated. The postulated higher levels are by no means definitive. How perception works at the various levels is unknown; thus wonders of perception remain to be uncovered. But there can be no doubt that we control our perceptions.

Willingness and ability to understand
You have been raised in a culture where everyone *knows* that we react to stimuli in our environment and control our actions. Now someone suggests that you don't react, you oppose disturbances. You don't control your actions, you control your perceptions. Your brain does not issue commands, it sets reference signals. What would you make of it?

Are you willing and capable of making the effort to grasp the model? Might you find the idea strange and obviously false?

Acceptance
The basic PCT model of how living organisms control their internal worlds will hopefully be widely accepted fifty years from now. Children most likely will take it for granted because they will learn the concept already in elementary school.

To learn more
People as Living Things; The Psychology of Perceptual Control by Philip J. Runkel introduces the theory and shows its implications for numerous aspects of human experience, thereby illustrating its significance and challenging crucial contemporary notions of how humans and human relationships can work. This is a very good place to start. The book refers to other PCT literature and points to web sites where you can download tutorials and simulations. See http://www.livingcontrolsystems.com.

Name index—letters only